PHOSPHORUS–NITROGEN COMPOUNDS

Cyclic, Linear, and High Polymeric Systems

PHOSPHORUS–NITROGEN COMPOUNDS

Cyclic, Linear, and High Polymeric Systems

H. R. ALLCOCK

Department of Chemistry
The Pennsylvania State University
University Park, Pennsylvania

ACADEMIC PRESS New York and London 1972

ACADEMIC PRESS, INC.
111 Fifth Avenue, New York, New York 10003

United Kingdom Edition published by
ACADEMIC PRESS, INC. (LONDON) LTD.
24/28 Oval Road, London NW1

LIBRARY OF CONGRESS CATALOG CARD NUMBER: 78-159613

PRINTED IN THE UNITED STATES OF AMERICA

To Noreen

CONTENTS

Chapter 12. **Phosphazenes as Brønsted–Lowry Bases**

Chapter 13. **Rearrangements**

Chapter 14. **Skeletal Bond Cleavage, Thermolysis, and Other Reactions**

Part III

POLYMER CHEMISTRY

Chapter 15. **Polymerization and Depolymerization**

Chapter 16. **Phosphazene Polymers**

PREFACE

One of the striking developments in chemistry during the last fifteen years has been the surge of interest in inorganic ring systems and polymers. The cyclo- and polyphosphazenes (phosphonitriles) and related phosphorus–nitrogen compounds have been in the forefront of these developments, and these systems have attracted considerable attention from the theoretical, synthetic, mechanistic, and technological points of view. Much of this interest has arisen since the mid-1950's when the demands of aerospace and polymer technology for new and unusual materials focused attention on the phosphazenes. This practical impetus was accompanied by a burst of activity in fundamental research, and this is continuing at the present time. In recent years nearly every modern chemical technique has been brought to bear on this subject, with the result that important information is now widely scattered throughout the literature. This book represents an attempt to bring together this diverse data in order to provide a unified picture of a rapidly advancing field.

This volume is concerned with the chemistry of compounds which contain alternating phosphorus–nitrogen atoms in the skeleton. The emphasis is placed on cyclic and high polymeric phosphazenes (phosphonitriles), although monophosphazenes (phosphinimines) and phosphazanes are also discussed, particularly with respect to the synthetic and bonding aspects. The book has been planned as an introduction to phosphorus–nitrogen chemistry, as a review of advances in the field, and as a reference work. These objectives are reflected in the layout of the volume. First, the structure and bonding in phosphorus–nitrogen compounds is a subject of considerable importance, and this is dealt with in Chapters 2 and 3. Second, by far the greatest research effort has been devoted to the synthetic and mechanistic aspects of phosphorus–nitrogen chemistry, and this is reflected in the space devoted to that subject.

Thus, Chapters 3 through 14 deal with the multiplicity of synthetic processes and reactions now known in this area. Third, Chapters 15 and 16 are concerned with the polymer chemistry of phosphazenes, with polymerization and depolymerization processes, syntheses, the characteristics of polymers, and their possible applications. Finally, the two appendices comprise compilations of physical constants and data of value to chemists working in this area. Appendix I is a compilation of X-ray crystallographic and related information, and Appendix II lists melting points, boiling points, infrared, and other useful data for a wide range of different derivatives. The main literature coverage extends to the middle of 1971, but several later 1971 references are also included.

It is hoped that this book will be useful to the research worker, the student, and to those contemplating research in this field. It is also hoped that this volume may help to bridge the narrowing gap between inorganic, organic, structural, and polymer chemistry.

In writing this book I have been greatly assisted by a number of people. Drs. R. L. Kugel, E. J. Walsh, and W. J. Birdsall, Mrs. M. T. Stein, and Mr. D. P. Mack were kind enough to read sections of the manuscript, and I am indebted to them for their helpful suggestions. I would also like to acknowledge the work of Miss Marilyn Matthews whose careful compilation of a large amount of numerical data from the literature provided the basis for Appendix II. Finally, I am indebted to those authors and publishers who gave permission for the use of several diagrams used in the text.

H. R. ALLCOCK

Part I

INTRODUCTION AND STRUCTURAL CHEMISTRY

Chapter 1

INTRODUCTION, HISTORICAL BACKGROUND, AND NOMENCLATURE

A. Introduction

This book reviews the chemistry of those ring and chain compounds which contain alternating phosphorus and nitrogen atoms in their skeleton. Three important groups of compounds fall into this category—the cyclo- and polyphosphazenes or phosphonitriles, the monophosphazenes or phosphinimines, and the phosphazanes.

The cyclo- or polyphosphazenes (phosphonitriles) are probably the best known and most intensively studied phosphorus–nitrogen derivatives. They are ring or chain compounds which have two substituents connected to each phosphorus atom, but no substituents on nitrogen, and they are characterized by a valence-unsaturated skeleton. Representative structures are the cyclic trimer (I), the cyclic tetramer (II), and the high polymer (III). The substituent,

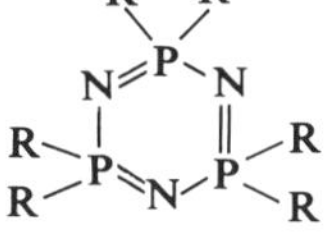

I

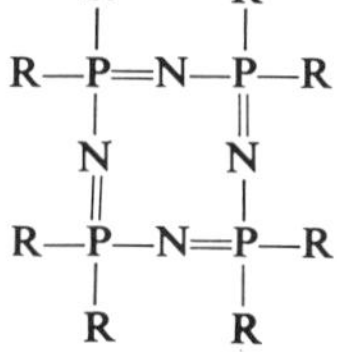

II

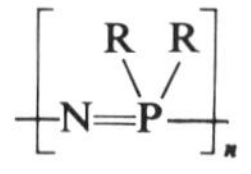

($n \approx 15{,}000$)

III

R, can be halogen, pseudohalogen, amino, azido, or a wide variety of organic groups such as alkoxy, aryloxy, alkylamino, arylamino, alkyl, or aryl. Mixed substitution is also possible. It is generally believed that the number of possible homologues in each polymeric series is very large, with the range extending from cyclic trimers in a continuous series to very high molecular weight cyclic or linear polymers. Stable monomers of structure, $N{\equiv}PR_2$, and cyclic dimers $(N{=}PR_2)_2$, have not been isolated in this series. Most phosphazene cyclic trimers and tetramers are white, crystalline solids which are soluble in organic liquids. The linear high polymers are elastomeric or thermoplastic. Nearly all these derivatives are stable to atmospheric oxygen and moisture. Phosphazene derivatives are also known in which short-chain segments are "end-capped" by the elements of an inorganic halide. Phosphorus pentachloride is an effective end-capping reagent, giving rise to compounds of the general type depicted in IV. These compounds are reminiscent of the monophosphazenes or phosphin-

$$Cl_4P\text{—}\left(N{=}\overset{Cl_2}{P}\right)_n\text{—}N{=}PCl_3$$

IV

imines to be introduced below. Many polymeric phosphazene derivatives are formed by the cross-linking of rings or chains.

Monophosphazenes or phosphinimines have the structure V. They show

$$RN{=}PR_3$$

V

many fundamental similarities to the cyclic and polymeric phosphazenes mentioned above. They are particularly interesting from the viewpoint of their bonding structure, and much of the physical-inorganic work carried out on these compounds has an important bearing on the skeletal bonding found in cyclic and polymeric phosphazenes.

The phosphaz*anes* constitute a diverse group of compounds which have the common characteristic of a formally saturated phosphorus–nitrogen skeletal bond. Several different types of phosphazane compounds are known, including the cyclotriphosphazanes or trimetaphosphimates depicted in VI and the cyclic

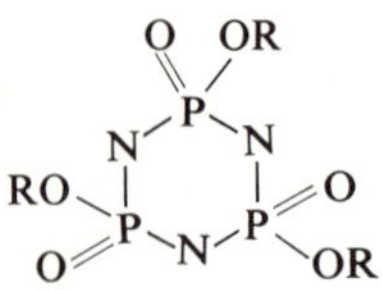

VI

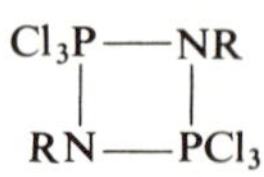

VII

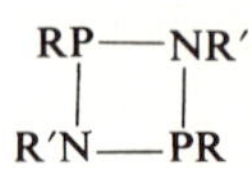

VIII

dimeric derivatives of the types shown in VII and VIII. It should be emphasized that other groups of phosphorus–nitrogen compounds are known, such as the phosphinic amides, $R_2(O)P\text{—}NR_2$, and aminophosphines, $R_2P\text{—}NH_2$. These latter compounds will not be reviewed in detail here.

Although the phosphazenes and phosphazanes are unique in their own right, certain formal similarities to other heteroatom systems can be discerned. Thus, cyclo- and polyphosphazenes may be compared to cyclo- or polysiloxanes and silthianes, and certain phosphazanes are reminiscent of silazanes. There are also formal structural analogies between phosphazenes, phosphazanes, and metaphosphates, or even with polyoxymethylenes. Such similarities are valuable for the rationalization of molecular conformations or for an understanding of the packing arrangements within a crystalline or polymeric lattice. However, bonding characteristics are nearly always more important than superficial structural resemblances, and the special properties of the phosphorus–nitrogen bond are largely responsible for the unique properties of phosphazenes and phosphazanes.

B. Historical Background

The earliest phosphazene compounds to be synthesized were the chlorophosphazenes, $(NPCl_2)_n$, which were prepared by the interaction of phosphorus pentachloride with ammonia. This reaction was first studied by Liebig and Wöhler[1] and by Rose[2] in 1834. The principal reaction product was phospham, $(NPNH)_n$, although Liebig and Wöhler also isolated a small amount of what appears to have been hexachlorocyclotriphosphazene, $(NPCl_2)_3$. This compound was a white, crystalline solid which could be steam-distilled, or heated with hot acid or base without decomposition. Ten years later Gerhardt[3,4] and Laurent[5] showed by analysis that the correct empirical formula was $NPCl_2$, and eighteen years thereafter Gladstone and Holmes[6–8] and subsequently Wichelhaus[9] used vapor density measurements to show that the molecular formula was, in fact, $(NPCl_2)_3$. A growing interest in phosphazene chemistry was evident in the closing years of the nineteenth century, when papers by Gladstone,[10–14] Besson,[15–17] Besson and Rosset,[18,19] Couldridge,[20,21] and particularly by Stokes[22–28] laid the foundations of the substitution, hydrolysis, and polymerization reactions of phosphazenes. Stokes first suggested the cyclic structure of $(NPCl_2)_3$. He also identified a series of higher cyclic homologues, $(NPCl_2)_{4-7}$, and described the thermal polymerization of chlorophosphazenes to "inorganic rubber" and the subsequent depolymerization of this at higher temperatures and reduced pressures. Stokes also studied the hydrolysis reactions of chlorophosphazenes to "metaphosphimic acids" and he was able to isolate salts of derivatives that we now call cyclophospha*zanes*. Ammonolysis reactions of chlorophosphazenes were also investigated by de Ficquelmont.[29]

From this point, in the early 1900s, it is possible to discern the beginnings of distinct lines of research which have led to increased specialization. Improvements to the early synthetic procedures were evident in the important work of Schenk and Römer[30] in 1924. They developed a much improved method for the synthesis of chlorophosphazenes and, even today, this remains the basis of the commercial production of these materials. The overall reaction scheme is

$$n\ PCl_5 + n\ NH_4Cl \xrightarrow[120^\circ C]{\text{Solvent}} (NPCl_2)_n + 4n\ HCl$$

The detailed mechanism of this reaction was not investigated until the early 1960s.[31–33] Meanwhile, fluorophosphazenes, $(NPF_2)_n$, had been prepared by an alternative route in 1956,[34] and bromophosphazenes, $(NPBr_2)_n$, were first reported in 1960.[35] The first direct syntheses of several organophosphazenes, $(NPR_2)_n$, were also described at this time.[36–42]

Since the mid-1950s there has been an almost explosive increase in research on the substitution reactions of halophosphazenes. Aminolysis reactions have been studied in particular detail,[43] but alcoholysis, phenolysis, metathetical replacements, Friedel-Crafts reactions, and many other reactions have been examined by different groups of investigators. Large sections of this book are devoted to a discussion of such reactions.

The polymer chemistry of phosphazenes dates back to the early work of Stokes. However, the sustained interest in phosphazenes as inorganic polymers has been partly responsible for the phenomenal growth of synthetic activity in this field since the 1950s. During this time, there has been a steadily rising demand by the aerospace industry and other organizations for new and unusual polymers. Many phosphazene-based high polymers have been developed, and this area promises to have an important technological future. Three important lines of research have evolved in the polymeric area—investigations of the polymerization and depolymerization reactions of linear and cyclic species,[44] the synthesis of linear, high molecular weight organophosphazene polymers,[45, 46] and the development of cross-linked cyclomatrix polymers.[47, 48]

The six-membered cyclic structure for hexachlorocyclotriphosphazene was first proposed by Stokes. However, it was many years later before a detailed definitive structural analysis was instigated. The first X-ray structural examination of a phosphazene was reported in 1936 by Meyer, Lotmar, and Pankow.[49] They measured the chain-repeating distance of the rubbery high polymer, $(NPCl_2)_n$, and from this value they deduced the P—N bond distance and the chain angles. Then in 1939 Ketelaar and de Vries[50] reported an X-ray structural determination of the chloro-substituted tetramer, $(NPCl_2)_4$, and this work provided accurate bond lengths and angles. The structure of the trimer, $(NPCl_2)_3$, was investigated by Brockway and Bright,[51] who reported the results of an electron-diffraction study in 1943. However, it was not until 1958, when Wilson and Carroll[52] described an X-ray structural analysis of $(NPCl_2)_3$, that the use of the X-ray method in this field gathered momentum. Since that time, a steadily increasing amount of structural data for phosphazenes and phosphazanes has become available, and this has markedly facilitated the development of structural and bonding theories.

Other physical techniques have proved to be powerful tools for the structural examination of phosphazenes. Infrared and Raman spectroscopy have been used extensively since 1954,[53] and ultraviolet spectroscopy, nuclear magnetic resonance, nuclear quadrupole resonance, electron spin resonance, thermochemistry, mass spectrometry, and many other techniques have now been applied to the study of phosphazene structures. As might be expected, the development of bonding theories has paralleled the structural investigations. The first reasonable theories of bonding in phosphazenes were put forward between 1958 and 1961.[54–56] Each of these theories invoked the postulate that π-bonding existed in the phosphazene framework as a result of involvement by the phosphorus $3d$ orbitals. This view will be discussed in Chapter 2.

It is also worthwhile to recall the parallel development of the related series of monomeric phosphazenes, the phosphinimines, $RN{=}PR_3$. The first examples were isolated by Staudinger and co-workers in the early 1920s.[57–59] Since that time, a considerable chemistry of these derivatives has been developed, much of which is complementary to that of the cyclic and polymeric phosphazenes. The relationships between these two groups of phosphorus–nitrogen compounds will be discussed in later sections of this book.

The expanding development of phosphorus–nitrogen chemistry has been paralleled by the appearance of several journal reviews during recent years. Reviews by Audrieth, Steinman, and Toy,[60] Becke-Goehring,[61] Gribova and Ban-yuan,[62] Shaw, Fitzsimmons, and Smith,[63] Shaw, Keat, and Hewlett,[64] Schmulbach,[65] Paddock,[66, 67] Haber,[68] Fluck,[69] and the writer,[70] illustrate the expansion of interest and change of emphasis during the last 25 years. Although it is difficult to predict the future course of any branch of science, there is much evidence that phosphazene polymer chemistry, reaction mechanism studies, structural investigations, and possibly biomedical applications will attract increasing attention during the future.

C. Nomenclature

In common with most other inorganic heterocyclic or chain systems, much confusion exists about the nomenclature of phosphorus–nitrogen compounds.[71] Three main naming systems have evolved in recent years. These are the "phosphazene," the "phosphonitrilic," and the "hydroazaphosphorine" notations. In addition, for low molecular weight linear species, names based on the "phosphinimine" or "phosphazo" notations are common. Of these systems, the phosphazene-type notation has much to offer from the viewpoint of systematics, and this is the nomenclature to be employed in this book. However, brief descriptions of the most important alternative naming methods are also given here.

1. The Phosphazene–Phosphazane Notation

For cyclic derivatives and high polymers, the phosphazene nomenclature system is the most systematic naming method available at this time. It has the added advantage of being readily applied to different members of the large polymeric series found in phosphorus–nitrogen chemistry. The system bears many resemblances to the nomenclature approach used in organosilicon chemistry and for similar structural reasons.

The basis of the phosphazene notation is that compounds which contain the repeating unit IX are known as phosphazenes, while those which have the general structure X are called phosphazanes.[63] Thus, phosphorus–nitrogen

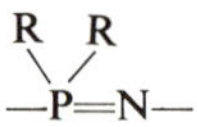

IX

X

compounds can be considered to be substituted derivatives of the hypothetical parent compounds $HN{=}PH_3$ (phosphazene), $H_2N{-}PH_4$ (phosphazane), or $N{\equiv}PH_2$ (phosphazyne or phosphazine).*

The degree of polymerization is indicated by the prefixes tri-, tetra-, penta-, poly-, etc. A *ring system* is denoted by the prefix cyclo-, and a linear chain is denoted by the absence of the cyclo prefix. Thus, compounds of structure XI are called cyclotriphosphazenes, those of type XII are cyclotetraphosphazenes, and polymers of type XIII are polyphosphazenes.

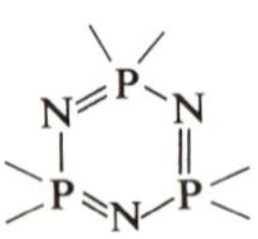

XI

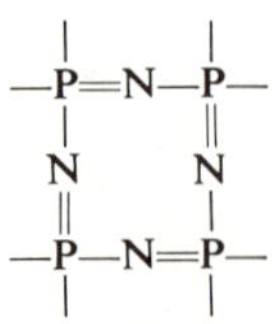

XII

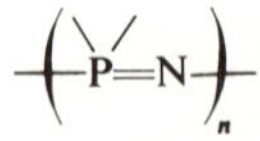

XIII

* It is equally valid to consider cyclo- and polyphosphazenes, $(N{=}PR_2)_n$, as addition polymers of $N{\equiv}PR_2$, in which case, a reasonable argument could be put forward for calling them cyclo- or polyphosphazynes or -phosphazines. A related complication arises when we attempt to distinguish between monophosphazenes of the phosphinimine type ($RN{=}PR_3$) and the (yet undetected) first member of a cyclophosphazene homologous series ($N{\equiv}PR_2$).

The designation of *substituents* attached to the phosphazene ring or chain is quite straightforward, provided agreement is reached about the skeletal numbering system. Some authors commence the numbering at phosphorus and some at nitrogen. We shall employ the former system and number from phosphorus. Compound XIV is, therefore, called 1,3,5-trichloro-1,3,5-tris-(dimethylamino)cyclotriphosphazene, while compound XV is 1,1-diphenylhexachlorocyclotetraphosphazene. It is sometimes simpler to refer to substituents as being geminal or non-geminal. Thus, XV could be described as the

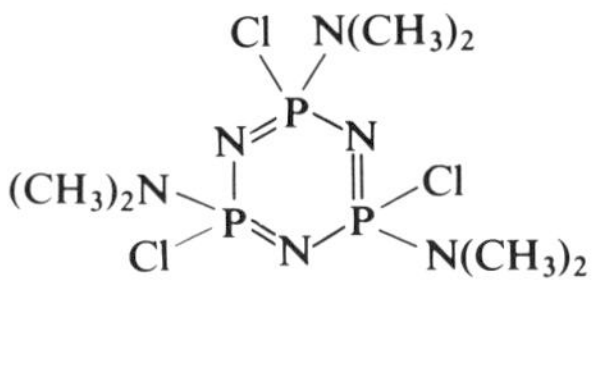

XIV

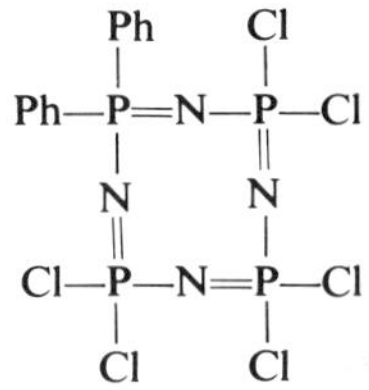

XV

gem-diphenyltetrachloro tetramer, and XIV as the *non-geminal* trichlorotris-(dimethylamino) trimer. Some authors expand the names to emphasize the number of classical double bonds in the skeletal structure.[63] For example, compounds of type XIV would be called cyclotriphosphaza*trienes*. We will not use this designation unless the rings are partially saturated.

Monomeric species of the $RN{=}PR_3$ type are called phosphinimines, phosphine imines, phosphinimides, phosphine imides or, when phenyl groups are present, phosphazoaryls. They can be named in a straightforward manner by the use of the monophosphazene notation. Thus, the compound, $Cl_3P{=}NCCl_3$, would be called 1-trichloro-2-(trichloromethyl)monophosphazene, or *P*-trichloro-*N*-trichloromethylmonophosphazene.

Some care must be exercised when naming compounds of the phospha*zane* type. When necessary, the classical valency of phosphorus can be specified by Roman numerals III or V as in phospha(III)azane ($H_2P{-}NH_2$), or phospha-(V)azane ($H_4P{-}NH_2$). When chains of alternating phosphorus and nitrogen atoms are present, the prefixes di, tri, etc., can be used as before, although it should be noted that such designations do not imply the identity of all repeating units [e.g., $H_2N{-}PH_3{-}NH{-}PH_4$, diphospha(V)azane]. The position of substituents in phosphazanes can be designated by a numbering system [e.g., 1,3,5-tris(methoxyoxo)-2,3,6-trimethylcyclotriphosphazane for XVI] or by

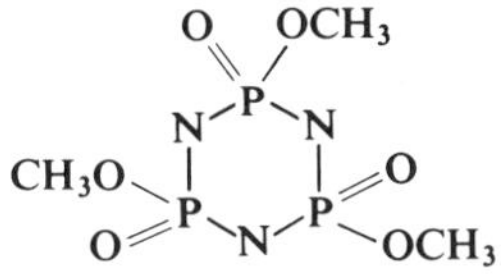

XVI

elemental designation, as in *P*-tris(methoxyoxo)-*N*-trimethylcyclotriphosphazane. XVII to XXIV demonstrate further the use of this nomenclature system. In these and later formulas and equations, the symbols Ph, Me, Et, Pr, and Bu, refer to phenyl, methyl, ethyl, propyl, and butyl groups, respectively.

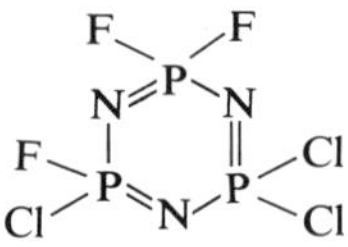

XVII

1,1,3-Trifluoro-3,5,5-trichlorocyclotriphosphazene

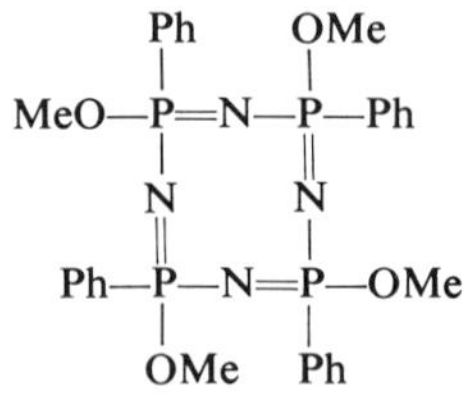

XVIII

1,3,5,7-Tetraphenyl-1,3,5,7-tetramethoxycyclotetraphosphazene

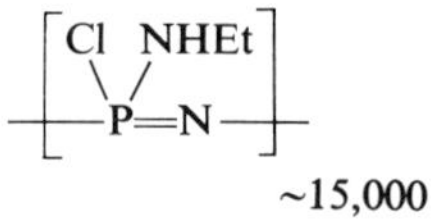

XIX

Non-geminal poly(ethylaminochlorophosphazene)

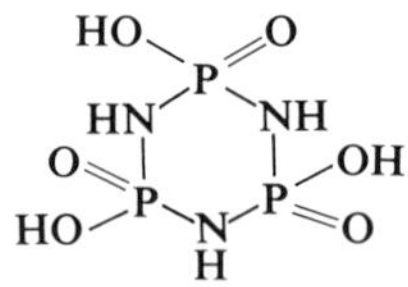

XX

1,3,5-Tris(hydroxyoxo)-2,4,6-trihydrocyclotriphospha(V)azane (Trimetaphosphimic acid)

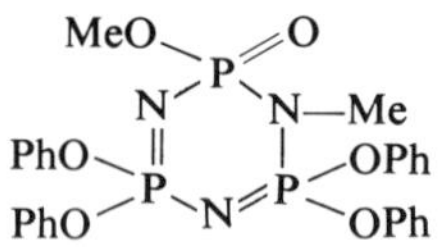

XXI

1,1-Methoxyoxo-3,3,5,5-tetraphenoxy-2-methylcyclotriphosphazadiene

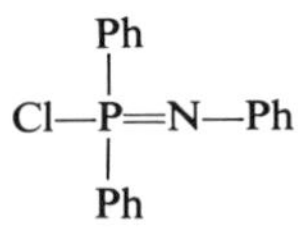

XXII

P-Diphenylchloro-*N*-phenylmonophosphazene

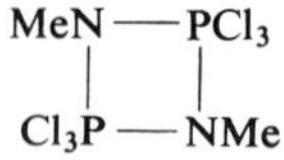

XXIII

1,1,1,3,3,3-Hexachloro-2,4-dimethylcyclodiphospha(V)azane

XXIV

1,3-Bis(phenylamino)-2,4-diphenylcyclodiphospha(III)azane

2. The Phosphonitrilic Notation

This is an older, more established naming system than the phosphazene notation. However, it suffers from the disadvantages of nonuniformity and the difficulty of adapting it to a systematic approach for different types of phosphorus–nitrogen compounds. It is also inaccurate since the term "phosphonitrile" implies similarities to the p_π–p_π triple bonding found in nitriles, a relationship which is at variance with the facts. Nevertheless it is widely used as a general, informal naming system for uncomplicated structures.

Thus, compounds XI, XIV, or XVII are known as phosphonitrilic trimers and XII, XV, or XVIII as phosphonitrilic tetramers. The well-known derivative $(NPCl_2)_3$ is known as phosphonitrilic chloride trimer or triphosphonitrilic chloride, whereas $[NP(OPh)_2]_4$ would be called octaphenoxyphosphonitrile tetramer. Compounds of structure $[NP(NH_2)_2]_n$ are called phosphonitrilamides. Many confusing modifications are found in the literature including terms such as nitrilophosphoric chloride, phosphoronitridic dichloride, chlorophosphinic nitride, and phosphorus chloronitride, for compounds of formula $(NPCl_2)_n$, and open-chain derivatives.

The system is not readily adaptable to include a classification of phosphazane or phosphinimine type compounds. Cyclophosphazanes, such as X or XVI, are often referred to as metaphosphimates by derivation from metaphosphimic acids, $[HN—P(O)OH]_n$.

3. The Hydroazaphosphorine Notation

Strict use of the standard IUPAC nomenclature system for heterocyclic compounds[72] leads to a rather curious naming system for phosphorus–nitrogen compounds. Thus, the compound $(NPF_2)_3$ would be called 2,2,4,4,6,6-hexafluoro-2,2,4,4,6,6-hexahydro-1,3,5,2,4,6-triazatriphosphorine, and $(NPPh_2)_4$ becomes 2,2,4,4,6,6,8,8-octaphenyl-2,2,4,4,6,6-octahydro-1,3,5,7,2,4,6,8-tetrazaphosphocine. The main defect of this type of nomenclature, apart from its unwieldiness, is the extreme difficulty involved in naming the higher cyclic and linear members of a polymeric series. The system has been used in *Chemical Abstracts* and in some organic-oriented journals, but it has not gained favor among those working in this area.

4. Stereochemical Nomenclature

Irrespective of the general nomenclature system employed, some provision must be made for the depiction of geometrical isomerism, which is a common feature of phosphorus–nitrogen chemistry. Geometrical isomerism can usually be described adequately in terms of the familiar *cis–trans* notation. Thus derivative XXV would be called *cis*-1,3-diphenyltetrachlorocyclotriphosphazene. More complexity is evident when geometrical isomerism exists in cyclic

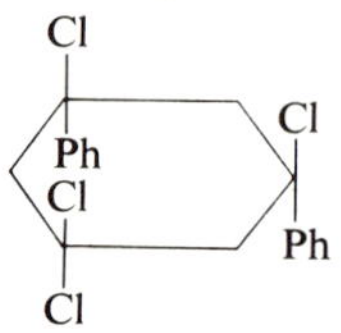

XXV

tetramers. One method which has been used to classify non-geminal substituents is to describe species as α, β, and γ-*trans* isomers.[73] Thus, three *trans*-type isomers are known for $(NPClPh)_4$, and these are depicted as XXVI, XXVII, and XXVIII.

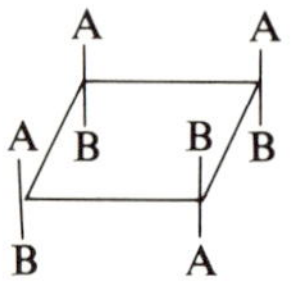

XXVI
α-*trans*

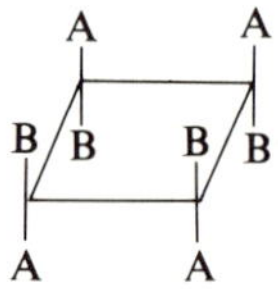

XXVII
β-*trans*

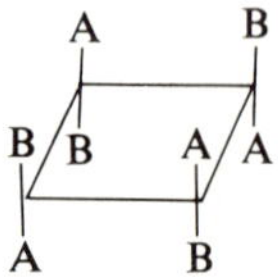

XXVIII
γ-*trans*

When puckered rings are present, additional opportunities exist for stereochemical isomerism. Cyclic trimers can exist as boat and chair forms (XXIX

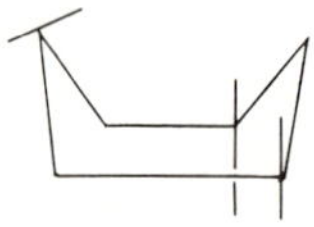

XXIX
Boat

XXX
Chair

and XXX), and cyclic tetramers may occupy tub (XXXI), chair (XXXII), or crown (XXXIII) conformations. Many more conformations are, of course,

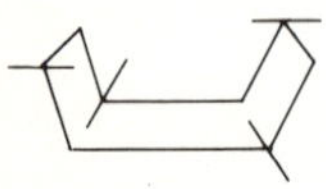

XXXI
Tub

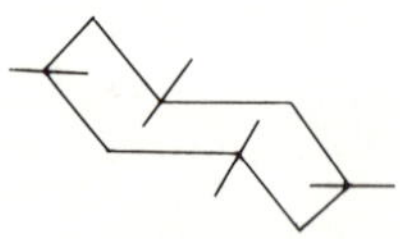

XXXII
Chair

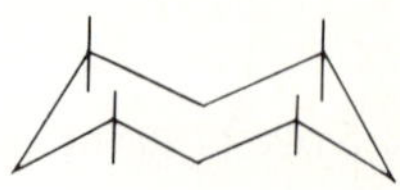

XXXIII
Crown

possible for the higher cyclic homologues. For high polymers, the theoretical number of conformations is very large but, in practice, three conformations are to be anticipated. These are the *trans*-planar (XXXIV), the *cis–trans* planar (XXXV), and the gauche or helical conformation (XXXVI).

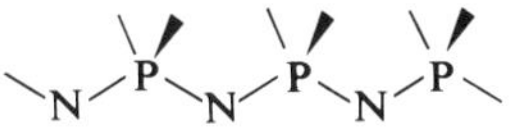

XXXIV
trans planar

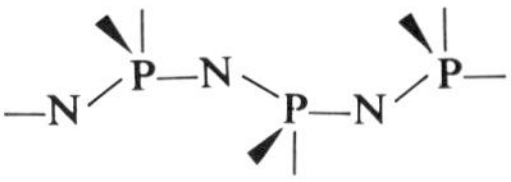

XXXV
cis–trans planar

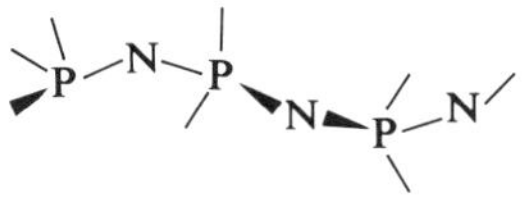

XXXVI
Helical

Optical isomerism is also a possibility for many phosphorus–nitrogen compounds, and at least one example has been reported.[74]

References

1. J. Liebig, *Ann. Chem.* **11**, 139 (1834).
2. H. Rose, *Ann. Chem.* **11**, 131 (1834).
3. C. Gerhardt, *Ann. Chim. Phys.* [3] **18**, 188 (1846).
4. C. Gerhardt, *C. R. Acad. Sci.* **22**, 858 (1846).
5. A. Laurent, *C. R. Acad. Sci.* **31**, 356 (1850).
6. J. H. Gladstone and J. D. Holmes, *J. Chem. Soc., London* **17**, 225 (1864).
7. J. H. Gladstone and J. D. Holmes, *Ann. Chim. Phys.* [4] **3**, 465 (1864).
8. J. H. Gladstone and J. D. Holmes, *Bull. Soc. Chim. Fr.* [2] **3**, 113 (1865).
9. H. Wichelhaus, *Chem. Ber.* **3**, 163 (1870).
10. J. H. Gladstone, *Ann. Chem.* **76**, 74 (1850).
11. J. H. Gladstone, *J. Chem. Soc., London* **2**, 121 (1850).
12. J. H. Gladstone, *J. Chem. Soc., London* **3**, 135 (1851).
13. J. H. Gladstone, *J. Chem. Soc., London* **3**, 353 (1851).
14. J. H. Gladstone, *Ann. Chem.* **77**, 314 (1851).
15. A. Besson, *C. R. Acad. Sci.* **111**, 972 (1890).
16. A. Besson, *C. R. Acad. Sci.* **114**, 1264 (1892).
17. A. Besson, *C. R. Acad. Sci.* **114**, 1479 (1892).
18. A. Besson and G. Rosset, *C. R. Acad. Sci.* **143**, 37 (1906).
19. A. Besson and G. Rosset, *C. R. Acad. Sci.* **146**, 1149 (1908).
20. W. Couldridge, *J. Chem. Soc., London* **53**, 398 (1888).
21. W. Couldridge, *Bull. Soc. Chim. Fr.* [2] **50**, 535 (1888).
22. H. N. Stokes, *Amer. Chem. J.* **17**, 275 (1895).
23. H. N. Stokes, *Chem. Ber.* **28**, 437 (1895).
24. H. N. Stokes, *Amer. Chem. J.* **18**, 629 (1896).

25. H. N. Stokes, *Amer. Chem. J.* **18**, 780 (1896).
26. H. N. Stokes, *Amer. Chem. J.* **19**, 782 (1897).
27. H. N. Stokes, *Amer. Chem. J.* **20**, 740 (1898).
28. H. N. Stokes, *Z. Anorg. Chem.* **19**, 36 (1899).
29. A. M. de Ficquelmont, *C. R. Acad. Sci.* **200**, 1045 (1935).
30. R. Schenck and G. Römer, *Chem. Ber.* **57B**, 1343 (1924).
31. L. G. Lund, N. L. Paddock, J. E. Proctor, and H. T. Searle, *J. Chem. Soc., London* p. 2542 (1960).
32. M. Becke-Goehring and E. Fluck, *Angew. Chem.* **74**, 382 (1962).
33. M. Becke-Goehring and W. Lehr, *Z. Anorg. Allg. Chem.* **327**, 128 (1964).
34. F. Seel and J. Langer, *Angew. Chem.* **68**, 461 (1956).
35. K. John and T. Moeller, *J. Amer. Chem. Soc.* **82**, 2647 (1960).
36. C. P. Haber, D. L. Herring, and E. A. Lawton, *J. Amer. Chem. Soc.* **80**, 2116 (1958).
37. R. A. Shaw and C. Stratton, *Chem. Ind.* (*London*) p. 52 (1959).
38. A. J. Bilbo, *Z. Naturforsch B* **15**, 330 (1960).
39. V. V. Korshak, I. A. Gribova, T. V. Artamonova, and A. N. Bushmarina, *Vysokomol. Soedin.* **2**, 377 (1960).
40. I. I. Bezman and J. H. Smalley, *Chem. Ind.* (*London*) p. 839 (1960).
41. D. L. Herring, *Chem. Ind.* (*London*) p. 717 (1960).
42. G. Tesi, C. P. Haber, and C. M. Douglas, *Proc. Chem. Soc., London* p. 219 (1960).
43. R. Keat and R. A. Shaw, *Angew. Chem., Int. Ed. Engl.* **7**, 212 (1968), and references cited therein.
44. H. R. Allcock, *Inorg. Chem.* **5**, 1320 (1966).
45. H. R. Allcock, R. L. Kugel, and K. J. Valan, *Inorg. Chem.* **5**, 1709 (1966).
46. H. R. Allcock and R. L. Kugel, *Inorg. Chem.* **5**, 1716 (1966).
47. C. A. Redfarn, U.S. Pat. 2,866,733 (1958).
48. R. G. Rice, B. H. Geib, and L. A. Kaplan, U.S. Pat. 3,121,704 (1964).
49. K. H. Meyer, W. Lotmar, and G. W. Pankow, *Helv. Chim. Acta* **19**, 930 (1936).
50. J. A. A. Ketelaar and T. A. de Vries, *Rec. Trav. Chim. Pays Bas* **58**, 1081 (1939).
51. L. O. Brockway and W. M. Bright, *J. Amer. Chem. Soc.* **65**, 1551 (1943).
52. A. Wilson and D. F. Carroll, *Chem. Ind.* (*London*) p. 155 (1958); *J. Chem. Soc., London* p. 2548 (1960).
53. L. W. Daasch, *J. Amer. Chem. Soc.* **76**, 3403 (1956).
54. D. P. Craig and N. L. Paddock, *Nature* (*London*) **181**, 1052 (1958); *J. Chem. Soc., London* p. 4118 (1962).
55. D. W. J. Cruickshank, *J. Chem. Soc., London* p. 5486 (1961).
56. M. J. S. Dewar, E. A. C. Lucken, and M. A. Whitehead, *J. Chem. Soc., London* p. 2423 (1960).
57. H. Staudinger and E. Hauser, *Helv. Chim. Acta* **4**, 861 and 887 (1921).
58. H. Staudinger and J. Meyer, *Helv. Chim. Acta* **2**, 635 (1919).
59. H. Staudinger and J. Meyer, *Chem. Ber.* **53**, 72 (1920).
60. L. F. Audrieth, R. Steinman, and A. D. F. Toy, *Chem. Rev.* **32**, 109 (1943).
61. M. Becke-Goehring, *Quart. Rev., Chem. Soc.* **10**, 437 (1956).
62. I. A. Gribova and U. Ban-yuan, *Russ. Chem. Rev.* **30**, 1 (1961).
63. R. A. Shaw, B. W. Fitzsimmons, and B. C. Smith, *Chem. Rev.* **62**, 247 (1962).
64. R. A. Shaw, R. Keat, and C. Hewlett, *Prep. Inorg. React.* **2**, 1 (1965).
65. C. D. Schmulbach, *Progr. Inorg. Chem.* **4**, 275 (1962).
66. N. L. Paddock and H. T. Searle, *Advan. Inorg. Chem. Radiochem.* **1**, 348–383 (1959).
67. N. L. Paddock, *Quart. Rev., Chem. Soc.* **18**, 168 (1964).
68. C. P. Haber, *Chem. Soc., Spec. Publ.* **15**, 115 (1961).

69. E. Fluck, *Top. Phosphorus Chem.* **4**, 291 (1967).
70. H. R. Allcock, *Chem. & Eng. News* **46**, 68 (1968).
71. H. R. Allcock, "Heteroatom Ring Systems and Polymers." Academic Press, New York, 1967.
72. *J. Amer. Chem. Soc.* **82**, 5566 (1960).
73. B. Grushkin, A. J. Berlin, J. L. McClanahan, and R. G. Rice, *Inorg. Chem.* **5**, 172 (1966).
74. C. D. Schmulbach, C. Derderian, O. Zeck, and S. Sahuri, *Inorg. Chem.* **10**, 195 (1971).

Chapter 2

THEORIES OF BONDING IN PHOSPHAZENES

A. Introduction

A characteristic trend in modern chemistry is the drive to correlate reactivity and physical properties with molecular structure. This trend is particularly obvious in phosphorus–nitrogen chemistry, where an impressive background of synthetic information can now be rationalized in terms of structural data. Molecular structural results generally fall into two closely related categories—information about molecular geometry, bond angles and distances, the existence of rings or chains, the presence of planar or puckered ring systems, etc., and information about the electronic structure of molecules, bond strengths, orbital hybridization, and the presence of π bonding. Theoretical models of molecular structure provide a valuable tool for understanding both types of results. In this chapter, several models for the electronic structure of phosphorus–nitrogen compounds will be considered, and the following chapter will review the experimental structural data which pertain to these theories.

B. Anomalies of the Phosphorus–Nitrogen Bond

A large number of compounds are known in which tetracoordinate phosphorus is linked to dicoordinate nitrogen. The monophosphazenes (I), cyclotriphosphazenes (II), cyclotetraphosphazenes (III), and high molecular weight polyphosphazenes (IV), are all characterized by this bonding arrangement. The formation of the σ-bond framework in such compounds utilizes four of the five valence shell electrons of phosphorus and two of the five valence electrons of nitrogen. Thus, one electron from phosphorus and three from nitrogen remain to be accounted for (I–IV). It is widely assumed that two of the

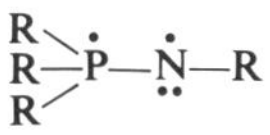

I

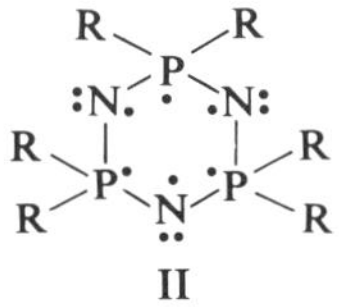

II

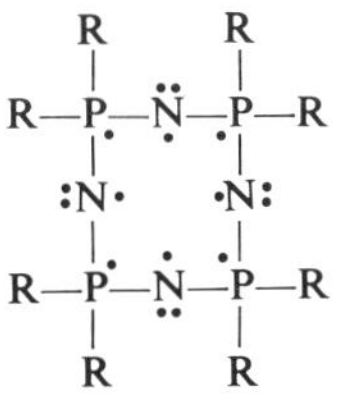

III

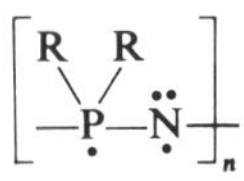

IV

remaining electrons at nitrogen occupy a nonbonding, lone-pair orbital, while the remaining electron from nitrogen interacts with the one from phosphorus to form some kind of π bond. We will examine this viewpoint later, but first it is instructive to briefly consider several bonding anomalies associated with this class of compounds.

A detailed examination of the structural evidence will be considered in Chapter 3, but the following is a brief summary of the pertinent facts:

(1) Phosphazene ring and chain structures are surprisingly stable.

(2) Phosphorus–nitrogen bond distances in phosphazenes are shorter than expected for pure covalent σ bonds. The shortest skeletal bond lengths are associated with the presence of the most electronegative ligands.

(3) The skeletal bond lengths are usually equal around the ring except when an unsymmetrical ligand arrangement exists.

(4) Skeletal bond angles at phosphorus in cyclo- and polyphosphazenes approximate to 120°, while those at nitrogen vary from 120° to 148.6°.

(5) Some cyclic phosphazenes are planar and some are puckered, but the presence or absence of puckering appears to have little effect on molecular stability.

(6) The skeletal nitrogen atoms in cyclophosphazenes function as basic coordination sites, especially when electron-supplying ligands are attached to phosphorus.

(7) Spectral effects normally associated with organic π-electron systems, such as the bathochromic ultraviolet shifts which accompany increased delocalization, are not found for cyclo- or polyphosphazenes.

(8) Unlike organic aromatic systems, the phosphazene skeleton is difficult to reduce electrolytically.

Thus, the phosphorus–nitrogen bond in mono-, cyclo-, or polyphosphazenes differs from a covalent σ bond and yet does not resemble the familiar σ–π bond of organic aromatic compounds.

The problem of understanding the bonding arrangement in phosphazenes is one which is common to a number of inorganic-type ring systems and polymers.[1] In particular, alternating heteroatom compounds with skeletal bonds between second- or third-row elements, such as aluminum, silicon, phosphorus, sulfur, germanium, arsenic, or selenium, and oxygen or nitrogen often show bonding anomalies similar to those listed above. Thus, an understanding of the bonding situation in phosphorus–nitrogen compounds has wide implications for other areas of inorganic and organometallic chemistry. Some of the electronic characteristics of nitrogen and phosphorus atoms will be summarized in the following sections, and possible bonding models will then be discussed.

C. Bonding Characteristics of Nitrogen

Compounds of nitrogen exist in which bonds are formed to one, two, three, or four other atoms. Nitriles, *s*-triazines, amines, and ammonium salts provide respective examples of these bonding states. This progression is accompanied by a change from a six-electron bond in nitriles, and a delocalized four-electron bond in *s*-triazines, to two-electron bonds in amines and ammonium salts, a trend which is reflected in an increase in the C—N bond distance and a narrowing of the angle at nitrogen (Table 2.1)[2–5]. It is also known that *s*-triazines form planar rings, that amines are pyramidal, and that ammonium ions are tetrahedral.

TABLE 2.1

BOND DISTANCES AND ANGLES IN NITROGEN COMPOUNDS

Compound	C—N bond length (Å)	Bond angles at nitrogen (°)	Electrons per C—N bond	Nonbonding valence electrons at nitrogen	References
C≡N	1.16	—	6	2 (s)	2
$-\!(\!-HC\!=\!=N\!-\!)\!-_3$	1.32	113	4	2 (sp^2)	3
Me_3N	1.47	108 ± 0.02	2	2 (sp^3)	4
$Me_4N^{\oplus}$	1.55	109.4	2	0	5

The disposition of the electrons in such molecules appears to be well established. In nitriles, one of the five valence electrons participates in σ-bond formation, two others contribute to two mutually orthogonal p_π–p_π bonds, and the two remaining electrons occupy a nonbonding $2s$ orbital. The situation in *s*-triazines is apparently similar to that found in phosphazenes. The C—N—C bond angle is close to 120°, and the hybridization at nitrogen appears to be sp^2, with an additional p_z orbital directed at right angles to the ring plane. Thus, two electrons from nitrogen occupy two of the sp^2 orbital lobes for σ bonding to carbon, two occupy the remaining sp^2 orbital lobe as a nonbonding lone pair directed outward in the ring plane, and the fifth uses the p_z orbital for p_π–p_π bonding to carbon. In amines, the hybridization is sp^3, with two electrons occupying a lone-pair orbital directed along one of the four tetrahedral axes. In ammonium-type ions, all the valence electrons are involved in forming four σ bonds through an sp^3-hybrid arrangement.

Thus, we may anticipate that, in phosphazenes, a 120° ring angle is suggestive of sp^2-type hybridization with a lone-pair orbital directed radially, and with an additional electron available for π-type interactions with phosphorus. A wider angle at nitrogen implies a change in hybridization whereby the lone-pair electrons assume more s character, the σ bonds to phosphorus take on more sp character, and one electron remains available for additional interactions with phosphorus through a nitrogen $2p_z$ orbital. In phosphazanes and in those phosphazene adducts which utilize the nitrogen lone-pair electrons, the hybridization scheme at nitrogen should resemble that of amines.

D. Bonding Characteristics of Phosphorus

Phosphorus forms stable compounds in which it is bonded to from three to six other atoms. The electronic arrangement in tricoordinate phosphorus compounds is fairly well understood. Three electrons participate in σ-bond formation to three other atoms, and the remaining two electrons form a nonbonding lone pair. Bond angle measurements suggest that the hybridization scheme lies somewhere between p^3 and sp^3.

1. The Possibility of *d*-Orbital Participation

For tetracoordinate compounds of phosphorus, and especially for penta- and hexacoordinate species, the suggestion has been made that the $3d$ orbitals may also participate in the bonding schemes.[6–23] First-row elements are unable to make use of *d* orbitals but, for second-row elements such as phosphorus, the energy of the $3d$ state is not prohibitively higher than that of the $3p$ state. The possibility exists, therefore, that $3spd$ hybrid orbital arrangements could exist in penta- or hexacoordinate structures and that, in tetracoordinate phosphorus compounds, such as phosphazenes, the 3*d* orbitals could be occupied for the purpose of π bonding. The latter situation would correspond to a stabilized excited state. Some dispute exists about whether or not phosphorus and the other second-row elements can make significant use of *d*-orbital arrangements. The following sections outline the arguments that have been used for and against this viewpoint.

2. Arguments for the Availability of 3*d* Orbitals

The evidence used to demonstrate that the second-row main group elements can utilize their 3*d* orbitals can be divided into the following categories: (a) the existence of stable penta- and hexacoordinate compounds; (b) evidence from reaction mechanisms; (c) evidence from "back-bonding" studies; and (d) arguments based on atomic and molecular orbital calculations.

a. *Penta- and Hexacoordinate Compounds*

Pentacoordinate phosphorus compounds such as PF_5, $MePF_4$, Me_2PF_3, PF_3Cl_2, PPh_5, and PCl_5 (vapor) exist as stable species. The structures are those of trigonal bipyramidal molecules in which the axial bonds are longer and weaker than the equatorial ones.[24–33] One explanation of this behavior is that the five valence electrons of phosphorus are accommodated in an sp^3d-hybrid arrangement. Since the 3*d* participation is restricted to the $3d_{z^2}$ orbital for σ-bond formation, the axial bonds should indeed be longer and weaker than the equatorial ones. This hybridization scheme could also explain why the most electronegative groups occupy the axial positions.[34–36]

Hexacoordinate phosphorus anions, such as $PhPF_5^-$,[37] PF_6^-, PCl_6^-, and $[P(O_2C_6H_4)_3]^-$[38] are also known. Their octahedral structures can be rationalized in terms of an sp^3d^2-hybrid orbital arrangement.

b. *Reaction Mechanisms*

Nucleophilic substitution reactions at tetracoordinate phosphorus probably proceed through the formation of pentacoordinate, trigonal bipyramidal transition states.[21, 39] Presumably, these correspond to sp^3d-hybrid arrangements.

c. *Back-Bonding*

Evidence exists that the base strength of trisilylamine is lower than expected,[40] that the S—O bonds in sulfoxides and sulfones are less polar than anticipated,[41] and that aryl groups in arylsilicon compounds, $ArSiR_3$, supply electron density to silicon via a π-type mechanism.[6, 41, 42] These results are ascribed to "back-bonding" whereby unpaired electrons on nitrogen or oxygen or the π electrons in an aromatic group are donated to the second-row element through a d_π–p_π coordinate π bond (Fig. 2.1).

Back-bonding is assumed to be partly responsible for the wide bond angles at oxygen in compounds which contain Si—O—Si or Si—N—Si bonds. By analogy it is assumed that phosphorus participates in a similar arrangement

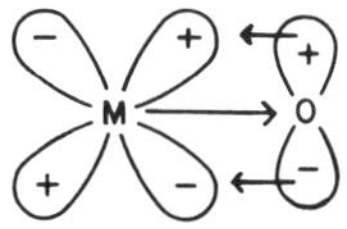

FIG. 2.1. Back-bonding through a d_π–p_π coordinate π bond.

when it is linked to oxygen,[19, 23] and evidence has been presented for P—C bond shortening when tetracoordinate phosphorus is linked to tricoordinate carbon.[43–45] Indeed, the phosphorus–oxygen bond in phosphoryl compounds, such as POF_3, is very strong[46, 47] presumably because of a back-bonding interaction.

d. *Atomic and Molecular Orbital Calculations*

The crucial question, which underlies all of this discussion, is whether the energy difference between phosphorus 3*p* and 3*d* orbitals is sufficiently low to permit appreciable occupation of the 3*d* level or its *spd* hybrids for ground-state molecules. The earliest analysis of the question was made on the basis of hydrogen-like orbitals, with energies computed from spectroscopic data. Figure 2.2 shows the energy-level sequence expected for atomic phosphorus. The promotional energy is calculated to be at least 100–150 kcal, which is probably too high to explain the involvement of the 3*d* orbitals. Furthermore, use of Slater-type equations for the calculation of 3*d*-orbital distribution functions show that the 3*d* orbitals are much too diffuse to overlap effectively with bonding orbitals from neighboring atoms in either σ- or π-type interactions.

Two types of argument can be used to circumvent this difficulty. In the first, the reasonable assumption is made that hydrogen-like orbitals differ greatly in radial density distribution from those found in phosphorus atoms. In the second, it is assumed that the presence of ligands in a molecule serves to lower the energy of the phosphorus 3*d* orbitals and to reduce their size and diffuseness. Craig and co-workers[48–53] have proposed SCF and valence state models, particularly for sulfur, in which the 3*d* functions are contracted and less diffuse than in Slater-type orbitals, but the problem of the high promotional energy still remains.[23]

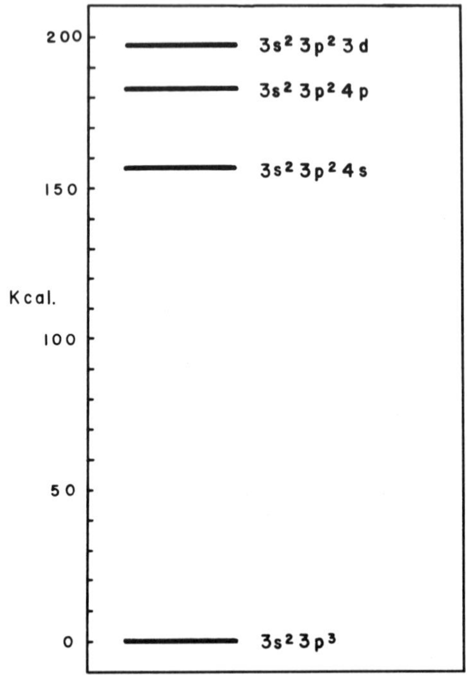

FIG. 2.2. Spectroscopic energy levels for the phosphorus atom, showing relative order and separation of 3*p*, 4*s*, 4*p*, and 3*d* orbitals. Data from Moore.[47a]

The second argument, based on ligand-induced contraction of the 3*d* orbitals, appears to have a wider application.[51] The theory is based on the observation that the highest valence states of the second-row elements are associated with the presence of the most electronegative ligands, such as fluorine, chlorine, oxygen, or nitrogen. It is assumed, therefore, that electronegative ligands serve to contract the *d* orbitals and lower their energy. Calculations by Craig and co-workers,[52, 53] and by Mitchell,[54, 55] for SF_6 and POF_3 have provided some evidence for this viewpoint.

3. Arguments against the Participation of 3*d* Orbitals

It will be clear from the preceding section that an analysis of the role of 3*d* orbitals in phosphorus chemistry really involves two separate questions—whether *d*-orbital hybrids are involved in stable five- and six-coordinate phosphorus structures, and whether phosphorus 3*d* orbitals can participate in d_π–p_π bonding. It is this latter question that concerns us most in phosphazene chemistry, but positive evidence about the first question would probably reinforce the arguments for d_π–p_π bonding.

Unfortunately, for nearly every argument that can be put forward in favor of *d*-orbital contributions, another argument can be raised against it. This is particularly true for the question of the structure of penta- and hexacoordinate phosphorus species. Reviews by Lucken,[56] Jorgensen,[57] and Mitchell[23] have noted these counterarguments, which are summarized as follows.

First, in spite of the ligand orbital-contraction hypothesis, the suspicion remains that the energy required to promote an electron from the 3*s* or 3*p* levels to the 3*d* level in phosphorus is too great to permit a significant *d*-orbital involvement.

Second, assuming that the ligand-field effect lowers the 3*d*-orbital energy appreciably, it is reasonable to suppose that the 4*s* and 4*p* levels would be reduced in energy to an extent which would ensure their participation also. In the free atom, the 4*s* and 4*p* levels lie below that of the 3*d* (Fig. 2.2). Ligand effects would be expected to affect the 3*d* and 4*p* orbitals most because of their directionality, but, nevertheless, all three sets of orbitals would be available for bonding. Thus, there is no profound energetic reason for assuming an exclusive involvement by the 3*d* orbitals. Indeed, it has been suggested[56,57] that, for the second-row and later elements, the formation of high coordination number compounds may reflect an involvement of several high-level orbitals, with the concept of *n* orbitals being required for formation of *n* bonds no longer valid.

Another alternative to the sp^3d-orbital formulation is the possibility that ionic resonance forms exist. For example, PCl_5 could be described as a resonance hybrid of the type $PCl_4^{\oplus}Cl^{\ominus}$. The fact that the higher coordination numbers are found with the most electronegative ligands is consistent with this view. Thus, *d*-orbital contributions are not necessary to explain these structures.

From a molecular orbital point of view, it has also been shown that the existence of pentacoordinate and hexacoordinate compounds of the second-row elements can be understood in terms of 3*s*- and 3*p*-orbital contributions only.[58,59]

Finally, perhaps the most damaging argument against *d*-orbital involvement results from the inconclusive data obtained from attempts to confirm the existence of *d* contributions by the use of physical techniques.[56,57] The results of nuclear quadrupole coupling constant experiments for silicon compounds, ^{31}P NMR experiments, electron spin resonance results for silicon and phosphorus compounds, and ultraviolet data for aryl derivatives of second-row elements are sufficiently ambiguous that no convincing proof for *d*-orbital participation has emerged.[56,57]

Thus, from the viewpoint of phosphorus–nitrogen chemistry, the concept of 3*d*-orbital contributions must be taken as yet unproved. In spite of this fact, the idea of σ-d_π–p_π bonds in phosphazenes has received wide attention and, at the present time, it provides the most satisfactory working hypothesis for understanding the structure and chemistry of these compounds. Two other models are also mentioned briefly in the following sections. These are the zwitterionic model and one based on the use of 4*s*, 4*p*, or mixed orbitals.

E. The Zwitterionic Phosphazene Model

The zwitterionic model constitutes one of the earlier attempts to rationalize the structure of phosphazenes. For a cyclotriphosphazene, the zwitterionic structure is formed by transfer of an electron from phosphorus to nitrogen (V).

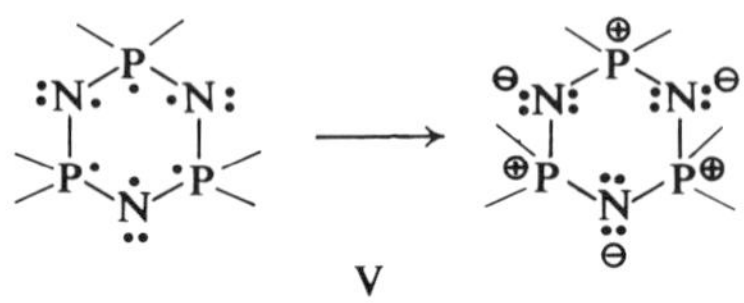

V

A model such as this would certainly explain the shortness of the phosphorus–nitrogen bonds in phosphazenes, and it is consistent with the electronegativity difference (0.9) between the contributing atoms. The model has the added attribute that phosphorus 3*d* orbitals need not be invoked.

However, the highly polar skeletal structure implies a high reactivity to ionic reagents, a prediction which is not consistent with the experimental facts. Furthermore, the electron transferred to nitrogen would presumably be accepted into the half-filled p_z orbital to yield a lone pair directed normal to the ring plane (Fig. 2.3). Repulsions between this lone pair and the radial lone-

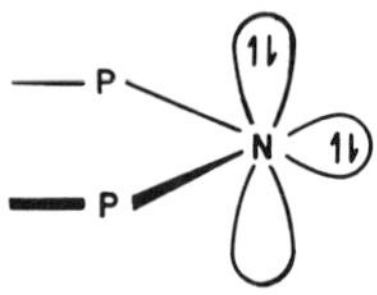

VI

FIG. 2.3. Zwitterionic phosphazene structure, with occupation of a nitrogen p_z orbital by the second set of nonbonding electrons.

pair orbital would favor a hybridization change to a ψ sp^3 tetrahedral arrangement (VII, Fig. 2.4) accompanied by a narrowing of the P—N—P bond toward 109.5°. This would be an untenable situation for cyclic trimers and higher cyclic species, and the existence of pure zwitterionic species seems unlikely from energy considerations. The chief value of this model lies in the fact that it describes one extreme in a range of structures which extend to a nonpolar d_π–p_π model.

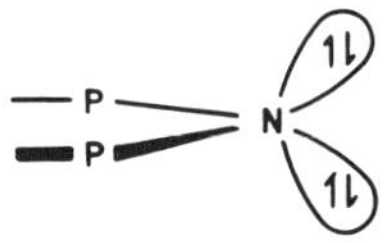

VII

FIG. 2.4. Rehybridized orbital arrangement at nitrogen for a phosphazene zwitterionic structure.

F. The Phosphorus 4s or 4p Model

As discussed earlier, the structures of phosphazenes are consistent with the view that the skeletal nitrogen atoms show sp^2 hybridization with two lobes forming σ bonds to phosphorus, and the third directed radially as a lone-pair orbital. The remaining electron then occupies a p_z orbital oriented normal to the plane (VIII, Fig. 2.5). If the ligands attached to phosphorus are sufficiently

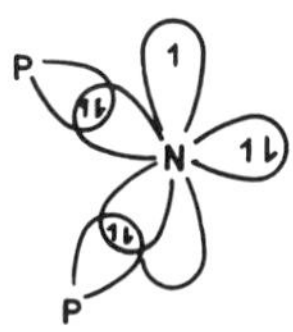

VIII

FIG. 2.5. Orbital hybridization scheme at nitrogen, showing p_z orbital available for π bonding.

electronegative, the 3*d*, 4*s*, and 4*p* orbitals may be contracted sufficiently to interact with the nitrogen $2p_z$ orbital. The role of the phosphorus 3*d* orbitals will be considered in the next section. The symmetry of the 4*p* orbital is appropriate for formation of a p_π–p_π bond (IX, Fig. 2.6).

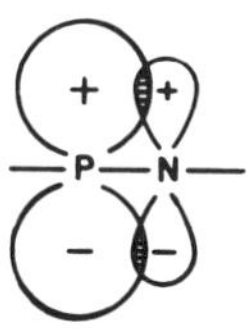

IX

FIG. 2.6. A π-bond formed from phosphorus $4p_z$ and nitrogen $2p_z$ orbitals.

The model would explain the short P—N bond length and the increased P—N bond strength as the ligand electronegativity is increased, but it does not explain why ring puckering has little affect on molecular stability, why organic-type electron delocalization effects are not observed, or why the barrier to torsion of a P—N bond is low.[60] It could be argued that the size and diffuseness of the $4p$ orbital would create a system insensitive to spectroscopic or physical tests, but this same diffuseness would lead to poor π overlap and a reduced influence on the bond strength.

A reasonable modification of this model makes use of the assumption that it is impossible to distinguish between discrete bonding contributions from $3s$, $3p$, $3d$, $4s$, and $4p$ orbitals on phosphorus, and that the actual bonding situation involves appreciable mixing of these states. In these terms, the distinction between σ and π bonds breaks down.

G. The d_π–p_π Model

1. General Features of the Model

The development of the d_π–p_π model for bonding in phosphazenes has been largely the result of publications by Craig and Paddock,[7–16] and by Dewar, Lucken, and Whitehead.[17] More recently, an outline of this theory has been presented in a review by Mitchell.[23] The basic assumptions behind the theory are as follows:

(1) The atomic orbitals can be separated into noninteracting sets in which the π orbitals are treated separately from the σ orbitals. Thus, as in the Hückel treatment of organic aromatic compounds, the σ orbitals form a localized framework while the π orbitals provide opportunities for electron delocalization and additional molecular stabilization.

(2) The σ-bond framework is made up of overlapping orbitals from phosphorus and nitrogen. The contributing orbitals from nitrogen are two $2sp^2$ lobes and those from phosphorus are two $3sp^3$ lobes. The remaining two sp^3

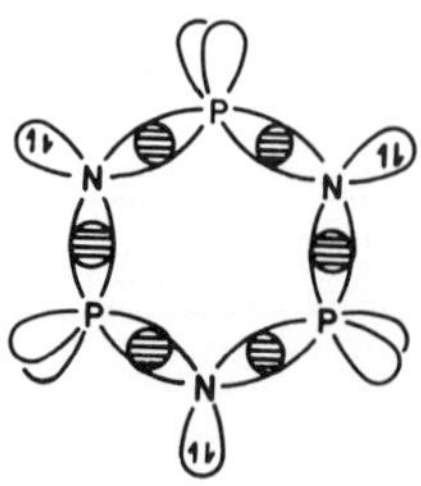

FIG. 2.7. σ-Bond framework in a cyclotriphosphazene.

orbitals from phosphorus form σ bonds to the ligands, whereas the remaining sp^2 orbital at nitrogen contains a lone pair of electrons (Fig. 2.7). Because the ring angle at phosphorus is nearly always closer to 120° than 109.5°, some distortion of the sp^3 tetrahedron is required.

(3) π Bonds are formed between phosphorus and nitrogen by promotion of the remaining valence electron at phosphorus to the 3d level. Overlap of a $3d_{xz}$ orbital with the nitrogen p_z orbital (which contains one electron) yields a d_π–p_π bond (Fig. 2.8). This orbital arrangement, which provides for an alternation of resonance integral sign around the ring, is known as a heteromorphic or π interaction. Such interactions could give rise to delocalization–stabilization effects similar to those found in organic aromatic molecules (Fig. 2.9).

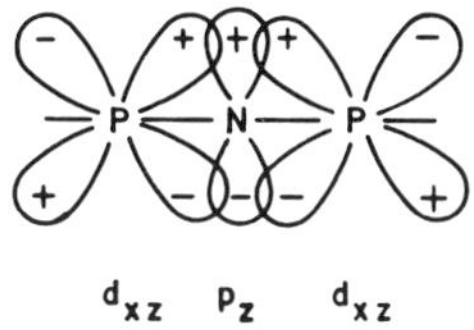

FIG. 2.8. d_π–p_π Bonds.

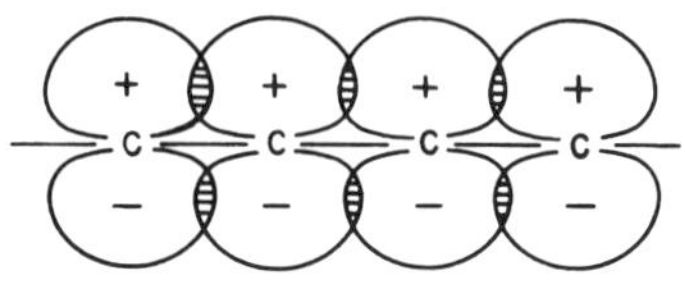

FIG. 2.9. p_π–p_π Bonds.

(4) An additional set of π bonds can be formed in the ring plane by donation of the lone-pair electrons at nitrogen into the unoccupied $d_{x^2-y^2}$ or d_{xy} orbitals of phosphorus (Fig. 2.10). This interaction is known as a π' system and it is symmetric to reflection in the molecular plane. Changes in the π' system should be detectable by changes in the availability of the nitrogen lone-pair electrons.

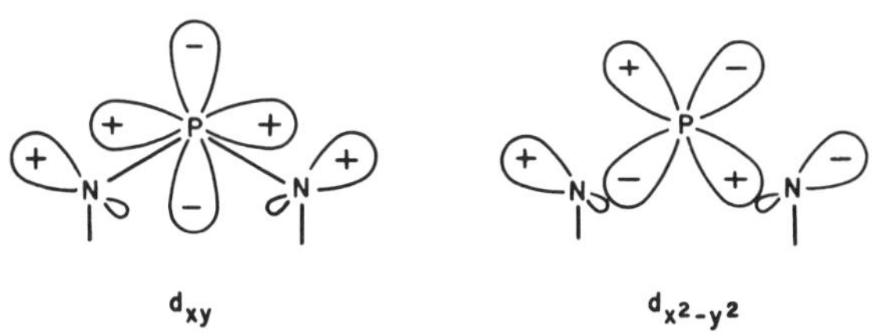

FIG. 2.10. π' Bonding.

Coordinate π' bonding of this kind is reminiscent of that believed to exist in siloxanes, silazanes, and phosphates.[1]

2. Theoretical Treatment

Consider first the situation which results from an interaction of a d_{yz} orbital with the nitrogen p_z orbitals as depicted in Fig. 2.11. A planar cyclic molecule

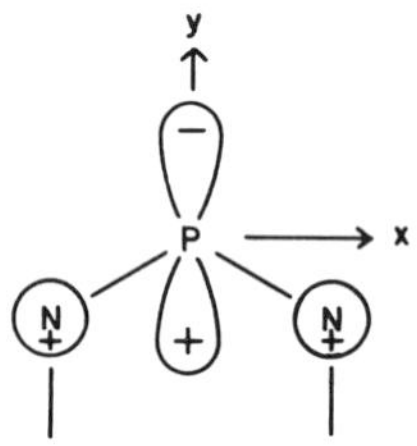

FIG. 2.11. Homomorphic interaction between a phosphorus d_{yz} and two nitrogen p_z orbitals.

belongs to the symmetry group D_{nh} and molecular orbital formation requires that the available atomic orbitals transform like representations of this group. A cyclotriphosphazene, $(NPX_2)_3$, for example, is assigned to group D_{3h}. The atomic sites around the ring are numbered according to the scheme shown in X,

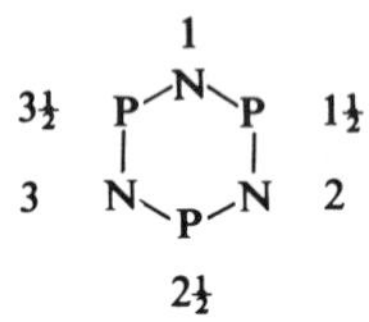

X

and since the P and N are not interconvertible by symmetry, they must be treated separately by the Hückel equations

$$\phi_l^{A} = n^{-1/2} \sum_{p=1}^{n} \exp^{(2\pi i l p/n)} A_p$$

$$\phi_l^{B} = n^{-1/2} \sum_{p=1}^{n} \exp^{[2\pi i l (p+\frac{1}{2})/n]} B_{p+1/2}$$

where A is nitrogen, B is phosphorus, n is the number of nitrogen or phosphorus atoms, A_p and $B_{p+1/2}$ are the atomic π orbitals at the centers, and l is a ring quantum number. If n is odd (a cyclic trimer, pentamer, etc.), l can assume values of $0, \pm 1, \ldots, \pm(n-1)/2$. If n is even (a cyclic tetramer, hexamer, etc.),

l takes values of $l = 0, \pm 1, \ldots, n/2$. The functions ϕ belong to irreducible representations of the appropriate group (D_{3h} for a trimer) and to the site group representations of the atomic orbitals. The molecular orbitals are formed by combining ϕ_l^{A} and ϕ_l^{B} to give minimum energy. For the p_π–d_{yz} interaction the energies are given by the roots of the two-row secular equations[10]

$$\begin{vmatrix} \alpha_{\mathrm{A}} - E & 2\beta \cos l\pi/n \\ 2\beta \cos l\pi/n & \alpha_{\mathrm{B}} - E \end{vmatrix} = 0$$

where α_{A} and α_{B} are the Coulomb integrals, and β is the resonance integral. This situation resembles those of p_π–p_π systems such as benzene or *s*-triazine, where Hückel's rule applies and where there is an alternation of stability as the ring size in increased (Fig. 2.12).

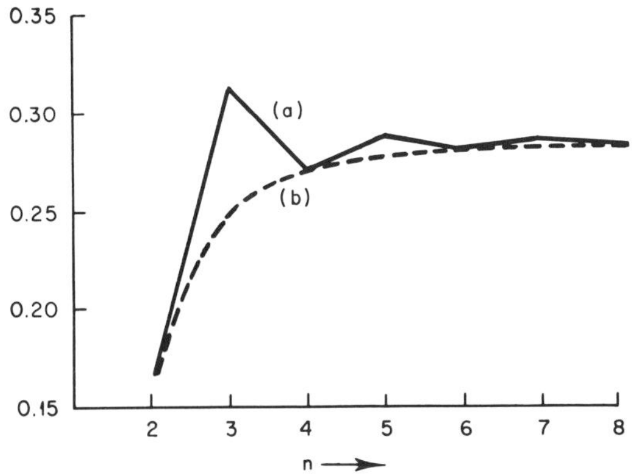

FIG. 2.12. Delocalization energies per electron (in units of β) for $(\mathrm{AB})_n$, assuming $\alpha_{\mathrm{B}} = \alpha_{\mathrm{A}} + \beta$: (a) homomorphic system, (b) heteromorphic system. From Mitchell, ref. 23.

However, for the heteromorphic system, which results from the interaction of a d_{xz} and a p_z orbital (Fig. 2.13), the secular equation is

$$\begin{vmatrix} \alpha_{\mathrm{A}} - E & 2\beta \sin l\pi/n \\ 2\beta \sin l\pi/n & \alpha_{\mathrm{B}} - E \end{vmatrix} = 0$$

and solution of this equation for cyclic trimers, tetramers, and higher homologues shows that the delocalization energy per π electron increases steadily to a limiting value with increasing ring size (Fig. 2.12). Thus, if the homomorphic and heteromorphic systems could be separated, two different types of behavior would be expected. In the limit, it might be supposed that preferential interaction with, say, the d_{xz} orbital, could give rise to a situation in which predominantly heteromorphic characteristics would be seen.

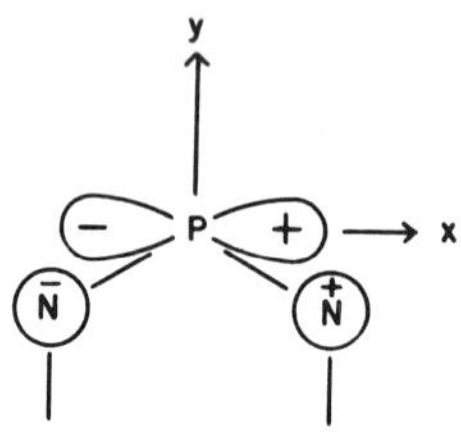

Fig. 2.13. Heteromorphic interaction between a phosphorus d_{xz} and two nitrogen p_z orbitals.

In practice, each nitrogen p_z orbital is expected to interact with both the d_{xz} and d_{yz} orbitals and, for this situation, the more general secular equation is used[8]

$$\begin{vmatrix} \alpha - E & 2i\beta_{xz}\sin l\pi/n & 2\beta_{yz}\cos l\pi/n \\ -2i\beta_{xz}\sin l\pi/n & \alpha_{xz} - E & 0 \\ 2\beta_{yz}\cos l\pi/n & 0 & \alpha_{yz} - E \end{vmatrix} = 0$$

where the Coulomb integrals, α, α_{xz}, and α_{yz} correspond to the p_π, d_{xz}, and d_{yz} orbitals respectively, and the resonance integrals β_{xz} and β_{yz} depend on the molecular geometry and on the binding energies of the interacting atomic orbitals.[14, 23] Implicit in this argument is the assumption that the d_{xz} and d_{yz} orbitals do not have equal electronegativities and do not therefore, contribute equally to the π bond.

The foregoing treatment is largely the result of theoretical investigations by Craig, Paddock, and Mitchell.[7–14] However, Dewar, Lucken, and Whitehead[17] have proposed the additional interpretation that no good reason exists for assuming that the d_{xz} orbital will have a greater electronegativity than the d_{yz} orbital. Thus, both orbitals should contribute equally to the π bond, and α_{xz} should equal α_{yz}, while β_{xz} should be the same as β_{yz}. Under these circumstances, a linear combination of the reoriented contributing *d* orbitals can be taken in such a way that

$$d_\pi{}^{\mathrm{a}} = 2^{-1/2}(d_{xz} - d_{yz})$$
$$d_\pi{}^{\mathrm{b}} = 2^{-1/2}(-d_{xz} - d_{yz})$$

and the π-molecular orbitals are now localized in such a way that the skeletal delocalization is interrupted at each phosphorus atom. Thus, this leads to the postulation of *islands* of π-character disposed as illustrated in Figs. 2.14 and 2.15.

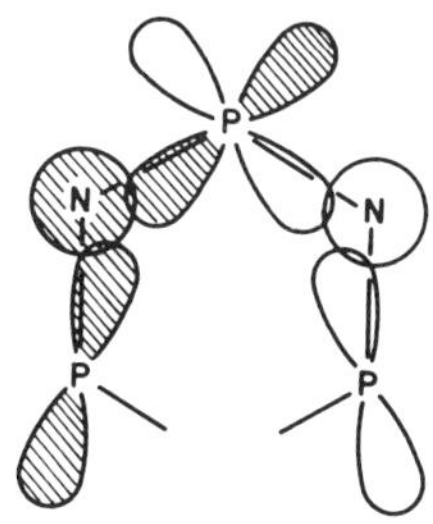

FIG. 2.14. Projection of atomic orbitals to be combined in molecular orbitals for three-center "island" delocalization. From Craig and Mitchell.[14]

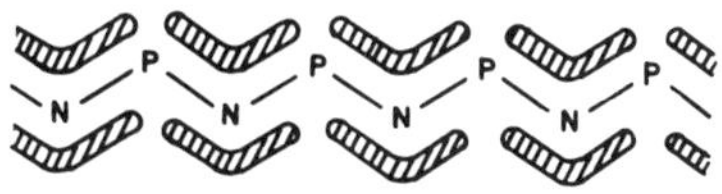

FIG. 2.15. "Islands" of d_π–p_π character in a macrocyclic or linear polymer.

3. SUMMARY OF *d*-ORBITAL BONDING POSSIBILITIES

There are five interconnected possibilities for an interpretation of the π-bonding arrangements of phosphazenes in terms of d orbitals.

(1) The homomorphic π system (Fig. 2.16b) utilizes the d_{yz} orbital. Such a system should resemble an aromatic p_π–p_π system, with an alternation of delocalization stability evident as a homologous series is ascended (Fig. 2.12). It would become manifest only if the d_{yz} orbital contributed to the π system to a greater extent than the d_{xz} orbital. Evidence for a homomorphic π system of this type might be obtainable from an examination of bond lengths and strengths for different homologues, and from ultraviolet, reduction potential, and electron spin resonance studies.

(2) The heteromorphic π system (Fig. 2.16a) utilizes the d_{xz} orbital. No alternation of stability would be expected as the homologous series is ascended (Fig. 2.12), but a steady increase in stability up to a limiting value would be anticipated. This stability difference would be particularly evident between cyclic trimers and tetramers. For this phenomenon to operate, ligand electronegativity effects would contract the d_{xz} orbital more than the d_{yz} orbital, or the preferential directional interaction with the nitrogen p_z orbital would serve the same purpose. Here too, structural evidence about bond lengths and strengths, and data from ultraviolet and ESR studies might be expected to throw light on the validity of this model.

(3) The island-type π orbital (Fig. 2.14) would result from equal participation by the d_{xz} and d_{yz} orbitals. Since the π orbitals are interrupted at each

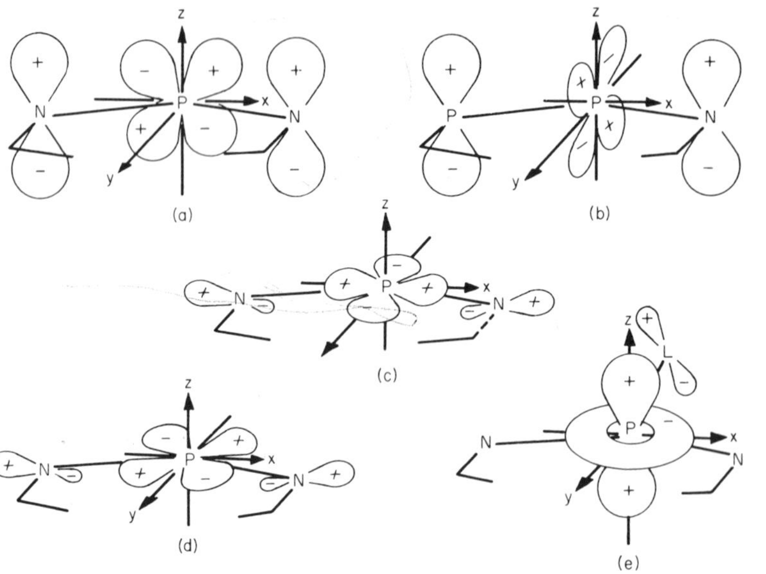

FIG. 2.16. Overlap schemes for π bonding of (a) d_{xz} and (b) d_{yz} orbitals with p_z orbitals; (c) π' bonding of $d_{x^2-y^2}$ and (d) d_{xy} orbitals with an s-p_y hybrid, and (e) π bonding of a d_{z^2} orbital with a ligand p orbital. From Paddock, ref. 16.

phosphorus atom, no change in delocalization stability should be evident when cyclic trimers are compared with higher cyclic homologues, and aromatic behavior would not be evident. The system would, in fact, resemble the "allylic" π systems of carbon chemistry. However, the optimum condition for "island"-type molecular orbital formation is a ring with a bond angle of 90° at phosphorus, a feature not found in cyclic phosphazenes. Thus, it is speculated that a whole spectrum of molecular conditions may exist in which the "island"-type contributions may vary from 100% of the heteromorphic π character down to less than 50% as the d_{xz} and d_{yz} orbitals become non-equivalent.[14] It should be noted that this model and the heteromorphic model predict that moderate ring puckering should have little influence on delocalization stability because of the diffuseness and symmetry characteristics of the contributing d orbitals.

(4) The possibility also exists that π bonding could occur between phosphorus d orbitals and the p orbitals of the ligands. In such a case, the $3d_{z^2}$ orbital probably has the optimum symmetry for this interaction (Fig. 2.16e), although the $3d_{xz}$ and $3d_{x^2-y^2}$ orbitals could also participate. Evidence for this interaction should be seen in a shortening of the bond distance between phosphorus and the ligand, particularly when phosphorus is bonded to an exo-

cyclic oxygen, nitrogen, or sulfur atom which can supply electrons from a lone-pair orbital.

(5) The π' system in the plane of the phosphazene ring utilizes the lone pair of electrons in a nitrogen sp^2 lobe for donation into empty d_{xy} or $d_{x^2-y^2}$ orbitals at phosphorus (Figs. 2.16c and 2.16d). Thus, studies of the availability of the lone-pair electrons at nitrogen should provide an insight into the working of this system.

H. Comparisons with p_π–p_π Aromatic Systems

Any attempt to search for physical evidence pertaining to the nature of bonding in phosphazenes must be tempered with the realization that the types of "evidence" which confirm the presence of π bonding or aromaticity in carbocyclic compounds are not necessarily applicable to phosphazenes. Thus, although p_π–p_π bonds in ethylenic compounds have barrier heights sufficient to permit the isolation of geometrical isomers, a d_π–p_π bond in a phosphazene high polymer, for example, may show virtually no barrier because of the diffuseness of the *d* orbitals and the opportunities for the nitrogen p_z orbital to bond with several different 3*d* orbitals as the bond undergoes torsion. Then again, the large resonance energy of benzene results from a coincidence of factors including the 120° (sp^2) ring angle, the ring planarity, and the fact that all the skeletal atoms have the same electronegativity. A comparison of benzene with *s*-triazine, which has a much lower resonance energy, illustrates how aromaticity is affected deleteriously by the presence of elements with different electronegativities.[1] Dramatic delocalization energies comparable to those of benzene or polyaromatic hydrocarbons cannot, therefore, be expected for phosphazenes, even if broadly delocalized π orbitals exist. Thus, the evidence about π bonding in phosphazenes is not nearly so clear-cut as is the comparable evidence for organic aromatic species. The nature of the available evidence is discussed in Chapter 3.

References

1. H. R. Allcock, "Heteroatom Ring Systems and Polymers." Academic Press, New York, 1967.
2. C. C. Costain, *J. Chem. Phys.* **29**, 864 (1958).
3. P. J. Wheatley, *Acta Crystallogr.* **8**, 224 (1955).
4. L. O. Brockway and H. O. Jenkins, *J. Amer. Chem. Soc.* **58**, 2036 (1936).
5. B. Morosin and E. C. Lingafelter, *Acta Crystallogr.* **12**, 611 (1959).
6. F. G. A. Stone and D. Seyferth, *J. Inorg. Nucl. Chem.* **1**, 112 (1955).
7. D. P. Craig, *Chem. Ind. (London)*, p. 3 (1958).
8. D. P. Craig, *Chem. Soc., Spec. Publ.* **12**, 343 (1958).

9. D. P. Craig and N. L. Paddock, *Nature* (*London*) **181**, 1052 (1958).
10. D. P. Craig, *Theor. Org. Chem., Pap. Kekule Symp., 1958* p. 20 (1959).
11. D. P. Craig, *J. Chem. Soc., London* p. 997 (1959).
12. D. P. Craig, M. L. Heffernan, R. Mason, and N. L. Paddock, *J. Chem. Soc., London* p. 1376 (1961).
13. D. P. Craig and N. L. Paddock, *J. Chem. Soc., London* p. 4118 (1962).
14. D. P. Craig and K. A. R. Mitchell, *J. Chem. Soc., London* p. 4682 (1965).
15. J. Emsley and N. L. Paddock, *J. Chem. Soc., A* p. 2590 (1968).
16. N. L. Paddock, *Quart. Rev., Chem. Soc.* **18**, 168 (1964).
17. M. J. S. Dewar, E. A. C. Lucken, and M. A. Whitehead, *J. Chem. Soc., London* p. 2423 (1960).
18. D. A. Brown and C. G. McCormack, *J. Chem. Soc., London* p. 5385 (1964).
19. D. W. J. Cruickshank, *J. Chem. Soc., London* p. 5486 (1961).
20. R. F. Hudson, *Pure Appl. Chem.* **9**, 371 (1964).
21. R. F. Hudson, "Structure and Mechanism in Organo-Phosphorus Chemistry." Chapter 3. Academic Press, New York, 1965.
22. K. A. R. Mitchell, *J. Chem. Soc., A* p. 2683 (1968).
23. K. A. R. Mitchell, *Chem. Rev.* **69**, 157 (1969).
24. K. W. Hansen and L. S. Bartell, *Inorg. Chem.* **4**, 1775 and 1777 (1965).
25. G. V. Romanov and V. P. Spiridonov, *Zh. Strukt. Khim.* **8**, 159 (1967); *J. Struct. Chem.* (*USSR*) **8**, 131 (1967).
26. L. O. Brockway and J. Y. Beach, *J. Amer. Chem. Soc.* **60**, 1836 (1938).
27. M. Roualt, *Ann. Phys.* (*Paris*) [11] **14**, 78 (1940); *C. R. Acad. Sci.* **207**, 620 (1938).
28. G. Wittig and M. Rieber, *Justus Liebigs Ann. Chem.* **562**, 177 and 187 (1949).
29. G. A. Rasuvaev and N. A. Osanova, *Dokl. Akad. Nauk SSSR* **104**, 552 and 773 (1955).
30. F. Ramirez and N. B. Desai, *J. Amer. Chem. Soc.* **85**, 3253 (1963).
31. P. J. Wheatley, *J. Chem. Soc., London* p. 2206 (1964).
32. F. A. Cotton, *J. Chem. Phys.* **35**, 228 (1961).
33. P. C. Van Der Voorn, K. F. Purcell, and R. S. Drago, *J. Chem. Phys.* **43**, 3457 (1965).
34. J. E. Griffiths, R. P. Carter, and R. R. Holmes, *J. Chem. Phys.* **41**, 863 (1964).
35. L. S. Bartell and K. W. Hansen, *Inorg. Chem.* **4**, 1777 (1965).
36. A. J. Downs and R. Schmutzler, *Spectrochim. Acta, Part A* **23**, 681 (1967).
37. R. Schmutzler and G. S. Reddy, *Inorg. Chem.* **4**, 191 (1965).
38. H. R. Allcock, *J. Amer. Chem. Soc.* **86**, 2591 (1964).
39. F. H. Westheimer, *Accounts Chem. Res.* **1**, 70 (1968).
40. K. Hedberg, *J. Amer. Chem. Soc.* **77**, 6491 (1955).
41. C. Eaborn, "Organosilicon Compounds," Chapter 3. Butterworth, London, 1960.
42. L. Goodman, A. H. Konstam, and L. H. Sommer, *J. Amer. Chem. Soc.* **87**, 1012 (1965).
43. F. S. Stephens, *J. Chem. Soc., London* pp. 5640 and 5658 (1965).
44. P. J. Wheatley, *J. Chem. Soc., London* p. 5785 (1965).
45. T. C. W. Mak and J. Trotter, *Acta Crystallogr.* **18**, 81 (1965).
46. F. Abel and E. Bretscher, *Helv. Chim. Acta* **12**, 450 (1929).
47. H. Siebert, *Z. Anorg. Allg. Chem.* **275**, 210 (1954).
47a. C. E. Moore, Atomic Energy Levels, Vol. 1. Natl. Bur. St. Publ. 467 (1949).
48. D. P. Craig and T. Thirunamachandran, *J. Chem. Phys.* **43**, 4183 (1965).
49. D. P. Craig and T. Thirunamachandran, *J. Chem. Phys.* **45**, 3355 (1966).
50. D. P. Craig and T. Thirunamachandran, *Proc. Roy. Soc., Ser. A* **303**, 233 (1968).
51. D. P. Craig, A. Maccol, R. S. Nyholm, L. E. Orgel, and L. E. Sutton, *J. Chem. Soc., London* p. 332 (1954).
52. D. P. Craig and E. A. Magnusson, *J. Chem. Soc., London* p. 4895 (1956).

53. D. P. Craig and C. Zauli, *J. Chem. Phys.* **37**, 601 and 609 (1962).
54. K. A. R. Mitchell, *J. Chem. Soc., A* 2676 and 2683 (1968).
55. K. A. R. Mitchell, *Can. J. Chem.* **46**, 3499 (1968).
56. E. A. C. Lucken, *in* "Structure and Bonding" (P. Hemmerick *et al.*, eds.), p. 1. Springer-Verlag, Berlin and New York. 1969.
57. C. K. Jorgensen, *in* "Structure and Bonding" (P. Hemmerick *et al.*, eds.), p. 94. Springer-Verlag, Berlin and New York, 1969.
58. R. E. Rundle, *Surv. Progr. Chem.* **1**, 81 (1963).
59. D. P. Santry, *J. Amer. Chem. Soc.* **90**, 3309 (1968).
60. H. R. Allcock, R. L. Kugel, G. F. Konopski, and E. G. Stroh, *Chem. Commun.* p. 985 (1970).

Chapter 3

THE STRUCTURE OF PHOSPHAZENES AND PHOSPHAZANES

A. Introduction

The question of the nature of the bonding in phosphorus–nitrogen compounds has attracted considerable attention, and numerous investigators have attempted to distinguish between one model or another by the application of physical techniques to the study of the structures of these compounds. This chapter will review the physicochemical information in the light of the alternative bonding theories. X-Ray diffraction results form the cornerstone of nearly all structural arguments, and these will be discussed first. Thermochemical data will be considered next, to be followed by vibrational and ultraviolet spectroscopic results and an analysis of polarographic and electron spin resonance data. Conductivity information, mass spectrometry, dipole moments, diamagnetic anisotropy, basicity results, and nuclear quadrupole resonance data follow, and a discussion of nuclear magnetic resonance spectroscopy completes the survey.

B. X-Ray Diffraction Data

X-Ray diffraction experiments have shown that those compounds which have the formula $(NPR_2)_{3\text{ to }6}$ or $(RNPR'_3)_2$ are cyclic, whereas the rubbery or thermoplastic derivatives with the formula $(NPR_2)_n$ are linear-type high polymers. X-Ray experiments also provide accurate data about bond lengths, bond angles, and molecular conformations, and these aspects will be considered in turn.

1. Bond Contraction

Few structure results are available which provide a measure of the length of a phosphorus–nitrogen single bond. The value quoted most frequently is 1.77 Å, as derived from X-ray studies of sodium phosphoramidate.[1, 2] This corresponds to covalent radii of 0.70 Å for nitrogen and 1.07 Å for phosphorus, values which are consistent with those derived from amines[3] and phosphine.[4] A similar single-bond value of 1.78 Å has been reported for two of the skeletal bonds in the cyclodiphosphazane $(MeN—PCl_3)_2$.[5, 6]

As shown in Table 3.1,[1, 2, 5–35] the phosphorus–nitrogen skeletal bond distances in cyclo- or polyphosphazenes are in the range of 1.47 to 1.62 Å. This represents an average contraction of 0.22 Å when compared to the single-bond distance. Bond contractions can be ascribed to π bonding or to high ionic character of the bond. X-Ray studies can only rarely define the location of electron density between atoms and no direct means exist for distinguishing between these alternatives. However, the data in Table 3.1 indicate that the shortest skeletal bonds are found when highly electronegative ligands, such as fluorine, are present, and the longest values are found when electron-supplying units, such as dimethylamino, are attached to phosphorus. This suggests that the shortness of the phosphorus–nitrogen bond in phosphazenes is not due to σ-bond polarization,

$$\overset{\delta+}{P} \rightarrow \overset{\delta-}{N}$$

since the effect of electron withdrawal from phosphorus by a ligand would be to reduce the polarity of the bond and increase its length. The alternative explanation is that the most electronegative ligands contract the phosphorus d orbitals and enhance the orbital overlap in the π system. Ligand electronegativity will also favor donation of the nitrogen lone-pair electrons into the phosphorus d_{xy} or $d_{x^2-y^2}$ orbitals via the π' system. Thus, the bond length contraction phenomenon favors the idea of d_π–p_π bonding in phosphazenes.

Some bond contraction is also evident in "saturated" phosphazanes. For example, two of the skeletal bonds in $(MeNPCl_3)_2$ (I) and $(MeNPF_2Ph)_2$ (II) display a contraction of ~0.13 Å, and the cyclic trimers and tetramers, III, IV, and V, show bond contractions in the region of 0.11 Å. Thus, the possibility exists that some coordinate π' character exists in the skeletal structures of these species.

For cyclophosphazenes, small P—N bond contractions are evident in the change from cyclic trimers to higher cyclic homologues. For example, a bond contraction of 0.05 Å can be discerned in the change from the planar, trimeric $(NPF_2)_3$ to the planar tetramer, $(NPF_2)_4$.[12, 13] The P—N bond length decreases

TABLE 3.1

SELECTED STRUCTURAL DATA FOR PHOSPHAZANES AND PHOSPHAZENES[a]

Compound	P—N bond length (Å)	N—P—N bond angle (°)	P—N—P bond angle (°)	R—P—R bond angle (°)	Skeletal structure	References
$NaPO_3NH_3$	1.77	—	—	—	—	1, 2
$(MeNPCl_3)_2$	1.77	81.1	98.9	~90	Four-membered ring	5, 6
	1.64			~110		
$[MeNP(S)Cl]_2$	1.67	84	96	111.5	Four-membered ring	7
$(MeNPF_2Ph)_2$	1.78	80.6	99.4	91.8	Four-membered ring	8
	1.64			87.0		
$[MeNP(O)OMe]_3$	1.66	105.9	121.7	117	Twisted boat	9
		103.8		113.4		
$[HNP(O)ONa]_3$	1.68	104.5	123	118	Chair	10
$[HNP(O)OH]_4$	1.66	107.3	125.6	116.1	Puckered ring	11
$(NPF_2)_3$	1.57	119.4	120.3	<99.9	Planar ring	12
$(NPF_2)_4$	1.51	122.7	147.4	99.9	Planar ring	13
$(NPF_2)_n$	1.52	119.0	136.0	95.0	*Cis–trans* planar chain	14
$N_3P_3FCl_5$	1.56	118.5	121.5	100.5	Planar ring	15
$(NPCl_2)_3$	1.58	118.4	121.4	101.4	Planar or slightly puckered ring	16, 17
$(NPCl_2)_4$	1.57	121.2	131.3	102.8	Puckered ring (tub) (K form)	18
$(NPCl_2)_4$	1.56	120.5	133.6	103.1	Puckered ring (chair) (T form)	19
			137.6			
$(NPCl_2)_5$	1.52	118.4	148.6	102.2	Nearly planar ring	19a
$(NPCl_2)_n$	1.60	119	127	—	Distorted *cis–trans* planar chain	20

$(NPBr_2)_3$	1.58	117	126.8	102.5	Puckered ring	21, 22
			118.4			
$[NP(NCS)_2]_3$	1.58	119	121	100	Slightly puckered ring	23
$(NPMe_2)_4$	1.60	119.8	131.9	104	Puckered ring	24
1,1-$N_4P_4Me_2F_6$	1.47	116.9	143.3	93.2	Puckered ring (saddle)	24a
	1.58	126.1	146.7	105.7		
1,1,5,5-$N_4P_4F_4Me_4$	1.54	117.5	134.6	96.4	Slightly puckered ring (saddle)	24a
	1.59	125.9		106.9		
$[NP(NMe_2)_2]_4$	1.58	120	133	104	Puckered ring	25
$[NP(NMe_2)_2]_6$	1.56	120	147.5	102.9	Puckered ring	26
$[NP(OPh)_2]_3$	1.58	117.3	121.9	94.1	Slightly puckered ring	26a
				100.1		
$(NPO_2C_6H_4\text{-}o)_3$	1.58	117.5	122.5	97	Slightly puckered ring	27
$(NPO_2C_{12}H_8\text{-}2,2')_3$	1.57	118.4	121	102.5	Slightly puckered ring	28
$[NP(OMe)_2]_4$	1.58	122	132	105	Puckered ring (saddle)	29
$[NP(OMe)_2]_8$	1.56	116.7	136.7	101.3	Puckered ring	30
		(121.2)				
$(NPClPh)_4$	1.57	121.2	137.4	102.9	Puckered ring	31
1,1-$N_3P_3F_4Ph_2$	1.56	—	—	—	Slightly puckered	32
	1.54					
1,1-$N_3P_3Cl_4Ph_2$	1.62	119.7	122	104.4	Slight chair puckering	33
	1.56	115.2	119.2	100.4		
1,1-$N_3P_3Cl_2Ph_4$	1.58	120.7	124.9	104.4	Slight boat puckering	34
	1.56	115.5	121	98.5		
$(NPPh_2)_3$	1.60	117.8	122.1	103.8	Slight chair puckering	35

[a] When different values are available, average or most recent figures are listed. In most cases, these data are mean values for closely related angles or distances. When appreciable differences are noted within the same molecule, a second value is given. A more complete listing of structural information is included in Appendix I.

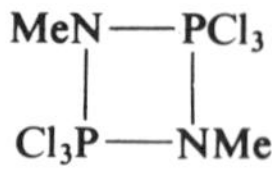

I

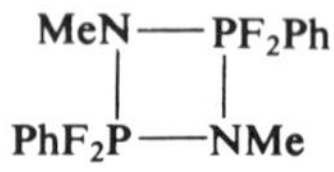

II

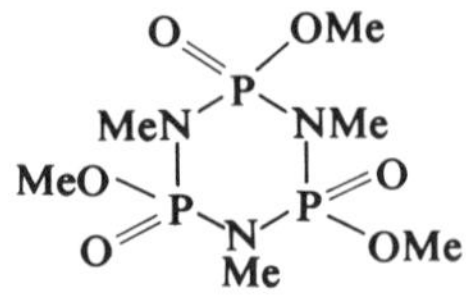

III

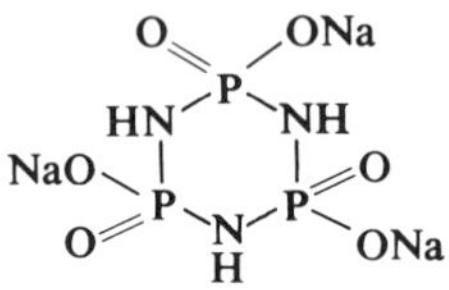

IV

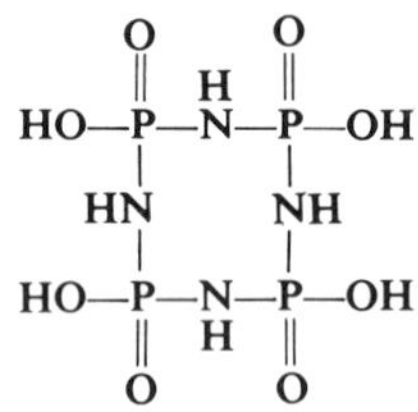

V

by 0.03 Å in moving from the planar trimer, $(NPCl_2)_3$, to the puckered tetramer, $(NPCl_2)_4$,[16-18] and a further 0.05 Å decrease accompanies the change from tetramer to planar pentamer, $(NPCl_2)_5$.[19a] Similarly, a small 0.02 Å contraction in skeletal P—N bond distance is found when the puckered tetramer $[NP(OMe)_2]_4$ is compared with the puckered octamer $[NP(OMe)_2]_8$.[29, 30] Thus, ring size appears to exert a more important influence on bond length than does skeletal puckering, and this is consistent with a minor broad delocalization effect. However, there is no evidence that the high polymers have appreciably shorter bond distances than trimers or tetramers,[14, 20] and a limiting plateau in the bond contraction must exist. The observed skeletal bond contractions which accompany increased ring size are very close to the standard deviations of the X-ray method. However, the trends are nearly all in the same direction, and this provides one of the strongest items of evidence in favor of a π-delocalization phenomenon.

2. Bond Equality around the Ring

Symmetrically substituted cyclotriphosphazenes of formula, $(NPR_2)_n$ or $(NPRR')_n$, show equal P—N bond lengths around the ring (Table 3.1). No separation occurs into alternating σ and σ–π bonds, even when ring puckering exists. This contrasts markedly with p_π–p_π systems, such as cyclooctatetraene, tetrathiazyl fluoride, or cyclotetraborazenes,[36] where ring puckering is accompanied by a delocalization breakdown to give alternating long and short bonds. For phosphazenes, ring puckering apparently has little effect on bond length and this probably reflects the ease with which alternative phosphorus 3*d* orbital lobes can participate in π bonding as the bond undergoes torsion and the ring puckers. Minor P—N bond length differences around a symmetrically substituted ring can usually be attributed to crystal-packing forces or to ring strain resulting from the formation of a particular conformer.[28]

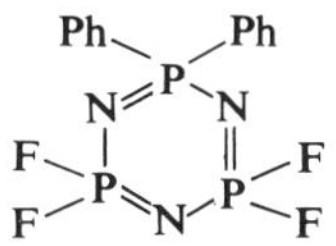

VI

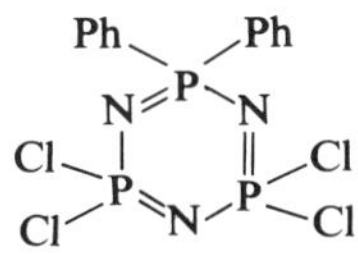

VII

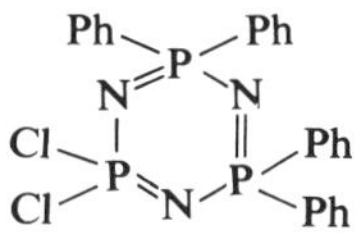

VIII

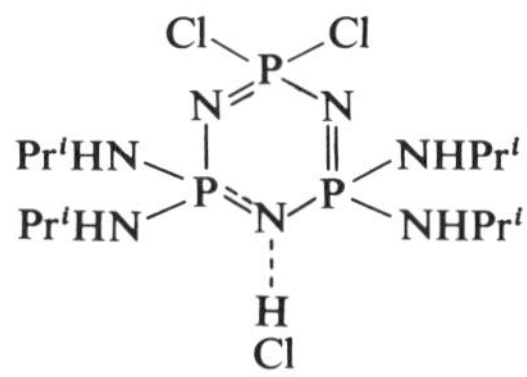

IX

However, appreciable bond length differences exist within a phosphazene ring when different ligands are disposed unsymmetrically. In compounds VI,[32] VII,[33] and VIII,[34] the shorter skeletal P—N bond distances (Table 3.1) are adjacent to the most electronegative ligands (fluorine or chlorine). These differences can be attributed to greater *d*-orbital contractions at the phosphorus atoms which bear the most electronegative ligands, with a corresponding enhancement of π overlap. For compound IX, the bonds adjacent to the PCl_2 unit are 1.56 Å long, the N—$P(NHPr^i)_2$ bonds are 1.58 Å long, and those close to the protonated nitrogen are 1.67 Å.[37] Thus, this molecule has both phosphazene and phosphazane characteristics.

Cyclodiphosphazanes I and II show alternating short and long bonds around the ring.[5, 6, 8] In both cases, the bonds are close to 1.77 Å and 1.64 Å, values which are characteristic of σ and σ-d_π–p_π bonds, respectively. The two skeletal bonds form axial and equatorial components of a trigonal bipyramid (X), with the axial skeletal bonds being the longer of the two. This is consistent with the idea of d_{z^2} involvement in the axial component of a distorted $3sp^3d$ hybrid at phosphorus, and with π' shortening of the P—N bond.

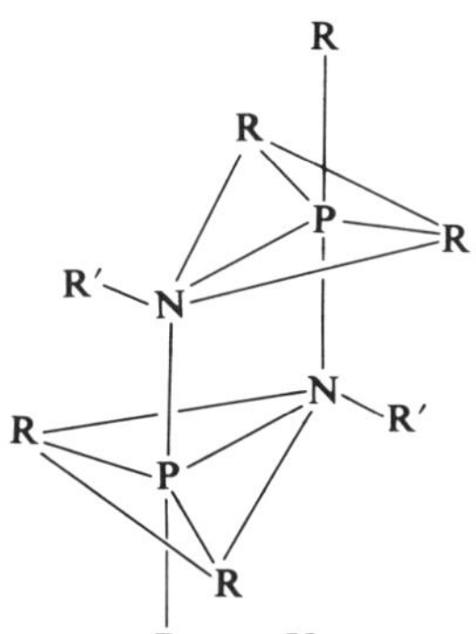

X

3. Exocyclic Bond Lengths

The exocyclic P—N bond lengths in $[NP(NMe_2)_2]_4$ and $[NP(NMe_2)_2]_6$ are in the region of 1.67 to 1.69 Å,[25, 26] a noticeable contraction from the single-bond value of 1.77 Å. This implies that exocyclic d_π–p_π bonding exists in these compounds, with the lone-pair electrons from the ligand nitrogen atoms being donated into the phosphorus d_{z^2} or other suitable orbitals. The exocyclic phosphorus–oxygen bonds in $[NP(OMe)_2]_8$ have mean lengths of 1.58 Å,[30] a 0.13 Å contraction from the theoretical single-bond value of 1.71 Å,[38] while the values of $[NP(OMe)_2]_4$ show an average contraction of 0.11 Å.[29] In both compounds, the P—O—C angle approximates to 120°, which is 8° to 9° wider than found in most ethers and organophosphates. This evidence suggests that the oxygen lone-pair electrons are partly donated to phosphorus in a d_π–p_π system.

4. Bond Angles at Phosphorus

The bond orientation around tetracoordinate phosphorus in phosphates and phosphine oxides corresponds to a distorted tetrahedral sp^3 arrangement. In cyclo- and polyphosphazenes, the tetrahedron is appreciably distorted, with ring angles at phosphorus generally approximating to 120°, and with the external R—P—R angle being close to 100°. Several reasons for this distortion can be advanced, including the effects of endocyclic ring strain, mutual repulsion by ring atoms, hybridization changes, and mutual repulsion or attraction by the ligands.

It is doubtful if endocyclic ring strain has any significant influence on the N—P—N bond angle. Although the 120° ring angle in cyclic trimers might be ascribed to the limitations imposed by a six-membered ring, approximately the same angle is found in cyclic tetramers, pentamers, hexamers, and even in linear high polymers where opportunities for N—P—N bond-angle narrowing should certainly exist. However, ring strain undoubtedly explains the nonexistence of cyclic dimeric phosphazenes.

It seems more reasonable to ascribe the 120° ring angle at phosphorus to mutual repulsion by the two adjacent nitrogen atoms. The repulsion may be Coulombic in nature (XI), a result of polarization of the skeletal σ bonds, or it could involve van der Waals repulsions, as shown by the van der Waals boundary contacts illustrated in Fig. 3.1. Contraction of the N—P—N bond angle

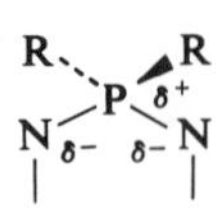

XI

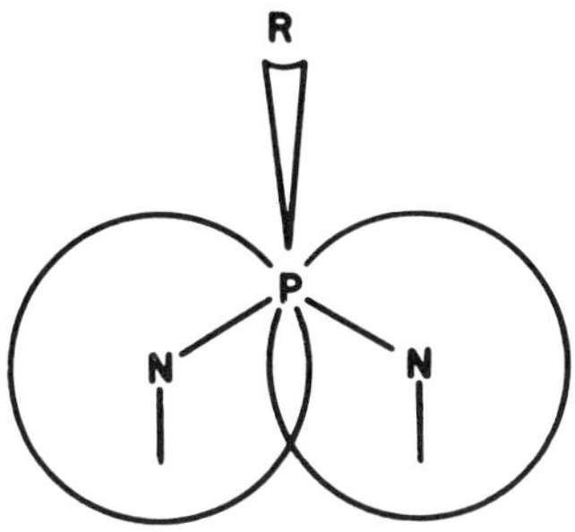

FIG. 3.1. Van der Waals boundary contacts in a phosphazene ring. A van der Waals radius of 1.55 Å is used for nitrogen[39] and a P—N bond length of 1.56 Å is employed.

below 117–120° would cause serious interpenetration of the nitrogen van der Waals boundaries. Thus, it is not surprising that N—P—N angles of less than 118° are only rarely found in phosphazenes. The lowest N—P—N angles reported are ~115°, when two phenyl groups are attached to the phosphorus atom, and this suggests that mutual repulsions by the two bulky phenyl groups may narrow the N—P—N angle. There are many precedents for this effect at tetrahedral carbon, where widening of one bond angle in the tetrahedron causes narrowing of the opposing angle in a "scissoring" or Thorpe-Ingold effect.[40, 41] Presumably the N—P—N and R—P—R bond angles in phosphazenes represent a subtle balance between the N···N repulsions on one hand and the R···R repulsions on the other. An additional factor, which would favor a wide ring angle at phosphorus, would be the stabilization which could result from a more favorable overlap of the nitrogen p_z orbitals with the d_{xz} orbital at phosphorus.

If the ligands form part of a five-membered ring system, as in XII, the exocyclic O—P—O angle at phosphorus is compressed to 97°.[27] This angle contraction is almost certainly induced by the restricting influence of the five-membered ring, since the related seven-membered exocyclic system in XIII

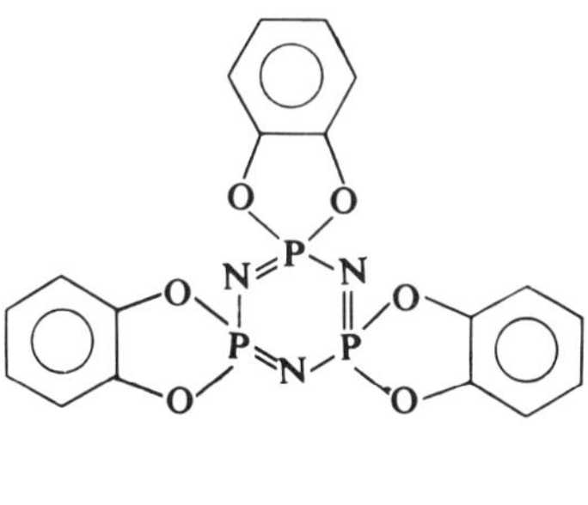

XII

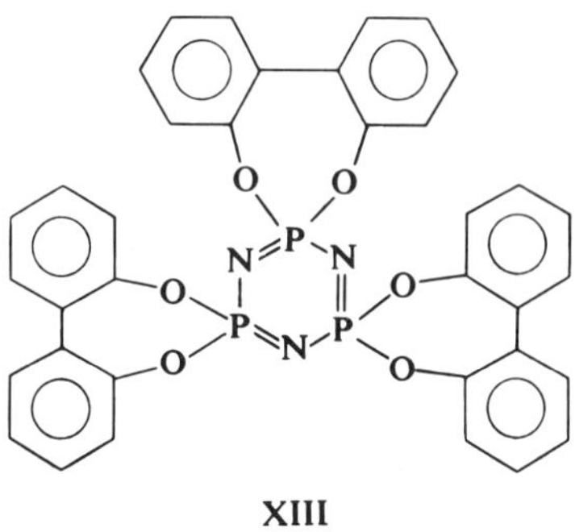

XIII

shows normal O—P—O angles of 102.5°.[28] In this latter case, side group twisting occurs to relieve the ring strain. However, one of the O—P—O angles in $[NP(OPh)_2]_3$ is also low (94.1°), although the others are normal (100.1°).[26a]

It is significant that four-membered cyclic phosphaz*anes* are found, but not four-membered cyclophosphaz*enes*. Two reasons for this can be discerned. Firstly, the general absence of *delocalized* heteromorphic d_π–p_π ring bonding in cyclophosphazanes removes the limitation of a wide (120°) ring angle at phosphorus and, secondly, the greater crowding of exocyclic ligands attached to the phosphorus can provide a steric inducement for low ring angles.[42] The fact that phosphorus forms part of an sp^3d, trigonal bipyramidal structure in these compounds, suggests the involvement of the d_{z^2} orbital in the longest (axial) phosphorus–nitrogen skeletal bond. The 90° ring angle at phosphorus is thus required. In compounds such as I, II, or the generalized structure X, ring bond angles near 90° also reduce the total transannular van der Waals repulsion. However, since the van der Waals radius of nitrogen (~1.55 Å) is lower than the value for phosphorus (~1.80 Å),[39] a distortion of the angles at phosphorus from 90° to 80–84° and those at nitrogen from 90° to 96–100° minimizes these repulsions. Significantly, in cyclic trimeric and tetrameric phosphazanes, such as III, IV, or V, where only four bonds exist at each phosphorus, the N—P—N angle of 104° to 107° is closer to the tetrahedral angle.

5. Bond Angles at Nitrogen

In cyclo- and polyphosphazenes, the dicoordinate skeletal nitrogen atoms appear to be $2sp^2$-hybridized (Fig. 3.2) (see Chapter 2). The orientation of the

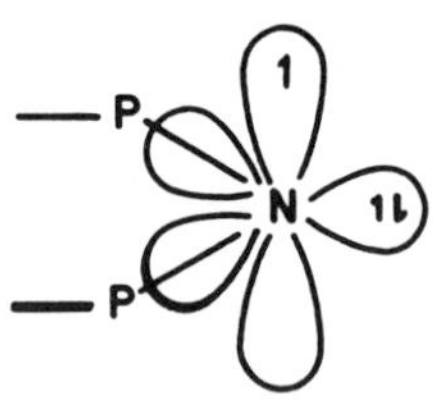

FIG. 3.2. sp^2–p_z Hybridization at skeletal nitrogen.

lone-pair orbital in the plane of the ring is important since it determines the symmetry of the π' system. Ideally, P—N—P bond angles in phosphazenes should be close to 120°. As shown in Table 3.1 the angle actually varies from 120° to 148°. Although it could be argued that a wide angle at nitrogen accompanies a depopulation of the lone-pair orbital when electron-withdrawing substituents (F or Cl) are attached to phosphorus, a wide angle is also found when strongly electron-supplying ligands such as NMe_2 or OMe are present. Thus, π' effects do not explain the trends.

It appears that ring size has a more important influence on the P—N—P bond angle in phosphazenes, for the widest angles are associated with tetrameric, pentameric, and hexameric rings. This reflects the high angular flexibility of bonds to nitrogen in phosphazenes and this, in turn, is an indication of the ease of rehybridization at that element.

In cyclodiphosphaz*anes* the ring angle at nitrogen is close to 100°, whereas in cyclotri- and cyclotetraphosphazanes it is much wider, in the range of 121–125°. Bond angles at tricoordinate nitrogen can vary from ~90° (a p^3 structure) to ~120° (an sp^2 structure) according to the amount of *s* character in the hybrid. Presumably, the wider angles in trimers and tetramers permit more effective π' coordinate bonding, but here again, the evidence suggests that the hybridization scheme responds to the requirements of ring size rather than vice versa.

6. Molecular Conformations

As illustrated in Table 3.1, both planar and puckered cyclic phosphazenes are found. Puckering of the phosphazene ring appears to have little effect on the stability of the molecule. Minor distortions from planarity do not reduce the d_π–p_π interactions because of the diffuseness and almost cylindrical symmetry of the *d*-orbital cloud. Similarly, the P—N—P skeletal angles can vary over a wide range before destabilization occurs, and the overall picture is one of a series of flexible bonds which do not themselves determine the molecular conformation. In these circumstances, nonbonding intramolecular forces can tip the balance between one molecular conformation or another. This is illustrated by the structural data of Wagner, Mani, and Vos,[19, 26, 37] which indicate that transannular Cl···Cl interactions distort one of the P—N—P angles in the chair form of $(NPCl_2)_4$[19] and that the wide P—N—P angles in $[NP(NMe_2)_2]_6$[26] and 1,1-$N_3P_3Cl_2(NHPr^i)_4 \cdot HCl$ are a response to side-group repulsions. A similar influence by the ligands on the ring conformation of $[NP(NMe_2)_2]_4$ has been noted by Bullen.[25] Paddock, Trotter, and Whitlow[30] have pointed out that the "step"-type structure of $[NP(OMe)_2]_8$ may represent a balance between nonbonding side-group interactions and the ring planarity induced by π bonding. Furthermore, for the cation $[NP(Me_2)_4H]^{\oplus}$, both tub and saddle conformations exist in the same crystal, and the π system appears to exert little influence on the molecular shape.[43] In the solid state, *inter*-molecular forces may also exert an appreciable influence on the ring conformation and, for this reason, considerable care must be exercised during the interpretation of X-ray data in terms of bonding differences.

The conformations of phosphazene linear high polymers can also be explained in terms of nonbonding steric interactions. X-Ray diffraction data indicate that poly(difluorophosphazene), $(NPF_2)_n$, occupies a *cis–trans* planar conformation[14] (Fig. 3.3) and several related polymers, such as $(NPCl_2)_n$,

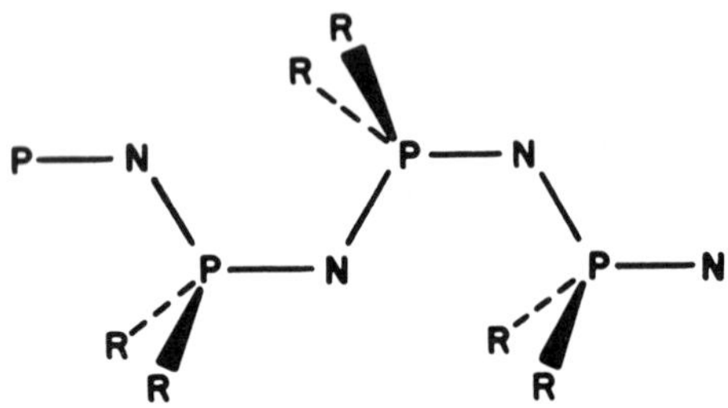

FIG. 3.3. A *cis–trans* planar chain conformation.

$[NP(OCH_2CF_3)_2]_n$, and $[NP(OPh)_2]_n$ crystallize in a similar conformation[20, 44] (see Chapter 17). The near-planarity of these molecules is not a result of π-type interactions, but rather a function of the ligand–ligand and ligand–chain, nonbonding interactions.[45, 46] In other words, when bulky side groups are attached to phosphorus, the *cis–trans* planar conformation is the arrangement which minimizes the nonbonding intramolecular repulsions. Even when the ligand is small, such as a chloro group, the combination of Cl···N attractions and Cl···Cl repulsions force the chain into a distorted *cis–trans* planar arrangement. In the absence of bulky substituents, the high polymers show evidence of an unusually flexible backbone structure. For example, for $(NPF_2)_n$, skeletal torsional motions persist at −80°C,[47] and this indicates that, if d_π–p_π bonding exists in this molecule, it certainly does not create significant torsional barriers.

C. Thermochemical Bond Strengths and π-Bond Energy

The relative shortness of the phosphorus–nitrogen skeletal bonds in phosphazenes is believed to be an indication of high bond strength, and therefore of d_π–p_π character. If this is true, there should be a direct correlation between bond shortness and thermochemical bond energy. Unfortunately, different methods of calculation yield different phosphorus–nitrogen bond energy values. Bond energy terms are listed in Table 3.2.[48–52]

The surprising fact is that the P—N bond energy values for cyclophosphazenes are only marginally higher (3–4 kcal) than those of the control P—N single bonds in $(Et_2N)_3P$, $(Me_2N)_3P$, $M^+(PO_3NH_3)^-$, $P_4(NMe)_6$ (cage), and in the long axial bond of $(PhN{—}PCl_3)_2$. Indeed, the P—N bonds in $[PhN{—}P(O)NHPh]_2$ and the short (axial) bond in $(PhN{—}PCl_3)_2$ have higher energies than the skeletal bonds in the cyclophosphazenes studied, and the P—N bond energy in $EtN{=}PPh_3$ is higher still. Even more surprising is the observation that $\bar{E}$ values, which specifically include an energy term for π bonding, suggest that the P—N bond energies in phosphazenes are actually lower than those in phosphoramidates. Thus, the results indicate that, in terms of stabilization

TABLE 3.2

PHOSPHORUS–NITROGEN BOND ENERGIES

Compound	Bond energy (kcal)	$\bar{\epsilon}^a$	References
$(Et_2N)_3P$	66.8 ($\bar{D}$)	—	48
$(Me_2N)_3P$	70 ($\bar{D}$)	—	49
$M^+(PO_3NH_3)^-$	~69 ($\bar{E}$)	83	49
$P_4(NMe)_6$	69	—	50
$[PhN—P(O)NHPh]_2$	79	—	50
$(PhN—PCl_3)_2$	69 (axial)	—	50
	77.5 (eq)	—	50
$(NPCl_2)_3$	72.3 ($\bar{E}$)	70	49, 51, 52
$(NPCl_2)_4$	72.5 ($\bar{E}$)	70	49, 52
$[NP(OC_6H_{11})_2]_3$	~72.4	76	49, 51
$(NPPh_2)_4$	~73	74	49, 51
$(NPMe_2)_3$	~72.4	68	49, 51
$EtN{=}PPh_3$	98.4 (E)	52	48
$EtN{=}PMe_3$	69.6 (E)	52	48

[a] $\bar{\epsilon} = \bar{E}_{(P-N\ tervalent)} + E'_{\pi\ (P-N)}$, where $E'_{\pi\ (P-N)}$ is a correction term for π bonding based on infrared force constant data.[49]

energy, d_π–p_π bonding does not exert a profound influence on the strength of the skeletal bonds in cyclophosphazenes. However, the unusually high P—N bond energy term for $EtN{=}PPh_3$ is consistent with strong d_π–p_π bonding and with appreciable delocalization. The fact that there is a sharp fall in P—N bond energy when $EtN{=}PPh_3$ is replaced by $EtN{=}PMe_3$, illustrates that the phenyl ligands participate strongly in strengthening the P—N bond in monophosphazenes.

TABLE 3.3

P—R BOND ENERGIES IN $(NPR_2)_n$ AND PR_3 (KCAL/MOLE)[a]

Compound	E(P—R) in $(NPR_2)_n$	E(P—R) in PR_3	E(PR + PN) excess/bond over single-bond values[b]
$(NPCl_2)_3$	80.3	76.2	9.3
$[NP(OC_6H_{11})_2]_3$	135.3	92.0	16.1
$(NPMe_2)_3$	68.3	61.6	12.4
$(NPPh_2)_4$	80.5	67.5	18.9

[a] From Bedford and Mortimer.[51]
[b] From Paddock.[53]

Bonds from phosphorus to various ligands show a higher energy in cyclophosphazenes than in phosphines. This is illustrated by the values determined by Bedford and Mortimer[51] and depicted in Table 3.3. Thus, there is some evidence that cyclophosphazenes are stabilized by exocyclic d_π–p_π bonding between phosphorus and the ligands, and that the *total* π bonding within the whole molecule [illustrated by the $E(\mathrm{PR} + \mathrm{PN})$ excess energy terms in Table 3.3[53]] may be significant, even though the ring π bonding alone may be weak.

D. Vibrational Spectra

Because most phosphazenes and phosphazanes are relatively complicated molecules, the characteristic group frequency concept has been applied widely to these compounds, and only the simplest structures have been analyzed in terms of symmetry considerations or by normal coordinate analysis.

1. Structure Determination

A few complete structures of phosphorus–nitrogen compounds have been elucidated from infrared and Raman data. These are listed in Table 3.4.[54–73a]

As a comprehensive structural tool, vibration–rotation spectroscopy is less satisfactory for solid phosphazenes and phosphazanes than is X-ray crystallography. However, all solid-state structural methods suffer from the disadvantage that the measured structural parameters and conformations may reflect the influence of crystal-packing forces as much as intramolecular interactions. Thus, detailed infrared and Raman studies offer the advantage that structure determinations performed on the solid, liquid, and vapor can provide more meaningful information about bonding and intramolecular conformational effects. This is illustrated by the results of Hisatsune's study with $(NPCl_2)_3$ and $(NPCl_2)_4$.[62, 63] The phosphazene ring in the trimer is slightly puckered in the solid state but apparently planar in the liquid and vapor states. As the tetramer changes from the solid, to the liquid, to the vapor state, the symmetry changes from S_4 or C_{2h} (for different polymorphs), through D_{2d}, to D_{4h}. This illustrates the dangers inherent in basing theoretical bonding arguments on minor structural differences determined for molecules in the solid state. Other conformational differences are found when the X-ray data in Table 3.1 and the vibrational data in Table 3.4 are compared for $(NPF_2)_4$ (planar in the solid by X-ray, but nonplanar in all phases by infrared–Raman), $(NPCl_2)_5$ (planar by X-ray, but nonplanar by infrared–Raman), and $(NPBr_2)_3$ (puckered in the solid state but planar in the vapor). However, it should be noted that the apparent planarity of a molecule in the vapor state could be a consequence of rapid thermal averaging between mirror-image puckered conformers.

TABLE 3.4

MOLECULAR STRUCTURES FROM INFRARED AND RAMAN DATA

Compound	Symmetry	Structure	Phase[a]	References
$(MeNPF_3)_2$	C_{2h} or C_i	Planar P—N ring with C atoms in ring plane	v, l	54
	C_{2h}	Planar P—N ring with C atoms in ring plane	v, l	55
$(MeNPCl_3)_2$	C_{2h} or C_i	Planar P—N ring with C atoms in ring plane	v, l	54
$(NPF_2)_3$	D_{3h}	Planar P—N ring	v, l, s, sln	56–58
$(NPF_2)_4$	C_{2h} or lower	Nonplanar ring	v, l, s, sln	56, 57
$(NPF_2)_5$	$\sim D_{5h}$	Slightly nonplanar	v, l, s, sln	57
$(NPCl_2)_3$	D_{3h}	Planar ring	v, l, s, sln	57, 59–61
	$\sim D_{3h}$	Slightly nonplanar	s	62
$(NPCl_2)_4$	D_{4h}	Planar ring	v	63
	D_{2d}	Nonplanar ring	l, sln	59, 63–65
	S_4	Nonplanar ring	s	63
	C_{2h}	Nonplanar ring	s	63
$(NPCl_2)_5$	Not D_{5h} or C_{5v}	Nonplanar ring	v, s, sln	57
$(NPCl_2)_n$	C_2	Distorted *cis–trans* planar chain	s	66
$(NPBr_2)_3$	D_{3h}	Planar ring	v, sln	64, 67–69
	$D_{3h} \rightarrow C_{3v}$	Slightly nonplanar	s	64
$(NPBr_2)_4$	D_{2d}	Nonplanar ring	sln	64
$(NPBr_2)_5$	$\sim D_{5h}$	Slightly nonplanar	s, sln	70
$N_3P_3Cl_{6-n}Br_n$	D_{3h}	Planar ring	s	71, 72
$[NP(NCS)_2]_3$	D_{3h}	Planar ring	s, sln	73
$[NP(N_3)_2]_3$	D_{3h}	Planar ring	s	73a

[a] v = Vapor phase, l = liquid, s = solid, and sln = solution.

2. GROUP FREQUENCY ASSIGNMENTS

The group frequency concept is based on the assumption that the vibrational–rotational frequencies of a structural unit are almost independent of the vibrations of the ligands attached to it. In the present context, this means that different characteristic infrared frequencies should be recognizable for P—N σ bonds in phosphazanes and for σ–π bonds in phosphazenes, with the frequency increasing as a σ bond gains π character and the force constant becomes larger. This is, in reality, a gross oversimplification, particularly for *cyclo*phosphazenes or *cyclo*phosphazanes, for which the observed infrared bands result from vibrations of N—P—N units and from ring distortion modes. The group frequency concept is also valid only if the ligands attached to phosphorus do not participate in combination vibrations with the ring or chain. Thus, the most useful information can be obtained for derivatives which have relatively small ligands, and the least valid conclusions are derived from compounds which have large, bulky ligands with vibrational frequencies close

TABLE 3.5

P=N STRETCHING FREQUENCIES FOR MONOPHOSPHAZENES (PHOSPHINIMINES)

Compound	ν (cm^{-1})	References
$Cl_3CN{=}PCl_3$	1360 (1450)	74
$Cl_3CC(Cl_2)N{=}PCl_3$	1375 (1440)	74
$(Cl_3C)_2C(Cl)N{=}PCl_3$	1465	74
$Me_3CN{=}PCl_3$	1370	74
$PhN{=}PPh_3$	1344	75
$p\text{-}NO_2C_6H_4N{=}PPh_3$	1373	75
$p\text{-}R_2NC_6H_4N{=}PPh_3$	1328	76
$MeN{=}PPh_3$	1230	75
$PhN{=}PClPh_2$	1357	75
$PhN{=}PEtPh_2$	1330	75
$PhN{=}P(OEt)Ph_2$	1330	75
$PhN{=}P(NHPh)Ph_2$	1304, 1316	75
$PhN{=}P(NPr_2)Ph_2$	1335, 1350	75
$PhN{=}PBu_3$	1339	75
$Me_3SiN{=}PMe_3$	1286	77
$Me_3SiN{=}PPh_3$	1302, 1315	77
$PhN{=}P(OAlk)_3$	1325–1385	78
$F_3CC(O)N{=}PCl_3$	1390	74, 79
$Cl_3CC(O)N{=}PCl_3$	1290	79
$PhC(O)N{=}PCl_3$	1295	79
$PhC(O)N{=}PPh_3$	1337	76
$MeO(O)N{=}Ph_3$	1370	80
$ArSO_2N{=}PPh_3$	1147	81
$ArSO_2N{=}PCl_3$	1199	75, 76, 81

to those of the skeleton. Provided these limitations are recognized, the group frequency approach can provide useful information about bonding.

Many monophosphazenes (phosphinimines) show a characteristic P=N stretching vibration in the 1325–1375 cm^{-1} region of the spectrum. Table 3.5 provides some examples.[74–81] These vibrations occur at much higher frequencies than those in phosphazanes, and this implies an appreciable degree of d_π–p_π bonding in monophosphazenes. In general, the highest frequencies are associated with the presence of phenyl or electronegative substituents. However, the last two entries in Table 3.5 illustrate that tosyl groups attached to nitrogen markedly reduce the frequency. This may be the result of π delocalization into the SO_2 unit or to combination vibrations involving the heavy sulfur atom. Linear phosphazenes which contain two phosphorus atoms linked to nitrogen, such as MeS—PMe_2=N—P(S)Ph_2,[82] or Ph_3P=N—P(O)Ph_2,[75] also show P=N vibrations at low frequencies (~1200 cm^{-1}), but alkyl and alkoxy derivatives, such as $(MeO)_3P$=N—P(O)Et_2,[83] show higher frequency vibrations near 1300 cm^{-1}.

Cyclo- and polyphosphazenes usually show two characteristic infrared bands. The strongest, in the 1200–1400 cm^{-1} region, corresponds to a P—N—P asymmetric vibration or a degenerate ring-stretching mode. A second characteristic band is found in the 700–950 cm^{-1} region for solids and it can be considered as a P—N—P symmetric stretch. Its position varies with the skeletal degree of polymerization, being close to 885 cm^{-1} for cyclic trimers, near 895 cm^{-1} for cyclic tetramers, and in the 750 cm^{-1} region for high polymers. Because this is a forbidden vibration, it is seen only when the selection rules break down, often in the solid or molten states. However, most of the structural interpretations have made use of the 1200–1400 cm^{-1} bands, and the following comments refer predominantly to that spectral region.

Table 3.6 lists characteristic P—N vibrational frequencies for a variety of ring sizes and substituent groups. In general, only rough trends should be derived from these data and not detailed comparisons. Four types of structural information can be derived from the observed trends. These are (1) information about relative changes in skeletal force constants as the degree of polymerization is increased, (2) data about the effects of ligand electronegativity on skeletal π character and bond strength, (3) side group steric effects, and (4) calculation of force constant values. These aspects will be considered in turn.

The change from cyclic trimer to cyclic tetramer is usually accompanied by an increase in the characteristic ring frequency. This is particularly noticeable where the ligand is F, Cl, OMe, OEt, OPr^n, OBu^n, OPh, or amino. Attempts have been made to ascribe this effect to a strengthening of the ring bonds in the tetramer by enhanced delocalization. However, further increases in ring size do not give rise to higher-frequency vibrations. The frequency usually remains nearly constant between $n = 4$ and $n = 6$ or higher, or it may decrease. These frequency shifts appear to be a consequence of the fact that the characteristic ring vibration is a degenerate ring-stretching mode and not a vibration of a monomerlike unit. It can be shown that, even if the P—N force constant is independent of ring size, the vibrational frequency should *decrease* with increasing ring size.[84] The frequency change observed when cyclic trimers are compared with tetramers is thus an anomaly which may have little significance from the point of view of bond energies.

TABLE 3.6

CHARACTERISTIC P—N STRETCHING VIBRATIONS FOR CYCLO- AND POLYPHOSPHAZENES[a, b]

Compound	n = 3	n = 4	n = 5	n = 6	n = ~15,000	References
$(NPF_2)_n$	1297	1419, 1438	1405[c]–1450[d]	1408	1290	84–88
$(NPFCl)_n$[e]	1255	1377	—	—	—	89
$(NPFBr)_n$[e]	1235	—	—	—	—	90
$(NPCl_2)_n$	1218	1315	1298, 1354	1325	1230, 1275	60, 66, 69, 72, 86, 88, 91–94
$(NPClBr)_n$	1190	—	—	—	—	72
$(NPBr_2)_n$	1175	1253–1280	1265–1330	—	—	64, 65, 67, 69, 70, 72, 87, 93
$[NP(NCS)_2]_n$	1218	—	—	—	1240	73, 95
$[NP(N_3)]_n$	1200	—	—	—	—	95a
$(NPMe_2)_n$	1180	1180	—	—	1160	93, 96, 97
$[NP(CF_3)_2]_n$	1205, 1300	1216, 1412	—	—	1335–1430	98, 99
$(NPEt_2)_n$	(1157) 1225	1320	—	—	—	93
$(NPPh_2)_n$	1190	1213	—	—	1200	93, 100

$(NPClPh)_n$	1180	—	—	—	1290	101, 102
$[NP(OMe)_2]_n$	1235, 1275	1337	1340	1335	1250	92–94, 103
$[NP(OEt)_2]_n$	1225–1240	1320	—	—	1240	92, 93, 103
$[NP(OPr^n)_2]_n$	1225–1240	1323	—	—	—	103
$[NP(OBu^n)_2]_n$	1225–1240	1323	—	—	—	103
$[NP(OCH_2CF_3)_2]_n$	1240, 1280	1272	—	—	1270	92, 104
$[NP(OCH_2C_2F_5)_2]_n$	1200	1200	—	—	1250	95, 104
$[NP(OCH_2C_3F_7)_2]_n$	1227	1220	—	—	1220	95, 104
$[NP(OPh)_2]_n$	1160–1200 1250–1280	1330–1350	—	—	1240	92, 103, 105
$[NP(SEt)_2]_n$	1150	—	—	—	—	105a
$[NP(NH_2)_2]_n$	1170	1240	—	—	—	93, 106
$[NP(NHMe)_2]_n$	1180	—	—	—	1255, 1200	107, 108
$[NP(NMe_2)_2]_n$	1195	1265	—	—	1240, 1280	93, 109
$[NP(NHPh)_2]_n$	—	—	—	—	1180–1220	109
$[NP(NHMe)(Ph)]_n$	1122, 1185	~1240	—	—	—	110, 111

[a] For a more comprehensive listing, see Appendix II.
[b] Some of these assignments are tentative, particularly when organic substituents are present.
[c] Solid.
[d] Vapor.
[e] The positional isomerism is not defined.

More meaningful comparisons can be derived from a study of the influence of different ligands on the characteristic vibrational bonds. In general, electronegative ligands raise the characteristic P—N vibrational frequency (Table 3.6[60, 64–67, 69, 70, 72, 73, 84–111]). For example, for cyclic trimers, vibrational frequencies above 1200 cm^{-1} are found when the ligands are F, Cl, NCS, CF_3, or OR, whereas frequencies below 1200 cm^{-1} are associated with Br, Me, Ph, NH_2, NHR, or NR_2 groups. Fluorine induces the highest-frequency skeletal vibrations, which are near 1300 cm^{-1} for the cyclic trimer, and over 1400 cm^{-1} for the tetramer, pentamer, and hexamer. These values are comparable to the highest frequencies observed for monophosphazenes (Table 3.5). Presumably, fluorine exerts an influence on the skeletal bonds by a strong withdrawal of electrons from phosphorus. This could facilitate a drift of electrons from the nitrogen lone pair to phosphorus in a π' system, or it could aid d_π–p_π bond

TABLE 3.7

P—N VIBRATIONAL FREQUENCIES IN ALKYLAMINOCYCLOPHOSPHAZENES (in cm^{-1})[a]

R	NH_2	MeNH	EtNH	Pr^nNH	Bu^nNH	n-$C_5H_{11}NH$	n-$C_6H_{13}NH$
$(NPR_2)_3$	1170	1175	—	1183	1195	1190	1192
$(NPR_2)_4$	1240	1215	1262	1266	1260	1265	1265

[a] From Paddock and Moeller *et al.*[112–114]

formation above and below the ring plane by contraction of the phosphorus d_{xz} and d_{yz} orbitals. In these terms, chloro, isothiocyano, and trifluoromethyl groups also strengthen the skeletal π bonding by a similar mechanism, but weakly electron withdrawing or electron supplying groups, such as bromo, methyl, phenyl, or amino, weaken the skeletal bonds by the opposite effect. The position of alkoxy and aryloxy substituents in this series is surprising, since nearly all the OR groups listed in Table 3.6 show the characteristics of skeletal bond-strengthening units. For the fluoroalkoxy or phenoxy ligands this is not surprising, but for methoxy, ethoxy, n-propoxy, and n-butoxy, the result is unexpected. The suspicion exists that the group frequency concept may not be applicable when relatively bulky substituents, such as these, are present.

When alkylamino substituents are present, the ligand exerts a small steric influence on the P—N stretching mode. This is illustrated by the data listed in Table 3.7.[112–114] Amino and methylamino substituents are associated with low P—N characteristic frequencies, as expected for electron-supplying groups. However, the longer-chain primary amino substituents induce a slight increase in skeletal vibrational frequency. It has been suggested that this phenomenon results from the inability of the bulkier substituents to assume the planar configuration required to supply electrons from *exocyclic* nitrogen to phosphorus. Thus, electrons are acquired from cyclic nitrogen with a strengthening of the π' ring system.

Infrared data for $(NPCl_2)_3$ and $(NPF_2)_3$ have been used by Chapman and Carroll to calculate the force constants of the P—N and P—halogen bonds.[115] A P—N force constant of 8.19 mdyne/Å was calculated for the chloroderivative and 8.95 mdyne/Å for the fluoride derivative. Equally interesting are the values of 4.38 mdyne/Å for the P—Cl bond and 5.79 mdyne/Å for the P—F

TABLE 3.8

P—N CHARACTERISTIC VIBRATIONAL FREQUENCIES IN CYCLOPHOSPHAZANES

Compound	Frequency (cm^{-1})	References
$(MeN—PCl_3)_2$	847	54, 116
$(MeN—PF_3)_2$	847 (614, 535)	54, 117
$(ClCH_2N—PCl_3)_2$	821	118
$(HN—PFPh_2)_2$	864	119
$(MeN—P(O)NMe_2)_2$	871	120
$(EtN—P(O)NMe_2)_2$	910	120
$(MeN—P(O)SMe)_2$	847, 855	120
$(EtN—P(O)Cl)_2$	886	120
$(PhN—P(O)Cl)_2$	893, 904, 949, 980	120

bond. Both of these values are sufficiently high, compared to the equivalent bonds in PCl_3 (2.83 mdyne/Å), PCl_5 equatorial (2.89 mdyne/Å), $POCl_3$ (6.18 mdyne/Å), to suggest that considerable exocyclic, P—halogen π character is present in the cyclophosphazenes. If this is so, the P—Cl π character is more pronounced than that of the P—F bond.

The phosphorus–nitrogen bond in phosphaz*anes* appears to be longer and weaker than in phosphazenes, and characteristic vibrational frequencies would be expected at lower frequencies. The data tabulated in Table 3.8[54, 116–120] confirm that this is so.

The cyclodiphosphazane ring-stretching mode appears in the region of 847–870 cm^{-1},[54] although assignments for this vibration in the 530–615 cm^{-1} and the 870–1053 cm^{-1} regions have also been made.[117] Presumably, there is a rough correlation between increasing coordinate π character and increasing vibrational frequency within this range, although the separation into long and short skeletal bonds, as discussed earlier in the X-ray section, will complicate the picture.

E. Ultraviolet Spectra

Ultraviolet–visible spectroscopy has proved to be a powerful tool for the study of π-electron interactions in organic systems. Attempts have been made, with considerably less success, to apply similar techniques to the study of bonding in cyclo- and polyphosphazenes. However, definitive information is available about the electronic spectra of monophosphazene-type derivatives.

1. Monophosphazenes and Related Derivatives

Bock[121–124] has examined the influence of the phosphorus–nitrogen linkage on the electronic spectrum of an adjacent azido grouping. For example, the spectra of the azophosphonic acid derivatives XIV and XV have been investigated. Compounds of structure XIV, where X is OPh, Ph, OK, or NR_2,

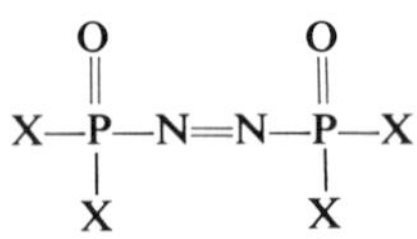

XIV

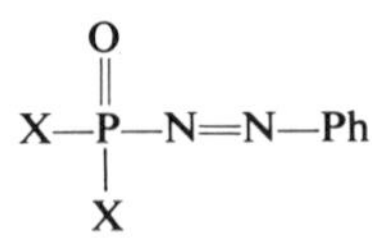

XV

are violet. The color results from an $n \rightarrow \pi^*$ transition at ~500 mμ, which is accompanied by an ultraviolet $\pi \rightarrow \pi^*$ transition near 300 mμ. Both transitions are derived from the P⋯N⋯N⋯P framework, with contributions from the phosphorus 3*d* orbitals being included. However, although the substituents, X, exert inductive influences on the visible band, the skeletal π delocalization is not transmitted through the phosphorus atoms. A similar behavior was observed for compounds of structure XV, where the $n \rightarrow \pi^*$ transition occurs at 500–600 mμ and the $\pi \rightarrow \pi^*$ transition appears near 300 mμ.

In the absence of chromophoric linkages, absorption bands appear only in the ultraviolet. The monophosphazene, $PhEt_2P{=}NPh$, absorbs at 243 and 283 mμ,[125] $Ph_3P{=}NH$ absorbs at 275, 303, 258, and 298 mμ,[126] and $[Ph_2(NH_2)P{=}N{-}P(NH_2)Ph_2]^{\oplus}Cl^{\ominus}$ absorbs at 225 and 261–273 mμ.[127] Paciorek[128] has noted that the series of linear phosphazenes listed in Table 3.9 absorb at the same wavelengths, irrespective of the number of P—N units linked together. Thus, there is no evidence for extensive skeletal delocalization, and this favors an "island"-type delocalization model (see Chapter 2). Indeed, the close correspondence between the spectra of $Ph_2P(O)OH$, $Ph_2P(O){-}NH{-}P(O)Ph_2$ and those of the linear phosphazenes suggests that

TABLE 3.9

ULTRAVIOLET ABSORPTION INTENSITIES (log ε) OF LINEAR PHOSPHAZENES[a]

Compound	247 mμ	260 mμ	266 mμ	273 mμ
$Ph_2P(O)OH$	2.98	3.06	3.14	3.07
$Ph_2P(O)$—NH—$P(O)Ph_2$	—	3.30	3.39	3.30
$Ph_2P(O)$—N=PPh_3	—	3.37	3.33	3.27
$Ph_2P(O)$—N=PPh_2—NH—$P(O)Ph_2$	—	3.45	3.48	3.35
$Ph_2P(O)$+N=PPh_2+$_2$—NH—$P(O)Ph_2$	—	3.64	3.61	3.41
$Ph_2P(O)$+N=PPh_2+$_4$—Ph	—	—	3.80	3.59

[a] From Paciorek.[128]

phenyl $\pi \rightarrow \pi^*$ and perhaps lone-pair $n \rightarrow \pi^*$ transitions are responsible for the spectra. Linear chlorophosphazenes of structure, $Cl_4P{+}NPCl_2{+}_n$—N=PCl_3, show no absorption maxima between 400 and 290 mμ, but they do show a continuous absorption increase as the wavelength is decreased.[128]

2. CYCLO- AND POLYPHOSPHAZENES

Most aryl-substituted cyclo- and polyphosphazenes are colorless compounds with near-ultraviolet spectra that can be explained satisfactorily in terms of side-group absorptions alone. Specific examples are shown in Table 3.10.[92, 109, 128–132]

Aryl-substituted cyclo- and polyphosphazenes have spectra which are very reminiscent of side-group derivatives. For example, the spectra of $[NP(OPh)_2]_3$ and $[NP(OPh)_2]_n$ resemble that of anisole,[129] the spectrum of $[NP(O_2C_6H_4\text{-}o)]_3$ resembles that of 1,2-dimethoxybenzene,[129] and the spectrum of $[NP(NHPh)_2]_n$ resembles that of aniline.[109] Thus, delocalization effects between the substituents and the phosphazene ring are not detected.[129] Those phosphazenes which contain aliphatic substituents show no absorption maxima at wavelengths below 200 mμ and this again reflects the behavior of the substituent groups. Cyclophosphazene rings themselves do not absorb in the near-ultraviolet region.

Perhaps more significant is the fact that high-polymeric organophosphazenes yield ultraviolet spectra which are almost identical to those of equivalent cyclic trimers and tetramers.[92, 109] This provides a strong argument against a broadly delocalized skeletal d_π–p_π bonding arrangement. Even when a phenyl group is bonded directly to phosphorus, open-chain species, such as $Ph_2P(O)$+N=PPh_2+$_4$Ph, show absorption maxima at the same wavelengths as the cyclic trimer, $(NPPh_2)_3$.[128]

Halocyclophosphazenes show absorption maxima at wavelengths near or below 200 mμ.[92, 132–139] Specific values are shown in Table 3.11.[132] Again, no bathochromic shifts are observed with an increase in ring size,[132, 134] but a shift to longer wavelengths is observed in the change from the fluoride, to the

TABLE 3.10

ULTRAVIOLET ABSORPTION MAXIMA OF ORGANOPHOSPHAZENES[a]

Compound	Absorption maxima	References
$(NPPh_2)_3$	266 (3.09), 273 (2.89)	128
$[NP(OPh)_2]_3$	206 sh (4.70), 259 (3.23), 266 (3.31), 272 (3.18)	129
$[NP(OPh)_2]_n$	264 (2.96/monomer unit)	92
$[NP(OC_6H_4NO_2\text{-}o)_2]_3$	254 sh	130
$[NP(OC_6H_4NO_2\text{-}m)_2]_3$	258	130
$[NP(OC_6H_4NO_2\text{-}p)_2]_3$	264	130
$[NP(O_2C_6H_4\text{-}o)]_3$	209 (4.38), 266 sh (3.68), 271 (3.79), 276 (3.72)	129
$[NP(O_2C_{12}H_8)]_3$[b]	244 (4.68), 269 sh (3.88), 277 sh (3.76)	131
$[NP(OMe)_2]_3$	<200	92, 132
$[NP(OMe)_2]_n$	<200	92
$[NP(OEt)_2]_3$	<200	92, 132
$[NP(OEt)_2]_n$	<200	92
$[NP(OCH_2CF_3)_2]_3$	<200	92
$[NP(OCH_2CF_3)_2]_n$	<200	92
$(NPMe_2)_3$	<200	132
$[NP(NCS)_2]_3$	196 (5.09), 250 (3.44)	132
$[NP(N_3)_2]_3$	197 (3.0), 210 (3.80), 260 (1.30)	132
$N_3P_3Cl_4(SEt)_2$	~240 sh (~3.0)	132
$[NP(NMe_2)_2]_3$	<200	132
$[NP(NMe_2)_2]_n$	<200	109
$[NP(NHEt)_2]_n$	<200	109

[a] λ_{max} in mμ with $\log_{10} \epsilon$ values in parentheses.
[b] 2,2′-Dioxybiphenyl derivative.

chloride, and to the bromide. The exact assignment of these absorption maxima is not clear. The low extinction coefficient suggests that the absorptions result from a nitrogen $n \rightarrow \pi^*$ transition, but protonation of the ring nitrogen atoms has little effect on the absorption.[137] The possibility also exists that halogen atom lone-pair excitation may be responsible. Finally, the absorptions could represent skeletal $\pi \rightarrow \pi^*$ transitions but, if so, only limited π-electron delocalization is implied.

TABLE 3.11[a]

ULTRAVIOLET ABSORPTION MAXIMA OF HALOGENOPHOSPHAZENES[b]

Compound	λ_{max}	
$(NPF_2)_3$	149.4	(4.0)
$(NPF_2)_4$	147.5	(4.0)
$(NPCl_2)_3$	175	(4.0)
$(NPCl_2)_4$	175	(4.0)
$(NPCl_2)_n$	<220	—
$N_3P_3Cl_5Br$	192	(4.0)
$N_3P_3Cl_4Br_2$	192	(4.0)
$(NPBr_2)_3$	200	(4.4)
$(NPBr_2)_4$	199	(4.5)

[a] From Lakatos *et al.*[132]
[b] λ_{max} in mμ with $\log_{10} \epsilon$ values in parentheses.

3. CHARGE-TRANSFER SPECTRA

Attempts have also been made to obtain information about the cyclophosphazene electronic arrangement by examination of the spectra of possible charge-transfer complexes.[136, 139, 140] No evidence could be found for the presence of charge-transfer bands or spectral shifts when $(NPCl_2)_3$ or $(NPF_2)_3$ were allowed to interact with hexamethylbenzene or tetracyanoethylene in cyclohexane,[139] or when $(NPCl_2)_{3 \text{ or } 4}$ was treated with iodine in carbon tetrachloride or heptane,[136, 140] or with benzene, 1,3,5-trinitrobenzene, naphthalene, pyridine, or chloranil.[136] Moreover, attempts to detect π complexes after the melting together of $(NPCl_2)_3$ and hexamethylbenzene, followed by ultraviolet and infrared examination of the mixture, provided no evidence for complex formation.[139]

F. Polarography

Specific organophosphazenes can be reduced polarographically by a one-electron process to form the radical-anion[141, 142]

$$(NPR_2)_n + e^- \rightarrow (NPR_2)_n^{\dot{-}}$$

The electron is presumably accepted into the lowest unfilled molecular orbital, which could be a phosphazene ring antibonding orbital, a ligand antibonding orbital, or a broadly delocalized combination of the two. In these terms, the least negative half-wave reduction potentials should be associated with the lowest unfilled antibonding orbital energy levels. Attempts have been made to reduce a number of different cyclo- and polyphosphazenes in the 0 to −3 V potential range in dimethylformamide solvent with the use of tetraalkylammonium halides as supporting electrolytes. The results are shown in Table 3.12. Figure 3.4 shows a polarogram for the reduction of $(NPPh_2)_3$.

TABLE 3.12
POLAROGRAPHIC REDUCTION POTENTIALS OF PHOSPHAZENES[a]

Compound	$E_{1/2}$[b] (V)	Compound	$E_{1/2}$[b] (V)
$(NPCl_2)_3$	n.r.[c]	$[NP(OCH_2CF_3)_2]_n$	n.r.
$(NPCl_2)_4$	n.r.	$[NP(OMe)_2]_3$	n.r.
$(NPCl_2)_n$	n.r.	$[NP(O_2C_6H_4\text{-}o)]_3$	n.r.
$(NPF_2)_3$	n.r.	$[NP(NHPh)_2]_3$	n.r.
$(NPBr_2)_3$	n.r.	$(NPPh_2)_3$	−2.65
$(NPClPh)_3$	n.r.	$(NPPh_2)_4$	−2.67
$N_3P_3Cl_4Ph_2$-1,1	n.r.	$[NP(Ph)(OCH_2CF_3)]_3$	−2.45
$[NP(OPh)_2]_3$	n.r.	$N_3P_3Ph_2(OCH_2CF_3)_4$-1,1	−2.57
$[NP(OPh)_2]_4$	n.r.	$[NP(OC_6H_4NO_2\text{-}p)_2]_3$	−1.24
$[NP(OPh)_2]_n$	n.r.	$[NP(2,3\text{-}O_2C_{10}H_6)]_3$[d]	−1.83
$[NP(OCH_2CF_3)_2]_3$	n.r.	$[NP(1,8\text{-}O_2C_{10}H_6)]_3$[e]	−2.15
$[NP(OCH_2CF_3)_2]_4$	n.r.	$[NP(2,2'\text{-}O_2C_{12}H_8)]_3$[f]	−2.33

[a] From Allcock and Birdsall.[141,142]
[b] Versus saturated calomel electrode.
[c] n.r. indicates no reduction observed in the 0 to −3 V potential range. Those compounds which contained halogen groups bonded directly to phosphorus showed evidence of decomposition during reduction.
[d] 2,3-Dioxynaphthyl derivative.
[e] 1,8-Dioxynaphthyl derivative.
[f] 2,2′-Dioxybiphenyl derivative.

The results indicate that phosphazenes which contain phenyl, *p*-nitrophenoxy, 2,3-dioxynaphthyl, 1,8-dioxynaphthyl, or 2,2′-dioxybiphenyl side groups can be reduced at potentials more positive than −3 V, whereas derivatives with fluoro, chloro, bromo, trifluoroethoxy, phenoxy, methoxy, *o*-dioxyphenyl, or phenylamino side groups cannot. On the basis of the half-wave potential values, the ease of reduction decreases in the order: $p\text{-}NO_2C_6H_4O >$ $2,3\text{-}O_2C_{10}H_6 > 1,8\text{-}O_2C_{10}H_6 > 2,2'\text{-}O_2C_{12}H_8 > (OCH_2CF_3)(Ph)$-*non-gem* > $1,1\text{-}Ph\text{-}2,2,4,4\text{-}OCH_2CF_3 > Ph$.

The following facts are deduced from these data[142]:

(1) Phosphazenes which contain only the most electronegative ligands, such

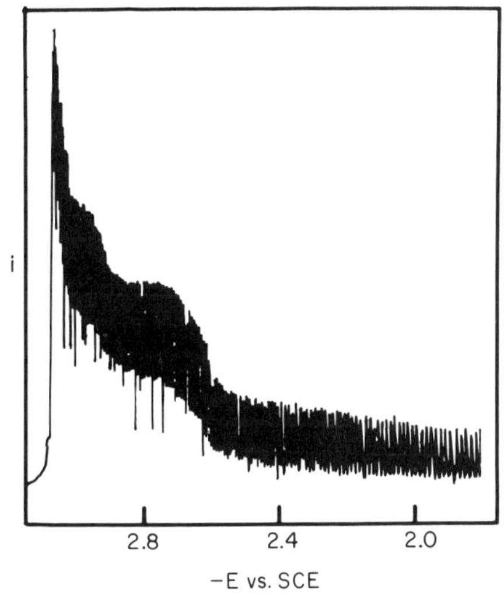

FIG. 3.4. Polarogram for the reduction of $(NPPh_2)_3$ (0.001 *M*) in dimethylformamide with the use of tetrabutylammonium iodide (0.1 *M*) as a supporting electrolyte. From Allcock and Birdsall.[142]

as F, Cl, Br, or OCH_2CF_3, cannot be reduced in this potential range. This suggests that, even when the most favorable conditions exist for the development of skeletal d_π–p_π bonding, the phosphazene ring itself cannot be reduced. However, it should be noted that benzene or anisole cannot be reduced at potentials more positive than −3 V, and the presence of phenyl substituents or fused aromatic units is required before the reduction potential is lowered into the 0 to −3 V range.

(2) Phosphazenes which contain an unsubstituted phenyl group linked through oxygen or nitrogen to phosphorus cannot be reduced under these conditions, but those with phenyl groups linked directly to phosphorus can. Presumably, phenoxy and phenylamino ligands resemble benzene and anisole in their reduction behavior, whereas the reducibility of the molecule is markedly enhanced when a phenyl group is directly bonded to phosphorus. The facile reducibility of phenylphosphazenes cannot reflect the reduction of a phenyl ligand, and it must be assumed that this behavior is a function of lowered antibonding orbital levels in a Ph—P unit, a Ph—P—Ph segment, or over larger sections of the molecule.

(3) Reduction of phenylphosphazenes is facilitated by the presence of trifluoroethoxy groups either on the same phosphorus atom as the phenyl group or on adjacent phosphorus atoms. Thus, XVI is more easily reduced than XVII, which itself is more easily reduced than XVIII. Presumably, electron withdrawal by trifluoroethoxy groups enhances the reducibility of Ph—P

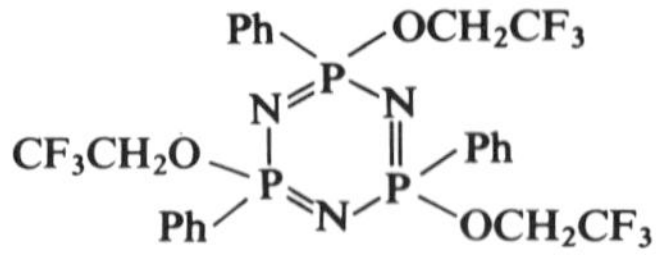

XVI

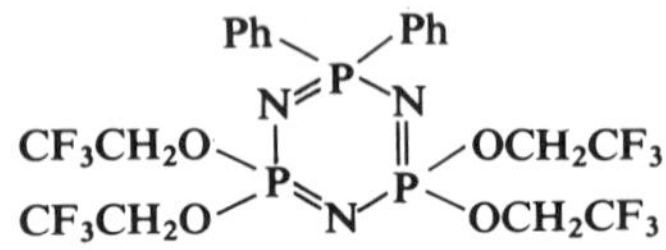

XVII

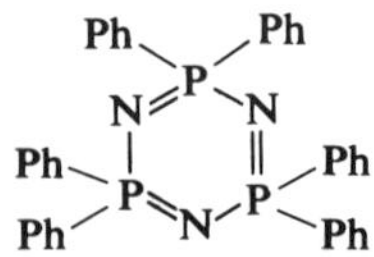

XVIII

units or of a larger molecular segment. However, limited delocalization within a Ph—P—Ph unit is inconsistent with the easier reducibility of XVI than of XVII or XVIII.

(4) Reduction of an aryloxyphosphazene is possible only if the aromatic group itself is independently reducible in the 0 to −3 V potential range. For example, compounds XIX, XX, XXI, and XXII are reduced in the 1.2 to −2.6 V range. For such compounds, it seems clear that the aromatic side units are the primary reduction sites, with the reduction potentials corresponding closely to those of derivatives of *p*-nitrophenol, naphthalenediols, or biphenylenediols.

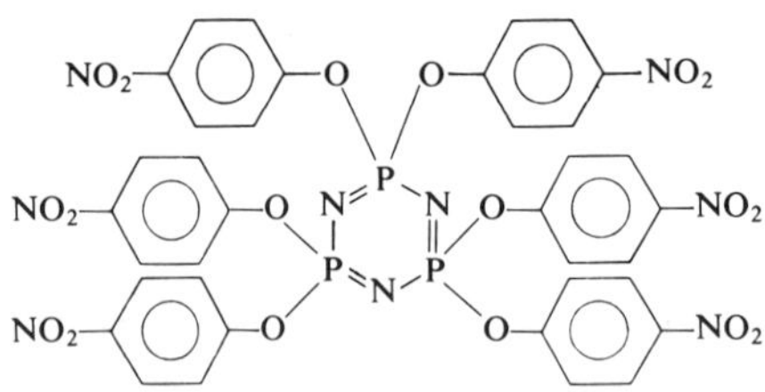

XIX

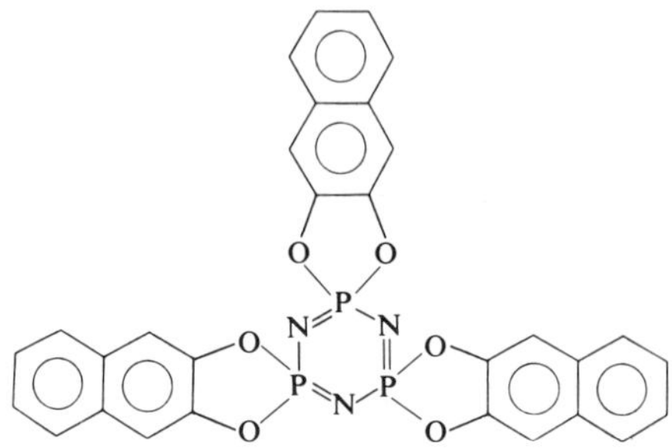

XX

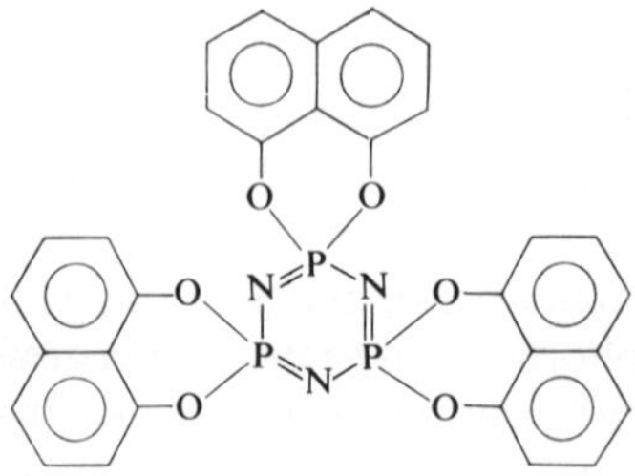

XXI

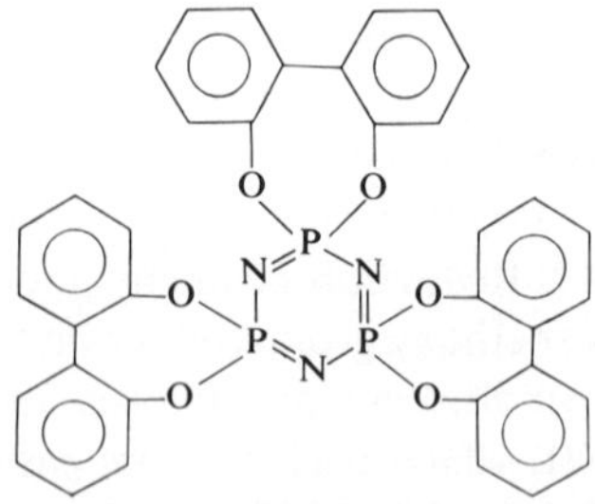

XXII

The evidence suggests that the side groups function as separate moieties, electronically insulated from the skeleton.

(5) Within the limited range of compounds studied, the skeletal degree of polymerization has little influence on the ease of reduction. Thus, the $E_{1/2}$ value of $(NPPh_2)_3$ is very close to that of $(NPPh_2)_4$ (Table 3.12) and the linking together of ~15,000 repeating units in $[NP(OPh)_2]_n$ or $[NP(OCH_2CF_3)_2]_n$ does not lower the reduction potential into the measurable range. Thus, there is no evidence that the degree of π-electron delocalization varies with ring size or chain length, and this provides an argument against a broadly delocalized π-electron model. A further argument against extensive delocalization is the fact that diphenylvinylphosphine oxide, $Ph_2(CH_2{=}CH)P{=}O$, and triphenylphosphine oxide, $Ph_3P{=}O$, are reduced at more positive potentials (−2.4 and −2.51 V, respectively) than $(NPPh_2)_{3\ \text{or}\ 4}$.[142]

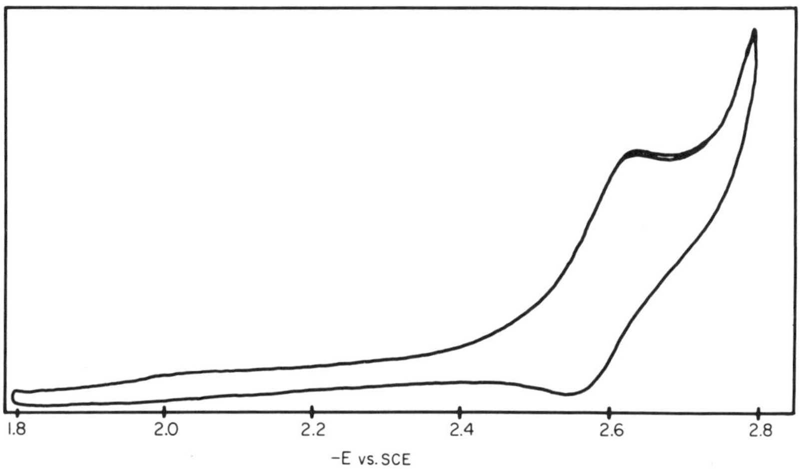

FIG. 3.5. Cyclic voltammogram of $(NPPh_2)_3$ (0.001 *M*) with tetrabutylammonium iodide (0.1 *M*) in dimethylformamide at a sweep rate of 0.05 cps. From Allcock and Birdsall.[142]

Cyclic voltammetric experiments have also been conducted on organophosphazenes.[141, 142] Figure 3.5 shows a cyclic voltammogram for $(NPPh_2)_3$ in dimethylformamide with tetra-*n*-butylammonium iodide used as a supporting electrolyte. The effects of variations in sweep rates indicated that the radicals formed from $(NPPh_2)_3$ (XVIII), $(NPPh_2)_4$, $[NP(Ph)(OCH_2CF_3)]_3$ (XVI), and $N_3P_3Ph_2(OCH_2CF_3)_4$ (XVII) were unstable at low sweep rates (<0.1 cps). This appeared to result from relaxation and electron exchange rather than from rapid chemical decomposition. Compounds XIX, XX, XXI, and XXII showed stable cyclic voltammograms at sweep rates from 1.0 to 60 cps.

Attempts to generate radical-cations of cyclophosphazenes by electrolytic oxidation at a mercury or platinum anode provided evidence for oxidation processes with $(NPCl_2)_3$, $(NPCl_2)_4$, and $(NPF_2)_3$.[142] However, the oxidation products could not be identified.

G. Electron Spin Resonance Spectra

Two attempts have been reported to detect electron spin resonance spectra from phosphazene radical-anions. In the first, Chapman, Glarum, and Massey[143] treated $(NPPh_2)_3$ with sodium or potassium in dimethoxyethane at −78°C and observed the ESR signal characteristic of the biphenyl radical-anion. Presumably, this was formed by cleavage of phenyl groups from phosphorus, followed by dimerization and reduction.

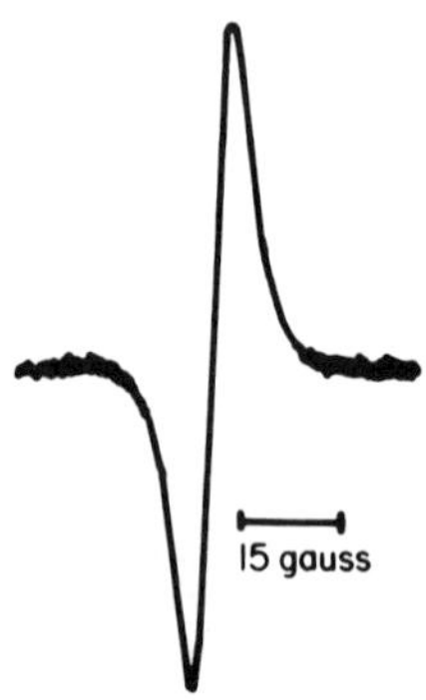

FIG. 3.6. Electron spin resonance spectrum of the product of reduction of $(NPPh_2)_{3 \text{ or } 4}$. From Allcock and Birdsall.[142]

In a later attempt, Allcock and Birdsall[142] reduced a number of cyclophosphazenes polarographically (see preceding section) with the samples contained within the cavity of an ESR spectrometer, and transient radical species were detected. A singlet ESR spectrum at $g = 2$ was detected from $(NPPh_2)_3$ (XVII), $(NPPh_2)_4$, and $N_3P_3Ph_2(OCH_2CF_3)_4$ (XVII) (Fig. 3.6). All attempts to resolve this singlet failed. Two explanations for the absence of hyperfine splitting are advanced. Firstly, if the unpaired electron spin density is delocalized within a Ph—P—Ph or a larger molecular unit, 150 or more hyperfine lines would be expected and these may not be resolvable. Secondly, rapid inter- or intramolecular electron exchange could broaden the line and destroy the spectral fine structure. The spectrum of the biphenyl radical-anion was not detected. Cyclophosphazene XVI yielded no ESR signal during reduction, a fact which suggests that the formation of moderately stable radicals requires the presence of two phenyl groups on each phosphorus.

The reduced species from $[NP(OC_6H_4NO_2\text{-}p)_2]_3$ (XIX) yielded a 13-line ESR spectrum and a blue-green color. Decomposition of the initial anion provides an explanation for the results. A five-line ESR spectrum is formed during reduction of XX, and a 15–20 line spectrum is obtained from XXI (Figs. 3.7 and 3.8). The hyperfine splitting schemes for these radicals suggest

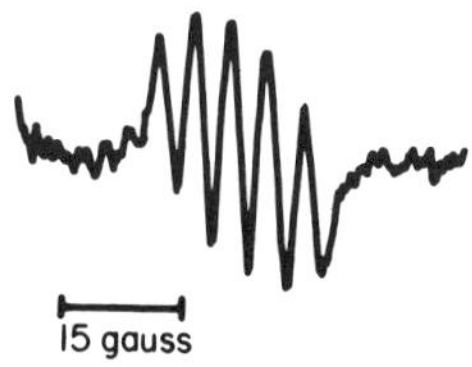

FIG. 3.7. Electron spin resonance spectrum of the product formed by electrolytic reduction of XX. From Allcock and Birdsall.[142]

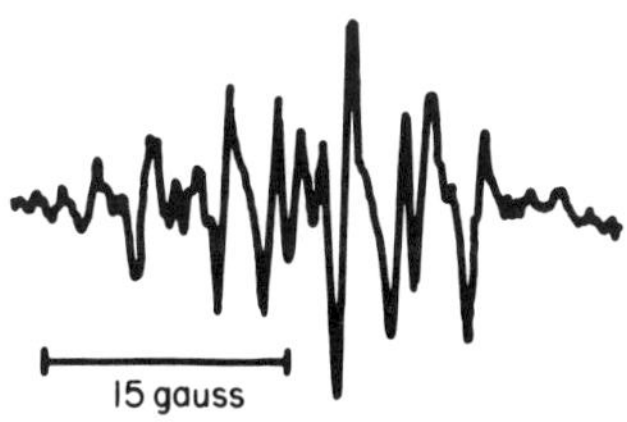

FIG. 3.8. Electron spin resonance spectrum of the product formed by electrolytic reduction of XXI. From Allcock and Birdsall.[142]

that the unpaired electron spin density is restricted to the aryl or arylenedioxy units, and that no interaction occurs with the phosphorus atoms or the phosphazene ring. This reinforces the conclusions derived from polarographic experiments.[142]

H. Electrical Conductance Experiments

Solid $(NPCl_2)_3$ and $(NPCl_2)_4$ show a small, temperature-variable conductance.[144] Eley and Willis[144] have shown that this is not the result of electronic (i.e. π-electron) conduction, but that it appears to result from ionization in the solid state.

$$(NPCl_2)_3 \rightarrow N_3P_3Cl_5^{\oplus} + Cl^{\ominus}$$

Thus, electrons can migrate through the crystalline lattice under the influence of an applied electric field, without undue interaction with the neutral molecules. This process is analogous to the movement of electrons in conventional semiconductors, with the crystal resistance decreasing exponentially with increasing temperature. Two types of conduction are envisaged. The first involves migration of chloride ions through the lattice, whereas the second requires transfer of a chloride ion from a neutral to an ionized molecule. The ionic nature of the conductivity is confirmed by direct current polarization and by the low mobility of the change carrier.

Allcock and Best[145] showed that the conductivity of *molten* $(NPCl_2)_3$ varies very little with temperature between 114° and 163°C (~1.7 mho $cm^{-1} \times 10^{11}$). However, above 200°C, the conductance rises dramatically with temperature (to 6250 mho $cm^{-1} \times 10^{11}$ at 253°C), and this is ascribed to ionization of chloride ion from phosphorus, with concurrent polymerization of the cyclic trimer. In contrast, the compound $[NP(OPh)_2]_3$, with side groups which are presumably less ionizable, shows only minor conductivity changes (1.5 to 17.7 mho $cm^{-1} \times 10^{11}$) as the temperature of the melt is raised from 120° to 252°C. Only at 350°C does this compound show evidence of phenoxide ion formation and electrical conduction.[145]

Conductance experiments have also been used to verify the ionic nature of linear, low molecular weight phosphazenes. For example, the compound, $[Ph_2(NH_2)\cdots N\cdots P(NH_2)Ph_2]^{\oplus}Cl^-$, acts as an electrolytic conductor in acetonitrile solution.[146]

I. Mass Spectrometry

Valuable information about the stability of various homologues in a cyclophosphazene series can be derived from an analysis of positive-ion mass-spectrometric data.[147–154] Ionization potentials can also be obtained. These two aspects will be considered separately.

1. Stability of Positive Ions

Mass-spectrometric examination of the fluorophosphazenes, $(NPF_2)_{3-16}$, indicates that the positive ions formed from the larger ring systems break down to form cations of smaller rings, probably by transannular interactions.[149] Thus, the cyclic hexamer yields trimer, the heptamer gives trimer and tetramer, the octamer gives tetramer, the nonamer yields trimer, the undecamer breaks down to pentamer and trimer, and the dodecamer yields trimer. From this information it appears that ring contraction does not proceed through monomer formation. A cyclic dimer, $(NPF_2)_2^{\oplus}$, can be recognized among the fragmentation species, but its concentration is low (0.1% or less), and this reflects the inherent instability of the four-membered ring. However, the cyclic dimeric cation is formed in greater quantities from fluorophosphazenes than from chlorophosphazenes, and this has been ascribed to greater skeletal π delocalization when the more electronegative fluorine ligands are present.[149] Some importance is attached to the observation that the higher cyclic species $(NPF_2)_n$, where n is greater than 6, yield a particularly stable $(NPF_2)_n^{\oplus}$ parent ion, and this has been cited as evidence for the greater delocalization stability of the higher homologues than of the lower cyclic species.[149] Linear fragmentation products are also detected from the mass spectra of fluorocyclophosphazenes.

The chlorophosphazenes, $(NPCl_2)_{3-8}$ have also been studied by mass spectrometry.[148, 149] Cyclic chlorophosphazene cations are particularly stable relative to chains, and the longer chains formed by fragmentation are less stable than the shorter chains. Thus, the cyclic species, $(NPCl_2)_{3-5}$, tend to retain the parent ring structure during fragmentation, but $(NPCl_2)_6$ fragments to the cyclic trimer, $(NPCl_2)_7$ gives trimer and tetramer, and $(NPCl_2)_8$, in spite of its relatively high stability, yields trimer, tetramer, and pentamer. The higher cyclic species, such as $(NPCl_2)_{6-8}$ yielded condensed cyclic cations also. It has also been pointed out[149] that, for $(NPCl_2)_{3\text{ and }4}$, the relative abundance of the even-electron ion is always greater than that of the odd-electron ion (excluding the parent ion). Also, the even-electron ion is more abundant than the odd-electron ions with one more or one less chlorine atom. An explanation of this has been made in terms of more efficient overlap by sp^2 phosphorus orbitals in the σ-bond system.[150]

Bromophosphazenes of structure $(NPBr_2)_{3-6}$ have also been investigated.[147] For $(NPBr_2)_{3-5}$, the most intense peak in the mass spectrum is $N_3P_3Br_5^{\oplus}$, $N_4P_4Br_7^{\oplus}$, or $N_5P_5Br_9^{\oplus}$ (the species formed by loss of a bromine atom from the parent ion). The hexamer $(NPBr_2)_6$ yields the ions $N_3P_3Br_5^{\oplus}$ and $N_6P_6Br_{11}^{\oplus}$. The data indicate that the trimeric ring is the most stable unit, and ring contraction occurs with the higher cyclic cations. This process takes place by loss of phosphorus–nitrogen units following removal of both bromine atoms from the labile phosphorus, but the loss of molecular bromine is not the most important reaction in this process. The cyclic hexamer, $(NPBr_2)_6$, gives rise to condensed ring species during mass spectrometry.

Information has also been obtained about the mass spectra of chlorobromocyclotriphosphazenes of formula, $N_3P_3Cl_xBr_{6-x}$.[151, 152] For these species, bromine atoms are lost 50 times more easily than chlorine atoms, and this undoubtedly reflects the greater strength of P—Cl than P—Br bonds. Apparently, replacement of chlorine atoms by bromine in a cyclic trimer has little effect on the fragmentation stability of the ring, and cyclic ions are always formed in preference to chains in a ratio of 72 to 17%.

Mass spectra of the isothiocyanophosphazenes, $[NP(NCS)_2]_3$ and $[NP(NCS)_2]_4$, at 70 eV indicate that the phosphazene rings are particularly stable.[154a] No linear trimers or tetramers are detected, and the fragmentation products are formed by removal of all or part of the substituent groups. Evidence for a ring contraction from cyclic trimer to cyclic tetramer is evident from the metastable peaks.

2. Ionization Potentials

Ionization potentials have been measured by conventional electron impact mass spectrometry and by photoelectron spectroscopy. Table 3.13 lists

TABLE 3.13

First Ionization Potentials of Cyclophosphazenes

Compound	First ionization potential (eV)	References	Compound	First ionization potential (eV)	References
$(NPF_2)_3$	11.4	154	$N_3P_3Cl_2Br_4$	9.60	152
$(NPF_2)_4$	10.7	154	$N_3P_3ClBr_5$	9.47	152
$(NPF_2)_5$	11.4	154	$(NPBr_2)_3$	9.56	152
$(NPF_2)_6$	10.9	154	$(NPBr_2)_4$	9.21	152
$(NPF_2)_7$	11.3	154	$[NP(OCH_2CF_3)_2]_3$	10.43	154
$(NPF_2)_8$	10.9	154	$[NP(OCH_2CF_3)_2]_4$	10.01	154
$(NPCl_2)_3$	10.26	154	$[NP(OMe)_2]_3$	9.29	154
$(NPCl_2)_4$	9.80	154	$[NP(OMe)_2]_4$	8.83	154
$(NPCl_2)_5$	9.83	154	$[NP(OPh)_2]_3$	8.83	154
$(NPCl_2)_6$	9.81	154	$[NP(OPh)_2]_4$	8.70	154
$(NPCl_2)_7$	9.80	154	$[NP(NMe_2)_2]_3$	7.85	154
$N_3P_3Cl_5Br$	9.38	152	$[NP(NMe_2)_2]_4$	7.45	154
$N_3P_3Cl_4Br_2$	9.80	152	$(NPMe_2)_3$	8.35	154
$N_3P_3Cl_3Br_3$	9.72	152	$(NPMe_2)_4$	8.35	154

measured values. The first ionization potentials apparently reflect the loss of an electron from the skeletal π system rather than from the ligands.[154] Paddock, Mitchell, and their co-workers[154] have interpreted the ionization potential data in terms of the dual d_π–p_π system. The following experimental facts need to be explained:

(1) The first ionization potentials for $(NPX_2)_{3 \text{ or } 4}$ increase in the order, $X = NMe_2 < Me < OPh < OMe < Br < Cl < OCH_2CF_3 < F$. This order roughly follows the increasing electronegativity of the ligands.

(2) Cyclic tetramers have lower ionization potentials than cyclic trimers.

Hence, the electron removed during cation formation must be derived from a bonding orbital rather than a nonbonding lone pair.

(3) In ascending the fluorophosphazene homologous series, $(NPF_2)_{3-8}$, there is an alternation of ionization potential, with those species with odd numbers of monomer units having higher values than those with even numbers of monomer units.

(4) No such alternation of ionization potentials occurs during the ascent of the chlorophosphazene series. The tetramer, pentamer, hexamer, and heptamer all have similar ionization potentials, and all show lower values than the trimer.

(5) Introduction of one bromine atom into $(NPCl_2)_3$ causes an appreciable decrease in ionization potential, although additional bromine atoms have a less dramatic influence.[152]

Consider first the situation with the fluorophosphazenes. By the use of photoelectron spectroscopy, the ionization potentials of the first to as many as the eleventh electron have been measured. The results are shown in Table 3.14.

TABLE 3.14[a]

VERTICAL IONIZATION POTENTIALS (eV) OF FLUOROCYCLOPHOSPHAZENES $(NPF_2)_n^{a,b}$

$n = 3$	4	5	6	7	8
—	10.7	—	10.9	—	10.9
11.4	11.5	11.4	11.7	11.3	—
—	12.5	12.5	12.7	12.3 ± 0.3	*c*
13.1	13.6	13.3	13.4	13.0 ± 0.2	*c*
—	14.9	14.3	14.1	13.5 ± 0.2	*c*
15.4	15.9	15.8	15.8	15.8	15.7
16.6	16.7	16.5	16.3	16.2 ± 0.2	*c*
17.2	17.4	17.2	17.2	17.2	17.2
18.2	18.2	18.1	18.0	17.9	18.0
19.6	19.4	20.0	20.0	19.9	19.8
20.1	19.9	—	—	—	—

[a] From Branton *et al.*[154]

[b] Estimated uncertainty ±0.1 eV unless otherwise stated.

[c] Unresolvable.

Ionization potentials between 20.1 and 16.6 eV appear to be almost independent of ring size and these are attributed to ionization of the fluorine lone-pair electrons and electrons of the P—F σ bonds. Ionization potentials near 14.95 eV are presumed to represent removal of electrons from P—N σ bonds, and values below 14 eV apparently correspond to the loss of electrons from π-bond orbitals. The problem arises when attempts are made to distinguish between heteromorphic (π) levels and homomorphic levels, and to explain the alternation of the first ionization potential as the homologous series is ascended. It is assumed that the highest energy level in $(NPF_2)_3$ (at 11.4 eV) corresponds to the coincidence in energy levels of the homomorphic and heteromorphic π-levels, whereas the highest level in $(NPF_2)_4$ (at 10.7 eV) represents ionization from the homomorphic system, and the next lower level (11.5 eV) represents the highest level of the heteromorphic system.[154] Coincidence of the highest-level homomorphic and heteromorphic orbitals is attributed to molecular flexibility effects. Evidence for this viewpoint was derived from the similarities of the experimentally observed energy-level sequence and those calculated by Hückel molecular orbital methods.[154]

If the highest level π-molecular orbital is assumed to be the homomorphic, π', system, the influence of ligand electronegativity can be understood, since the most electronegative substituents should bind the nitrogen lone-pair electrons more strongly because of the enhanced interaction with a contracted phosphorus $3d_{x^2-y^2}$ orbital.

The absence of an alternation of ionization potentials in the chlorophosphazene series can be ascribed to the fact that the tetramer and possibly the hexamer and heptamer are puckered. Ring puckering tends to blur the distinction between π and π' interactions and it favors removal of an electron from a molecular orbital composed only of nitrogen atomic orbitals in the first ionization step, rather than from a d_π–p_π molecular orbital. Similar arguments may perhaps explain the general differences between planar trimers and puckered tetramers, and the pattern of ionization potentials in chlorobromocyclotriphosphazenes.

J. Dipole Moments

Dipole moments can provide an indication of the planarity or nonplanarity of a ring system, and this technique has been used to examine $(NPCl_2)_3$, $(NPCl_2)_4$, $(NPF_2)_4$, and $[NP(OPh)_2]_3$.[145,155] Table 3.15 lists representative dipole moments determined by the dielectric constant technique.[145,155,156,156a]

The low dipole moments of $(NPCl_2)_3$ and $(NPF_2)_4$ are consistent with a planar ring, but the low value for the tetramer, $(NPCl_2)_4$, is surprising in view of the puckered structure found in the solid state. Presumably, the ring is sufficiently flexible that, in solution, the dipoles from the puckered structure average out. Confirmation for this view is found in infrared spectral data.[63] The high dipole moment of $[NP(OPh)_2]_3$) suggests the presence of a relatively inflexible, puckered ring. The use of dipole moments to distinguish between different isomers is also illustrated in Table 3.15.

TABLE 3.15

DIPOLE MOMENTS OF CYCLOPHOSPHAZENES

Compound	Phase	Temperature (°C)	Dipole moment (D)	References
$(NPCl_2)_3$	Melt	120	0.98 ± 0.2	145
$(NPCl_2)_3$	Benzene	70	0.92 ± 0.1	145
$(NPCl_2)_3$	Benzene	30	0.83 ± 0.1	145
$(NPCl_2)_4$	Benzene	*a*	0.20	155
$(NPF_2)_4$	Benzene	*a*	0.10	155
$[NP(OPh)_2]_3$	Melt	120	3.00	145
$[NP(OPh)_2]_3$	Benzene	70	3.28	145
$[NP(OPh)_2]_3$	Benzene	30	2.84	145
$(NPBrPh)_3$-*trans*	Benzene	25	2.36	110
$(NPBrPh)_3$-*cis*	Benzene	25	5.27	110
$N_3P_3Cl_4(NMe_2)_2$-*gem*	Benzene	24	3.12	156
$N_3P_3Cl_4(NMe_2)_2$-*trans*	Benzene	25	2.61	156
$N_3P_3Cl_4(NMe_2)_2$-*cis*	Benzene	25	4.3	156
$N_3P_3Cl_3(NMe_2)_3$-*gem*	Benzene	24	2.01	156
$[NP(Cl)(NC_4H_8)]_3$-*trans*	Benzene	25	2.44	156a
$[NP(Cl)(NC_4H_8)]_3$-*cis*	Benzene	25	5.64	156a
$N_3P_3Cl_4(NC_5H_{10})_2$-*trans*	Benzene	25	3.02	156a
$N_3P_3Cl_4(NC_5H_{10})_2$-*cis*	Benzene	25	4.61	156a

[a] Temperature not specified.

K. Diamagnetic Anisotropy

For organic aromatic compounds, the measurement of the magnetic susceptibility of a crystal along two axes can be correlated with the density of delocalized π electrons in a particular crystal plane. The magnetic anistropy value for $(NPCl_2)_3$ has been measured, and attempts have been made to correlate the data with d_π–p_π delocalization effects.

The diamagnetic anisotropy of $(NPCl_2)_3$ is -10.5×10^{-6} cgs units, compared to a value of -60×10^{-6} cgs units for benzene. The low value for $(NPCl_2)_3$ seriously complicates the data interpretation. Thus, as pointed out by Craig, Heffernan, Mason, and Paddock,[157] it is practically impossible to assign the observed anisotropy value to a particular ring current effect, since the influence of σ electrons and π atomic orbitals also contributes to the overall value. Serious approximations must be used to estimate the magnitude of these latter effects. The average susceptibilities per $NPCl_2$ unit in species of formula $(NPCl_2)_{3-7}$ are -49.8, -49.2, -49.6, -50.1, and -48.9×10^{-6} cgs units, respectively. Cyclophosphazenes are expected to show paramagnetic ring current effects. The experimental data are probably consistent with this view, although the ring current contribution to the average susceptibility is small.

L. Basicity Measurements

If the π' system in cyclo- and polyphosphazenes is weakly developed, the lone-pair electrons will be contained in the nitrogen $2sp^2$ lone-pair orbitals. The skeletal nitrogen atoms will then show basic character. However, if these electrons are delocalized toward phosphorus in a π' system under the influence of electronegative ligands, the basicity of the skeletal nitrogen atoms should decrease. This subject is reviewed in detail in Chapter 12. Here, it is sufficient to note that the effect is observed experimentally. Strongly electron withdrawing substituents, such as chloro or trifluoroethoxy, reduce the skeletal basicity, whereas electron-supplying ligands, such as amino, enhance it.[158–164]

M. Nuclear Quadrupole Resonance

Since the nuclear spin quantum number for ^{35}Cl is 3/2, chlorophosphazenes can be studied in the solid state by NQR techniques. Three chlorophosphazene systems have been studied in detail: $(NPCl_2)_3$; the metastable K form of $(NPCl_2)_4$; and the stable T form of $(NPCl_2)_4$.[165–170] For the first two species, the NQR data tend to confirm the results of X-ray crystallographic studies on these compounds.

Thus, $(NPCl_2)_3$ gives a four-line NQR spectrum with intensities of 2:1:1:2. This is in complete accord with the X-ray structure determinations which suggest that the 24 chlorine atoms in the unit cell can be distinguished into four sets of 8, 4, 4, and 8.[165, 167, 169] The NQR results for $(NPCl_2)_3$ suggest a nonplanarity of the ring and a slightly wider Cl—P—Cl angle (104.5°)[167] than was indicated by X-ray data.

Two crystallographic forms exist for $(NPCl_2)_4$. The metastable K form has a tublike structure according to X-ray data.[18] Each unit cell contains two molecules and there are two distinguishable chlorine environments. The NQR spectrum consists of two lines of equal intensity, but studies of the influence of temperature on the spectrum suggest that the nonequality of the chlorine atoms may result from the thermal motions of the molecules.[169] Above 63 ± 1.5°C the K form is converted to the stable T form.[170] This transformation is accompanied by a symmetry change from $\bar{4}$ to $\bar{1}$ (chair shape).[19] The NQR spectrum of the T form consists of four lines with intensities in the ratio of 2:3:2:3.[168, 169]

N. Nuclear Magnetic Resonance

NMR spectroscopy is an important structural tool in phosphorus–nitrogen chemistry. The ^{31}P nucleus with a spin of $\frac{1}{2}$ can be studied readily, and the presence of hydrogen or fluorine nuclei ($I = \frac{1}{2}$) in the side groups permits a high degree of sophistication to be reached in structural interpretations. Two types of structural information can be obtained. Firstly, it is possible to identify positional and *cis–trans* isomers by chemical shift and spin–spin coupling data and secondly, a careful interpretation of NMR data permits conclusions to be drawn about skeletal flexibility and the electronic arrangements within the molecule. Both of these aspects will be considered for each type of NMR technique.

1. Identification of Positional and *Cis–Trans* Isomers (General)

The use of NMR spectra to identify positional and *cis–trans* isomers can be illustrated by the work of Allen, Tsang, and Moeller[171, 172] with the three diaryltetrafluoro cyclotriphosphazenes depicted as XXIII, XXIV, and XXV. All three are formed by the interaction of $(NPF_2)_3$ with phenyllithium.

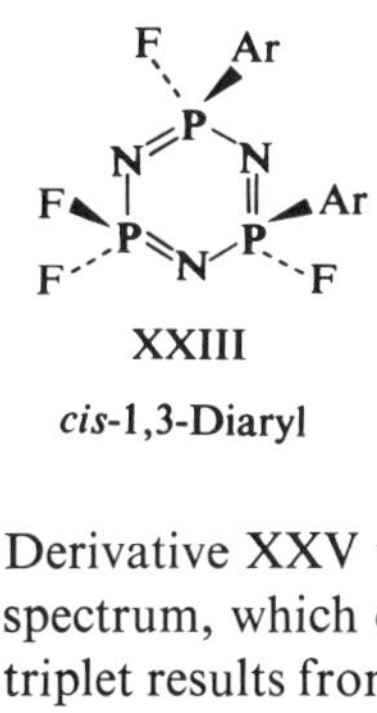

XXIII

cis-1,3-Diaryl

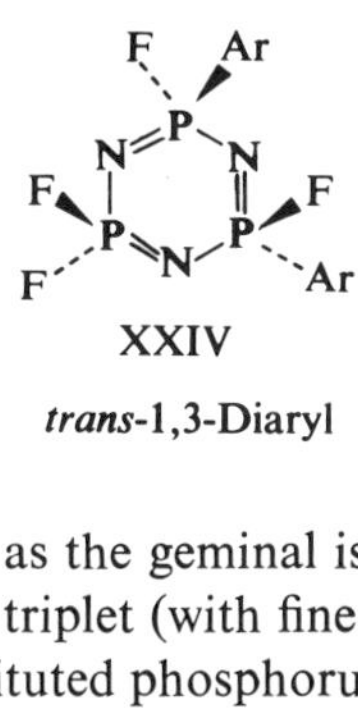

XXIV

trans-1,3-Diaryl

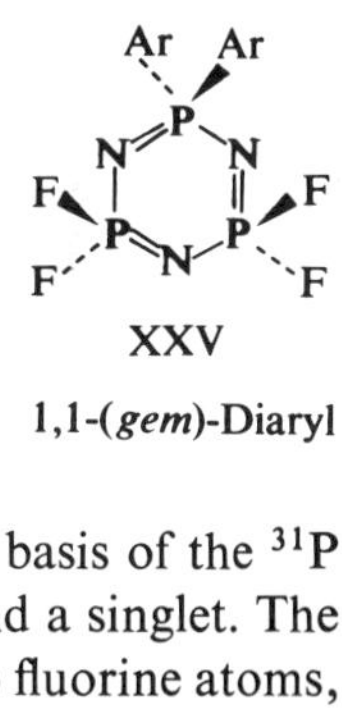

XXV

1,1-(*gem*)-Diaryl

Derivative XXV was identified as the geminal isomer on the basis of the ^{31}P spectrum, which consisted of a triplet (with fine structure) and a singlet. The triplet results from fluoro-substituted phosphorus split by two fluorine atoms, and the singlet is derived from the aryl-substituted phosphorus. The *trans*-non-geminal isomer, XXIV, gave a ^{31}P spectrum which consisted of a triplet (J_{PF} in PF_2) further split to a triplet of triplets (J_{PP}), plus a doublet (J_{PF} in PFPh) further split to a doublet of doublets (J_{PP}). The ^{19}F spectrum of this isomer consisted of two large doublets (J_{PF} in PF_2 and PFPh). The *cis*-non-geminal isomer, XXIII, gave a complex ^{19}F spectrum because of the non-equivalence of the geminal fluorine atoms.

NMR experiments have been used to identify the structures of a large number of other cyclophosphazenes. In particular, dimethylamino-substituted derivatives have been studied in detail by ^1H and ^{31}P methods,[173–179] and

TABLE 3.16

SELECTED ^{31}P NMR SHIFT DATA FOR PHOSPHAZENES

Group	Compound	^{31}P shift[a] (ppm)	References
PF_2	$(NPF_2)_3$	−13.9	172
PF_2	$(NPF_2)_4$	+17.0	185a
PF_2	$N_3P_3F_4Ph_2$-*gem*	−12.1	172
PF_2	$N_3P_3F_3Ph_3$-*gem*	−8.6	172
PF_2	$N_3P_3F_2Ph_4$-*gem*	−6.2	172
PF_2	$N_3P_3F_4Ph_2$-*non-gem, cis*	−12.4	171
PF_2	$N_3P_3F_4Ph_2$-*non-gem, trans*	−10.8	171
PFCl	$N_3P_3FCl_5$	−14.4	177
PFPh	$N_3P_3F_3Ph_3$-*gem*	−31.2	172
PFPh	$N_3P_3F_4Ph_2$-*non-gem, cis*	−38.4	171
PFPh	$N_3P_3F_4Ph_2$-*non-gem, trans*	−38.3	171
$P(F)NMe_2$	$N_3P_3F_5(NMe_2)$	−24.9	185b
$P(F)NMe_2$	$N_4P_4F_7(NMe_2)$	−4.0	185b
$P(F)NMe_2$	$N_5P_5F_9(NMe_2)$	−4.5	185b
$P(F)NMe_2$	$N_6P_6F_{11}(NMe_2)$	−4.4	185b
PCl_2	$(NPCl_2)_3$	−20	92, 138, 177, 183
PCl_2	$(NPCl_2)_4$	+7.4	138, 183
PCl_2	$(NPCl_2)_5$	+17.0	138
PCl_2	$(NPCl_2)_6$	+16.0	138
PCl_2	$(NPCl_2)_7$	+18.0	138
PCl_2	$(NPCl_2)_8$	+18.0	138
PCl_2	$(NPCl_2)_n$	+17.4	92
PCl_2	$N_3P_3FCl_5$	−23.0	177
PCl_2	$N_3P_3BrCl_5$	−17.7	186
PCl_2	$N_3P_3Br_2Cl_4$	−16.1	186
PCl_2	$N_3P_3Br_3Cl_3$	−14.0	186
PCl_2	$N_3P_3Br_4Cl_2$	−13.9	186
PCl_2	$N_3P_3Cl_5(OMe)$	−22.5	177
PCl_2	$N_3P_3Cl_5(OEt)$	−21.3	177
PCl_2	$N_3P_3Cl_5(OPr^i)$	−21.7	177
PCl_2	$N_3P_3Cl_5(OCH_2CF_3)$	−22.7	177
PCl_2	$N_3P_3Cl_4(SPh)_2$	−18.6	183
PCl_2	$N_4P_4Cl_4(SEt)_4$	+9.7	183
PCl_2	$N_3P_3Cl_5NH_2$	−20.4	187
PCl_2	$N_3P_3Cl_4(NH_2)_2$	−18.3	187
PCl_2	$N_3P_3Cl_5(NMe_2)$	−20.5	176, 177
PCl_2	$N_3P_3Cl_4(NMe_2)_2$-*non-gem, cis*	−21.6	176
PCl_2	$N_3P_3Cl_4(NMe_2)_2$-*non-gem, trans*	−20.5	176

[a] Relative to aqueous 85% phosphoric acid or triethylphosphate.

TABLE 3.16 (*Continued*)

Group	Compound	^{31}P shift[a] (ppm)	References
PCl_2	$N_3P_3Cl_4(NPCl_3)_2$-*gem*	−17.5	187–189
PClBr	$N_3P_3ClBr_5$	+14.0	186
PClBr	$N_3P_3Cl_5Br$	+7.8	186
PClPh	$(NPClPh)_3$-*cis*	−29.4	190
PClPh	$(NPClPh)_3$-*trans*	−30.3, −32.7	190
PCl(OMe)	$N_3P_3Cl_5(OMe)$	−16.7	177
PCl(OEt)	$N_3P_3Cl_5(OEt)$	−13.6	177
$PCl(OPr^i)$	$N_3P_3Cl_5(OPr^i)$	−12.6	177
$PCl(OCH_2CF_3)$	$N_3P_3Cl_5(OCH_2CF_3)$	−16.5	177
$PCl(NH_2)$	$N_3P_3Cl_5(NH_2)$	−19.0	187
PCl(NHMe)	$N_3P_3Cl_5(NHMe)$	(−22.2)	188, 191
$PCl(NMe_2)$	$[NPCl(NMe_2)]_3$-*cis*	−27.6	176, 192
$PCl(NMe_2)$	$[NPCl(NMe_2)]_3$-*trans*	−26.1	176
$PCl(NMe_2)$	$N_3P_3Cl_5(NMe_2)$	−22.7	176, 177
$PCl(NMe_2)$	$N_3P_3Cl_4(NMe_2)_2$-*non-gem, cis*	−24.9	176
$PCl(NMe_2)$	$N_3P_3Cl_4(NMe_2)_2$-*non-gem, trans*	−24.5	176
PCl(NMePh)	$[NPCl(NMePh)]_4$	+1.1	114
$PCl(NC_5H_{10})$	$[NPCl(NC_5H_{10})]_4$	−1.7	114
$PCl(N{=}PCl_3)$	$N_3P_3Cl_5(N{=}PCl_3)$	+2.20	187
PCl_3	$N_3P_3Cl_5(N{=}PCl_3)$	+3.29	187
PCl_3	$Cl_3CC(O)N{=}PCl_3$	−25.1	79
PCl_3	$PhC(O)N{=}PCl_3$	−13.2	79
PCl_3	$MeOC(O)N{=}PCl_3$	−8.9	79
PCl_2Ph	$Cl_3CC(O)N{=}PCl_2Ph$	−31.6	79
$PClPh_2$	$Cl_3C(O)N{=}PClPh_2$	−40.0	79
PBr_2	$(NPBr_2)_3$	+45.4 (+49.5)	178, 192a
PBr_2	$(NPBr_2)_4$	+71.8	192a
PBr_2	$N_3P_3Cl_4Br_2$	+38.6	186
PBr_2	$N_3P_3Br_4(NMe_2)_2$-*non-gem, trans*	+36.6	178
PBr_2	$N_3P_3Br_5(NMe_2)$	+39.3	178
PBrPh	$(NPBrPh)_3$-*cis*	−16.4	181
PBrPh	$(NPBrPh)_3$-*trans*	−20.1, −18.0	181
$PBr(NMe_2)$	$N_3P_3Br_5(NMe_2)$	−4.5	178
$PBr(NMe_2)$	$[NPBr(NMe_2)]_3$	−15.4	178
$PBr(NMe_2)$	$N_3P_3Br_4(NMe_2)_2$-*non-gem, trans*	−10.0	178
$P(NCS)_2$	$[NP(NCS)_2]_3$	−30	188
PMe_2	$S{=}PMe_2{-}N{=}PPh_2{-}SMe$	−44.7	82
PPh_2	$(NPPh_2)_3$	−14.3	188
PPh_2	$N_3P_3F_4Ph_2$-*gem*	−30.4	172
PPh_2	$N_3P_3F_2Ph_4$-*gem*	−27.3	172
PPh_2	$N_3P_3Cl_4Ph_2$-*gem*	−20.6	193
PPh_2	$N_3P_3Cl_2Ph_4$-*gem*	−19.1 (−17.2)	193, (188)

TABLE 3.16 (*Continued*)

Group	Compound	^{31}P shift[a] (ppm)	References
PPh_2	$N_3P_3ClPh_5$	−17.6	193
PPh_2	$HO\text{+}PPh_2{=}N\text{+}_3PPh_2{=}O$	−15.4, −2.4	194
PPh_2	$S{=}PPh_2{-}N{=}PPh_2{-}OMe$	−41.8	195
PPh_2	$O{=}PPh_2{-}N{=}PPh_2{-}SMe$	−26.7	195
PPh_2	$S{=}PPh_2{-}NH{-}PPh_2{=}S$	−55.1	195
PPh_3	$Cl_3CC(O)N{=}PPh_3$	−23.8	79
PPh_3	$PhC(O)N{=}PPh_3$	−14.9	79
$P(NH_2)_2$	$[NP(NH_2)_2]_3$	−18.8	188
$P(NH_2)_2$	$N_3P_3Cl_4(NH_2)_2$-*gem*	−9.0	187, 188
$P(NHMe)_2$	$N_3P_3Cl_4(NHMe)_2$-*gem*	−12.3	188, 191
$P(NH_2)(N{=}PPh_3)$	$N_3P_3Cl_4(NH_2)(N{=}PPh_3)$-*gem*	+1.2 to +4.1	188, 196
$P(NMe_2)_2$	$N_3P_3Cl_2(NMe_2)_4$-*non-gem*	−23.9	176
$P(NMe_2)_2$	$[NP(NMe_2)_2]_n$	−5.3, +0.7	109
$P(NC_5H_{10})_2$	$[NP(NC_5H_{10})_2]_n$	+7.7	109
$P(NHPh)_2$	$[NP(NHPh)_2]_n$	+14.0	109
$P(NH_2)(Ph)$	$[NP(NH_2)(Ph)]_3$	−23.0	131
$P(OMe)_2$	$[NP(OMe)_2]_3$	−21.7	92
$P(OMe)_2$	$[NP(OMe)_2]_n$	+3.5	92
$P(OEt)_2$	$[NP(OEt)_2]_3$	∼−15.3	92
$P(OEt)_2$	$[NP(OEt)_2]_n$	+7.3	92
$P(OCH_2CF_3)_2$	$[NP(OCH_2CF_3)_2]_3$	−17.7	92
$P(OCH_2CF_3)_2$	$[NP(OCH_2CF_3)_2]_n$	+7.5	92
$P(OPh)_2$	$[NP(OPh)_2]_3$	−9	92
$P(OPh)_2$	$[NP(OPh)_2]_n$	+18.6	92
$PO_2C_6H_4$-*o*	$[NP(O_2C_6H_4\text{-}o)]_3$	−11.8	131
$PO_2C_{12}H_8$-2,2′	$[NP(O_2C_{12}H_8\text{-}2,2')]_3$	−9.5	131
$P(OEt)_3$	$(EtO)_2P(O){-}N{=}P(OEt)_3$	−3	197
$P(SEt)_2$	$[NP(SEt)_2]_3$	−45.7	183
$P(SEt)_2$	$N_3P_3Cl_4(SEt)_2$-*gem*	−51.7	183
$P(SEt)_2$	$N_4P_4Cl_4(SEt)_4$-*gem*	−28.6	183
$P(SPh)_2$	$N_3P_3Cl_4(SPh)_2$-*gem*	−46.8	183
$PH(OPh)$	$N_3P_3Ph_4(OPh)H$-*gem*	−5.8	198
PHMe	$N_3P_3Ph_4MeH$-*gem*	−6.9	198
PHEt	$N_3P_3Ph_4EtH$-*gem*	−12.5	198

phenyl,[171, 172, 175, 179–182] methoxy,[177] ethoxy,[177, 182] trifluoroethoxy,[177] isopropoxy,[177] phenoxy,[175] mercapto,[183] amino,[184] and fluoro[177, 185] derivatives have been structurally identified by NMR techniques. The following sections review the principles used to identify different isomers.

2. ^{31}P NMR Shifts

Tabulation of Data. A large body of ^{31}P NMR shift data has been reported for cyclo- and polyphosphazenes, and low molecular weight linear phosphazenes. Selected values for typical compounds are listed in Table 3.16,[79, 82, 92, 109, 114, 131, 138, 171, 172, 176–178, 183, 185a–198] with chemical shifts recorded in parts per million relative to aqueous 85% phosphoric acid or triethyl phosphate.

Utility as a "Fingerprinting" Technique. Proton chemical shifts are often sufficiently characteristic of various groups that they can be used directly for identification purposes. The fact that this is less true for ^{31}P shifts in phosphazenes is illustrated by the tabulation in Table 3.17 of the range of chemical

TABLE 3.17

^{31}P NMR Chemical Shift Ranges

Unit	^{31}P Shift range (ppm)	Unit	^{31}P Shift range (ppm)
PF_2	−6 to −14	$P(NCS)_2$	−30
PFCl	−14	PMe_2	(−45)
PFPh	−27 to −38	PPh_2	−14 to −30
PCl_2 (trimers)	−14 to −23	PPh_2 (linear)	−2.4 to −55
PCl_2 (tetramers)	+7.4	$-PPh_3$	−15 to −24
PCl_2 (pentamers to polymers)	+16 to +18	$P(NH_2)_2$	−9 to −19
PClBr	+8 to +14	P(NHMe)	−12
PClPh	−29 to −33	$P(NH_2)(N{=}PPh_3)$	+1 to +4
PCl(OAlkyl)	−13 to −17	$P(NMe_2)_2$ trimer	−24
$PCl(NH_2)$	−19	$P(NMe_2)_2$ polymer	0 to −5
$PCl(NMe_2)$	−23 to −28	$P(NC_5H_{10})_2$ polymer	+8
$PCl(N{=}PCl_3)$	+2	$P(NHPh)_2$ polymer	+14
$-PCl_3$	−9 to +3	P(OAlkyl) trimer	−15 to −22
$-PCl_2Ph$	−32	P(OAlkyl) polymer	+3 to +8
$-PClPh_2$	−40	P(OAryl)trimer	−9 to −12
PBr_2	+37 to +45	P(OAryl) polymer	+19
PBrPh	−16 to −20	$P(SEt)_2$	−29 to −46
$PBr(NMe_2)$	−4 to −15	PHR	−6 to −13

shifts reported for each PR_2 unit in different molecular environments. The chemical shift values for most cyclic trimers fall in the range of 0 to −46 ppm, with considerable overlap of the ranges being evident for each type of PR_2 unit. Of the cyclic trimers, bromophosphazenes are almost unique in showing positive chemical shifts. Positive shifts also appear to be characteristic of cyclic tetramers, pentamers, hexamers, heptamers, octamers, and of linear high polymers. Small chemical shift differences exist between *cis* and *trans* isomers of cyclophosphazenes. However, insufficient evidence is available to enable general diagnostic trends to be recognized.

Influence of Different Ligands Attached to Phosphorus. The influence of a group, R, in the unit, –N—PR_2–, on the phosphorus chemical shift might be expected to provide information about the nature of the bonds attached to phosphorus. However, the influence of other units in the molecule on this phosphorus atom is usually so profound that meaningful comparisons can be made only between closely related groups of compounds. Consider first the series depicted by the general formula XXVI. Heatley and Todd[177] have shown that

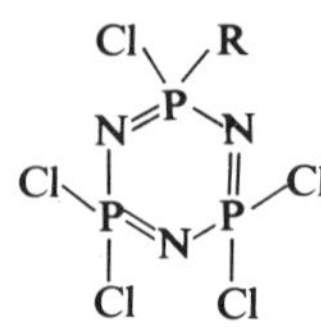

XXVI

the shifts become more negative for the phosphorus attached to R in the order $R = OPr^i > OEt > F > OCH_2CF_3 > OMe > Cl > NMe_2$. However, if the comparison is made between the *difference* in chemical shifts between the PClR and PCl_2 units, in each molecule, the difference becomes enhanced in the order, $NMe_2 < Cl < OMe < OCH_2CF_3 < OEt < F < OPr^i$. There is a rough correlation between this order and the electronegativity of the group directly connected to phosphorus, and it is suggested that electron withdrawal by R from phosphorus favors delocalization of the skeletal nitrogen lone-pair electrons to phosphorus, thus inducing a chemical shift to higher (i.e., more positive) field.[177] On the same scale, an NH_2 group is roughly equivalent to a chloro group in its shielding effect, whereas a phenyl group (δPClPh ≈ -30 ppm) is strongly electron supplying even when compared with dimethylamino. Bromine, on the other hand (δ PClBr $= +7.8$), shows the most positive shift of all the groups,[186] a fact which is inconsistent with this mechanism. From these facts it seems clear that the disposition of the skeletal π electrons has a more profound effect on the shielding at phosphorus than does the primary increment of electrons within the P—R σ bond.

Consider next compounds of structure XXVII. For changes in R, the ^{31}P

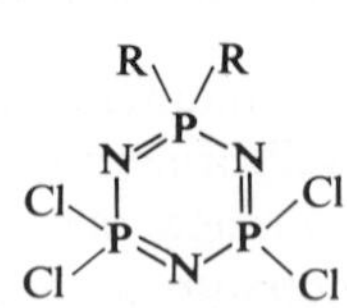

XXVII

shifts become more negative in the order Br > NH_2 > NHMe > Cl ≈ Ph. In this case, phenyl behaves as a deshielding ligand, while bromine anomalously functions as a highly shielding substituent. It must be concluded, therefore, that the influence of ligand electronegativity on the ^{31}P chemical shift is complex, with interactions within the R—P—R framework and the asymmetry of the π-orbital envelope at phosphorus also exerting profound influences. The possibility also exists that bromine substituents may shield the phosphorus by a neighbor-anisotropy effect.

Influence of Ligands Attached to Other Phosphorus Atoms. The range of chemical shifts for each PR_2 unit listed in Table 3.17 is a consequence of remote influences by other substituent groups elsewhere on the ring. For example, the PF_2 ^{31}P shift in XXVIII varies over the range, −8.6 to −13.9 ppm, according to whether X is fluoro or phenyl, and the PPh_2 shift in XXIX varies from −14.3 ppm to −30 ppm depending on the distribution of fluoro, chloro, or phenyl groups between the other substituent sites.

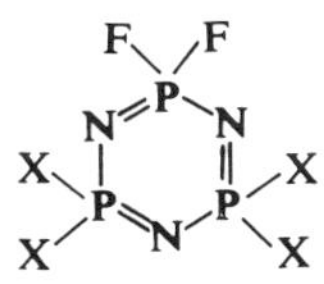

XXVIII

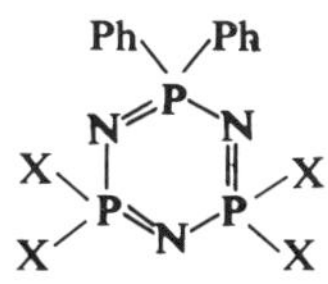

XXIX

For XXVIII, replacement of fluorine as X by phenyl groups, results in shifts to higher field. The same is true when fluorine substituents in XXIX are replaced by chloro and phenyl groups, and when non-geminal $N_3P_3Cl_2(NMe_2)_4$ is replaced by $[NP(NMe_2)_2]_3$. Similar effects are observed with monophosphazenes, where replacement of $Cl_3CC(O)N{=}PPh_3$ by $PhC(O)N{=}PPh_3$ causes a change in chemical shift from −23.8 ppm to −14.9 ppm.[79] The general implication is that replacement of a strongly electron withdrawing group by a less electron withdrawing unit at a particular phosphorus atom increases the shielding of a phosphorus at the *meta* position. It is reasonably well established that the chemical shift of a quadruply connected phosphorus atom depends on σ-bond electronegativity influences on the phosphorus *s*- and *p*-orbital electrons and on π-bond influences on the 3*d*-orbital electrons.[199] Presumably the influence of a remote ligand is through the skeletal π electrons since 1:4 inductive effects should be minimal. It could be argued, therefore, that remote substituents deshield by drawing the π-electron cloud away from the deshielded phosphorus. The process can be viewed as a competition between the substituents R and X (XXX) for the π electrons of the P—N—P unit. In this sense, the remote substituent effects provide some evidence in favor of delocalized π orbitals.

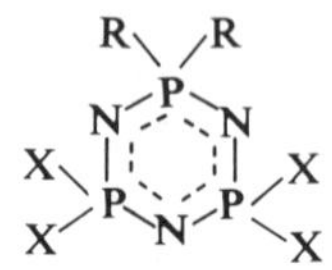

XXX

Influence of Ring or Chain Size. Most cyclotriphosphazenes show negative ^{31}P shifts, whereas cyclic tetramers and larger rings or high polymers show positive shifts. In the chlorophosphazene series, $(NPCl_2)_n$, the chemical shift of the pentamer through octamer and the high polymer are all in the range of +16 to +18 ppm. Positive shifts with increasing degree of polymerization are also evident in the series $(NPF_2)_{3\text{ and }4}$, $(NPBr_2)_{3\text{ and }4}$, $[NP(F)NMe_2]_{3\text{ to }6}$, $[NP(OMe)_2]_{3\text{ and }n}$, $[NP(OEt)_2]_{3\text{ and }n}$, $[NP(OCH_2CF_3)_2]_{3\text{ and }n}$ and $[NP(OPh)_2]_{3\text{ and }n}$. One factor which distinguishes most tetramers and higher homologues from cyclic trimers is the greater molecular flexibility of tetramers and polymers. Skeletal bond torsion will have the effect of improving the π-bond symmetry around phosphorus and thereby increasing the shielding.[192] Indeed, the ^{31}P shift of the high polymer $[NP(OCH_2CF_3)_2]_n$ becomes more positive as the temperature is raised from −80° to +80°C (δ changes from +4.7 ppm to +8.2 ppm),[92] presumably because the torsional mobility of the skeletal bonds is much enhanced at the higher temperatures.

Effect of Skeletal Saturation. Table 3.18[200–206] lists ^{31}P chemical shifts for a number of cyclophosphazanes. One interesting feature of the data is that the chemical shift is very little affected by the presence of a negative charge, as

TABLE 3.18

SELECTED ^{31}P NMR SHIFT DATA FOR CYCLOPHOSPHAZANES

Group	Compound	^{31}P shift (ppm)	References
PO_2^-	$(HN—PO_2)_3^{3-}$ 3 Na^+	1.5	200
PO_2^-	$(HN—PO_2)_4^{4-}$ 4 K^+	3.7, 4.1	201
PO_2^-	$(EtN—PO_2)_3^{3-}$	−5.3	201
P(O)OEt	$[EtN—P(O)OEt]_3$	−5.3	202
P(O)Cl	$[MeN—P(O)Cl]_2$	5.3	203
P(O)Cl	$N_6P_4O_2Cl_2Me_6$	−4.0	203
P(O)NHPh	$[PhN—P(O)NHPh]_2$	12.0	201
$P(Cl)(NMe–)_2$	$N_6P_4O_2Cl_2Me_6$	73	203
P(S)Cl	$[MeN—P(S)Cl]_2$	−51.5	203
PCl_3	$(MeN—PCl_3)_2$	78.2	204
PF_3	$(MeN—PF_3)_2$	71.5, 69.5	205, 206

evidenced by the identical shifts found for $(EtN—PO_2)_3^{3-}$ and $[EtN—P(O)OEt]_3$. Furthermore, for species of formula, $(HN—PO_2)_n^{n-}$, only a very small shift difference exists between the cyclic trimer and tetramer. It is also noteworthy that in halocyclodiphosphazanes, such as $(MeN—PF_3)_2$ or $(MeN—PCl_3)_2$, the phosphorus atoms are much more highly shielded than in the corresponding cyclotriphosphazenes, $(NPF_2)_3$ or $(NPCl_2)_3$, or the linear species, $RC(O)N{=}PCl_3$. Sulfur attached to phosphorus appears to be strongly deshielding relative to oxygen, as indicated by the strongly negative shift observed when $[MeN—P(S)Cl]_2$ replaces $[MeN—P(O)Cl]_2$.

3. Proton Chemical Shifts

The chemical shift associated with a proton in a phosphazene side group is characteristic of the environment within the side group. Thus, protons in an NMe_2 ligand show characteristic chemical shifts in the region of $\tau = 7.27$ to 7.79 for cyclic trimers.[176] The other substituents exert minor influences on the chemical shift value of the NMe_2 protons, and small shift differences can be discerned between *cis* and *trans* isomers. The phosphazene ring size also exerts a small influence, with τ values falling from 7.20 for $[NP(NMe_2)_2]_3$ to 7.08 for $[NP(NMe_2)_2]_7$.[207] Thus, the gross chemical shift of a proton in a side unit is largely unaffected by the presence of a phosphazene ring. This is also illustrated by the following substituent τ values (in parentheses) relative to tetramethylsilane: NC_5H_{10}, αH (6.8 to 7.0),[208] NHMe, NH protons (7.3 to 7.6),[191] OCH_3 (6.29 to 6.46),[92, 207] OC_2H_5, αH (5.73 to 6.08),[177, 182, 207] OCH_2CF_3 (5.4),[92, 177] OPh (2.8 to 3.2),[92, 207, 209] and Ph (2.2 to 2.7).[182] The proton chemical shifts are usually sufficiently well separated that the data can be used for "fingerprint"-type identification.

4. ^{19}F Chemical Shifts

Although fluorine atoms in organic substituents which are attached to the ring show the expected chemical shifts, fluorine atoms connected directly to skeletal phosphorus show shifts which vary according to the molecular environment. Typical values are listed in Table 3.19.[47, 171, 185b, 210–212] Perhaps the most interesting feature is the variation of the ^{19}F chemical shift with increases in ring size. Chivers, Oakley, and Paddock[185b] have shown that, for the fully fluorinated derivatives, $(NPF_2)_{3\ \mathrm{to}\ 6}$, the chemical shift decreases slightly from tetramer to hexamer. However, when a fluorine atom is attached to the same phosphorus as a second substituent (Cl, Br, or NCS), the chemical shift increases sharply from trimer to tetramer, and thereafter decreases from tetramer through hexamer. This may be connected with changes in the availability of the ring π electrons for shielding of the fluorine nucleus.[185b]

TABLE 3.19

^{19}F SHIFT VALUES FOR FLUOROPHOSPHAZENES[a]

Group	Compound	Chemical shift	References
PF_2	$(NPF_2)_3$	71.9	185b
PF_2	$(NPF_2)_4$	71.9	185b
PF_2	$(NPF_2)_5$	69.0	185b
PF_2	$(NPF_2)_6$	68.6	185b
PF_2	$(NPF_2)_n$	70.5	47
P(F)Ph	$N_3P_3F_4Ph_2$-*non-gem, cis*	49.3	171
P(F)Ph	$N_3P_3F_4Ph_2$-*non-gem, trans*	51.5	171
P(F)NCS	$N_3P_3F_5(NCS)$	49.6	185b
P(F)NCS	$N_4P_4F_7(NCS)$	53.9	185b
P(F)NCS	$N_5P_5F_9(NCS)$	52.3	185b
P(F)NCS	$N_6P_6F_{11}(NCS)$	51.7	185b
PFCl	$N_3P_3F_5Cl$	30.7	185b
PFCl	$N_4P_4F_7Cl$	38.2	185b
PFCl	$N_5P_5F_9Cl$	33.9	185b
PFCl	$N_6P_6F_{11}Cl$	32.8	185b
PFBr	$N_3P_3F_5Br$	19.6	185b
PFBr	$N_4P_4F_7Br$	30.6	185b
PFBr	$N_5P_5F_9Br$	25.0	185b
PFBr	$N_6P_6F_{11}Br$	23.5	185b
$P(F)NMe_2$	$N_3P_3F_5(NMe_2)$	63.0	210
$P(F)NMe_2$	$N_3P_3F_4(NMe_2)_2$-*non-gem, cis*	60.2	211
$P(F)NMe_2$	$N_3P_3F_4(NMe_2)_2$-*non-gem, trans*	62.2	211
P(F)NSO	$N_3P_3F_5(NSO)$	70.0	212

[a] δ_F values relative to $CFCl_3$.

5. SPIN–SPIN COUPLING DATA

The identification of a cyclic phosphazene can often be accomplished by a conventional analysis of combined chemical shift and peak intensity data from ^{31}P, ^{1}H, and ^{19}F NMR experiments. However, an unambiguous assignment of a structure may require an investigation of the spin–spin coupling behavior. The long-range spin–spin coupling characteristics may also provide clues about electron delocalization within the skeleton.

In general, spin–spin coupling can occur by at least three mechanisms. Short-range σ-bond coupling for light elements occurs by polarization of the bonding electrons, and the magnitude of the coupling (as indicated by J, the coupling constant) falls off rapidly as the number of bonds between the two nuclei increases. Short-range coupling is usually negligible for nuclei which are separated by more than three bonds. Compounds which contain π electrons often show long-range coupling effects which are transmitted via the delocalized π-electron cloud. Heavier elements can also give rise to coupling effects through space by the interaction of nonbonding electron pairs.

Phosphorus–Phosphorus Spin Coupling. Phosphorus–phosphorus coupling is often observed in high-resolution spectra of cyclophosphazenes. The coupling constants depend on the substituents attached to each phosphorus, and typical values are listed in Table 3.20.[176, 177, 183, 185, 213, 214]

TABLE 3.20

PHOSPHORUS–PHOSPHORUS COUPLING CONSTANTS FOR

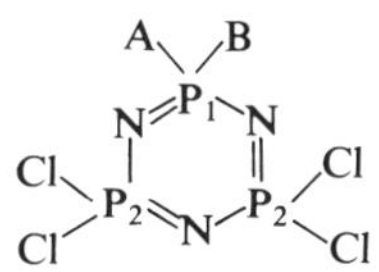

A	B	$J_{P_1P_2}$	References
SEt	SEt	4.8	183
SPh	SPh	4.2	183
NMe_2	NMe_2	~48	176
NHPh	NHPh	48	213
NMe_2	Cl	49.7	177
OPh	Cl	60.8	214
$OCHMe_2$	Cl	62.7	177
OMe	Cl	63.3	177
OEt	Cl	63.3	177
OCH_2CF_3	Cl	66.2	177
F	Cl	78.3	177
F	F	100	185

The $J_{P_1P_2}$ term shows a rough trend toward higher values as the electronegativity of the substituents A and B increases, and particularly as the electronegativity of the *atom* attached to phosphorus increases.[177, 215] It has been suggested that the contact term provides a predominant contribution to this coupling, and that the coupling is transmitted through the σ-bond framework.[209] The influence of the ligands on this coupling is then assumed to be related to the *s* character of the phosphorus hybrids which are involved in σ-bond formation to nitrogen.[215]

Phosphorus–Hydrogen Spin Coupling. In a system such as XXXI, where the group R contains hydrogen atoms, it can be anticipated that these hydrogens may couple with the nearly phosphorus atom (P_1). Such short-range coupling

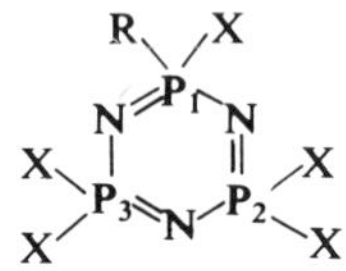

XXXI

is, in fact, detected when the group R is NHMe,[174, 175] NMe_2,[173, 176, 179] *O*-Alkyl,[177, 182] or *S*-Alkyl.[183] For example, the methyl protons in *cis*-$[NPCl(NMe_2)]_3$ yield a doublet signal from coupling with the nearby ^{31}P nucleus[176, 216] (Fig. 3.9), with a coupling constant, *J*, of 17.5 cps.

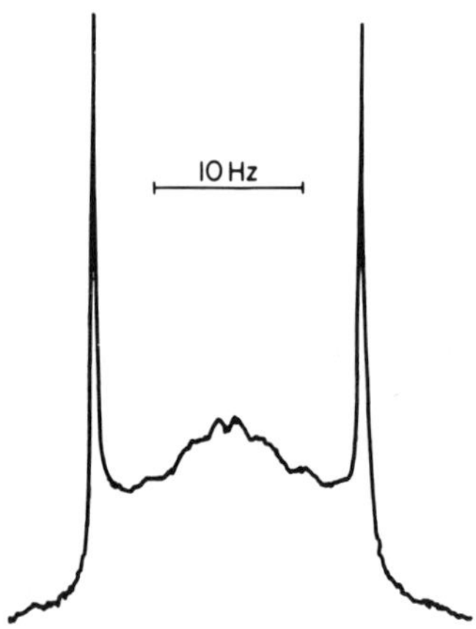

FIG. 3.9. Proton NMR spectrum of *cis*-$[NPCl(NMe_2)]_3$ ($\delta_{Me} = 2.68$ ppm). From Finer *et al.*[216]

Similarly, the methylene protons in ethoxy- and ethylthiophosphazenes couple with the supporting phosphorus atom, but the methyl protons, which are separated by four bonds from phosphorus, generally do not.[182, 183] In one case, $N_3P_3Ph_4(OEt)_2$-*gem*, a weak coupling is observed between the methyl protons and ^{31}P, but the coupling constant, *J*, is only 0.8 cps.

The short-range phosphorus–hydrogen coupling constants depend on the nature of the side groups and on the type of substituent which occupies the other ligand to the same phosphorus. For protons which are separated from phosphorus by three bonds, through carbon and nitrogen, the coupling constant is usually in the range of 11 to 18 cps,[173–176] with the higher values being associated with the presence of the more electronegative co-substituents on phosphorus. Analysis of the coupling constants can aid in the structural identification. In ethoxyphosphazenes, the methylene hydrogens show J_{PH} coupling constants in the region of 8.5 cps,[182] and in ethylthiophosphazenes, the value is near 18 cps.[183]

In many cases, the short-range coupling effects are complicated by the presence of NMR absorptions from long-range splittings which may involve atoms P_2 or P_3 (XXXI). These long-range effects may give rise to distinct hyperfine lines, or they may yield broad diffuse absorptions such as the one shown in Fig. 3.9. In a few cases, the long-range coupling scheme is sufficiently characteristic that it can be used for the identification of *cis–trans* isomers. Hewlett and Shaw[182] have noted that the methylene-proton spectrum of hexakis(ethoxy)cyclotriphosphazene is particularly complex, and an explanation has been formulated in terms of virtual coupling of these protons to all three phosphorus nuclei. Although attempts have been made to calculate phosphorus–phosphorus coupling constants from proton spectra, it was found that linewidth variations destroyed the primary requirement for a valid band shape analysis.[177, 216]

Phosphorus–Fluorine Coupling. Strong spin–spin coupling occurs between fluorine and the phosphorus to which it is attached. Coupling constants in the range of 840 to 1056 cps have been reported.[89, 171, 172, 185b, 211] In general, the lowest coupling constants are associated with the presence of the most electronegative groups attached to the same phosphorus atom. The coupling constant for PF_2 in $(NPF_2)_3$ is 868 cps,[185b] whereas the value for PFBr in $N_3P_3F_5Br$ is 1056 cps.[185b] Phenyl and chloro substituents in PFPh and PFCl groups induce coupling constants in the 960 to 1000 cps range. Long-range coupling between fluorine and the phosphorus atom in the *meta* position is also detected. The coupling constant in $N_3P_3Cl_5F$ (J_{FP_2}) is ± 11 cps[177] and for $N_3P_3Cl_4F_2$-*gem* it is -14 cps,[185] values which are sufficiently large to imply that the delocalized skeletal π electrons may transmit the spin characteristics.

6. Dynamic Nuclear Polarization

An attempt has been made to examine the relative π-electron delocalization in $(NPCl_2)_{3\ \text{to}\ 7}$ by ^{31}P dynamic nuclear polarization, with the use of four different free-radical species.[217] The ^{31}P enhancement follows a declining trend for $(NPCl_2)_n$ in the order $n = 3 > 4 > 6 > 7$. With three of the radicals, the pentamer breaks the sequence with a higher value than the tetramer but, with the diphenylpicrylhydrazyl radical, the value for the pentamer was lower than that of the hexamer. No explanation for this discrepancy has been made.

7. Restricted Rotation about the Phosphorus–Nitrogen Bond

The presence of p_π–p_π bonding in C═C or B═N bonds can be detected by the existence of *cis–trans* isomers formed as a result of the high barrier to torsion of the bond. This barrier may be derived directly from the symmetry requirements of the π-type orbital overlap, or it may be a consequence of steric hindrance associated with nearby groups in the molecule. NMR experiments can be used to study torsional barrier heights. The technique depends on the "freezing out" of torsional isomers as the sample temperature is lowered, with the concurrent separation of individual NMR peaks. The peak coalescence temperature is a function of the torsional barrier height. This technique has also been used to study the torsional barrier of the phosphorus–nitrogen bond in monophosphazanes. Table 3.21[218–220] lists barrier heights determined by this method.

TABLE 3.21

Torsional Barriers for Phosphorus–Nitrogen Compounds

Compound	^1H NMR coalescence temperature (°C)	Barrier height (kcal/mole)	References
PhClP—NMe_2	−50 ±2	12	218
$(ClCH_2)$ClP—NMe_2	−50 to −60	—	219
(Cl_2HCCF_2)ClP—NMe_2	−50 to −60	—	219
Pr^iClP—NMe_2	−50 to −60	—	219
$(ClCH_2)$FP—NMe_2	<−100	—	219
$(F_3C)_2$P—N(H)Me	−120	~8.5	220
Cl_2P—NMe_2	−120	8.4	220
$(F_3C)_2$P—NMe_2	−120	9.0	220

These barriers of 9 to 12 kcal/mole are probably too large to reflect nonbonding intramolecular interactions between the groups attached to phosphorus and nitrogen. Two explanations have been put forward. The first supposes that phosphorus–nitrogen d_π–p_π interactions are responsible,[219] and the second suggests that interaction between the nonbonding lone pairs on phosphorus and nitrogen gives rise to the barrier.[218] The ^{19}F NMR coalescence temperature for the high polymer $(NPF_2)_n$ is also known to be below −70°C,[47] and the proton spectrum of $[NP(OCH_2CF_3)_2]_n$ broadens suddenly at temperatures below −66°C.[95] This results from a "freezing out" of backbone torsional motions at this temperature, with the resultant formation of nonequivalent proton sites. The temperature at which this effect occurs (the polymer glass transition temperature) agrees well with values obtained from thermomechanical measurements.[92]

O. Conclusions

Use of the physical methods outlined in this chapter has enabled the structures of a great many phosphorus–nitrogen compounds to be elucidated. There is now no question about the presence of rings or chains or about the detailed structure of derivatives which show positional or *cis–trans* isomerism. However, serious questions still remain about the disposition of the electrons within the phosphorus–nitrogen skeleton. It seems likely that some form of π-electron interaction serves to stabilize the skeletal structure and transmit nuclear spin information, but the powerful conjugative influences which are characteristic of organic-type delocalized systems are certainly absent. Monophosphazenes appear to show more of the expected d_π–p_π bonding properties than do cyclo- or polyphosphazenes. This is particularly evident in the P—N bond strength data and from the analyses of infrared and electronic spectra. The cyclo- and polyphosphazenes appear to occupy a half-way position between the monophosphazenes on one hand and the phosphazanes on the other. The π character of cyclo- and polyphosphazenes is only weakly developed at best, and it confers considerably less delocalization stability on the molecules than in p_π–p_π systems, such as benzene or *s*-triazines. If a choice must be made between the various π bonding models for phosphazenes, a combined π' and discontinuous "island" type π orbital arrangement is perhaps the most compatible with the experimental data. However, none of the current theories satisfactorily explain all the experimental facts.

References

1. E. Hobbs, D. E. C. Corbridge, and B. Raistrick, *Acta Crystallogr.* **6**, 621 (1953).
2. D. W. J. Cruickshank, *Acta Crystallogr.* **17**, 671 (1964).
3. L. O. Brockway and H. O. Jenkins, *J. Amer. Chem. Soc.* **58**, 2036 (1936).
4. L. S. Bartell and R. C. Hirst, *J. Chem. Phys.* **31**, 449 (1959).
5. L. G. Hoard and R. A. Jacobson, *J. Chem. Soc., A* p. 1203 (1966).
6. H. Hess and D. Forst, *Z. Anorg. Allg. Chem.* **342**, 240 (1966).
7. J. Weiss and G. Hartmann, *Z. Naturforsch. B* **21**, 891 (1966).
8. J. W. Cox and E. R. Corey, *Chem. Commun.* p. 123 (1967).
9. G. B. Ansell and G. J. Bullen, *J. Chem. Soc., A* p. 3026 (1968).
10. R. Olthof, T. Migchelsen, and A. Vos, *Acta Crystallogr.* **19**, 596 (1965).

11. T. Migchelsen, R. Olthof, and A. Vos, *Acta Crystallogr.* **19**, 603 (1965).
12. M. W. Dougill, *J. Chem. Soc., London* p. 3211 (1963).
13. H. McD. McGeachin, and F. R. Tromans, *J. Chem. Soc., London* p. 4777 (1961).
14. H. R. Allcock, R. L. Kugel, and E. G. Stroh, unpublished work (1970).
15. R. Olthof, *Acta Crystallogr., Sect. B* **25**, 2040 (1969).
16. A. Wilson and D. F. Carroll, *J. Chem. Soc., London* p. 2548 (1960).
17. G. J. Bullen, *J. Chem. Soc., A* p. 1450 (1971).
18. R. Hazekamp, T. Migchelsen, and A. Vos, *Acta Crystallogr.* **15**, 539 (1962).
19. A. J. Wagner and A. Vos, *Acta Crystallogr., Sect. B* **24**, 707 (1968).
19a. A. W. Schlueter and R. A. Jacobson, *J. Chem. Soc., A* p. 2317 (1968).
20. E. Giglio, F. Pompa, and A. Ripamonti, *J. Polym. Sci.* **59**, 293 (1962).
21. P. deSantis, E. Giglio, and A. Ripamonti, *J. Inorg. Nucl. Chem.* **24**, 469 (1962).
22. E. Giglio and R. Puliti, *Acta Crystallogr.* **22**, 304 (1967).
23. J. B. Faught, T. Moeller, and I. C. Paul, *Inorg. Chem.* **9**, 1656 (1970).
24. M. W. Dougill, *J. Chem. Soc., London* p. 5471 (1961).
24a. W. C. Marsh, T. N. Ranganathan, J. Trotter, and N. L. Paddock, *Chem. Commun.* p. 815 (1970).
25. G. J. Bullen, *J. Chem. Soc., London* p. 3193 (1962).
26. A. J. Wagner and A. Vos, *Rec. Trav. Chim. Pays-Bas* **84**, 603 (1965); *Acta Crystallogr., Sect. B* **24**, 1423 (1968).
26a. W. C. Marsh and J. Trotter, *J. Chem. Soc., A* p. 169 (1971).
27. L. A. Siegel and J. H. van den Hende, *J. Chem. Soc., A* p. 817 (1967).
28. H. R. Allcock, M. T. Stein, and J. A. Stanko, *Chem. Commun.* p. 944 (1970).
29. G. B. Ansell and G. J. Bullen, *Chem. Commun.* p. 430 (1966).
30. N. L. Paddock, J. Trotter, and S. H. Whitlow, *J. Chem. Soc., A* p. 2227 (1968).
31. G. J. Bullen and P. A. Tucker, *Chem. Commun.* p. 1185 (1970).
32. C. W. Allen, I. C. Paul, and T. Moeller, *J. Amer. Chem. Soc.* **89**, 6362 (1967).
33. N. V. Mani, F. R. Ahmed, and W. H. Barnes, *Acta Crystallogr.* **19**, 693 (1965).
34. N. V. Mani, F. R. Ahmed, and W. H. Barnes, *Acta Crystallogr.* **21**, 375 (1966).
35. F. R. Ahmed, P. Singh, and W. H. Barnes, *Acta Crystallogr., Sect. B* **25**, 316 (1969).
36. H. R. Allcock, "Heteroatom Ring Systems and Polymers." Academic Press, New York, 1967.
37. N. V. Mani and A. J. Wagner, *Chem. Commun.* p. 658 (1968); *Acta Cryst.* **B27**, 51 (1971).
38. D. W. J. Cruickshank, *J. Chem. Soc., London* p. 5486 (1961).
39. A. Bondi, *J. Phys. Chem.* **68**, 441 (1964).
40. E. L. Eliel and G. S. Hammond, *in* "Steric Effects in Organic Chemistry" (M. S. Newman, ed.), pp. 117–120 and 460–470. Wiley, New York, 1956.
41. P. von R. Schleyer, *J. Amer. Chem. Soc.* **83**, 1368 (1961).
42. H. R. Allcock, *J. Macromol. Sci., Rev. Macromol. Chem.* **4**, 149 (1970).
43. J. Trotter and S. H. Whitlow, *J. Chem. Soc., A* p. 460 (1970).
44. H. R. Allcock and E. G. Stroh, unpublished work (1970).
45. H. R. Allcock, *Inorg. Chem.* **5**, 1320 (1966).
46. H. R. Allcock, J. E. Lehnsen, and J. J. Meister, unpublished work (1968).
47. H. R. Allcock, G. K. Konopski, R. L. Kugel, and E. G. Stroh, *Chem. Commun.* p. 985 (1970).
48. P. A. Fowell and C. T. Mortimer, *J. Chem. Soc., London* p. 2913 (1959).
49. S. B. Hartley, W. S. Holmes, J. K. Jacques, M. F. Mole, and J. C. McCoubrey, *Quart. Rev., Chem. Soc.* **17**, 204 (1963).
50. H. Fleig and M. Becke-Goehring, *Z. Anorg. Allg. Chem.* **376**, 215 (1970).
51. A. F. Bedford and C. T. Mortimer, *J. Chem. Soc., London* p. 4649 (1960).

52. S. B. Hartley, N. L. Paddock, and H. T. Searle, *J. Chem. Soc., London* p. 430 (1961).
53. N. L. Paddock, *Quart. Rev., Chem. Soc.* **18**, 168 (1964).
54. M. P. Yagupsky, *Inorg. Chem.* **6**, 1770 (1967).
55. A. J. Downs, *Chem. Commun.* p. 628 (1967).
56. H. J. Becher and F. Seel, *Z. Anorg. Allg. Chem.* **305**, 148 (1960).
57. A. C. Chapman and N. L. Paddock, *J. Chem. Soc., London* p. 635 (1962).
58. T. R. Manley and D. A. Williams, *Spectrochim. Acta, Part A* **23**, 1221 (1967).
59. L. W. Daasch, *J. Amer. Chem. Soc.* **76**, 3403 (1954).
60. S. Califano, *J. Inorg. Nucl. Chem.* **24**, 483 (1962).
61. S. Califano and A. Ripamonti, *J. Inorg. Nucl. Chem.* **24**, 491 (1962).
62. I. C. Hisatsune, *Spectrochim. Acta* **21**, 1899 (1965).
63. I. C. Hisatsune, *Spectrochim. Acta* **25A**, 301 (1969).
64. T. R. Manley and D. A. Williams, *Spectrochim. Acta, Part A* **23**, 149 (1967).
65. T. R. Manley and D. A. Williams, *Spectrochim. Acta, Part A* **24**, 1661 (1968).
66. T. R. Manley and D. A. Williams, *Polymer* **10**, 307 (1969).
67. E. Steger and R. Stahlberg, *Z. Naturforsch. B* **17**, 780 (1962).
68. E. Steger and R. Stahlberg, *Z. Anorg. Allg. Chem.* **326**, 243 (1964).
69. R. Stahlberg and E. Steger, *Spectrochim. Acta, Part A* **23**, 627 (1967).
70. G. E. Coxon and D. B. Sowerby, *Spectrochim. Acta, Part A* **24**, 2145 (1968).
71. E. Steger and R. Stahlberg, *J. Inorg. Nucl. Chem.* **28**, 688 (1966).
72. E. Steger and R. Stahlberg, *Spectrochim. Acta, Part A* **23**, 2057 (1967).
73. R. Stahlberg and E. Steger, *Spectrochim. Acta, Part A* **23**, 2185 (1967).
73a. F. Rauchle and M. Gayoso, *Anal. Real. Soc. Espanola, Ser. A*, **66**, 241 (1970).
74. E. S. Kozlov, A. A. Kisilenko, A. I. Sedlov, and A. V. Kirsanov, *Zh. Obshch. Khim.* **37**, 1611 (1967).
75. W. Wiegräbe and H. Bock, *Chem. Ber.* **101**, 1414 (1968).
76. H. Bock, M. Schnoller, and H. T. Dieck, *Chem. Ber.* **102**, 1363 (1969).
77. H. Schmidbaur and W. Wolfsberger, *Chem. Ber.* **100**, 1000 (1967).
78. M. I. Kabachnik, V. A. Gilyarov, and E. N. Tsvetkov, *Izv. Akad. Nauk SSSR, Otd. Khim. Nauk* p. 2135 (1959).
79. V. A. Shokol, A. A. Kisilenko, and G. I. Derkach, *Zh. Obshch. Khim.* **39**, 874 (1969).
80. V. I. Shevchenko, A. S. Shtepanek, and A. V. Kirsanov, *Zh. Obshch. Khim.* **32**, 2595 (1962).
81. W. Wiegräbe, H. Bock, and W. Lüttke, *Chem. Ber.* **99**, 3737 (1966).
82. A. Schmidpeter, H. Brecht, and J. Ebeling, *Chem. Ber.* **101**, 3902 (1968).
83. I. M. Filatova, E. L. Zaitseva, A. P. Simonov, and A. Ya. Yakubovich, *Zh. Obshch. Khim.* **38**, 1304 (1968).
84. A. C. Chapman, N. L. Paddock, D. H. Paine, H. T. Searle, and D. R. Smith, *J. Chem. Soc., London* p. 3608 (1960).
85. R. Rätz and C. Grundmann, *J. Inorg. Nucl. Chem.* **16**, 60 (1960).
86. H. J. Becher and F. Seel, *Z. Anorg. Allg. Chem.* **305**, 148 (1960).
87. T. R. Manley and D. A. Williams, *Spectrochim. Acta, Part A* **23**, 1221 (1967).
88. A. C. Chapman and N. L. Paddock, *J. Chem. Soc., London* p. 635 (1962).
89. J. Emsley and N. L. Paddock, *J. Chem. Soc., A* p. 2590 (1968).
90. E. Steger and D. Klemm, *J. Inorg. Nucl. Chem.* **29**, 1812 (1967).
91. J. Emsley, *J. Chem. Soc., A* p. 109 (1970).
92. H. R. Allcock, R. L. Kugel, and K. J. Valan, *Inorg. Chem.* **5**, 1709 (1966).
93. R. A. Chittenden and L. C. Thomas, *Spectrochim. Acta* **22**, 1449 (1966).
94. F. Rallo, *Chimica* **8**, 1134 (1965).
95. H. R. Allcock and G. F. Konopski, unpublished work (1967).

95a. P. Pulay, B. Lakatos, G. Toth, A. Hesz, and Z. Vetessy, *Acta Chim. Acad. Sci. Hung.* **60**, 333 (1969).
96. H. T. Searle, *Proc. Chem. Soc., London* p. 7 (1959).
97. H. Sisler, S. E. Frazier, R. G. Rice, and M. G. Sanchez, *Inorg. Chem.* **5**, 326 (1966).
98. G. Tesi and C. M. Douglas, U.S. Pat. 3,065,266 (1962).
99. G. Tesi and C. M. Douglas, *J. Amer. Chem. Soc.* **84**, 549 (1962).
100. D. D. Magnelli, G. Tesi, J. U. Lowe, and W. E. McQuiston, *Inorg. Chem.* **5**, 457 (1966).
101. B. Grushkin, M. G. Sanchez, and R. G. Rice, *Inorg. Chem.* **3**, 623 (1964).
102. R. H. Kratzer and K. L. Paciorek, *Inorg. Chem.* **4**, 1767 (1965).
103. B. W. Fitzsimmons and R. A. Shaw, *J. Chem. Soc., London* p. 1735 (1964).
104. H. R. Allcock, R. Caputo, A. Kalmus, C. W. Roberts, and E. T. McBee, *U.S. Govt. Res. Rep.* **AD 209,669** (1959).
105. H. R. Allcock, *J. Amer. Chem. Soc.* **86**, 2591 (1964).
105a. A. P. Carroll and R. A. Shaw, *J. Chem. Soc., A* p. 914 (1966).
106. D. B. Sowerby and L. F. Audrieth, *Chem. Ber.* **94**, 2670 (1961).
107. C. T. Ford, F. E. Dickson, and I. I. Bezman, *Inorg. Chem.* **4**, 890 (1965).
108. H. R. Allcock and D. P. Mack, unpublished results (1969).
109. H. R. Allcock and R. L. Kugel, *Inorg. Chem.* **5**, 1716 (1966).
110. P. Nannelli and T. Moeller, *Inorg. Chem.* **2**, 896 (1963).
111. R. A. Shaw and C. Stratton, *J. Chem. Soc., London* p. 5004 (1962).
112. N. L. Paddock, *Quart. Rev., Chem. Soc.* **18**, 168 (1964).
113. S. G. Kokalis, K. John, T. Moeller, and L. F. Audrieth, *J. Inorg. Nucl. Chem.* **19**, 191 (1961).
114. K. John, T. Moeller, and L. F. Audrieth, *J. Amer. Chem. Soc.* **83**, 2608 (1961).
115. A. C. Chapman and D. F. Carroll, *J. Chem. Soc., London* p. 5005 (1963).
116. H.-G. Horn and M. Becke-Goehring, *Z. Anorg. Allg. Chem.* **367**, 165 (1969).
117. A. J. Downs, *Chem. Commun.* p. 628 (1967).
118. T. Moeller and A. H. Westlake, *J. Inorg. Nucl. Chem.* **29**, 957 (1967).
119. R. Schmutzler, *Z. Naturforsch. B* **19**, 1101 (1964).
120. M. Green, R. N. Haszeldine, and G. S. A. Hopkins, *J. Chem. Soc., A* p. 1766 (1966).
121. H. Bock and E. Baltin, *Chem. Ber.* **98**, 2844 (1965).
122. H. Bock and G. Rudolph, *Chem. Ber.* **98**, 2273 (1965).
123. H. Bock, *Angew. Chem., Int. Ed. Engl.* **4**, 457 (1965).
124. H. Bock, *Chimica* **21**, 35 (1967).
125. B. I. Stepanov, T. G. Edelman, and I. M. Kolbasova, *Zh. Obshch. Khim.* **37**, 1883 (1967).
126. B. I. Stepanov and T. G. Edelman, *Zh. Obshch. Khim.* **39**, 1549 (1969).
127. F. G. Sherif and C. D. Schmulbach, *Inorg. Chem.* **5**, 322 (1966).
128. K. L. Paciorek, *Inorg. Chem.* **3**, 96 (1964).
129. H. R. Allcock, *J. Amer. Chem. Soc.* **86**, 2591 (1964).
130. H. R. Allcock and E. J. Walsh, unpublished work (1969).
131. H. R. Allcock and R. L. Kugel, *Inorg. Chem.* **5**, 1016 (1966).
132. B. Lakatos, A. Hesz, Z. Vetessy, and G. Horvath, *Acta Chim. Acad. Sci. Hung.* **60**, 309 (1969).
133. J. Nagy, I. Barta, and J. Reffy, *Period. Polytech., Chem. Eng.* **11**, 33 (1967).
134. H. J. Krause, *Z. Elektrochem.* **59**, 1004 (1955).
135. N. B. Jurinski, C. C. Thomson, and P. A. DeMaine, *J. Inorg. Nucl. Chem.* **27**, 1571 (1965).
136. N. B. Jurinski and P. A. DeMaine, *J. Inorg. Nucl. Chem.* **27**, 1591 (1965).
137. N. L. Paddock and H. T. Searle, *Advan. Inorg. Chem. Radiochem.* **1**, 347 (1959).

138. L. G. Lund ,N. L. Paddock, J. E. Proctor, and H. T. Searle, *J. Chem. Soc., London* p. 2542 (1960).
139. H. R. Allcock and P. Ostrowski, unpublished work (1968).
140. T. Moeller and S. G. Kokalis, *J. Inorg. Nucl. Chem.* **25**, 875 (1963).
141. H. R. Allcock and W. J. Birdsall, *J. Amer. Chem. Soc.* **91**, 7541 (1969).
142. H. R. Allcock and W. J. Birdsall, *Inorg. Chem.* **10** (1971).
143. D. Chapman, S. H. Glarum, and A. G. Massey, *J. Chem. Soc., London* p. 3140 (1963).
144. D. D. Eley and M. R. Willis, *J. Chem. Soc., London* p. 1534 (1963).
145. H. R. Allcock and R. J. Best, *Can. J. Chem.* **42**, 447 (1964).
146. I. Y. Ahmed and C. D. Schmulbach, *J. Phys. Chem.* **71**, 2358 (1967).
147. G. E. Coxon, T. F. Palmer, and D. B. Sowerby, *J. Chem. Soc., A* p. 1568 (1967).
148. C. E. Brion and N. L. Paddock, *J. Chem. Soc., A* p. 388 (1968).
149. C. E. Brion and N. L. Paddock, *J. Chem. Soc., A* p. 392 (1968).
150. C. D. Schmulbach, A. G. Cook, and V. R. Miller, *Inorg. Chem.* **7**, 2463 (1968).
151. G. E. Coxon, T. F. Palmer, and D. B. Sowerby, *Inorg. Nucl. Chem. Lett.* **2**, 215 (1966).
152. G. E. Coxon, T. F. Palmer, and D. B. Sowerby, *J. Chem. Soc., A* p. 358 (1969).
153. C. E. Brion, D. J. Oldfield, and N. L. Paddock, *Chem. Commun.* p. 226 (1966).
154. G. R. Branton, C. E. Brion, D. C. Frost, K. A. R. Mitchell, and N. L. Paddock, *J. Chem. Soc., A* p. 151 (1970).
154a. A. J. Wagner and T. Moeller, *J. Chem. Soc., A* p. 596 (1971).
155. G. Corfield, *J. Chem. Soc., London* p. 4258 (1962).
156. H. F. Koopman, F. J. Spruit, F. van Deursen, and J. Bakker, *Rec. Trav. Chim. Pays-Bas*, **84**, 341 (1965).
156a. I. Yu. Kokoreva, Ya. K. Syrkin, A. A. Kropacheva, N. M. Kashnikova, and L. E. Mukhina, *Dokl. Akad. Nauk SSSR*, **166**, 155 (1966).
157. D. P. Craig, M. L. Heffernan, R. Mason, and N. L. Paddock, *J. Chem. Soc., London* p. 1376 (1961).
158. S. K. Ray and R. A. Shaw, *Chem. Ind.* (*London*) p. 1173 (1961).
159. D. Feakins, W. A. Last, and R. A. Shaw, *J. Chem. Soc., London* p. 4464 (1964).
160. D. Feakins, W. A. Last, N. Neemuchwala, and R. A. Shaw, *J. Chem. Soc., London* p. 2804 (1965).
161. D. Feakins, W. A. Last, S. N. Nabi, and R. A. Shaw, *J. Chem. Soc., A* p. 1831 (1966).
162. D. Feakins, S. N. Nabi, R. A. Shaw, and P. Watson, *J. Chem. Soc., A* p. 10 (1968).
163. S. K. Das, R. Keat, R. A. Shaw, and B. C. Smith, *J. Chem. Soc., A* p. 1677 (1966).
164. T. Moeller and S. G. Kokalis, *J. Inorg. Nucl. Chem.* **25**, 875 (1963).
165. H. Negita and S. Satou, *J. Chem. Phys.* **24**, 621 (1956).
166. K. Torizuka, *J. Phys. Soc. Jap.* **11**, 84 (1956).
167. E. A. C. Lucken, *Proc. Colloq. AMPERE* (*At. Mol. Etud. Radio Elec.*) p. 678 (1962).
168. M. A. Whitehead, *Can. J. Chem.* **42**, 1212 (1964).
169. K. Kaplansky and M. A. Whitehead, *Can. J. Chem.* **45**, 1669 (1967).
170. M. Dixon, H. D. B. Jenkins, J. A. S. Smith, and D. A. Tong, *Trans. Faraday Soc.* **63**, 2852 (1967).
171. C. W. Allen and T. Moeller, *Inorg. Chem.* **7**, 2177 (1968).
172. C. W. Allen, F. Y. Tsang, and T. Moeller, *Inorg. Chem.* **7**, 2183 (1968).
173. C. T. Ford, F. E. Dickson, and I. I. Bezman, *Inorg. Chem.* **3**, 177 (1964).
174. C. T. Ford, F. E. Dickson, and I. I. Bezman, *Inorg. Chem.* **4**, 890 (1965).
175. C. T. Ford, J. M. Barr, F. E. Dickson, and I. I. Bezman, *Inorg. Chem.* **5**, 351 (1966).
176. R. Keat, S. K. Ray, and R. A. Shaw, *J. Chem. Soc., London* p. 7193 (1965).
177. F. Heatley and S. M. Todd, *J. Chem. Soc., A* p. 1152 (1966).
178. G. Engelhardt, E. Steger, and R. Stahlberg, *Z. Naturforsch. B* **21**, 586 (1966).

179. B. Grushkin, J. L. McClanahan, and R. G. Rice, *J. Amer. Chem. Soc*, **86**, 4204 (1964).
180. B. Grushkin, A. J. Berlin, J. L. McClanahan, and R. G. Rice, *Inorg. Chem.* **5**, 172 (1966).
181. T. Moeller and P. Nannelli, *Inorg. Chem.* **2**, 659 (1963).
182. C. Hewlett and R. A. Shaw, *J. Chem. Soc., A* p. 56 (1966).
183. N. Boden, J. W. Emsley, J. Feeney, and L. H. Sutcliffe, *Chem. Ind. (London)* p. 1909 (1962).
184. G. R. Feistel and T. Moeller, *J. Inorg. Nucl. Chem.* **29**, 1731 (1967).
185. M. L. Heffernan and R. F. M. White, *J. Chem. Soc., London* p. 1382 (1961).
185a. G. Allen, M. Barnard, J. Emsley, N. L. Paddock, and R. F. M. White, *Chem. Ind. (London)* p. 952 (1963).
185b. T. Chivers, R. T. Oakley, and N. L. Paddock, *J. Chem. Soc., A* p. 2324 (1970).
186. G. Englehardt, E. Steger, and R. Stahlberg, *Z. Naturforsch. B* **21**, 1231 (1966).
187. G. R. Feistel and T. Moeller, *J. Inorg. Nucl. Chem.* **29**, 2731 (1967).
188. H. P. Latscha, *Z. Anorg. Allg. Chem.* **362**, 7 (1968).
189. W. Lehr, *Z. Anorg. Allg. Chem.* **350**, 18 (1967).
190. B. Grushkin, M. G. Sanchez, and R. G. Rice, *Inorg. Chem.* **3**, 623 (1964).
191. W. Lehr, *Z. Anorg. Allg. Chem.* **352**, 27 (1967).
192. A. Schmidpeter, C. Weingand, and E. Hafner-Roll, *Z. Naturforsch. B* **24**, 799 (1969).
192a. K. John and T. Moeller, *J. Inorg. Nucl. Chem.* **22**, 199 (1961).
193. B. Grushkin, M. G. Sanchez, M. V. Ernest, J. L. McClanahan, G. E. Ashby, and R. G. Rice, *Inorg. Chem.* **4**, 1538 (1965).
194. K. L. Paciorek and R. H. Kratzer, *Inorg. Chem.* **5**, 538 (1966).
195. A. Schmidpeter, H. Brecht, and H. Groeger, *Chem. Ber.* **100**, 3063 (1967).
196. M. Feldt and T. Moeller, *J. Inorg. Nucl. Chem.* **30**, 2351 (1968).
197. V. V. Sheluchenko, I. M. Filatova, E. L. Zaitseva, and A. Ya. Yakubovich, *Zh. Obshch. Khim.* **39**, 194 (1969).
198. A. Schmidpeter and J. Ebeling, *Angew. Chem., Int. Ed. Engl.* **1**, 209 (1968).
199. M. M. Crutchfield, C. H. Dungan, J. H. Letcher, V. Mark, and J. R. Van Wazer, *Top. Phosphorus Chem.* **5**, (1967).
200. S. J. Fitch, *J. Amer. Chem. Soc.* **86**, 61 (1964).
201. M. L. Nielsen, J. V. Pustinger, and J. Strobel, *J. Chem. Eng. Data* **9**, 167 (1964).
202. M. L. Nielsen and J. V. Pustinger, *J. Phys. Chem.* **68**, 152 (1964).
203. M. Becke-Goehring, L. Leichner, and B. Scharf, *Z. Anorg. Allg. Chem.* **343**, 154 (1966).
204. E. Fluck, *Z. Anorg. Allg. Chem.* **320**, 64 (1963).
205. R. K. Harris and C. M. Woodman, *Mol. Phys.* **10**, 437 (1966).
206. R. Schmutzler, "International Review on Halogen Chemistry." Academic Press, New York, 1967.
207. G. Allen, D. J. Oldfield, N. L. Paddock, F. Rallo, J. Serregi, and S. M. Todd, *Chem. Ind. (London)* p. 1032 (1965).
208. R. Keat and R. A. Shaw, *J. Chem. Soc., A* p. 908 (1966).
209. D. Dell, B. W. Fitzsimmons, R. Keat, and R. A. Shaw, *J. Chem. Soc., A* p. 1680 (1966).
210. O. Glemser, E. Niecke, and H. W. Roesky, *Chem. Commun.* p. 282 (1969).
211. B. Green and D. B. Sowerby, *J. Chem. Soc., A* p. 987 (1970).
212. E. Niecke, O. Glemser, and H. Thamm, *Chem. Ber.* **103**, 2864 (1970).
213. H. Lederle, G. Ottmann, and E. Kober, *Inorg. Chem.* **5**, 1818 (1966).
214. R. F. M. White, quoted by E. G. Finer, *J. Mol. Spectrosc.* **23**, 104 (1967).
215. E. G. Finer, *J. Mol. Spectrosc.* **23**, 104 (1967).
216. E. G. Finer, R. K. Harris, M. R. Bond, R. Keat, and R. A. Shaw, *J. Mol. Spectrosc.* **33**, 72 (1970).

217. R. A. Dwek, N. L. Paddock, J. A. Potenza, and E. H. Poindexter, *J. Amer. Chem. Soc.* **91**, 5436 (1969).
218. A. H. Cowley, M. J. S. Dewar, and W. R. Jackson, *J. Amer. Chem. Soc.* **90**, 4185 (1968).
219. H. Goldwhite and D. G. Rowsell, *Chem. Commun.* p. 713 (1969).
220. A. H. Cowley, M. J. S. Dewar, W. R. Jackson, and W. B. Jennings, *J. Amer. Chem. Soc.* **92**, 1085 (1970).

Part II

SYNTHESIS AND REACTIONS

Chapter 4

SYNTHESIS OF THE PHOSPHORUS–NITROGEN SKELETON

A. General Survey of Synthetic Routes

A large number of phosphorus–nitrogen compounds have been described in the literature. Some are prepared by direct synthesis, and others are obtained by substitution reactions performed on suitable precursors. In this chapter we will consider the direct synthesis routes, and substitution processes will be discussed in Chapters 5 through 14. Although many direct synthetic routes to phosphorus–nitrogen compounds have been described, most of these methods fall into a few general categories, and these are considered below in turn. One reaction, the treatment of halophosphoranes with ammonium halides, has been studied so extensively that, in addition to being mentioned in the general survey, it is considered in greater detail in Section B.

1. Synthesis of Cyclo- and Polyphosphazenes

a. *Interaction of Halophosphoranes with Ammonium Halides*

Halophosphoranes react with ammonium halides to yield cyclophosphazenes, according to the general equation[1–71]

$$n\,R_2PX_3 + n\,NH_4X \rightarrow (NPR_2)_n + 4n\,HX$$

The group R can be chloro, bromo, or an organic unit, and the halogen atom, X, can be chloro or bromo. Typically, the reaction takes place in a boiling halogenated solvent such as *s*-tetrachloroethane or chlorobenzene. Cyclic trimers and tetramers are usually formed in the greatest yields, although smaller amounts of cyclic pentamers, hexamers, etc., and linear species can also be isolated.

The most extensively studied reaction is the one between phosphorus pentachloride and ammonium chloride to yield the stable, white, crystalline hexachlorocyclotriphosphazene, $(NPCl_2)_3$, and octachlorocyclotetraphosphazene, $(NPCl_2)_4$.[1–35] The overall reaction scheme is

$$n\,PCl_5 + n\,NH_4Cl \xrightarrow{-4n\,HCl} (NPCl_2)_3 + (NPCl_2)_4 + \cdots$$

This is one of the most important reactions in phosphazene chemistry, since it comprises the cheapest and most convenient first step to the synthesis of the majority of cyclo- and polyphosphazenes. It is also the basis of a manufacturing process for chlorocyclophosphazenes.[13–22]

A similar reaction is employed for the preparation of bromophosphazenes.[40–51] The general equation is

$$n\,PBr_3 + n\,Br_2 + n\,NH_4Br \xrightarrow{-4n\,HBr} (NPBr_2)_3 + (NPBr_2)_4 + \cdots$$

The lower bromocyclophosphazenes are stable, white, crystalline solids. They are more sensitive to atmospheric moisture than are the chloro derivatives. Although fewer cyclic bromophosphazene homologues have been reported than for the $(NPCl_2)_n$ or $(NPF_2)_n$ series, species up to the cyclic hexamer have been described[48, 49] and higher cyclic homologues undoubtedly can be isolated.

Mixed chloro–bromo cyclophosphazenes are prepared by the use of an appropriate mixture of phosphorus halide and ammonium halide, as illustrated by the following equation[43]:

$$3\,PCl_3 + 3\,Br_2 + 3\,NH_4Br \xrightarrow[110°C]{sym\text{-}C_2H_2Cl_4} N_3P_3Cl_2Br_4 + 12\,HX$$

Phosphorus tribromide, phosphorus pentachloride, and ammonium chloride give $N_3P_3Cl_4Br_2$, while phosphorus trichloride, bromine, and ammonium bromide yield $N_3P_3Cl_2Br_4$.[43]

Organocyclophosphazenes are synthesized by similar methods.[53–69] For example, 1,3,5-triphenyl-1,3,5-trichlorocyclotriphosphazene is prepared from phenyltetrachlorophosphorane and ammonium chloride, according to the reaction

$3\,PhPCl_4 + 3\,NH_4Cl \longrightarrow$ 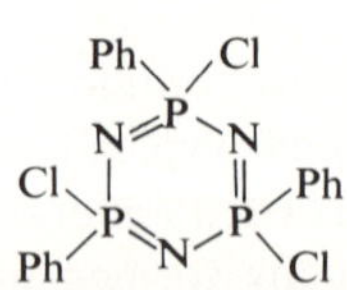 $+ 12\,HCl$

(Refs. 57, 60, 65)

Similarly, diphenyltrichlorophosphorane and ammonium chloride yield fully phenylated cyclophosphazenes by the overall process

$$n\,Ph_2PCl_3 + n\,NH_4Cl \xrightarrow{-4n\,HCl} (NPPh_2)_3 + (NPPh_2)_4 + \cdots$$

(Refs. 54, 55)

Other related reactions are

$$n\,PhPBr_2 + n\,Br_2 + n\,NH_4Br \xrightarrow{-4n\,HBr} (NPBrPh)_3 + (NPBrPh)_4$$

(Refs. 63, 64)

$$n\,Me_2PCl_3 + n\,NH_4Cl \xrightarrow{-4n\,HCl} (NPMe_2)_3 + (NPMe_2)_4$$

(Ref. 67)

$$n\,Et_2PCl_3 + 4n\,NH_3 \longrightarrow (NPEt_2)_n + 3n\,NH_4Cl$$

(Ref. 68)

$$3\,Me_2NPCl_4 + 12\,NH_3 \longrightarrow [NP(Cl)NMe_2]_3 + 9\,NH_4Cl$$

(Ref. 39)

$$n(C_3F_7)_2PCl_3 + nNH_4Cl \xrightarrow{-4nHCl} [NP(C_3F_7)_2]_3 + [NP(C_3F_7)_2]_4$$

(Ref. 69a)

A more detailed consideration of this reaction is given in Section B of this chapter.

b. *Cyclization of Linear Phosphorus–Nitrogen Compounds*

Cyclophosphazenes are formed when a linear phosphazene with terminal amino groups reacts with a halophosphorane.[72–77] There are a number of variations to this type of synthesis, but typical examples are shown in the following scheme:

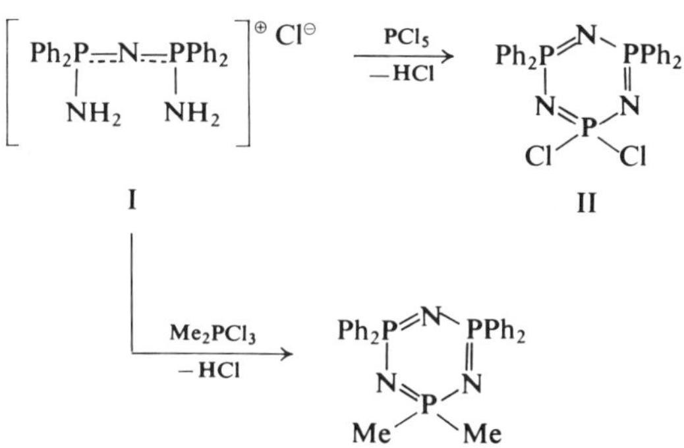

(Ref. 73)

(Ref. 76)

In one experiment, reported by Herring and Douglas,[74] the diaminophosphazene salt, I, was prepared by the reaction of diphenyltrichlorophosphorane, Ph_2PCl_3, with ammonia in chloroform solution.[59] Compound I is a white, crystalline solid, which melts at 245°–246°C. It is surprisingly stable in the atmosphere. A 1:1 molar mixture of I and phosphorus pentachloride was heated in vacuum at 135°C for 16 hours. The product was recrystallized from petroleum ether to give a 25% yield of white crystals of 1,1-dichloro-3,3,5,5-tetraphenylcyclotriphosphazene, II. Interestingly enough, the reactions of I with diphenyltrichlorophosphorane, Ph_2PCl_3, phenyltetrachlorophosphorane, $PhPCl_4$, or phosphorus pentachloride yield significant quantities of the appropriate cyclo*tetra*phosphazene.

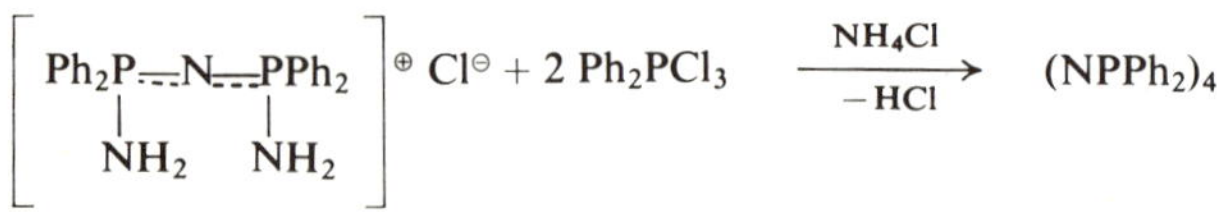

(Ref. 74)

I

The ammonium chloride required for the synthesis is thought to be formed during side reactions.

One of the compounds prepared by this route, 1,1-dichloro-*trans*-3,5-bis-(4-methylphenyl)-3,5-diphenylcyclotriphosphazene, is the first reported optically active cyclotriphosphazene. The compound was synthesized from the optically active phosphazene salt shown.[72]

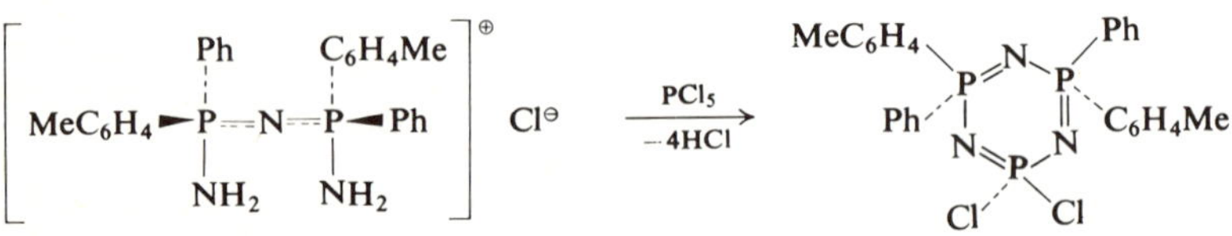

(Ref. 72)

d-isomer

A related reaction involves the interaction of a linear phosphazene with a phosphonite, as shown in the following equation:

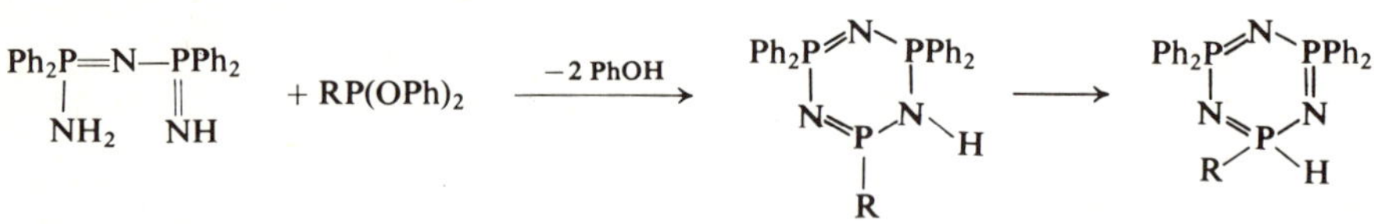

(Ref. 75)

where R is OPh, Me, or Et.

If compound I reacts with the halide of an element other than phosphorus, mixed heteroaromatic rings can be prepared. For example, with boron trichloride, the following reaction takes place:

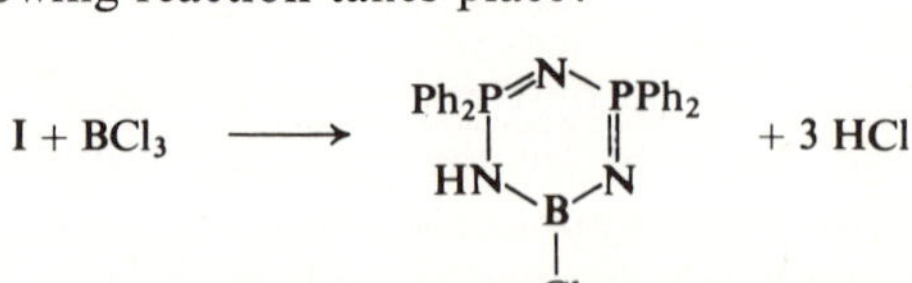

(Ref. 77)

c. *Thermal Decomposition of Azidophosphines*

This reaction route utilizes the metathetical exchange between a halophosphine and a metal azide, followed by careful decomposition of the resultant azidophosphine.[78–85] The overall reaction scheme is as follows:

$$R_2PCl + NaN_3 \xrightarrow{-NaCl} R_2PN_3 \longrightarrow (R_2PN)_n + n\,N_2$$

This reaction can be used to prepare a variety of cyclo- and polyphosphazenes, as illustrated by the following equations:

$$(CF_3)_2PCl + LiN_3 \xrightarrow{-LiCl} (CF_3)_2PN_3 \xrightarrow[30\text{ mm}]{50°\text{–}60°C} [(CF_3)_2PN]_n + n\,N_2 \quad \text{(Ref. 79)}$$

$$PBr_3 + NaN_3 \xrightarrow[165°\text{–}167°C]{-NaBr} [Br_2PN_3] \longrightarrow (Br_2PN)_n + n\,N_2 \quad \text{(Ref. 78)}$$

$$PhPCl_2 + NaN_3 \xrightarrow[170°\text{–}175°C]{-NaCl} [PhClPN_3] \rightarrow (PhClPN)_n + n\,N_2 \quad \text{(Ref. 78)}$$

$$Ph_2PCl + NaN_3 \xrightarrow[165°\text{–}167°C]{-NaCl} [Ph_2PN_3] \rightarrow (Ph_2PN)_4 + 4\,N_2 \quad \text{(Ref. 78)}$$

$2\,Ph_2PCl + 2\,PhPCl_2 + 4\,NaN_3 \longrightarrow$ 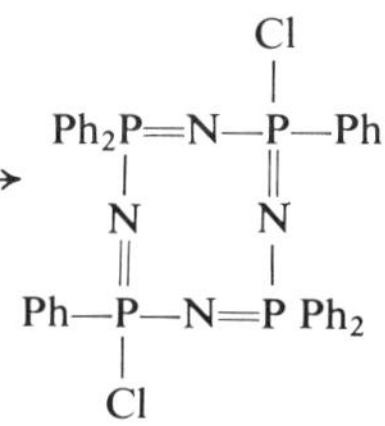 $+ 4\,N_2$ (Ref. 81)

$$Ph_2PCl + Me_3SiN_3 \xrightarrow[25°C]{-Me_3SiCl} Ph_2PN_3 \rightarrow (Ph_2PN)_3 + 3\,N_2 \quad \text{(Ref. 83)}$$

Some idea of the required reaction conditions can be obtained from the procedure reported by Tesi, Haber, and Douglas.[79] Bis(trifluoromethyl)chlorophosphine, $(CF_3)_2PCl$, was treated with lithium azide at 0°C for 24 hours. Distillation *in vacuo* then yielded about 70% of the highly explosive bis(trifluoromethyl)azidophosphine, $(CF_3)_2PN_3$. This material is a violent detonator even at liquid nitrogen temperatures. However, slow decomposition of the azide at 50°–60°C at 37 mm pressure caused elimination of nitrogen and formation of a white, waxy polyphosphazene of formula $[(CF_3)_2PN]_n$. As shown by Herring,[78] isolation of the intermediate azidophosphine is unnecessary in other reactions since, for example, direct addition of small quantities of sodium azide to molten diphenylchlorophosphine gives a mixture of diphenylphosphazenes directly.

d. *Dehydrohalogenation of Aminochlorophosphoranes*

Aminodichlorophosphoranes, $R_2P(Cl_2)NH_2$, can be dehydrohalogenated in the presence of a tertiary amine to yield a cyclo- or polyphosphazene.[86–90]

Appropriate aminodichlorophosphoranes for this reaction can be synthesized either by oxidative chlorination of aminophosphines (phosphinamides or phosphinous amides), or by chlorination of phosphinic amides. The general reaction sequence is

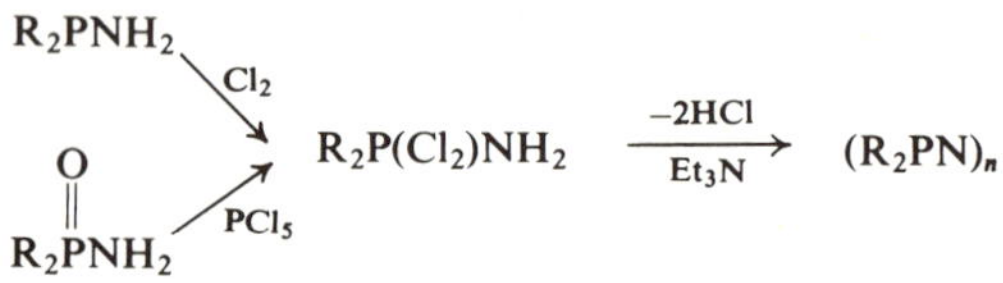

In this reaction, the aminodichlorophosphorane apparently functions as a phosphinimine hydrochloride with the structure $[R_2P(Cl)NH_2]^{\oplus}Cl^{\ominus}$. The scope of this synthetic route is illustrated by the following reaction sequences:

$$(CF_3)_2PNH_2 + Cl_2 \rightarrow (CF_3)_2P(Cl_2)NH_2 \xrightarrow{-2\,HCl} [(CF_3)_2PN]_{3,4,n}$$

(Refs. 86, 87)

$$(C_3F_7)_2PNH_2 + Cl_2 \rightarrow (C_3F_7)_2P(Cl_2)NH_2 \xrightarrow{-2\,HCl} [(C_3F_7)_2PN]_3$$

(Refs. 86, 87)

$$(C_6F_5)_2PNH_2 + Cl_2 \rightarrow (C_6F_5)_2P(Cl_2)NH_2 \xrightarrow{-2\,HCl} [(C_6F_5)_2PN]_3$$

(Ref. 90)

$$Ph_2P(O)NH_2 + PCl_5 \xrightarrow{-POCl_3} Ph_2P(Cl_2)NH_2 \xrightarrow{-2\,HCl} (Ph_2PN)_{3\text{ and }4}$$

(Ref. 89)

Dehydrohalogenation can be effected with a strong base such as triethylamine, or, as in the latter reaction, simply by heating in vacuum.

e. *Interaction of Chlorophosphines with Chloramine or Ammonia and Chloramine*

Sisler and co-workers[91–93] have shown that diarylchlorophosphines can be converted to cyclophosphazenes by treatment with chloramine (scheme 1),[92] or with ammonia and chloramine (scheme 2).[91] Two different reaction path-

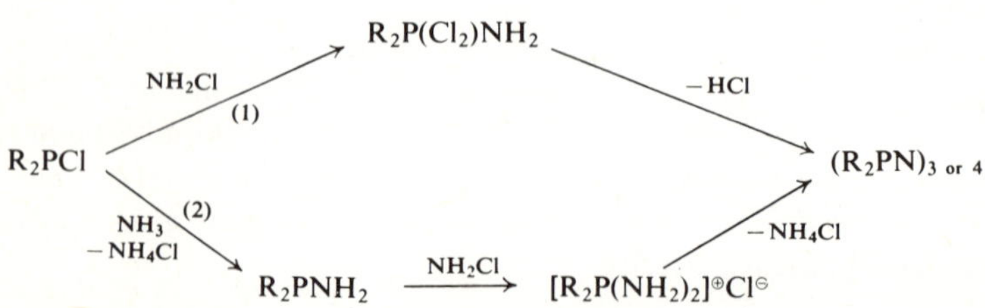

ways apparently exist. Both routes have been used to prepare the phenylcyclophosphazenes, $(NPPh_2)_{3\text{ and }4}$. Reaction scheme 1 resembles that discussed above in Section d. However, the second step of route 1 proceeds through the intermediate formation of linear phosphazanes, such as $Ph_2P(NH_2)(Cl)NHP(Cl_2)Ph_2$, and ammonolysis leads to the formation of $[Ph_2P(NH_2)NP(NH_2)Ph_2]^{\oplus}Cl^{\ominus}$. This latter compound was also isolated from reaction 2. Thus, there is a strong relationship between sequence 2 and the reactions discussed in Section b above. The chloramine–ammonia mixture used in route 2 was prepared by the gas-phase reaction of chlorine with excess ammonia, and this mixture was bubbled through a solution of diphenylchlorophosphine in tetrachloroethane.

f. *Pyrolysis of a Diorganophosphonium Halide*

This route utilizes part of the sequence outlined in route 2 of Section e above. Thus, intermediate molecular weight poly(dimethylphosphazenes) containing ~120 repeating units have been prepared by the pyrolysis of dimethyldiaminophosphonium chloride.[94] The overall reaction scheme is

$$n[Me_2P(NH_2)_2]^{\oplus}Cl^{\ominus} \xrightarrow{200^{\circ}C} (Me_2PN)_n + n\ NH_4Cl$$

The mechanism probably proceeds through the prior formation of $[Me_2P(NH_2)\text{---}N\text{---}P(NH_2)Me_2]^{\oplus}Cl^{\ominus}$ followed by cyclization to cyclophosphazenes and subsequent polymerization.

g. *Dehydrohalogenation of a Cyclophosphazane*

Cyclo*di*phosphazanes (prepared by methods to be described later) can be dehydrohalogenated to yield cyclotri- or higher cyclophosphazenes,[95] as shown in the following examples. The degree of polymerization of the latter product was not reported.

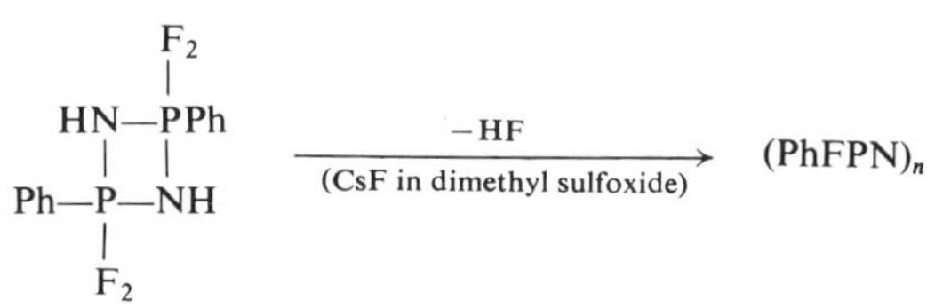

h. *Miscellaneous Syntheses*

It has been reported that chlorocyclophosphazenes can also be prepared by the treatment of phosphorus pentachloride with ammonobasic mercuric chloride, NH_2HgCl, and by the reaction of phosphoryl nitride with chlorine at temperatures above 800°C.[96–99] Poly(chlorofluorophosphazene) (III) can be prepared by ammonolysis of $F_2P(O)—O—P(O)F_2$ via the following synthesis route[99a]:

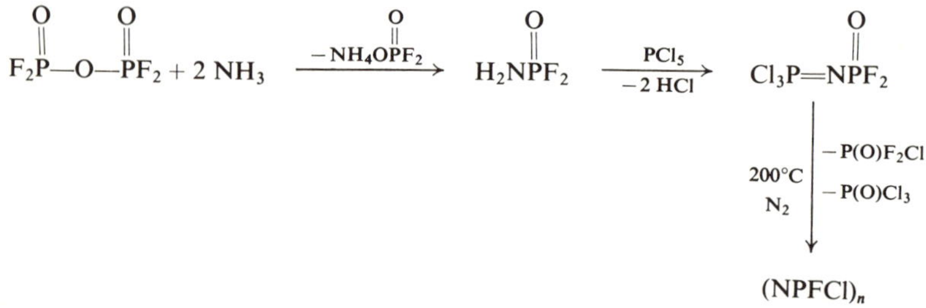

III

2. Synthesis of Cyclophosphazanes and Monophosphazanes

a. *By Rearrangement of Hydroxy- or Alkoxyphosphazenes*

The hydrolysis of halocyclotri- and tetraphosphazenes results in the formation of hydroxyphosphazenes (see Chapter 5). Such compounds can rearrange spontaneously to cyclophosphazanes (IV and V),[100–115] as shown below.

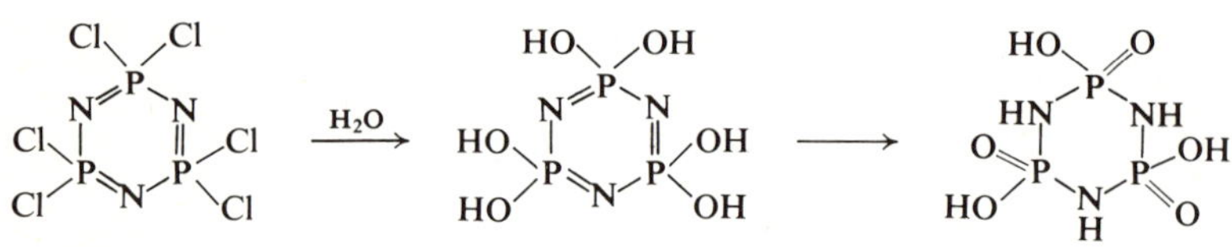

IV

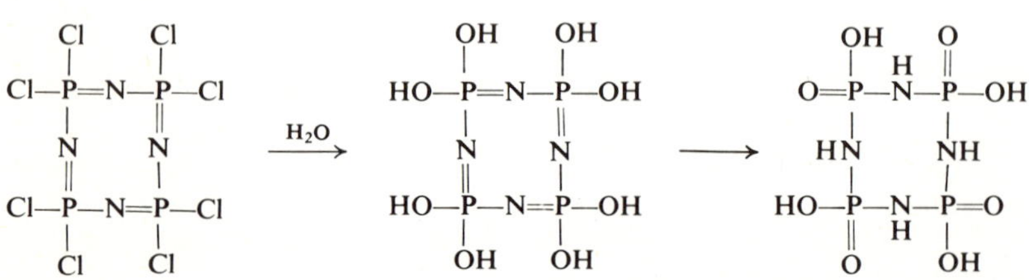

V

Stable salts of the hydroxyphosphazanes can also be isolated.

Similarly, alkyl group migration occurs from oxygen to nitrogen when alkoxycyclophosphazenes are heated at 160°–200°C. Alkyl halides are catalysts for this reaction.

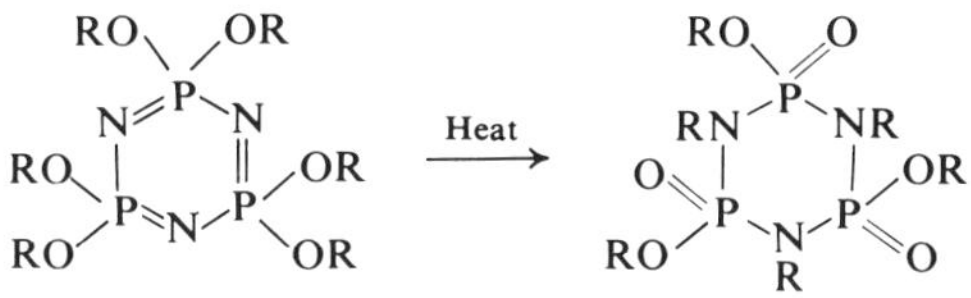

The rearrangement takes place when R is Me, Et, *n*-Pr, *i*-Pr, or benzyl, but it does not occur when R is phenyl or trifluoroethyl.[116] The reaction is considered in more detail in Chapter 13.

b. *Reactions of Halophosphorus Compounds with Amines*

Both the cyclic dimeric phosphazane, VI, and the triple ring structure, VII, have been isolated from the reaction of phosphorus pentachloride with methylammonium chloride.[117, 118] Triple ring compounds similar to VII, but with

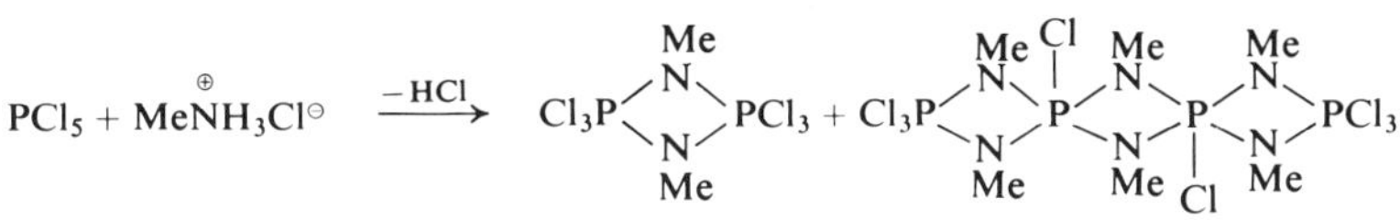

VI VII

CO or SO_2 units in place of the terminal PCl_3 groups, have also been described.[119, 119a]

In the presence of traces of water, phosphorus pentachloride reacts with methylammonium chloride to yield a compound which is believed to have a cage structure (VIII).[120] The ^{31}P NMR spectrum showed a singlet at +74.3 ppm, which is consistent with the cage structure. Compound VI can also be

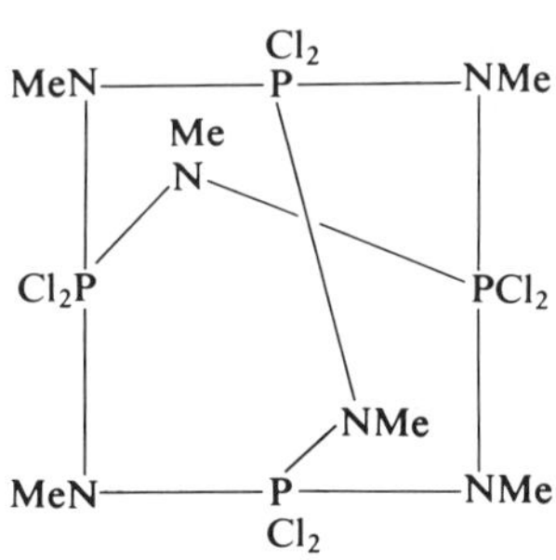

VIII

obtained from the reaction of phosphorus pentachloride with methylamine. It is a crystalline solid which melts at 178°C.

The same type of reaction occurs with arylchlorophosphoranes. For example, phenyltetrachlorophosphorane reacts with an alkylammonium chloride to yield a cyclodiphosphazane (IX).[121] These materials are colorless,

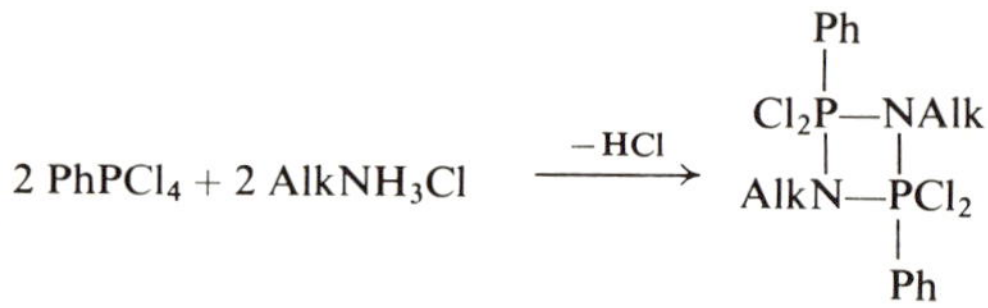

IX

crystalline substances which are readily hydrolyzed by atmospheric moisture. They cannot be depolymerized to the monophosphazene in boiling benzene. However, diphenyltrichlorophosphorane reacts with amines to form saltlike compounds of formula $[Ph_2P(Cl){-}NHR]^{\oplus}Cl^{\ominus}$.[121a]

Phosphorus pentafluoride reacts with secondary amines to yield first an adduct and then a monophosphazane.[122]

$$PF_5 + HNR_2 \rightarrow F_5P \cdot NHR_2 \xrightarrow[-HF]{\text{Heat}} F_4P{-}NR_2$$

A monophosphazane is also formed when phenyltetrafluorophosphorane interacts with a primary amine.[123]

$$PhPF_4 + 2\,MeNH_2 \rightarrow PhF_3P{-}NHMe + Me\overset{\oplus}{N}H_3Cl^{\ominus}$$

Phosphorus pentachloride reacts with aminomethanesulfonic acid to give a solid cyclic dimeric phosphazane (X) by the process[124]

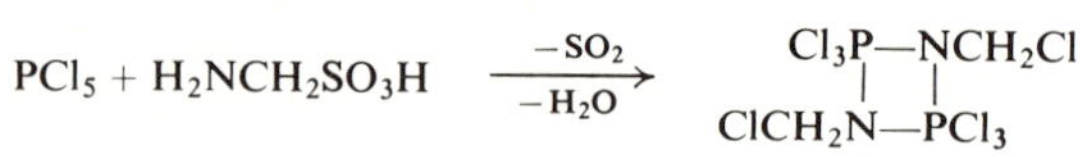

X

This product is a colorless, crystalline compound, m.p. 140°–141°C, which is rapidly decomposed by traces of moisture. Compound VI is formed by the interaction of phosphorus pentachloride with *N,N′*-dimethylsulfamide.

The cyclohexyl ester of dichlorophosphoric acid reacts with cyclohexylamine

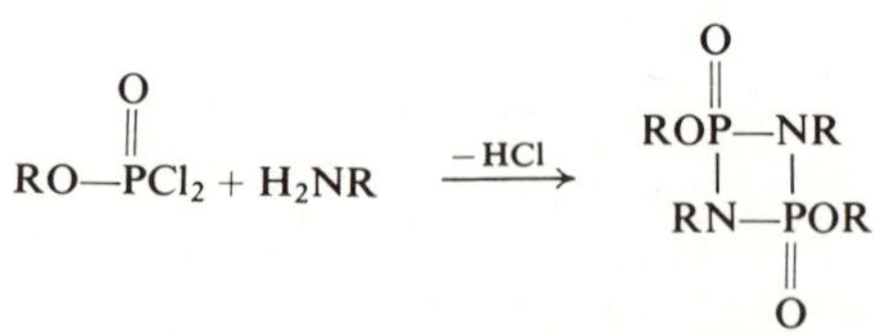

XI

to yield a cyclodiphosphazane with structure XI.[125] Similarly, thiophosphoryl chloride interacts with aniline hydrochloride to yield XII.[126] Finally, phos-

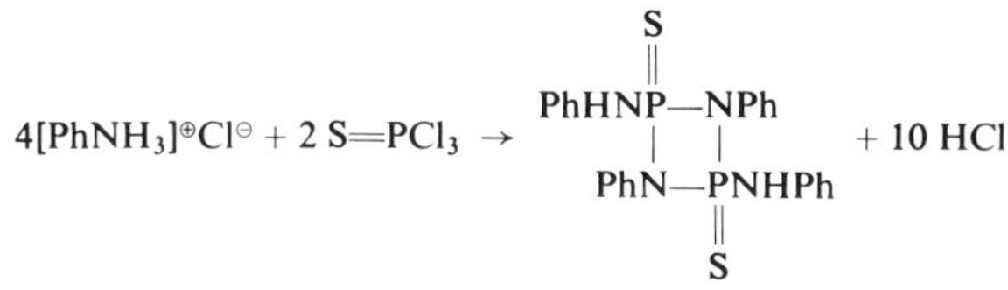

XII

phorus trichloride reacts with primary amines, such as aniline, to yield cyclodiphosphazanes such as XIII and XIV.[127, 128] It should be noted that similar

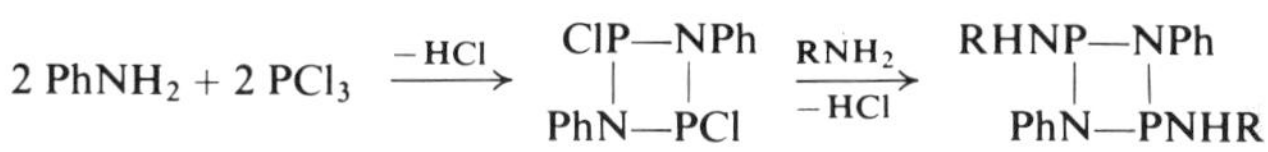

XIII XIV

reactions are employed for the synthesis of cyclodiphosphazanes and monophosphazenes (phosphinimines). Whether dimerization of the monophosphazene occurs appears to depend partly on steric factors.

c. *Reactions of Phosphorus Halides with Silicon–Nitrogen Compounds*[128–133]

Schmutzler and MacDiarmid have shown that the silicon–nitrogen bond in silylamines or silazanes can be cleaved by fluorophosphoranes to yield monophosphazanes or cyclodiphosphazanes. The following equations illustrate this method.

$$PF_5 + Me_2NSiMe_3 \rightarrow F_4P{-}NMe_2 + Me_3SiF \qquad \text{(Ref. 129)}$$

$$PF_5 + Et_2NSiMe_3 \rightarrow F_4P{-}NEt_2 + Me_3SiF \qquad \text{(Refs. 95, 130)}$$

$$PhPF_4 + Me_2NSiMe_3 \rightarrow PhF_3P{-}NMe_2 + Me_3SiF \qquad \text{(Ref. 130)}$$

(Refs. 129, 131)

(Ref. 131)

The four-membered ring systems are sufficiently stable to be distilled or sublimed. They are also hydrolytically stable.[131] However, cyclization does not occur if two phenyl groups are present on phosphorus, and a monophosphazene is formed instead.

$$Ph_2PF_3 + MeN(SiMe_2)_2 \rightarrow Ph_2FP{=}NMe + 2\ Me_3SiF \qquad \text{(Ref. 95)}$$

^1H, ^{31}P, and ^{19}F NMR spectroscopy have been used to confirm the structures of most of the above compounds.

A related cyclotriphosphazane synthesis involves the interaction of phosphorus trichloride with a disilazane,[131] as in the equation

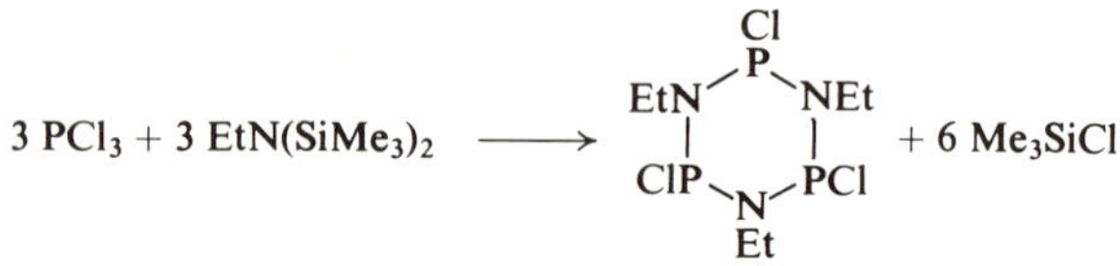

This compound is unstable in moist air.

In a closely related process, dichlorotrifluorophosphorane reacts with the bis(trimethylsilyl) derivative of *N*,*N'*-dimethylsulfamide (XV) to yield the monocyclic and bicyclic phosphazane compounds XVI and XVII by the following sequence[133]:

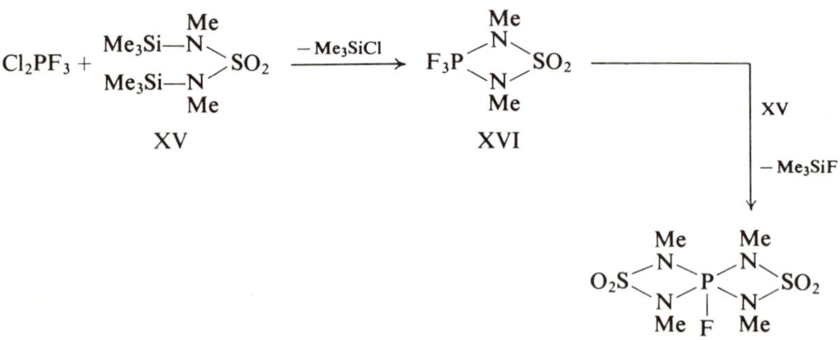

XV XVI XVII

d. *Pyrolysis of Amino-Phosphorus Compounds*

Some very early phosphazane syntheses fall into this category. For example, the pyrolytic condensation of $Cl_2P(O)NHPh$ to the cyclodiphosphazane XVIII was reported by Stokes in 1893.[134] Further reaction of this product with amine

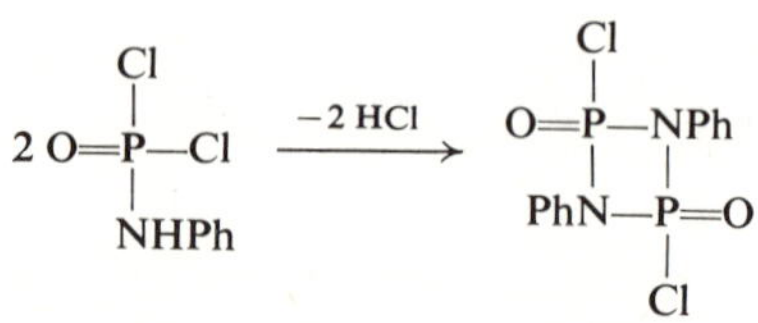

XVIII

hydrochloride results in the replacement of the halogen atoms and the formation of species such as $[R'NH(O)P—NPh]_2$.[135, 136]

More recently, it has been reported that the pyrolysis of pentafluorophenyldiaminophosphines results in the elimination of pentafluorobenzene and the formation of a possible cyclodiphosphazane (XIX).[137] However, this

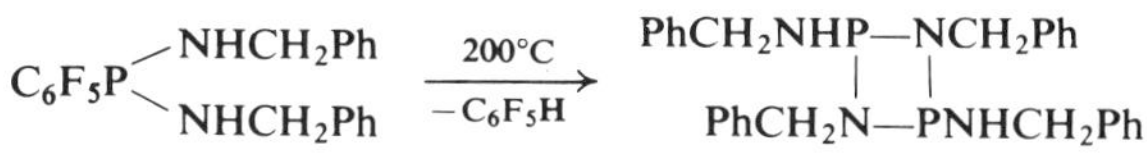

XIX

material is unstable in solution. The thermal decomposition of organophosphorus triamides also results in the formation of cyclophosphazane dimers.[138, 138a]

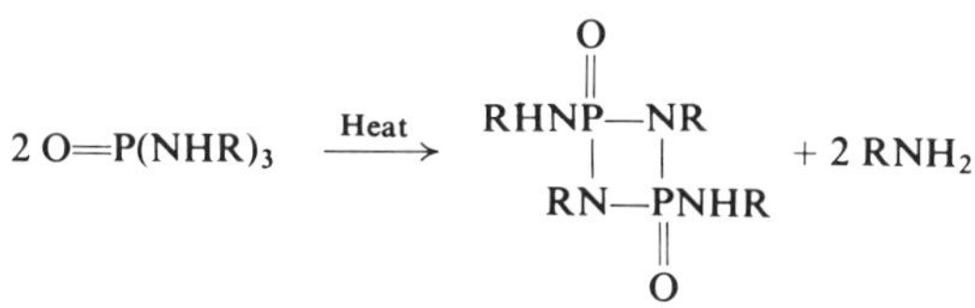

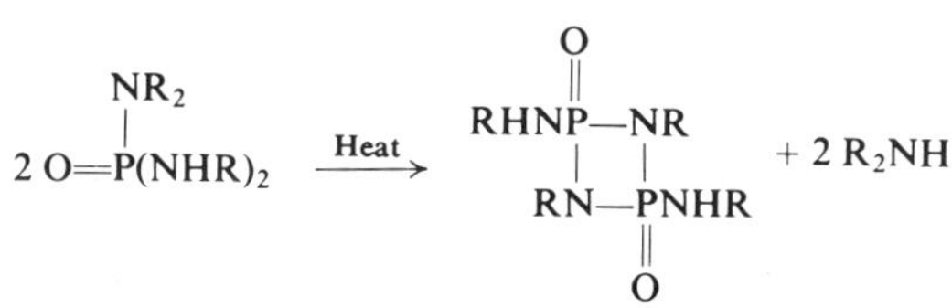

e. *Miscellaneous Phosphazane Syntheses*

Bis(trifluoromethyl)chloramine or the bromo analogue react with tris-(trifluoromethyl)phosphine to yield a monophosphazane by the process[139]

$$(CF_3)_2NCl + (CF_3)_3P \rightarrow (CF_3)_2N{-}P(CF_3)_2 + CF_3Cl$$

A monophosphazane is also formed by the following reaction[140]:

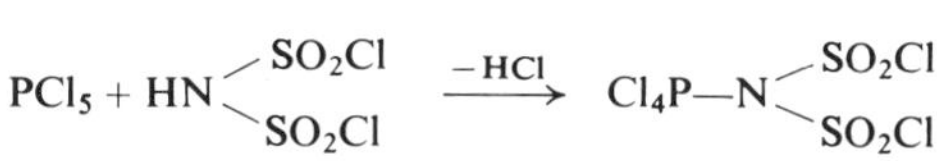

This is a moisture-sensitive compound. Finally, tris(diethylamino)phosphine interacts with aniline or toluidine to give the cyclodiphosphazane $(PhHNP{-}NPh)_2$.[141]

3. Synthesis of Monophosphazenes (Phosphinimines) and Higher Linear Phosphazenes

Many different reactions have been described for the preparation of monophosphazenes and related linear compounds.[142–208] The following sections separate these approaches into broad reaction classes, although it should be emphasized that certain monophosphazene syntheses follow processes similar to those mentioned previously for the preparation of cyclophosphazenes or cyclophosphazanes.

a. *Reaction of Chlorophosphoranes with Ammonium Chloride*

The interaction of phosphorus pentachloride with ammonium chloride usually yields linear phosphazenes as well as cyclophosphazenes.[25, 142–150] The mechanistic significance of this fact will be discussed later. Several linear species have been isolated by Becke-Goehring and co-workers,[25, 142–147] but the primary reaction is as follows:

$$3\ PCl_5 + NH_4Cl \rightarrow [Cl_3P{=}N{-}PCl_3]^{\oplus}[PCl_6]^{\ominus} + 4\ HCl$$

The synthesis is carried out in a *sym*-tetrachloroethane–nitrobenzene solvent system at 80°–140°C, and the phosphazene salt crystallizes from the solvent as white needles which melt at 310°–315°C. The compound is exceptionally sensitive to hydrolysis. The ^{31}P NMR spectrum shows two peaks (at –21.4 and +305.0 ppm relative to 85% H_3PO_4) with an intensity ratio of 2:1.

Longer-chain species, such as $[Cl_3P{=}N{-}PCl_2{=}N{-}PCl_3]^{\oplus}[PCl_6]^{\ominus}$, are formed by successive chain-building steps with ammonium chloride and phosphorus pentachloride,[145] and these compounds can also be identified by ^{31}P NMR spectra. For example, the triphosphazene salt shown above yields a triplet spectrum; one of these lines is resolvable to a doublet (PCl_3), one to a triplet (PCl_2), and one is unresolved ($PCl_6^{\ominus}$).[25] As discussed later (Section B), linear chlorophosphazenes of this type are believed to be intermediates formed during the synthesis of cyclophosphazenes, since ammonolysis of the linear compounds results in the formation of cyclophosphazenes. It is also known that liquid ammonia reacts with phosphorus pentachloride to yield the linear aminophosphazene XX by the process[59, 142, 147]

$$2\ PCl_5 + 16\ NH_3 \rightarrow \underset{\textbf{XX}}{[(NH_2)_3P{=}N{-}P(NH_2)_3]^{\oplus}Cl^{\ominus}} + 9\ NH_4Cl$$

One of the products isolated from the reaction between phosphorus pentachloride and ammonia is tetraaminophosphonium chloride (XXI).[151] This product can be condensed with phosphorus pentachloride to yield a branched chlorophosphazene salt.

A modification of the phosphorus pentachloride–ammonium chloride reaction takes place when metal halides such as $SbCl_3$, $SbCl_5$, $AlCl_3$, $AlBr_3$, BCl_3, BF_3, $TiCl_4$, $ZnCl_2$, $FeCl_3$, CuCl, or TlCl are present.[152, 153] Linear phosphazenes are obtained which are end-capped by the metal halide component. For example, compounds of composition $(N{=}PCl_2)_nSbCl_3$ are formed when antimony trichloride, phosphorus pentachloride, and ammonium chloride

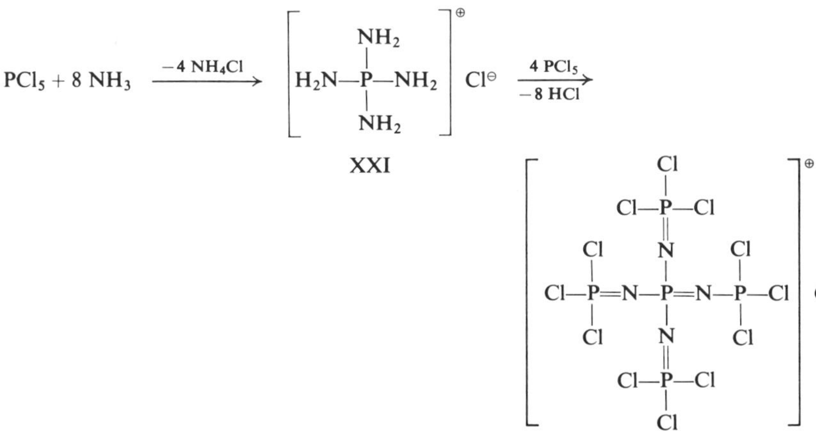

XXI

react in trichlorobenzene. Compounds of this type are thermally stable oils which are apparently resistant to polymerization above 500°C. They are very easily hydrolyzed, however.

Linear, phenyl-substituted phosphazenes can be isolated from the reactions of phenylchlorophosphoranes with ammonium chloride. Thus, the following reactions are known to occur:

$$2\,Ph_2PCl_3 + NH_4Cl \rightarrow [Ph_2ClP{=}N{-}PClPh_2]^{\oplus}Cl^{\ominus} + 4\,HCl \quad \text{(Refs. 58, 154)}$$

$$2\,PhPCl_4 + NH_4Cl \rightarrow [PhCl_2P{=}N{-}PCl_2Ph]^{\oplus}Cl^{\ominus} + 4\,HCl \quad \text{(Ref. 155)}$$

The interaction of diphenyltrichlorophosphorane with ammonium chloride takes place in tetrachloroethane at 130°C. The tetraphenyldiphosphazene salt forms colorless crystals in 90% yield. The reaction of phenyltetrachlorophosphorane with ammonium chloride also takes place in hot tetrachloroethane, and the linear diphenylphosphazene salt can be obtained in 60% yield as white crystals, m.p. 210°C. Similar products are obtained when phenyldichlorophosphine interacts with nitrogen trichloride.[155]

$$PhPCl_2 \xrightarrow{NCl_3} [PhCl_2P{=}N{-}PCl_2Ph]^{\oplus}Cl^{\ominus} + [PhCl_2P{=}N{-}P(ClPh){=}N{-}PCl_2Ph]^{\oplus}Cl^{\ominus} + [PhCl_2P{=}N{-}P(ClPh){=}N{-}P(ClPh){=}N{-}PCl_2Ph]^{\oplus}Cl^{\ominus} \quad \text{etc.}$$

Related linear phosphazenes are formed when tetraphenyldiphosphine reacts with a gaseous mixture of ammonia and chloramine[156] according to the following reaction scheme:

$$Ph_2P{-}PPh_2 \xrightarrow{NH_2Cl\ +\ NH_3} [PH_2(NH_2)P{=}N{-}P(NH_2)PH_2]^{\oplus}Cl^{\ominus} + Ph_2(NH_2)_2P^{\oplus}Cl^{\ominus}$$

or when diphenylchlorophosphine interacts with hydrazine hydrochloride.[157]

$$Ph_2PCl \xrightarrow{H_2NNH_2HCl} [Ph_2(NH_2)P{=}N{-}PPh_2{=}N{-}P(NH_2)Ph]^{\oplus}Cl^{\ominus}$$

This latter product is a white solid, m.p. 285°C. When heated in vacuum at 280°–305°C at reduced pressure it forms hexaphenylcyclotriphosphazene, $(NPPh_2)_3$.

b. *Treatment of Cyclophosphazenes with Inorganic Halides*

Phosphorus pentachloride reacts with hexachlorocyclotriphosphazene or octachlorocyclotetraphosphazene to yield linear phosphazenes by the process[13, 18, 158, 159]

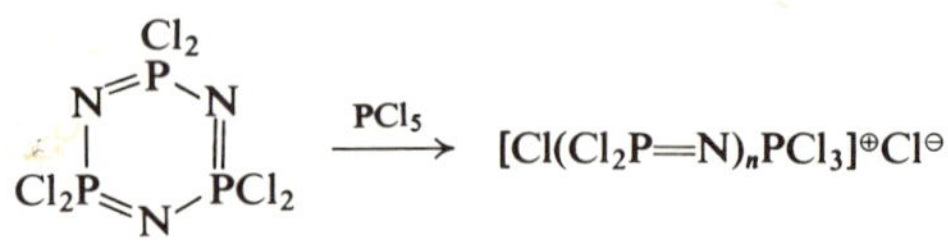

For example, Moran[159] has reported that the compound $[Cl(Cl_2P{=}N)_3PCl_3]^{\oplus}$-$Cl^{\ominus}$ can be prepared by heating $(NPCl_2)_3$ with PCl_5 in an approximately 1.0 to 1.9 molar ratio at 250°C for 11 hours. The yellow product is insoluble in benzene and hexane and it melts at 95°–96°C. Another yellow solid compound, $[Cl(Cl_2P{=}N)_4PCl_3]^{\oplus}Cl^{\ominus}$, is obtained after 100 hours reaction. When the $(NPCl_2)_3$ to PCl_5 ratio is 5:1, oily higher linear polymers are formed, and these polymerize to high polymers when heated above 300°C. A mechanism proposed for the initial attack by PCl_5 on $(NPCl_2)_3$ includes an electrophilic attack at nitrogen by the $[PCl_4]^+$ component of the $[PCl_4]^+[PCl_6]^-$ complex, followed by ring cleavage of the resultant cyclophosphazene cation.[159]

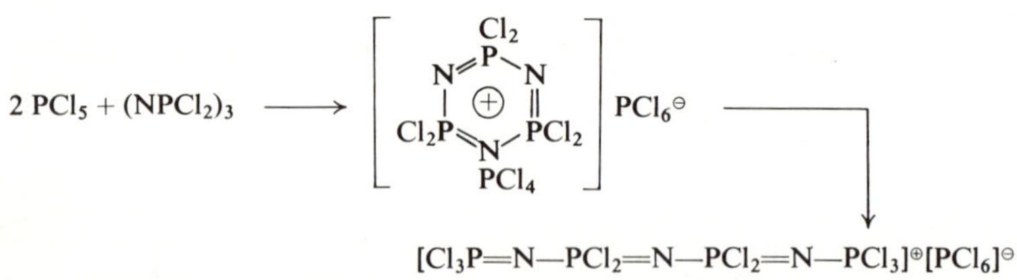

Treatment of these linear species with aluminum trichloride or boron trichloride results in the replacement of the $[PCl_6]^-$ anion by $[AlCl_4]^-$ or $[BCl_4]^-$ anions. This treatment enhances the thermal stability of the linear phosphazenes.[159]

It is also worthwhile to note that chlorocyclophosphazenes apparently undergo ring cleavage with aluminum chloride and other metal halides,[18, 160] and that $(NPF_2)_3$ appears to be cleaved by cesium fluoride.[160a] Octaphenyl-cyclotriphosphazene, $(NPPh_2)_4$, also reacts with phosphorus pentachloride to give linear derivatives.[161]

c. *Reaction of Amino Compounds with Phosphorus Pentachloride*[162–184]

This synthesis route, generally known as the Kirsanov reaction, can be formulated in general terms by the equation

$$PCl_5 + RNH_2 \rightarrow Cl_3P{=}NR + 2\,HCl \qquad \text{(Ref. 162)}$$

The group R can be varied over a considerable range, as illustrated by the following equations:

$$PCl_5 + PhNH_2 \rightarrow Cl_3P{=}NPh + 2\,HCl \qquad \text{(Refs. 166, 167)}$$

$$PCl_5 + RCONH_2 \rightarrow Cl_3P{=}NCOR + 2\,HCl \qquad \text{(Ref. 168)}$$

$$PCl_5 + PhSO_2NH_2 \rightarrow Cl_3P{=}N{-}SO_2{-}Ph + 2\,HCl \qquad \text{(Ref. 169)}$$

$$PCl_5 + NH_2OH \rightarrow Cl_3P{=}NOH + 2\,HCl \qquad \text{(Ref. 170)}$$

$$2\,PCl_5 + NH_2SO_3H \rightarrow Cl_3P{=}N{-}SO_2Cl + 3\,HCl + POCl_3 \qquad \text{(Refs. 171, 172)}$$

$$PCl_5 + H_2NP(O)(OPh)_2 \rightarrow Cl_3P{=}N{-}P(O)(OPh)_2 + 2\,HCl \qquad \text{(Ref. 173)}$$

$$PCl_5 + H_2NP(S)(OPh)_2 \rightarrow Cl_3P{=}N{-}P(S)(OPh)_2 + 2\,HCl \qquad \text{(Ref. 173)}$$

$$2\,PCl_5 + SO_2(NH_2)_2 \rightarrow Cl_3P{=}N{-}SO_2{-}N{=}PCl_3 + 4\,HCl \qquad \text{(Ref. 174)}$$

$$3\,PCl_5 + H_2NP(O)(OH)_2 \rightarrow Cl_3P{=}N{-}P(O)Cl_2 + 4\,HCl + 2\,POCl_3 \qquad \text{(Ref. 175)}$$

$$2\,PCl_5 + RP(O)(OAlk)NH_2 \rightarrow Cl_3P{=}NP(O)ClR + 2\,HCl + POCl_3 + AlkCl \qquad \text{(Ref. 176)}$$

$$F_3PCl_2 + H_2NSO_2F \rightarrow F_3P{=}NSO_2F + 2\,HCl \qquad \text{(Ref. 177)}$$

$$F_3PCl_2 + H_2NPOF \rightarrow F_3P{=}NPOF + 2\,HCl \qquad \text{(Ref. 177)}$$

$$F_3PCl_2 + H_2NP(S)F_2 \rightarrow F_3P{=}NP(S)F_2 + 2\,HCl \qquad \text{(Ref. 177)}$$

$$Ph_3PCl_2 + H_2NPh \rightarrow Ph_3P{=}NPh + 2\,HCl \qquad \text{(Refs. 178–180)}$$

$$Ph_2PCl_3 + H_2N{-}SO_2{-}C_6H_4Me \rightarrow Ph_2ClP{=}N{-}SO_2{-}C_6H_4Me + 2\,HCl \qquad \text{(Ref. 181)}$$

$$Ph_3PCl_2 + H_2NOSC_6H_4NO_2\text{-}o \rightarrow Ph_3P{=}NSOC_6H_4NO_2 + POPh_3 + 2\,HCl \qquad \text{(Ref. 182)}$$

$$Ph_3PCl_2 + H_2NC(S)NHPh \rightarrow Ph_3P{=}NC(S)NPhH + 2\,HCl \qquad \text{(Ref. 183)}$$

In a closely related process, phosphorus pentachloride, phosphorus trichloride, and hydroxylamine hydrochloride react according to the scheme[184]

$$2\,PCl_3 + PCl_5 + 2[H_3NOH]^{\oplus}Cl^{\ominus} \rightarrow Cl_3P{=}NP(O)Cl_2 + 4\,HCl + 4\,NH_4Cl + POCl_3$$

It should be noted that the majority of aromatic amines react with phosphorus pentachloride to form dimers of structure $(Cl_3P{-}NAr)_2$ (i.e., cyclodiphosphazanes) rather than monomers. Only weakly basic amines form stable monophosphazenes.[167] However, replacement of one chlorine atom in $Cl_3P{=}NAr$ by an organic unit sharply reduces the tendency for dimerization. A number of phosphazenes that were originally thought to be monomeric have subsequently been shown to be cyclic dimers.

d. *Interaction of a Tertiary Phosphine or a Halophosphine with an Azide*

This reaction route was used by Staudinger for the early synthesis of monophosphazenes.[185, 186] It requires the reaction of a tertiary phosphine with an organic azide in ether, according to the following general sequence:

$$R_3P + R'N_3 \rightarrow R_3PN{-}N{=}N{-}NR' \rightarrow R_3P{=}NR' + N_2$$

R and R′ can be aromatic or aliphatic. When aliphatic groups are present on nitrogen, the compounds are very sensitive to moisture, and the appropriate phosphine oxide and amine are formed by hydrolysis. However, when aromatic groups are attached to nitrogen, the derivatives show more stability to water[187] but can still be hydrolyzed by dilute acid. It is of interest that $Ph_3P{=}NPh$ is pale yellow in color, whereas $Ph_3P{=}NH$, $Ph_3P{=}NEt$, $Ph_3P{=}NCOPh$, and many other monophosphazenes are colorless.

Phosphorus ylides also react with organic azides to yield monophosphazenes according to the process[188]

$$Ph_3P{=}CHR + 2\ PhN_3 \rightarrow Ph_3P{=}NPh + PhN{=}CHR + 2\ N_2$$

It has recently been shown that silyl azides can also be employed for syntheses of this type.[189, 190] Thus, tertiary phosphines react with trimethylsilyl azide to yield monophosphazenes by the process

$$R_3P + Me_3SiN_3 \xrightarrow{-N_2} R_3P{=}NSiMe_3 \xrightarrow[-20°C]{MeOH{-}H_2SO_4} R_3P{=}NH + MeOSiMe_3$$

N-Silylated aminophosphazenes are formed when trimethylsilyl azide reacts with aminophosphines.[191] Phenyl azide interacts with phosphorus trichloride or with phenylchlorophosphines to give *N*-phenyl monophosphazenes,[192]

$$PCl_3 + PhN_3 \rightarrow Cl_3P{=}NPh + N_2$$

$$PhPCl_2 + PhN_3 \rightarrow PhCl_2P{=}NPh + N_2$$

$$Ph_2PCl + PhN_3 \rightarrow Ph_2ClP{=}NPh + N_2$$

and diphenylphosphinyl azide undergoes a reaction with diphenylchlorophosphine to form a linear phosphazene of structure $Ph_2(O)P[N{=}PPh_2]_3Cl$.[193] This is a white, crystalline substance, m.p. 145°–146°C. Diphenylphosphinyl azide also reacts with diphenylphosphine according to the sequence[194]

$$Ph_2PH + Ph_2P(O)N_3 \xrightarrow{-N_2} Ph_2P(O)NHPPh_2 \xrightarrow[-N_2]{Ph_2P(O)N_3} Ph_2P(O)NHPPh_2{=}NP(O)Ph_2$$

The product of this reaction tautomerizes to $Ph_2P(O)[N{=}PPh_2]_2{-}OH$.

An additional modification of the aryl azide route involves the reaction of this reagent with a bis(dialkylamino)chlorophosphine (XXII). The phos-

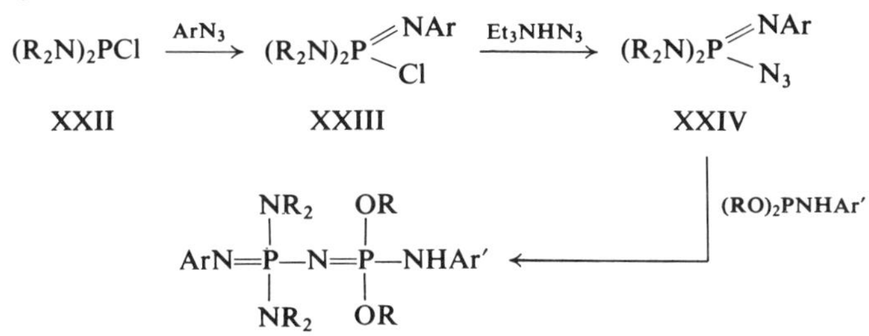

XXII XXIII XXIV XXV

phorodiamidamidic chloride (XXIII) is then converted to the azide (XXIV) with triethylammonium azide, and subsequent treatment with a dialkylaryl-phosphoramidite yields the linear diphosphazene XXV. Azides of type XXIV are reported to be light yellow liquids which are sufficiently stable to be vacuum-distilled or heated at 100°C for 3 hours without decomposition.[195]

e. *The Reaction of Tertiary Phosphines with Other Reagents*

Chloramine T, *O,N*-dibenzoylhydroxylamine, 2- or 4-bromoaniline, and chloramine followed by *n*-butyllithium, react with *tert*-phosphines to give phosphazenes according to the following equations:

$$R_3P + Na\,ClNSO_2C_6H_4Me \rightarrow R_3P{=}N{-}SO_2C_6H_4Me + NaCl \qquad \text{(Ref. 196)}$$

$$Ph_3P + PhCON(H)OCOPh \xrightarrow{100^\circ C} Ph_3P{=}NCOPh + PhCOOH \qquad \text{(Ref. 197)}$$

$$Ph_3P + H_2NC_6H_4Br \rightarrow [Ph_3P{-}NHPh]^{\oplus}Br^{\ominus} \xrightarrow[-NaBr]{NaOH} Ph_3P{=}NPh \qquad \text{(Ref. 198)}$$

$$C_{12}H_{25}PMe_2 + NH_2Cl \rightarrow [C_{12}H_{25}PMe_2NH_2]^{\oplus}Cl^{\ominus} \xrightarrow{n\text{-BuLi}}$$

$$C_{12}H_{25}PMe_2{=}NHLiCl \xrightarrow[-LiCl]{Heat} C_{12}H_{25}PMe_2{=}NH \qquad \text{(Ref. 191)}$$

The dimethyldodecylphosphazene formed in the latter reaction decomposes with water to give ammonia and the phosphine oxide.

f. *Miscellaneous Routes to Linear Phosphazenes*

The following are a number of reactions which do not fall readily into the above categories.

(1) Phosphazenes are formed quantitatively when *N*-acylamidotriphenyl-phosphiniminium bromides are dehydrohalogenated.[199]

$$[Ph_3PNHNHC(O)R]^{\oplus}Br^{\ominus} \xrightarrow[Et_3N]{-HBr} Ph_3P{=}NNHC(O)R$$

(2) Tetrasulfur tetranitride reacts with phenyldichlorophosphine to yield a linear diphosphazene[200] together with longer-chain species.

$$S_4N_4 + PhPCl_2 \rightarrow [PhCl_2P{=}N{-}PCl_2Ph]^{\oplus}[PCl_6]^{\ominus} + PhP(S)Cl_2 + PhCl_2P{=}N{-}P(S)ClPh$$

(3) Phosphorus trichloride and dinitrogen tetroxide react to form a phosphazene.[201]

$$PCl_3 + N_2O_4 \rightarrow Cl_3P{=}N{-}P(O)Cl_2 + PCl_5 + NOCl$$

(4) Monothiophosphoric acid triamide reacts with phosphorus pentachloride to yield phosphazene salts, as in the equation[202]

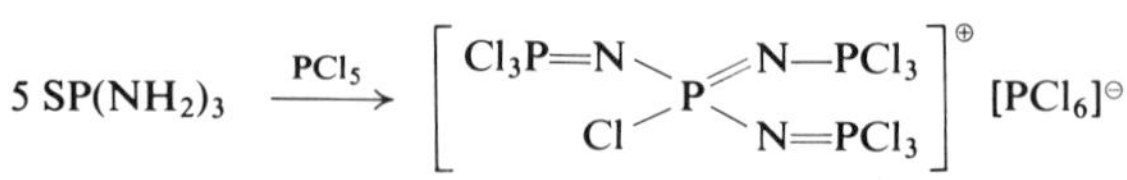

(5) Organometallic phosphazenes can be prepared by the following synthetic sequence,[203–205]

$$(Me_3C)_2PCl + 2\,NH_3 \xrightarrow{-NH_4Cl} (Me_3C)_2P{-}NH_2 \xrightarrow{n\text{-}BuLi} (Me_3C)_2PNHLi$$

$$(Me_3C)_2PNHLi \xrightarrow{Me_3MCl} (Me_3C)_2P{-}NH{-}MMe_2$$

$$(Me_3C)_2P{-}NH{-}MMe_3 \xrightarrow{n\text{-}BuLi} (Me_3C)_2P{-}NLi{-}MMe_3$$

where M is Si or Ge, and M′ is Si, Ge, or Sn.

(6) A monophosphazene is formed when diphenyltrifluorophosphorane reacts with *N*-methyl-hexamethyldisilazane.[95]

$$Ph_2PF_3 + MeN(SiMe_3)_2 \rightarrow Ph_2FP{=}NMe + 2\,Me_3SiF$$

(7) Monophosphazenes can also be isolated from the reaction of phosphorus pentachloride with an aryl-*N*-dichlorosulfonamide.[206]

$$PCl_5 + Cl_2NSO_2Ar \rightarrow 2\,Cl_2 + Cl_3P{=}NSO_2Ar$$

(8) Triaryl phosphites are converted to monophosphazenes by the following process[207]:

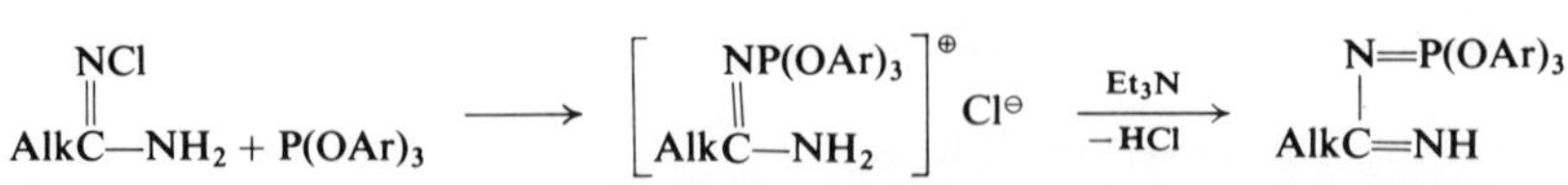

(9) It has been shown that the reaction between thiophosphoric tris(isothiocyanate) and chlorine leads to the formation of *N*-trichloromethyltrichlorophosphazene.[208]

$$S{=}P(NCS)_3 + Cl_2 \rightarrow Cl_3P{=}N{-}CCl_3 + SCl_2 + S_2Cl_2 + Cl{-}S{-}CCl_3$$

(10) Sulfur oxide (fluorosulfonyl) imide reacts with phosphorus pentabromide to yield *P*-tribromo-*N*-fluorosulfonylmonophosphazene.[208a]

$$FSO_2N{=}S{=}O + PBr_5 \rightarrow FSO_2N{=}PBr_3 + SOBr_2$$

(11) Finally, aminophosphines of formula $R_2P{-}NHR'$ (where R=Ph or OEt, and R'=Ph or Pr^n) react rapidly with carbon tetrachloride or bromotrichloromethane to give monophosphazenes.[209]

$$R_2P{-}NHR' + CCl_4 \rightarrow Ph_2\overset{\oplus}{P}(Cl)NHPh\cdot\overset{\ominus}{C}Cl_2 \rightarrow Ph_2P(Cl){=}NPh + CHCl_2$$

4. Synthesis of Some Miscellaneous Phosphorus–Nitrogen Compounds

Phosphorus pentachloride reacts with *N*,*N'*-dimethylurea to yield a heterocyclic molecule which contains phosphorus, nitrogen, and carbon.[210]

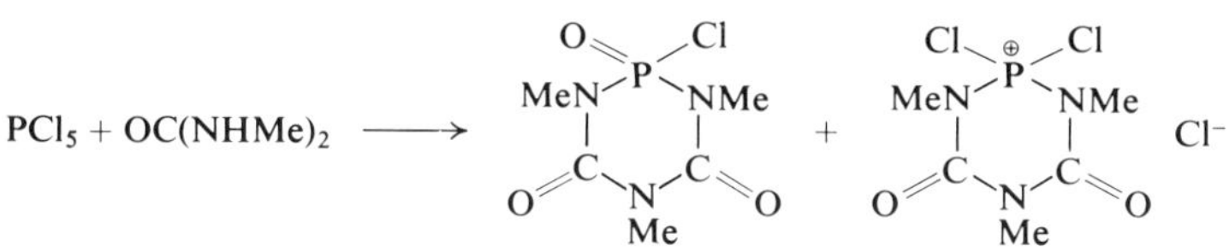

XXVI XXVII

Another interesting series of heterocyclic systems is formed when phosphoric dichlorides interact with *N*,*N'*-dimethylalkylenediamines.[211]

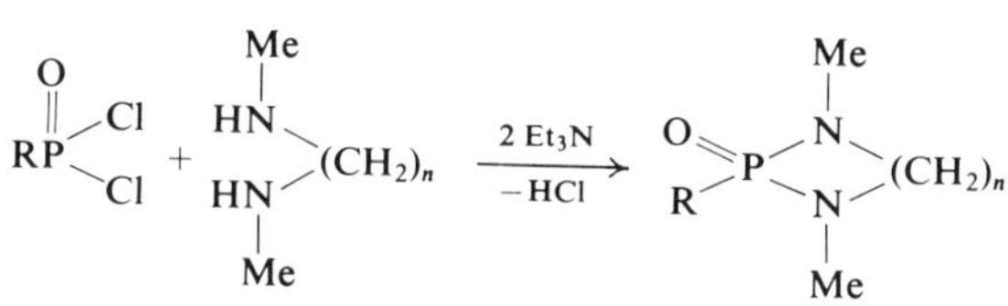

(n = 1 or 2).

Bis(diphenylphosphino)amines react with chloramine and ammonia in the manner shown by the equation[212]

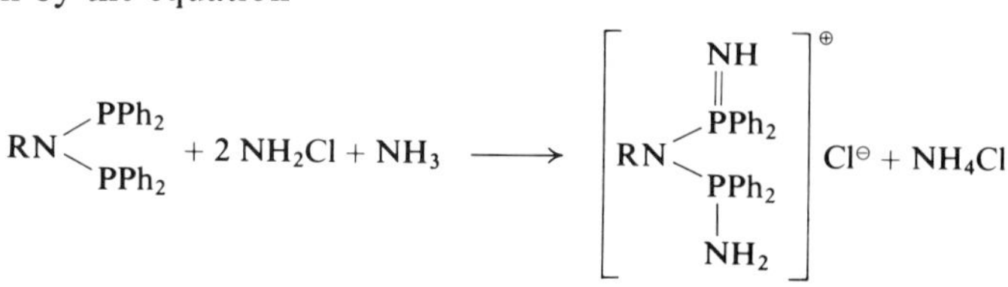

It is also worthwhile to note that the linear phosphazene formed by the interaction of phosphorus pentachloride with sulfuryl amide can be cyclized with hexamethyldisilazane or ammonia to yield two novel heteroaromatic systems, XXVIII and XXIX[213]

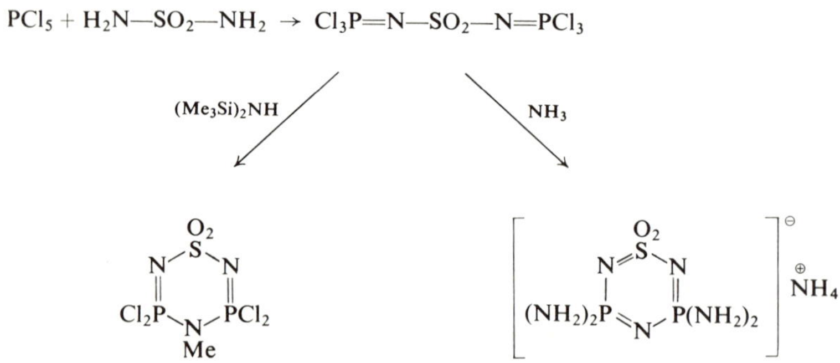

XXVIII XXIX

A number of other interactions which involve the synthesis of phosphorus–nitrogen–sulfur ring systems have been reported.[214]

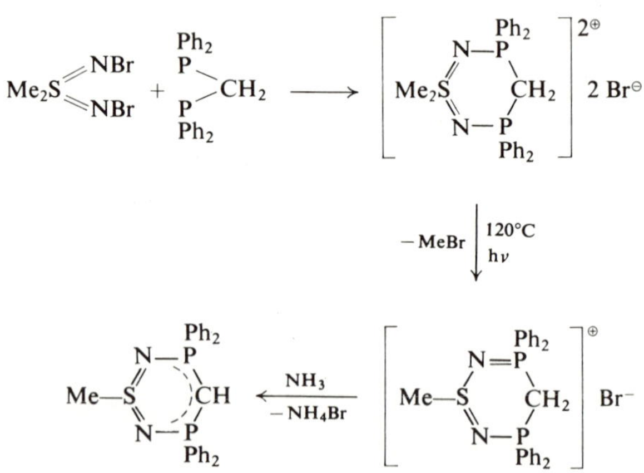

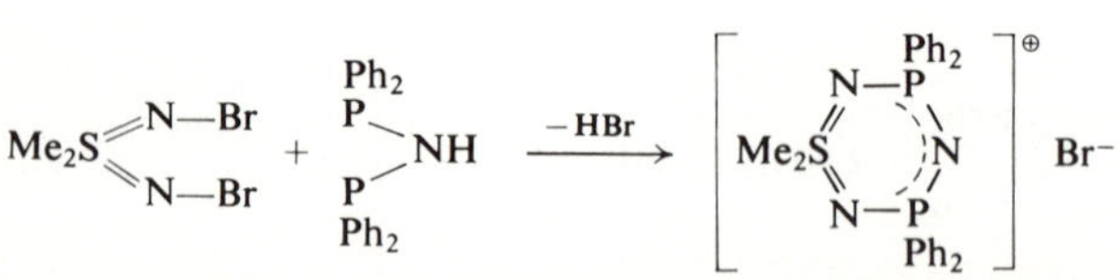

Schmidpeter and co-workers have reported a series of syntheses which give rise to diphospha-*s*-triazines (XXX).[215–221] These reactions are outlined below.

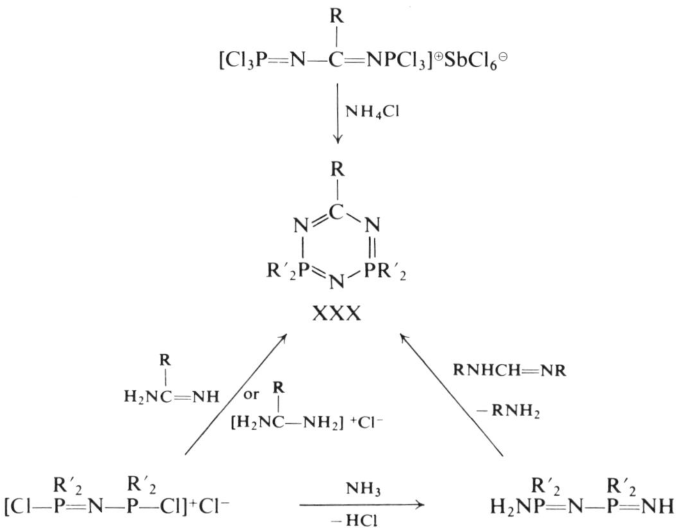

R′=Ph or Cl

A related eight-membered ring system (XXXI) is formed by the following reaction sequence[222]:

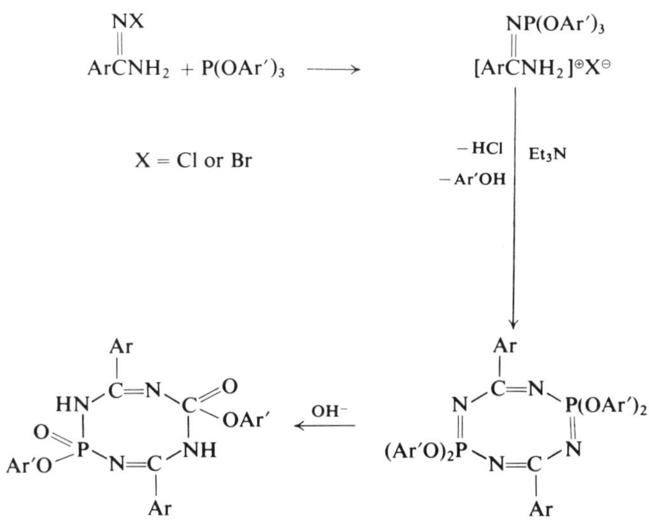

XXXI

Finally, attention is directed to the "phosphams" which are phosphorus–nitrogen compounds of composition PN_2H.[223] They are formed during the ammonolysis of phosphorus pentachloride, or by pyrolysis of aminocyclophosphazenes.

$$[(NH_2)_2P\text{—}N]_3 \rightarrow [(NPNH)_3]_n + 3\,NH_3$$

Phosphams can be decomposed thermally to triphosphorus pentanitride, $(P_3N_5)_n$, and phosphorus nitride, $(PN)_n$.

B. The Reaction between Halophosphoranes and Ammonium Halides

1. Experimental and Manufacturing Considerations

It was mentioned earlier that the reaction between phosphorus pentachloride and ammonium chloride to yield chlorophosphazenes comprises one of the most important synthetic reactions in this field. For this reason, the following detailed experimental outline is given.

a. *Synthesis of* $(NPCl_2)_3$, $(NPCl_2)_4$, *and Higher Homologues* (Standard laboratory procedure based on ref. 13)

A 2-liter flask is fitted with a reflux condenser and is charged with phosphorus pentachloride (625.5 g, 3.0 moles), ammonium chloride (176.5 g, 3.3 moles), and *sym*-tetrachloroethane (1.0 liter). The mixture is heated at the reflux temperature of the solvent for 7.5 hours, during which time hydrogen chloride (12.0 moles) is evolved. Hydrogen chloride fumes can be removed by a water-scrubber device connected to the condenser outlet. After completion of the reaction, the warm solution is filtered to remove ammonium chloride (20.0 g), and the solvent is removed by distillation at reduced pressure. The residue after distillation is a brown "buttery" mixture of oil and crystals (326.5 g, 93.9%). Repeated extraction of this mixture with petroleum (b.p. 40°–60°C) yields a mixture of cyclic $(NPCl_2)_n$ oligomers (235.5 g, 66.9%), and the residue is a light brown mixture of oily linear species (88.0 g, 27%). The lower cyclic chlorophosphazenes are quite stable to atmospheric moisture, whereas the linear analogues hydrolyze slowly.

In a typical preparation, the cyclic oligomer fraction consists of about 37% $(NPCl_2)_3$, 28% $(NPCl_2)_4$, and 35% higher cyclic $(NPCl_2)_n$ species. Several methods are available for the separation of this mixture. Fractional sublimation or distillation permits a facile separation of trimer from higher homologues. Fractional crystallization from light petroleum or *n*-heptane provides a laborious but effective means of separation of the lower homologues. An additional separation method reported by Paddock and co-workers[13] is as follows.

A mixture of predominantly cyclic chlorophosphazenes (7750 g) is extracted with 16.5 liters of 40°–60°C b.p. petroleum to obtain 19.5 liters of a solution of cyclic species (5700 g consisting of 35% trimer and 22% tetramer) and insoluble linear species (2050 g). The petroleum solution is extracted in batches with 98.7% sulfuric acid (2300 ml) in a three-stage fractionation scheme. Trimer is dissolved preferentially by the acid and is recovered from it by diluting the acid solution to approximately 60% sulfuric acid and back-extracting the precipitated trimer with fresh petroleum. Fractional crystallization from this solution then gives trimer (1060 g), m.p. 112°, and a mixed polymer fraction (690 g). Crystallization of the sulfuric acid-insoluble portion yields tetramer (712 g), m.p. 122°, 515 g of a 69% trimer, 20% tetramer mixture, 808 g of a mixture of oil and crystals, and 1915 g of a light yellow oil. The oily fractions are distilled at reduced pressure to yield pentamer, hexamer, heptamer, and octamer. After final recrystallizations from petroleum or *n*-heptane, the trimer melts at 112.8° and the tetramer at 122.8°.

b. *Modified Synthesis for Optimization of Cyclic Chlorophosphazene Formation*

Paddock and co-workers[13] have also shown that the yield of cyclic chlorophosphazenes can be improved dramatically at the expense of the oily linear species by maintaining an excess of ammonium chloride in the reaction mixture. For example, phosphorus pentachloride (62.5 g, 0.3 mole) and ammonium chloride (176.5 g, 3.3 moles) were heated in boiling *sym*-tetrachloroethane, 1.0 liter). The refluxing solvent was returned to the reaction vessel through a bed of phosphorus pentachloride (563 g, 2.7 moles) which was slowly extracted into the reaction mixture during a period of 6 hours. After a total reaction time of 7.5 hours, 11 moles (92%) of hydrogen chloride had been evolved and a 21.7 g excess of ammonium chloride remained. The solution yielded 322.3 g (92.6%) of cyclic species which consisted of approximately 54% of trimer, 11.4% of tetramer, and 29% of higher cyclic species. Similar results were obtained when warm phosphorus pentachloride solution was added slowly to the reaction mixture. It should be noted that an excess of phosphorus pentachloride in the reaction mixture favors the formation of linear species rather than cyclic compounds and that treatment of cyclic chlorophosphazenes with phosphorus pentachloride gives linear compounds of formula $(NPCl_2)_nPCl_5$.

Emsley and Udy[224] have pointed out that the ammonium chloride particle size has a greater influence on the course of the reaction than the rate of addition of phosphorus pentachloride. The use of finely divided ammonium chloride increases the reaction rate and favors the formation of cyclic chlorophosphazenes. Reaction times as short as 2.5 hours have been reported. This influence is a result of the fact that ammonium chloride is virtually insoluble in *s*-tetrachloroethane and surface effects are important. Reactions carried out at high dilutions favor the formation of $(NPCl_2)_3$ at the expense of linear products.

c. *Solvents and Catalysts*

sym-Tetrachloroethane has been used as a solvent for the phosphorus pentachloride–ammonium chloride reaction since the pioneering work of Schenk and Römer in 1924.[8] The advantages of this material as a reaction medium include its high boiling point, its relative inertness to reactants and products, and the fact that it is a solvent for phosphorus pentachloride and chlorophosphazenes. However, *sym*-tetrachloroethane is relatively expensive, highly toxic, and inconvenient to remove by distillation. For these reasons, attempts have been made to find more suitable reaction media and to discover catalysts for the synthesis.

Monochlorobenzene is less toxic than tetrachloroethane, it is cheaper, and it has a lower boiling point. However, at least 25 to 30 hours are required before the reaction is complete at the boiling point of this solvent.[14] It has been shown that the addition of 1–10% of anhydrous metal chlorides reduces the reaction time to 5 to 10 hours in chlorobenzene or 2 to 4 hours in *sym*-tetrachloroethane. Suitable chlorides include stannic chloride, magnesium chloride, zinc chloride, and titanium tetrachloride.[14, 31] Molybdenum pentachloride has a marked catalytic effect.[38] Quinoline also acts as a catalyst in chlorobenzene, and reaction times of 8 hours have been reported.[36] It has also been reported that phosphorus oxychloride (or reagents such as water, which will give rise to $POCl_3$ in the reaction mixture) reduce the reaction time in *sym*-tetrachloroethane to 3.5 hours.[32] Nitrobenzene can also be used as a medium for the synthesis at temperatures in the 140°–150°C region. The reaction apparently requires 5–6 hours for completion under these conditions.[35]

d. *Solid-State Reactions*

In 1942 Steinman, Schirmer, and Audrieth reported a convenient synthesis of chlorophosphazenes using no solvent at all.[10] The technique involved the heating of an intimate mixture of phosphorus pentachloride and ammonium chloride in a large diameter (35–50 mm) Pyrex glass test tube. The outer section of the tube contained ammonium chloride only. After 4–6 hours at 145°–160°C, a 90–95% conversion to cyclic chlorophosphazenes was obtained. This reaction has been reinvestigated more recently in the USSR, where yields of up to 85% were reported.[34, 37] In view of these reports, solid-state reactions would appear to provide a convenient small-scale route for the preparation of chlorophosphazenes.

e. *Safety Considerations*

Although small-scale laboratory preparations of chlorophosphazenes offer no particular hazards, adequate care should be taken to prevent inhalation of reaction products and solvents. *sym*-Tetrachloroethane is a powerful inhalation poison and adequate ventilation is required for large-scale handling of this material. The considerable quantities of hydrogen chloride evolved from these syntheses present a serious problem, and water extractors and scrubbers are necessary to prevent discomfort to personnel and severe corrosion of fume hoods, etc. The lower chlorophosphazenes (particularly the trimer) have a relatively high vapor pressure at room temperature, and they can be detected by their rather pleasant "organic" odor. However, precautions should be taken to avoid inhalation of the vapor, or exposure of the eyes to it. Even small concentrations cause prolonged irritation of the eye membranes, and many researchers have complained of "conjunctivitis" several days after exposure to the trimer. Temporary throat and lung irritation has also been reported following inhalation of these compounds. It follows that chlorophosphazenes should never be heated and volatilized in the open laboratory. Large quantities of chlorophosphazenes should be handled with adequate ventilation and with the use of protective clothing.

f. *The Manufacture of Chlorophosphazenes*

Several companies have undertaken the manufacture of hexachlorocyclotriphosphazene, and a number of patents covering commercial processes have been issued.[14–22] In general, the manufacturing process appears to follow the same procedures as the laboratory-scale preparations described above. The patent literature stresses the need for a slow addition of phosphorus pentachloride to an excess of ammonium chloride in *sym*-tetrachloroethane in order to improve the yield of cyclic rather than linear species.[14, 15, 16] The use of metal halides as catalysts and the use of chlorobenzene as a solvent have also been described in patent specifications.[14] A further modification involves the use of chlorine and phosphorus trichloride in place of phosphorus pentachloride.[20]

No serious technological problems are involved in scaling up the laboratory reaction, and large-scale syntheses can be carried out in the laboratory or pilot plant. The most serious problem is connected with the isolation and purification of the products. For this reason, crude mixtures of chlorophosphazenes can be produced quite cheaply, but the pure cyclic trimer is expensive. One patent[21] has suggested that purification may be effected by thermal depolymerization of the crude mixture to yield distillable trimer and tetramer. Differential partitioning of the chlorophosphazenes between an inert organic solvent and a strong mineral acid is also suggested as a purification process.[17]

All the above factors are of some consequence in view of the commercial potential of many of the phosphazene derivatives derived from hexachlorocyclotriphosphazene. In fact, the lack of an inexpensive route for the manufacture of chlorophosphazenes has been the principal bottleneck for a number of important technological processes.

g. *Bromophosphazenes*

Bromophosphazenes are synthesized with the use of the same general technique as described above for chlorophosphazenes. However, since phosphorus pentabromide is unstable, the synthesis is usually conducted by addition of bromine to a mixture of phosphorus tribromide and ammonium bromide in a hot solvent. Loss of phosphorus bromides by sublimation is avoided by slowly raising the reaction temperature to ~140°C. The following experimental procedure is based on the one reported by John and Moeller.[46]

A mixture of phosphorus tribromide (300 g, 1.1 mole), bromine (200 g, 1.25 mole), and ammonium bromide (300 g, 3.06 mole) in *sym*-tetrachloroethane (600 ml) is heated to 140°–142°C over a period of 5 days. The temperature is maintained at 142°C for 2 days, and additional bromine (150 g, 0.94 mole) is added. After completion of the reaction, unreacted ammonium bromide is filtered off and the solvent is removed by distillation to leave 255.2 g of crude bromophosphazenes. Extraction with hot benzene removes cyclic trimer and tetramer, and solvent evaporation followed by sublimation yields 119 g (52.5%) of a trimer–tetramer mixture. A modification of the reaction by addition of a phosphorus tribromide–bromine mixture to ammonium bromide and tetrachloroethane at 132°–134°C yields 89.9 g (36.9%) of a trimer–tetramer mixture. It should be noted that the monophosphazene $NPBr_2PBr_5$ is formed in substantial yields in these reactions. The cyclic trimer–tetramer mixture can be separated into its components by fractional sublimation or recrystallization from petroleum ether. Bromocyclophosphazenes are less stable to atmospheric moisture than the chloro-derivatives, and purification procedures should be conducted without excessive delay. Bromocyclophosphazenes are used in the manufacture of halogen lamps.[224a]

sym-Tetrabromoethane can be employed as a solvent in place of tetrachloroethane.[46] In fact, evidence exists that halogen exchange occurs in the latter solvent to give mixed chlorobromophosphazenes.[49] As mentioned earlier, hybrid halocyclophosphazenes are formed by related reactions. Mixed chlorobromo species such as these can be separated effectively by vapor-phase chromatography using a silicone oil substrate.

h. *Organophosphazenes*

The preparation of organocyclophosphazenes from organohalophosphoranes and ammonium halides follows the same general techniques as outlined above for halophosphazene synthesis. For example, the following preparation of $(NPClPh)_{3 \text{ and } 4}$ was reported by Grushkin, Sanchez, and Rice.[65]

A solution of $PhPCl_2$ (358 g, 210 moles) in chlorobenzene (2 liters) is chlorinated until it is bright yellow in color, and nitrogen is bubbled through until the yellow color has disappeared. This solution of $PhPCl_4$ is then added slowly over 12 hours to a suspension of ammonium chloride (214 g, 4.0 moles) in boiling chlorobenzene (8 liters). Hydrogen chloride is then removed by a stream of nitrogen. After a further 12 hours of reaction, the ammonium chloride is removed by filtration. Removal of approximately 75% of the solvent from the filtrate causes a slow precipitation of cyclic trimers and tetramers. These are separated by a rather arduous series of crystallizations from *n*-pentane. It should be noted that such preparations are complicated considerably if geometric isomers as well as species with different ring sizes are present, as in the present case. Separation of *cis* and *trans* isomers can often be accomplished by careful recrystallization sequences. The compounds are quite stable to atmospheric moisture.

2. Reaction Mechanisms

The syntheses outlined above can be summarized by the general equation

$$n\,R_2PX_3 + n\,NH_4X \rightarrow (R_2PN)_n + 4n\,HX$$

where R is halogen or an organic group, and X is halogen. In the subsequent discussion it will be assumed that the general reaction mechanisms are similar irrespective of the nature of R or X. It must be pointed out, however, that most of the mechanistic information about this reaction has been derived from systems where R is chlorine.

The following experimental facts are known about this reaction:

(1) Although the main reaction leads to the formation of cyclic phosphazenes of structure $(NPX_2)_n$, linear phosphazene species are nearly always formed at the same time. For example, the reaction between PCl_5 and NH_4Cl in *sym*-tetrachloroethane yields identifiable oily linear species of structures $Cl_3P{=}N{-}PCl_2{=}N{-}PCl_4$,[142] $Cl_3P(NPCl_2)_{\sim 10}{=}N{-}PCl_4$,[13] $[Cl_3P{=}N{-}PCl_3]^{\oplus}[PCl_6]^{\ominus}$ and $[Cl_3P{=}N{-}PCl_2{=}N{-}PCl_3]^{\oplus}[PCl_6]^{\ominus}$.[25, 26, 143, 146] Similarly, the reaction between phosphorus tribromide, bromine, and ammonium bromide yields species which correspond to the structures $[Br_3P{=}N{-}PBr_3]^{\oplus}$-$Br^{\ominus}$ and Br_2PNPBr_5. Furthermore, a compound, which may have the structure $[PhBr_2P{=}N{-}PBr_2Ph]^{\oplus}Br^{\ominus}$ was isolated from the reaction of phenyldibromophosphine and bromine with ammonium bromide,[64] and species of the types $[Ph_2ClP{=}N{-}PClPh_2]^{\oplus}Cl^{\ominus}$ and $[PhCl_2P{=}N{-}PCl_2Ph]^{\oplus}Cl^{\ominus}$ are formed when ammonium chloride reacts with diphenyltrichlorophosphorane,

Ph_2PCl_3, or with phenyltetrachlorophosphorane, $PhPCl_4$.[58, 154, 155] Linear phosphazenes such as these are believed to be intermediates in the synthesis of cyclophosphazenes. Becke-Goehring and co-workers have provided evidence for the ionic structures of many of these linear phosphazenes. These structures are comparable to the $PCl_4^{\oplus}PCl_6^{\ominus}$ formulation for phosphorus pentachloride.

(2) Cyclic phosphazenes are formed when linear species of the type described above are heated in *sym*-tetrachloroethane, particularly in the presence of ammonium halide or ammonia. For example, $(NPCl_2)_3$ can be formed directly from $[Cl_3P{=}N{-}PCl_2{=}N{-}PCl_3]^{\oplus}[PCl_6]^{\ominus}$, and $(NPCl_2)_4$ is formed from $[Cl_3P{=}N{-}PCl_3]^{\oplus}Cl^{\ominus}$ and ammonium chloride.[26]

(3) As discussed earlier, if the reaction between phosphorus pentachloride and ammonium chloride is conducted in such a way that ammonium chloride is always in excess, cyclic chlorophosphazenes are formed in high yields. If phosphorus pentachloride is in excess, the linear species are formed in greater amounts.[13]

(4) Linear phosphazenes can be formed by the treatment of cyclophosphazenes with phosphorus pentachloride.[13] For example, an intimate mixture of $(NPCl_2)_3$ and PCl_5 yields a mixture of linear phosphazenes at 350°C. Linear species of structure $Cl_3P{-}(NPCl_2)_nNH$ are formed when phosphorus pentachloride and ammonium chloride react in an autoclave in the absence of solvent.[24]

(5) Ammonia can replace ammonium chloride in several of the synthetic reactions with halophosphoranes. For instance, diphenyltrichlorophosphorane, Ph_2PCl_3, reacts with liquid ammonia at –40°C to yield diphenylcyclophosphazenes, $(NPPh_2)_n$, plus ammonium chloride.[54] Similarly, Me_2NPCl_4 reacts with ammonia to form $(NP(Cl)NMe_2)_3$.[39]

The above facts strongly favor a mechanism in which linear phosphazenes are formed as intermediates, prior to formation of the cyclic species. A mechanism which has gained wide acceptance was proposed by Becke-Goehring and co-workers.[25, 26] For convenience, three steps can be formulated for the overall reaction. These are monophosphazene formation, linear polymerization, and cyclization.

a. *Monophosphazene Formation*

Phosphorus pentachloride and ammonium chloride are assumed to function as the ionized and dissociated forms, respectively.

$$PCl_5 \rightleftharpoons \overset{\oplus}{P}Cl_4 + \overset{\ominus}{P}Cl_6$$

$$NH_4Cl \rightleftharpoons NH_3 + HCl$$

Ammonia then reacts with the PCl_4 cation according to the sequence

$$[PCl_4]^{\oplus}[PCl_6]^{\ominus} + NH_3 \rightarrow Cl_3P{=}NH + 2\,HCl + PCl_5$$

Paddock and Searle[12] suggested that the trichloromonophosphazene could also be produced through the intermediate formation of ammonium hexachlorophosphate, $[NH_4]^{\oplus}[PCl_6]^{\ominus}$, which would be formed directly from PCl_5 and NH_4Cl.

b. *Linear Polymerization*

The monophosphazene is then believed to react first with $[PCl_4]^{\oplus}[PCl_6]^{\ominus}$ and then with ammonia in a chain-lengthening sequence.

$$Cl_3P{=}NH + [PCl_4]^{\oplus}[PCl_6]^{\ominus} \rightarrow [Cl_3P{=}N{-}PCl_3]^{\oplus}[PCl_6]^{\ominus} + HCl$$

$$[Cl_3P{=}N{-}PCl_3]^{\oplus}[PCl_6]^{\ominus} + NH_3 \rightarrow Cl_3P{=}N{-}PCl_2{=}NH + 3\,HCl + PCl_5$$

A variety of chain-growing steps are then possible, including

$$[Cl_3P{=}N{-}PCl_3]^{\oplus} + Cl_3P{=}NH \rightarrow [Cl_3P{=}N{-}PCl_2{=}N{-}PCl_3]^{\oplus} + HCl$$

$$Cl_3P{=}N{-}PCl_2{=}NH + [PCl_4]^{\oplus} \rightarrow [Cl_3P{=}N{-}PCl_2{=}N{-}PCl_3]^{\oplus} + HCl$$

$$Cl_3P{=}N{-}PCl_2{=}NH + [Cl_3P{=}N{-}PCl_3]^{\oplus} \rightarrow [Cl_3P{=}(NPCl_2)_2{=}N{-}PCl_3]^{\oplus} + HCl$$

Thus, growth of the chains to contain three, four, five, six, or a higher number of phosphorus atoms can be readily understood. In practice, the compound $[Cl_3P{=}N{-}PCl_3]^{\oplus}[PCl_6]^{\ominus}$ is the first species detected,[225,226] and Emsley and Udy[225] have shown that it is formed in over 80% yield within 80 minutes of the start of the reaction and that it is almost insoluble in the medium. Because of this heterogeneity, the next higher members of the series, $[Cl_3P{=}N{-}PCl_2{=}N{-}PCl_3]^{\oplus}[PCl_6]^{\ominus}$ and $[Cl_3P{=}N{-}PCl_2{=}N{-}PCl_2{=}N{-}PCl_3]^{\oplus}[PCl_6]^{\ominus}$, are formed only slowly.

c. *Cyclization*

Cyclization of the chain is possible when a terminal NH group is present. Such a situation can arise whenever a cation–anion complex reacts with ammonia or ammonium chloride.

$$[Cl_3P{=}N{-}PCl_2{=}N{-}PCl_3]^{\oplus}[PCl_6]^{\ominus} + NH_4Cl \rightarrow Cl_3P{=}N{-}PCl_2{=}N{-}PCl_2{=}NH + 3\,HCl + PCl_5$$

$$Cl_3P{=}N{-}PCl_2{=}N{-}PCl_2{=}NH \xrightarrow{-HCl}$$

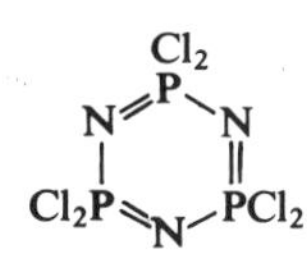

Thus, cyclization would be favored by an excess of ammonium halide and retarded by an excess of phosphorus pentachloride.

An alternative and probably more likely cyclization mechanism is one in which cyclic trimer is formed from the linear *tetramer* by loss of a $PCl_4^{\oplus}$ cation.[225]

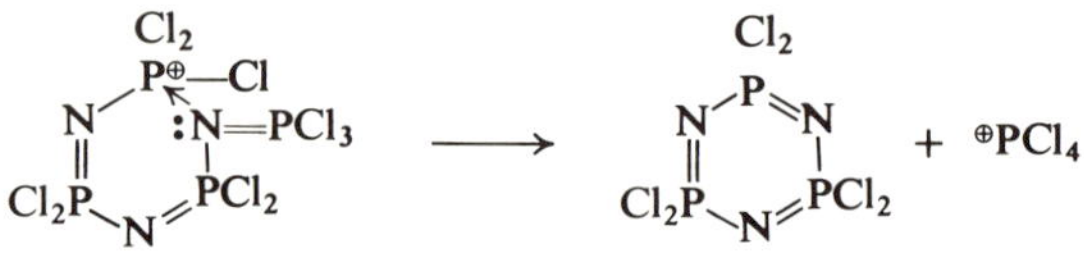

The relative proportions of cyclic trimer, tetramer, pentamer, etc., in the final reaction mixture depend partly on the amounts of linear precursors present and, in turn, on the efficiency of the chain growth. Removal of some linear species from the reaction phase because of their insolubility must also affect the final ratios of cyclic species. Furthermore, the relative amounts of the various cyclic compounds will depend on the steric and statistical ease of cyclization of the linear precursor. This undoubtedly provides an explanation for the fact that cyclic dimers are not isolated, and that the cyclic trimer and tetramer are usually formed in greater amounts than higher cyclic compounds. Steric factors involving the side group (R or halogen) may influence the relative ease of cyclization to trimers and tetramers. It also seems reasonable to suppose that cyclization is favored by the presence of solvents in view of the known effects of dilution on the cyclization constants of many polymers.

References

1. J. Liebig, *Ann. Chem.* **11**, 139 (1834).
2. C. Gerhardt, *Ann. Chim. Phys.* [3] **18**, 188 (1846); *C. R. Acad. Sci.* **22**, 856 (1846).
3. A. Laurent, *C. R. Acad. Sci.* **31**, 356 (1850).
4. H. N. Stokes, *Amer. Chem. J.* **17**, 275 (1895).
5. H. N. Stokes, *Chem. Ber.* **28**, 437 (1895).
6. H. N. Stokes, *Amer. Chem. J.* **19**, 782 (1897).
7. A. Besson and G. Rosset, *C. R. Acad. Sci.* **143**, 37 (1906).
8. R. Schenck and G. Römer, *Chem. Ber.* **57B**, 1343 (1924).
9. H. Moureu and P. Rocquet, *Bull. Soc. Chim. Fr.* [5] **3**, 821 (1936).
10. R. Steinman, F. B. Schirmer, and L. F. Audrieth, *J. Amer. Chem. Soc.* **64**, 2377 (1942).
11. L. F. Audrieth, R. Steinman, and A. D. F. Toy, *Chem. Rev.* **43**, 109 (1943).
12. N. L. Paddock and H. T. Searle, *Advan. Inorg. Chem. Radiochem.* **1**, 347 (1959).
13. L. G. Lund, N. L. Paddock, J. E. Proctor, and H. T. Searle, *J. Chem. Soc., London* p. 2542 (1960).
14. Albright and Wilson, Mfg. Ltd., Brit. Pats. 905,314 and 905,315 (1962).
15. Albright and Wilson, Mfg. Ltd., Brit. Pats. 1,011,237 (1965); 1,017,375 (1966).
16. Albright and Wilson, Mfg. Ltd., Aust. Pats. 230,487 and 233,600 (1958).
17. Albright and Wilson, Mfg. Ltd., U.S. Pat. 3,008,799 (1961).
18. Albright and Wilson, Mfg. Ltd., Can. Pat. 614,267 (1961).

19. French Pat. 1,382,954 (1964); *Chem. Abstr.* **63**, 11,797g (1965).
20. Albright and Wilson, Mfg. Ltd., and Hooker Chemical Corp. Fr. Pat. 1,331,078 (1965).
21. M. C. Taylor, U.S. Pat. 2,872,283 (1959).
22. Compagnie Francaise de Matieres Colorantes, Brit. Pat. 774,694 (1957); U.S. Pat. 2,782,133 (1957).
23. F. G. R. Gimblett, *Chem. Ind.* (*London*) p. 365 (1958).
24. M. Becke-Goehring and G. Koch, *Chem. Ber.* **92**, 1188 (1959).
25. M. Becke-Goehring and E. Fluck, *Angew. Chem.* **74**, 382 (1962).
26. M. Becke-Goehring and W. Lehr, *Z. Anorg. Allg. Chem.* **327**, 128 (1964).
27. M. Yokoyama, *J. Chem. Soc. Jap., Pure Chem. Sect.* **80**, 1189 (1959).
28. H. Saito and M. Kajiwara, *J. Chem. Soc. Jap., Ind. Chem. Sect.* **66**, 618 (1963).
29. E. Kobayashi, *J. Chem. Soc. Jap., Ind. Chem. Sect.* **69**, 618 (1966).
30. E. Kobayashi, *J. Chem. Soc. Jap., Pure Chem. Sect.* **87**, 135 (1966).
31. E. Kobayashi, *J. Chem. Soc. Jap., Ind. Chem. Sect.* **70**, 628 (1967).
32. J. Emsley and P. B. Udy, *Chem. Commun.* p. 633 (1967).
33. S. M. Zhivukhin, *Zh. Neorg. Khim.* **6**, 2414 (1961).
34. S. M. Zhivukhin, V. B. Tolstoguzov, V. V. Kireev, and K. G. Kuznetsova, *Russ. J. Inorg. Chem.* **10**, 178 (1965).
35. M. A. Glushkova, M. M. Ershova, and Yu. A. Buslaev, *Russ. J. Inorg. Chem.* **10**, 1060 (1965).
36. S. M. Zhivukhin, V. B. Tolstoguzov, and M. M. Levitskii, *Zh. Neorg. Khim.* **6**, 2414 (1961).
37. S. M. Zhivukhin, V. V. Kireev, G. S. Kolesnikov, V. P. Popilin, and E. A. Filippov, *Russ. J. Inorg. Chem.* **14**, 548 (1969).
38. R. W. Jenkins and S. Lanoux, *J. Inorg. Nucl. Chem.* **32**, 2453 (1970).
39. A. Schmidpeter, C. Weingand, and E. Hafner-Roll, *Z. Naturforsch. B* **24**, 799 (1969).
40. A. Besson, *C. R. Acad. Sci.* **114**, 1479 (1892).
41. H. Bode, *Z. Anorg. Allg. Chem.* **252**, 113 (1943).
42. H. Bode, *Angew. Chem.* **61**, 438 (1949).
43. R. G. Rice, L. W. Daasch, J. R. Holden, and E. J. Kohn, *J. Inorg. Nucl. Chem.* **5**, 190 (1958).
44. K. John and T. Moeller, *J. Amer. Chem. Soc.* **82**, 2647 (1960).
45. N. E. Bean and R. A. Shaw, *Chem. Ind.* (*London*) p. 1189 (1960).
46. K. John and T. Moeller, *J. Inorg. Nucl. Chem.* **22**, 199 (1961).
47. T. Moeller and P. Nannelli, *Inorg. Chem.* **2**, 659 (1963).
48. G. E. Coxon, D. B. Sowerby, and G. C. Tranter, *J. Chem. Soc., London* p. 5697 (1965).
49. G. E. Coxon and D. B. Sowerby, *J. Chem. Soc., A* p. 1566 (1967).
50. E. Steger and J. Rost, *J. Inorg. Nucl. Chem.* **25**, 732 (1963).
51. H. Rotzche, R. Stahlberg, and E. Steger, *J. Inorg. Nucl. Chem.* **28**, 687 (1966).
52. G. E. Coxon, T. F. Palmer, and D. B. Sowerby, *Inorg. Nucl. Chem. Lett.* **2**, 215 (1966).
53. H. Bode and H. Bach, *Chem. Ber.* **75B**, 215 (1942).
54. C. P. Haber, D. L. Herring, and E. A. Lawton, *J. Amer. Chem. Soc.* **80**, 2116 (1958).
55. C. P. Haber, E. A. Lawton, and C. G. Fitzgerald, U.S. Pat. 2,853,517 (1958).
56. R. A. Shaw and C. Stratton, *Chem. Ind.* (*London*) p. 52 (1959).
57. R. A. Shaw and C. Stratton, *J. Chem. Soc., London* p. 5004 (1962).
58. V. V. Korshak, I. A. Gribova, T. V. Artamonova, and A. N. Bushmarina, *Vysokomole. Soedin.* **2**, No. 3, 377 (1960).
59. I. I. Bezman and J. H. Smalley, *Chem. Ind.* (*London*) p. 839 (1960).
60. F. S. Humiec and I. I. Bezman, *J. Amer. Chem. Soc.* **83**, 2210 (1961).
61. I. I. Bezman, U.S. Pat. 3,098,871 (1963).

62. T. Moeller and P. Nannelli, *Inorg. Chem.* **1**, 721 (1962).
63. T. Moeller and P. Nannelli, *Inorg. Chem.* **2**, 659 (1963).
64. P. Nannelli and T. Moeller, *Inorg. Chem.* **2**, 896 (1963).
65. B. Grushkin, M. G. Sanchez, and R. G. Rice, *Inorg. Chem.* **3**, 623 (1964).
66. B. Grushkin, A. J. Berlin, J. L. McClanahan, and R. G. Rice, *Inorg. Chem.* **5**, 172 (1966).
67. H. T. Searle, *Proc. Chem. Soc., London* p. 7 (1959).
68. A. J. Bilbo, *Z. Naturforsch. B* **15**, 330 (1960).
69. W. R. Grace and Co., Brit. Pat. 981,821 (1965).
69a. V. N. Prons, M. P. Grinblat, and A. L. Klebanskii, *J. Gen. Chem. USSR* **40**, 2108 (1970).
70. H. Saito and M. Kajiwara, *J. Chem. Soc. Jap., Ind. Chem. Sect.* **66**, 618 (1963).
71. M. Yokoyama, A. Fujishiro, H. Saito, and T. Otsuka, *Kogyo Kagaku Zasshi* **68**, 185 (1965).
72. C. D. Schmulbach, C. Derderian, and S. Sahuri, *Inorg. Chem.* **10**, 195 (1971).
73. C. D. Schmulbach and C. Derderian, *J. Inorg. Nucl. Chem.* **25**, 1395 (1963).
74. D. L. Herring and C. M. Douglas, *Inorg. Chem.* **3**, 428 (1964).
75. A. Schmidpeter and J. Ebeling, *Angew. Chem., Int. Ed. Engl.* **7**, 209 (1968).
76. M. Bermann and K. Utvary, *J. Inorg. Nucl. Chem.* **31**, 271 (1969).
77. F. G. Sherif and C. D. Schmulbach, *Inorg. Chem.* **5**, 322 (1966).
78. D. L. Herring, *Chem. Ind. (London)* p. 717 (1960).
79. G. Tesi, C. P. Haber, and C. M. Douglas, *Proc. Chem. Soc., London* p. 219 (1960).
80. G. Tesi, C. M. Douglas, and C. P. Haber, U.S. Pat. 3,087,937 (1963).
81. D. L. Herring and C. M. Douglas, *Inorg. Chem.* **4**, 1012 (1965).
82. K. L. Paciorek, *U.S. Govt. Res. Rep.* **AD 276–920** (1962).
83. R. H. Kratzer and K. L. Paciorek, *Inorg. Chem.* **4**, 1767 (1965).
84. K. L. Paciorek and R. H. Kratzer, U.S. Pat. 3,297,751 (1966).
85. W. J. Birdsall, Ph.D. Thesis, The Pennsylvania State Univ., 1971, p. 59.
86. G. Tesi and C. M. Douglas, *J. Amer. Chem. Soc.* **84**, 549 (1962).
87. G. Tesi and C. M. Douglas, U.S. Pat. 3,065,266 (1962).
88. G. Tesi and C. M. Douglas, *U.S. Govt. Res. Rep.* **AD 236–914** (1961).
89. M. Becke-Goehring, *U.S. Govt. Res. Rep.* **AD 642–394** (1965).
90. D. D. Magnelli, G. Tesi, J. U. Lowe, and W. E. McQuiston, *Inorg. Chem.* **5**, 457 (1966).
91. H. H. Sisler, H. S. Ahuja, and N. L. Smith, *Inorg. Chem.* **1**, 84 (1962).
92. I. T. Gibson and H. H. Sisler, *Inorg. Chem.* **4**, 273 (1965).
93. W. R. Grace and Co., Ger. Pat. 1,189,077 (1965).
94. H. H. Sisler, S. E. Frazier, R. G. Rice, and M. G. Sanchez, *Inorg. Chem.* **5**, 326 (1966).
95. R. Schmutzler, *Z. Naturforsch. B* **19**, 1101 (1964).
96. J. H. Gladstone and J. D. Holmes, *J. Chem. Soc., London* **17**, 225 (1864).
97. J. H. Gladstone and J. D. Holmes, *Ann. Chim. Phys.* [4] **3**, 465 (1864).
98. J. H. Gladstone and J. D. Holmes, *Bull. Soc. Chim. Fr.* [2] **3**, 113 (1865).
99. H. Moureu and G. Wetroff, *C. R. Acad. Sci.* **204**, 51 (1937).
99a. S. Kongpricha and W. C. Preusse, *Inorg. Chem.* **6**, 1915 (1967).
100. J. H. Gladstone and J. D. Holmes, *J. Chem. Soc., London* **17**, 225 (1864).
101. H. Schiff, *Ann. Chem. Pharm.* **103**, 168 (1857).
102. H. N. Stokes, *Amer. Chem. J.* **18**, 629 (1896).
103. H. N. Stokes, *Amer. Chem. J.* **18**, 780 (1896).
104. H. N. Stokes, *Amer. Chem. J.* **20**, 740 (1898).
105. H. N. Stokes, *Z. Anorg. Chem.* **19**, 36 (1899).
106. A. M. de Ficquelmont, M. Magat, and L. Ochs, *C. R. Acad. Sci.* **208**, 1900 (1939).
107. M. Yokoyama, H. Cho, and M. Sakuma, *Kogyo Kagaku Zasshi* **66**, 422 (1963).
108. A. Narath, F. H. Lohman, and O. T. Quimby, *J. Amer. Chem. Soc.* **78**, 4493 (1956).

109. A. Narath, F. H. Lohman, and O. T. Quimby, *J. Amer. Chem. Soc.* **82**, 1009 (1960).
110. F. H. Pollard, G. Nickless, and A. M. Bigwood, *J. Chromatogr.* **11**, 534 (1963).
111. F. H. Pollard, G. Nickless, and R. W. Warrender, *J. Chromatogr.* **9**, 485 (1962).
112. F. H. Pollard, G. Nickless, and R. W. Warrender, *J. Chromatogr.* **9**, 493 (1962).
113. M. L. Nielsen and T. J. Morrow, *Inorg. Syn.* **6**, 97 (1960).
114. D. E. C. Corbridge and E. J. Low, *J. Chem. Soc., London* p. 4555 (1954).
115. K. Lunkwitz and E. Steger, *Z. Anorg. Allg. Chem.* **358**, 111 (1968).
116. B. W. Fitzsimmons, C. Hewlett, and R. A. Shaw, *J. Chem. Soc., London* p. 4459 (1964).
117. A. C. Chapman, W. S. Holmes, N. L. Paddock, and H. T. Searle, *J. Chem. Soc., London* p. 1825 (1961).
118. M. Becke-Goehring, L. Leichner, and B. Scharf, *Z. Anorg. Allg. Chem.* **343**, 154 (1966).
119. M. Becke-Goehring and H. J. Wald, *Z. Anorg. Allg. Chem.* **371**, 88 (1969).
119a. M. Becke-Goehring and M. R. Wolf, *Z. Anorg. Allg. Chem.* **373**, 245 (1970).
120. M. Becke-Goehring and L. Leichner, *Angew. Chem., Int. Ed. Engl.* **3**, 590 (1964).
121. I. N. Zhmurova, U. I. Dolgushina, and A. V. Kirsanov, *Zh. Obshch. Khim.* **37**, 1797 (1967).
121a. W. Haubold and M. Becke-Goehring, *Z. Anorg. Allg. Chem.* **372**, 273 (1970).
122. N. H. Brown, G. W. Fraser, and N. W. A. Sharpe, *Chem. Ind. (London)* p. 367 (1964).
123. R. Schmutzler and G. S. Reddy, *Inorg. Chem.* **4**, 191 (1965).
124. T. Moeller and A. H. Westlake, *J. Inorg. Nucl. Chem.* **29**, 957 (1967).
125. Rottweiler Kunstseidefabrik A. G., Ger. Pat. 1,139,497 (1962).
126. A. Michaelis and W. Kärsten, *Chem. Ber.* **28**, 1237 (1895).
127. A. Michaelis and G. Schroeter, *Chem. Ber.* **27**, 490 (1894).
128. H. W. Gimmel, A. Guenther, and J. F. Morgan, *J. Amer. Chem. Soc.* **68**, 539 (1946).
129. G. C. Demitras, R. A. Kent, and A. G. MacDiarmid, *Chem. Ind. (London)* p. 1712 (1964).
130. R. Schmutzler, *Angew. Chem., Int. Ed. Engl.* **3**, 753 (1964).
131. R. Schmutzler, *Chem. Commun.* p. 19 (1965).
132. E. W. Abel and G. Willey, *Proc. Chem. Soc., London* p. 308 (1962).
133. M. Becke-Goehring and H. Weber, *Z. Anorg. Allg. Chem.* **365**, 185 (1969).
134. H. N. Stokes, *Amer. Chem. J.* **15**, 198 (1893).
135. P. Otto, *Chem. Ber.* **28**, 616 (1895).
136. A. Michaelis and E. Silberstein, *Chem. Ber.* **29**, 716 (1896).
137. M. G. Barlow, M. Green, R. N. Haszeldine, and H. G. Higson, *J. Chem. Soc., A* p. 1592 (1966).
138. M. Green, R. N. Haszeldine, and G. S. A. Hopkins, *J. Chem. Soc., A* p. 1766 (1966).
138a. H. Bock and W. Wiegräbe, *Chem. Ber.* **99**, 1068 (1966).
139. H. G. Ang and H. J. Emeleus, *J. Chem. Soc., A* p. 1334 (1968).
140. G. Peiffer, A. Guillemonat, and J. C. Traynard, *C. R. Acad. Sci., Ser. C* **266**, 400 (1968).
141. J. K. Ruff, *Inorg. Chem.* **6**, 2108 (1967).
142. M. Becke-Goehring, *Angew. Chem.* **73**, 246 (1961).
143. M. Becke-Goehring and W. Lehr, *Chem. Ber.* **94**, 1591 (1961).
144. M. Becke-Goehring, E. Fluck, and W. Lehr, *Z. Naturforsch. B* **17**, 126 (1962).
145. M. Becke-Goehring and W. Lehr, *Z. Anorg. Allg. Chem.* **325**, 287 (1963).
146. M. Becke-Goehring and E. Fluck, *Inorg. Syn.* **8**, 94 (1966).
147. M. Becke-Goehring and B. Scharf, *Z. Anorg. Allg. Chem.* **353**, 320 (1967).
148. W. Lehr and M. Schwartz, *Z. Anorg. Allg. Chem.* **363**, 43 (1968).
149. A. Ya. Yakubovich, N. I. Shvetsov, I. V. Lebedeva, and V. S. Yakubovich, *Russ. J. Inorg. Chem.* **8**, 953 (1963).
150. E. Fluck, *Z. Anorg. Allg. Chem.* **320**, 64 (1963).

151. A. Schmidpeter and C. Weingand, *Angew. Chem., Int. Ed. Engl.* **8**, 615 (1969).
152. J. A. Parkins, WADC Tech. Rep. 59–251 (1959).
153. G. M. Nichols, *Rep. Conf. High Temp. Polym. Fluid Res.*, 1962 ASD-TDR-62-372 (1962); U.S. Pat. 3,249,397 (1966).
154. E. Fluck and F. L. Goldmann, *Chem. Ber.* **96**, 3091 (1963).
155. E. Fluck and R. M. Reinisch, *Chem. Ber.* **96**, 3085 (1963).
156. S. E. Frazier and H. H. Sisler, *Inorg. Chem.* **5**, 925 (1966).
157. E. F. Moran and D. P. Reider, *Inorg. Chem.* **8**, 1550 (1969).
158. E. Fluck, *Z. Anorg. Allg. Chem.* **315**, 191 (1962).
159. E. F. Moran, *J. Inorg. Nucl. Chem.* **30**, 1405 (1968).
160. H. R. Allcock and P. Ostrowski, unpublished work (1968).
160a. W. M. Douglas, M. Cooke, M. Lustig, and J. K. Ruff, *Inorg. Nucl. Chem. Lett.* **6**, 409 (1970).
161. C. M. Douglas, *U.S. Govt. Res. Rep.* **AD 276–920** (1962).
162. A. V. Kirsanov, *Izv. Akad. Nauk SSSR, Otd. Khim. Nauk* p. 646 (1954); *Chem. Abstr.* **49**, 13163 (1955).
163. I. N. Zhmurova, A. P. Martynyuk, and A. V. Kirsanov, *J. Gen. Chem. USSR* **37**, 1789 (1967).
164. V. A. Shokol, V. F. Gamaleya, and V. P. Kukhar, *J. Gen. Chem. USSR* **40**, 520 (1970).
165. H. W. Roesky and L. F. Grimm, *Chem. Ber.* **102**, 2319 (1969).
166. I. N. Zhmurova and A. V. Kirsanov, *J. Gen. Chem. USSR* **31**, 3440 (1961).
167. I. N. Zhmurova and A. V. Kirsanov, *J. Gen. Chem. USSR* **32**, 2540 (1962).
168. V. A. Shokol, V. F. Gamaleya, and G. I. Derkach, *J. Gen. Chem. USSR* **38**, 1815 (1968).
169. H. J. Bestmann, H. Buckschewski, and H. Leube, *Chem. Ber.* **92**, 1345 (1959).
170. M. Becke-Goehring, W. Gehrmann, and W. Goetze, *Z. Anorg. Allg. Chem.* **326**, 127 (1963).
171. F. Ephraim and M. Gurewitsch, *Chem. Ber.* **43**, 138 (1910).
172. A. V. Kirsanov, *Izv. Akad. Nauk SSSR, Otd. Khim. Nauk* p. 426 (1950); *Chem. Abstr.* **45**, 1503 (1951).
173. A. V. Kirsanov and I. N. Zhmurova, *Zh. Obshch. Khim.* **28**, 2478 (1958); *Chem. Abstr.* **53**, 3118 (1959).
174. A. V. Kirsanov, *Zh. Obshch. Khim.* **22**, 1346 (1952); *Chem. Abstr.* **47**, 5836 (1953).
175. M. Becke-Goehring, T. Mann, and H. D. Euler, *Chem. Ber.* **94**, 193 (1961).
176. V. A. Shokol, G. A. Golik, and G. I. Derkach, *Zh. Obshch. Khim.* **38**, 871 (1968).
177. M. Lustig, *Inorg. Chem.* **8**, 443 (1969).
178. L. Horner and H. Hoffmann, *Angew. Chem.* **69**, 478 (1956).
179. L. Horner and H. Oediger, *Ann. Chem.* **627**, 142 (1959).
180. H. Zimmer and G. Singh, *J. Org. Chem.* **28**, 483 (1963).
181. H. Bock and W. Wiegräbe, *Angew. Chem.* **74**, 327 (1962).
182. I. N. Zhmurova and A. P. Martynyuk, *J. Gen. Chem. USSR* **37**, 2577 (1967).
183. I. N. Zhmurova, A. P. Martynyuk, and G. I. Derkach, *J. Gen. Chem. USSR* **38**, 161 (1968).
184. M. Becke-Goehring and E. Fluck, *Inorg. Syn.* **8**, 92 (1966).
185. H. Staudinger and J. Meyer, *Helv. Chim. Acta* **2**, 635 (1919).
186. H. Staudinger and E. Hauser, *Helv. Chim. Acta* **4**, 861 (1921).
187. I. N. Zhmurova, A. A. Tukhar, and R. I. Yurchenko, *J. Gen. Chem. USSR* **39**, 2150 (1969).
188. H. Staudinger and J. Meyer, *Helv. Chim. Acta* **2**, 619 (1919).
189. L. Birkofer and S. M. Kim, *Chem. Ber.* **97**, 2100 (1964).
190. T. W. Rave and H. R. Hayes, *J. Org. Chem.* **31**, 2894 (1966).

191. O. J. Scherer and G. Schieder, *J. Organometal. Chem.* **19**, 315 (1969).
192. R. K. Bunting and C. D. Schmulbach, *Inorg. Chem.* **5**, 533 (1966).
193. K. L. Paciorek, *Inorg. Chem.* **3**, 96 (1964).
194. R. H. Kratzer and K. L. Paciorek, *J. Org. Chem.* **32**, 853 (1967).
195. V. A. Gilyarov, R. V. Kudryavtsev, and M. I. Kabachnik, *J. Gen. Chem. USSR* **38**, 351 (1968).
196. F. G. Mann, *J. Chem. Soc., London* p. 1209 (1940).
197. H. H. Wassermann and R. C. Koch, *Chem. Ind. (London)* p. 1014 (1956).
198. L. Horner and H. Hoffmann, *Angew. Chem.* **69**, 478 (1956).
199. C. C. Walker and H. Shechter, *J. Amer. Chem. Soc.* **90**, 5626 (1968).
200. E. Fluck and R. M. Reinisch, *Z. Anorg. Allg. Chem.* **328**, 165 (1964).
201. M. Becke-Goehring, A. Debo, E. Fluck, and W. Goetze, *Chem. Ber.* **94**, 1383 (1961).
202. H. P. Latscha, W. Haubold, and M. Becke-Goehring, *Z. Anorg. Allg. Chem.* **339**, 82 (1965).
203. O. J. Scherer and G. Schieder, *Angew Chem., Int. Ed. Engl.* **7**, 75 (1968).
204. O. J. Scherer and P. Klusmann, *Angew. Chem., Int. Ed. Engl.* **7**, 541 (1968); *Z. Anorg. Allg. Chem.* **370**, 171 (1969).
205. O. J. Scherer and G. Schieder, *Chem. Ber.* **101**, 4184 (1968).
206. E. S. Levchenko and A. V. Kirsanov, *Zh. Obshch. Khim.* **29**, 1813 (1959).
207. G. I. Derkach and M. V. Kolotilo, *Zh. Obshch. Khim.* **36**, 1437 (1966).
208. E. Fluck and F. L. Goldmann, *Z. Anorg. Allg. Chem.* **356**, 307 (1968).
208a. H. W. Roesky, *Z. Anorg. Allg. Chem.* **367**, 151 (1969).
209. R. F. Hudson, R. J. G. Searle, and F. H. Devitt, *J. Chem. Soc., C* p. 1001 (1966).
210. H. P. Latscha, *Z. Anorg. Allg. Chem.* **346**, 166 (1966).
211. H. Ulrich, B. Tucker, and A. A. R. Sayigh, *J. Org. Chem.* **32**, 1360 (1967).
212. J. M. Kanamueller and H. H. Sisler, *Inorg. Chem.* **6**, 1765 (1967).
213. M. Becke-Goehring, K. Bayer, and T. Mann, *Z. Anorg. Allg. Chem.* **346**, 143 (1966).
214. R. Appel, D. Hänssgen, and B. Ross, *Z. Naturforsch. B* **22**, 1354 (1967).
215. A. Schmidpeter and J. Ebeling, *Angew. Chem.* **79**, 100 (1967).
216. A. Schmidpeter and J. Ebeling, *Angew. Chem.* **79**, 534 (1967).
217. A. Schmidpeter and J. Ebeling, *Chem. Ber.* **101**, 2602 (1968).
218. A. Schmidpeter and N. Schindler, *Z. Anorg. Allg. Chem.* **362**, 281 (1968).
219. A. Schmidpeter and R. Boehm, *Z. Anorg. Allg. Chem.* **362**, 65 (1968).
220. A. Schmidpeter and J. Ebeling, *Chem. Ber.* **101**, 3883 (1968).
221. A. Schmidpeter and J. Ebeling, *Chem. Ber.* **101**, 815 (1968).
222. M. V. Kolotilo, *J. Gen. Chem. USSR* **35**, 1007 (1965).
223. M. C. Miller and R. A. Shaw, *J. Chem. Soc., London* p. 3233 (1963).
224. J. Emsley and P. B. Udy, *J. Chem. Soc., A* p. 768 (1971).
224a. J. M. Rees, *Lighting Res. Technol.* **2**, 257 (1970).
225. J. Emsley and P. B. Udy, *J. Chem. Soc., A* p. 3025 (1970).
226. G. Wunsch, R. Schiedermaier, V. Kiener, E. Fluck, and G. Heckmann, *Chem. Z.* **94**, 832 (1970).

Chapter 5

HYDROLYSIS OF PHOSPHAZENES AND PHOSPHAZANES

A. General Features

The phosphorus–nitrogen skeleton in cyclo- and polyphosphazenes is relatively stable to hydrolysis, but the cyclophospha*zane* framework is usually broken down more readily during hydrolytic attack. As might be expected, halophosphazenes, $(N{=}PF_2)_n$, $(N{=}PCl_2)_n$, and $(N{=}PBr_2)_n$, are hydrolyzed by water or aqueous base, but the reactions are slower than those of the analogous phosphorus halides. On the other hand, organophosphazenes, such as $(N{=}PR_2)_n$, $[N{=}P(OR)_2]_n$, or $[N{=}P(NHR)_2]_n$, are much more stable to hydrolysis, and it appears that hydrolytic stability in a phosphazene often depends on whether the side-group structure is stable to aqueous media. It should also be noted that, when hydrolysis of a cyclo- or polyphosphazene does occur, a phospha*zane* is usually formed as an intermediate species, and this then undergoes skeletal breakdown. The hydrolytic reactions are of practical importance in view of the potential use of phosphorus–nitrogen structures in polymers and in biologically active compounds.

B. Hydrolysis of Halophosphazenes

1. Hydrolysis of Fluorophosphazenes

The fluorophosphazenes, $(NPF_2)_3$ and $(NPF_2)_4$, react slowly with water, although the tetramer reacts faster than the trimer.[1] For example, in methanolic alkali, the tetramer is hydrolyzed rapidly at room temperature whereas comparable hydrolysis is achieved with the trimer only after 1 hour in a sealed tube at 100°C.[1] Hydrolysis of $(NPF_2)_4$ in a water–ether interfacial system yields octahydroxycyclotetraphosphazene (tetraphosphonitrilic acid), which can be isolated as the solid dihydrate $[NP(OH)_2]_4 \cdot 2\ H_2O$. Poly(difluorophosphazene) is a rubbery elastomer which hydrolyzes slowly with liquefaction when exposed to the atmosphere for several hours.[2]

2. Hydrolysis of Chlorophosphazenes

a. *Chlorocyclophosphazenes*

The lower chlorocyclophosphazenes are white, crystalline materials which, in the solid state, are quite stable to atmospheric moisture and to water. However, in solution they hydrolyze quite rapidly. The hydrolysis reactions of $(NPCl_2)_3$ and $(NPCl_2)_4$ were studied first in the 1890s by Stokes,[3-6] and more recent work has extended these studies.[7-10a] Hydrolysis occurs in acidic, basic, and neutral solution, with the basic reaction being the fastest.[9] The initial step involves hydrolysis of the phosphorus–chlorine bonds to give a hydroxycyclophosphazene (I), which then undergoes proton migration to yield the appro-

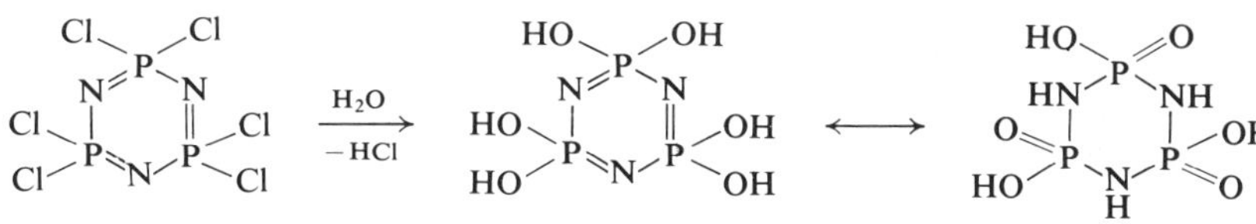

I II

priate hydroxyoxophosphazane (II). Subsequent hydrolysis of the phosphazane yields phosphoric acid and ammonia.

$$[HN\text{—}P(O)OH]_3 \xrightarrow{H_2O} 3\ NH_3 + 3\ H_3PO_4$$

In practice, hexachlorocyclotriphosphazene can be converted to an unstable, partially hydrolyzed product, $N_3P_3Cl_4(OH)_2$, when an ethereal solution of $(NPCl_2)_3$ is shaken with water. The dihydroxy compound remains in the ether, while hydrochloric acid and the hexahydroxy derivative, $[NP(OH)_2]_3$, are found in the aqueous layer.[3,4] The hexahydroxyphosphazene (I) is quite unstable in aqueous acidic media. Proton migration occurs to give II, followed by ring cleavage and skeletal degradation.[5,9]

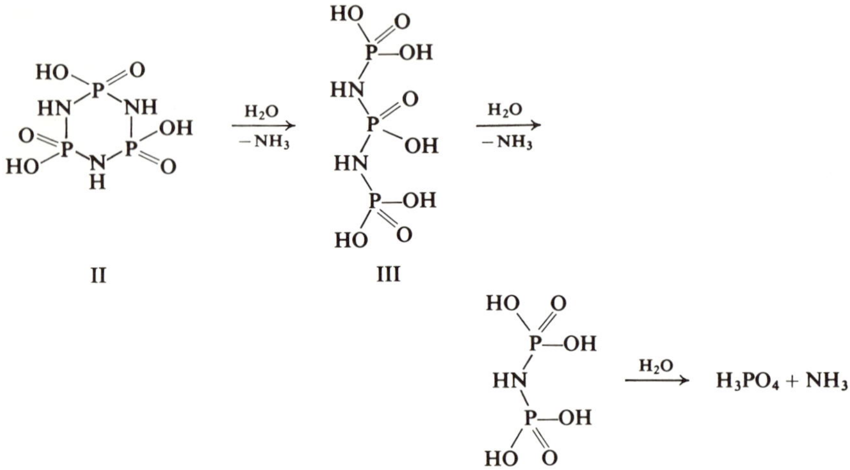

II III IV

Both the intermediate hydrolysis products (III and IV) can be isolated as their salts.[5] In alkaline aqueous media, salts of II are formed. For example, the trisodium salt is isolated when an ethereal solution of $(NPCl_2)_3$ is shaken with an aqueous solution of sodium acetate.[5] The isolation of trimetallic salts of II provides evidence in favor of the hydroxyoxophosphazane structure.

Octachlorocyclotetraphosphazene, $(NPCl_2)_4$ (V), is hydrolyzed more rapidly than the trimer.[3, 4, 6] The product is the surprisingly stable octahydroxycyclotetraphosphazene (VI) or the corresponding hydroxyoxophos-

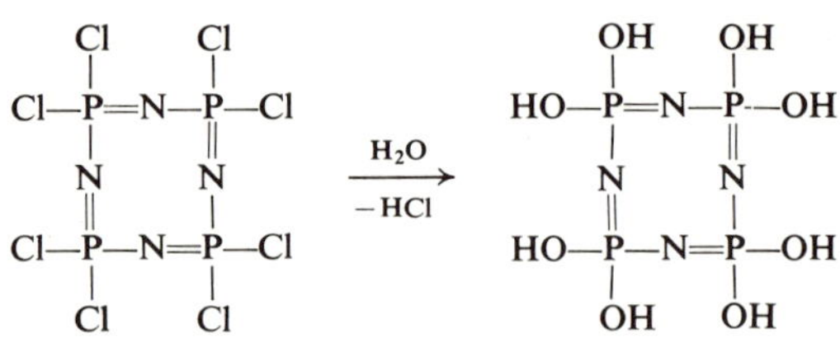

V VI

phazane. This derivative can be isolated as the colorless, crystalline dihydrate by treatment of an ethereal solution of V with water. Unlike the trimeric hydroxyphosphazene, VI is unusually stable to treatment with aqueous acid, although prolonged reaction under forcing conditions will bring about degradation to ammonia and phosphoric acid.[6] A similar, two-step degradation sequence occurs during the hydrolysis of $(NPCl_2)_{5\text{ to }11}$.[11]

The hydrolysis of chlorophosphazenes in the presence of aqueous pyridine has also been studied.[8, 10] Hydrolysis is so rapid that the reaction can be used as an analytical method for the determination of chlorine in phosphazenes.[10] Labile pyridine salts of the hydroxyphosphazene are formed, along with the hydrohalide pyridine salt.[8]

b. *Organochlorocyclophosphazenes*

Organochlorocyclophosphazenes undergo similar hydrolysis reactions. Pentaphenylchlorocyclotriphosphazene[12] (VII) and non-geminal hexaphenyldichlorocyclotetraphosphazene[13] (VIII) are hydrolyzed by pyridine–water

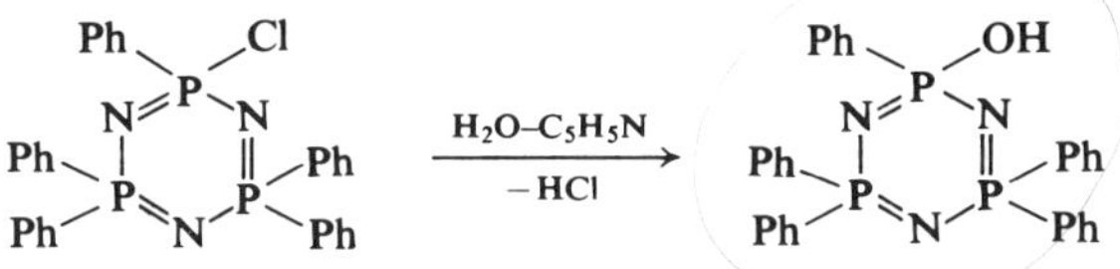

VII

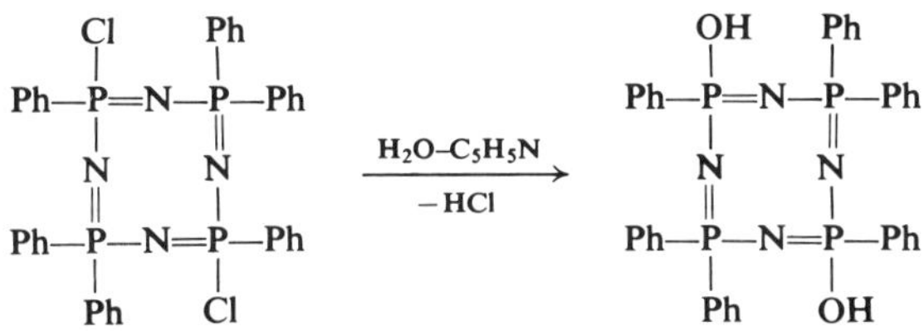

VIII

mixtures to the phenylhydroxyphosphazenes. The product in both cases is the phosphaz*ene* rather than the phosphaz*ane*-type species. However, it has been reported that the hydrolysis of pentaphenoxychlorocyclotriphosphazene (IX) in aqueous ethanolic potassium hydroxide solution yields the phosphazadiene (X).[14]

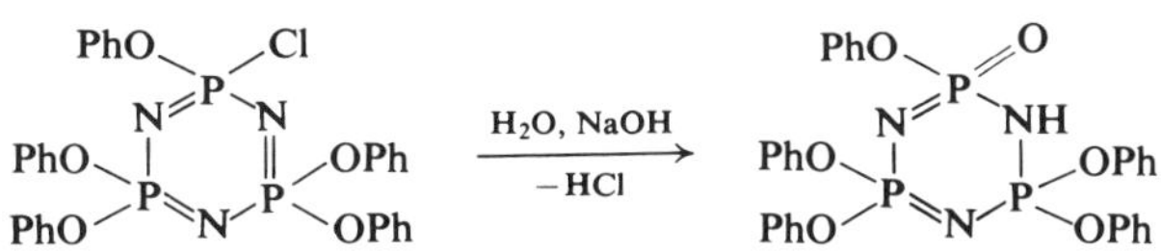

IX X

A kinetic study has been made by Schmulbach and Miller[15] of the hydrolysis of VII in pyridine–water. Pseudo-first-order kinetics were observed in which the rates were independent of the water concentration, but dependent on the pyridine concentration. Furthermore, it was found that VII, ammonium fluoride, water, and pyridine react in acetone to form $N_3P_3Ph_5F$ and $Ph_5P_3N_3ON_3P_3Ph_5$. A mechanism was proposed for the hydrolysis which required the formation of a pyridine–phosphazene ionic intermediate, and the subsequent hydrolysis of this species.

$$Ph_5ClP_3N_3 + C_5H_5N \underset{k_2}{\overset{k_1}{\rightleftharpoons}} Ph_5P_3N_3\cdot C_5H_5N^{\oplus} + Cl^{\ominus}$$

$$Ph_5P_3N_3\cdot C_5H_5N^{\oplus} + H_2O \xrightarrow{k_3} Ph_5(OH)P_3N_3 + C_5H_5NH^{\oplus}$$

An alternative mechanism, in which ionization of chloride ion from phosphorus precedes attack by a pyridine–water nucleophile, was considered to be inconsistent with the experimental data.

Cleavage of the phosphazene ring occurs when compounds such as XI or XII are heated with aqueous acid.[16–18] The products include diphenyl-

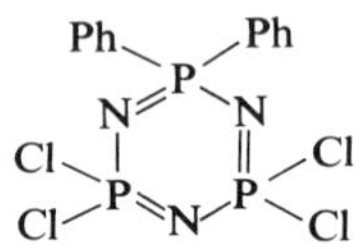

XI

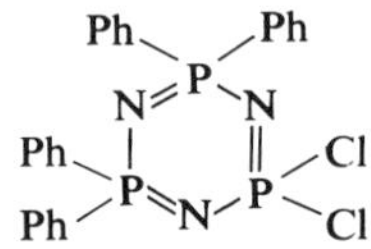

XII

phosphinic acid, $Ph_2P(O)OH$, phosphoric acid, and an ammonium salt. Such degradations have been used to prove the geminal structure of XI and XII.

Examples are also known in which both chloro and amino (NH_2) groups are cohydrolyzed from the cyclophosphazene skeleton. The compound $N_3P_3Cl_4$-$(NH_2)_2$ is hydrolyzed by aqueous sodium acetate to yield the trisodium salt of II; and $N_4P_4Cl_6(NH_2)_2$, $N_4P_4Cl_4(NH_2)_4$, and $[NP(NH_2)_2]_4$ react with water to form ammonium salts of the hydroxyphosphazene, VI.[19]

c. *High-Polymeric Chlorophosphazenes*

Poly(dichlorophosphazene), $(NPCl_2)_n$, is a rubbery elastomer which hydrolyzes slowly in contact with atmospheric moisture. The ultimate products are phosphates, ammonia, and hydrochloric acid. The hydrolysis process is accompanied by a gradual loss of elasticity of the material, and it is assumed that the initial reaction is a hydrolytic cross-linking process. Subsequent steps

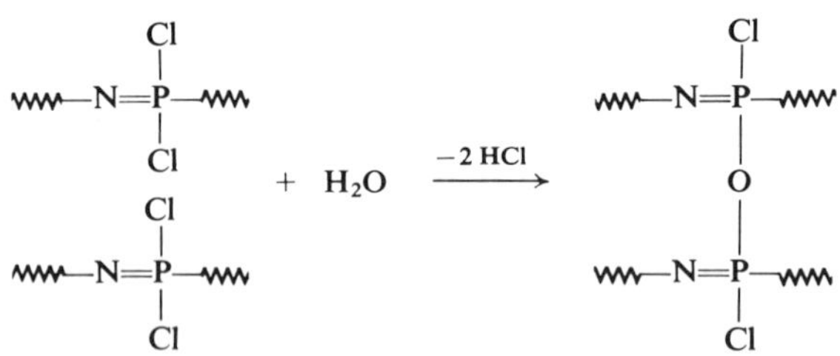

involve further hydrolysis of phosphorus–chlorine bonds and cleavage of the chains. Experiments have been conducted on the hydrolysis of poly(dichlorophosphazene), with the polymer swelled in a 5% aqueous acetone medium at 25°C. The first 58% of the chlorine is removed by a first-order-type process. Thereafter, the hydrolysis rate slows considerably. However, at elevated temperatures, total hydrolysis can be effected.[20] Partial hydrolysis of chlorophosphazene low polymers has been reported to give rise to rubbery polymeric materials.[21, 22]

d. *Chloromonophosphazenes*

The monophosphazene XIII reacts explosively with small amounts of water to form phosphoric acid, hydrochloric acid, sulfuric acid, and ammonium

$$\underset{\textbf{XIII}}{Cl_3P{=}N{-}SO_2{-}N{=}PCl_3} + 10\,H_2O \rightarrow 2\,H_3PO_4 + 4\,HCl + H_2SO_4 + 2\,NH_4Cl$$

chloride.[23] With a large excess of water, the reaction pathway is different, as shown by the equation

$$Cl_3P{=}N{-}SO_2{-}N{=}PCl_3 + 16\,H_2O \rightarrow 2\,H_3PO_4 + 6\,HCl + H_2N{-}SO_2{-}NH_2$$

It appears, that, in molecules of this type, the phosphorus–nitrogen bond is less stable hydrolytically than the sulfur–nitrogen bond, and this is confirmed by the acidic hydrolytic behavior of XIV and XV.[23] However, this instability is

$$\underset{\textbf{XIV}}{Cl_3P{=}N{-}SO_2{-}Cl} + 5\,H_2O \rightarrow 4\,HCl + H_3PO_4 + H_2N{-}SO_2{-}OH$$

$$\underset{\textbf{XV}}{Cl_3P{=}N{-}SO_2{-}Ph} + 4\,H_2O \rightarrow 3\,HCl + H_3PO_4 + H_2N{-}SO_2{-}Ph$$

associated with the presence of phosphorus–halogen bonds, since the phenoxymonophosphazene XVI undergoes sulfur–nitrogen rather than phosphorus–nitrogen cleavage during hydrolysis with water.[23]

$$\underset{\textbf{XVI}}{(PhO)_3P{=}N{-}SO_2{-}OPh} + 3\,H_2O \rightarrow (PhO)_2P(O){-}NH_2 + 2\,PhOH + H_2SO_4$$

C. Hydrolysis of Cyclophosphazanes

Hydroxyoxophosphazanes (metaphosphimates) are formed as intermediate species during the hydrolysis of several cyclophosphazenes. As discussed earlier, the cyclic tetramer, VI, is relatively stable to acid hydrolysis, whereas the trimer, II, undergoes rapid hydrolytic degradation in acidic solutions. The mechanism of this reaction has been investigated and a number of different hydrolysis products have been detected.[5, 24–27] The rather complex mixtures of reaction products were separated and identified by paper chromatography.[24–27] A likely reaction scheme for acid hydrolysis is shown in Scheme I, with the main reaction sequence indicated by solid arrows.

One of the more interesting facts about the observed reaction is that the main hydrolysis sequence involves the replacement of NH units in the phosphazane ring by oxygen bridges, rather than direct ring cleavage followed by chain breakdown. However, the introduction of oxygen into the ring progressively

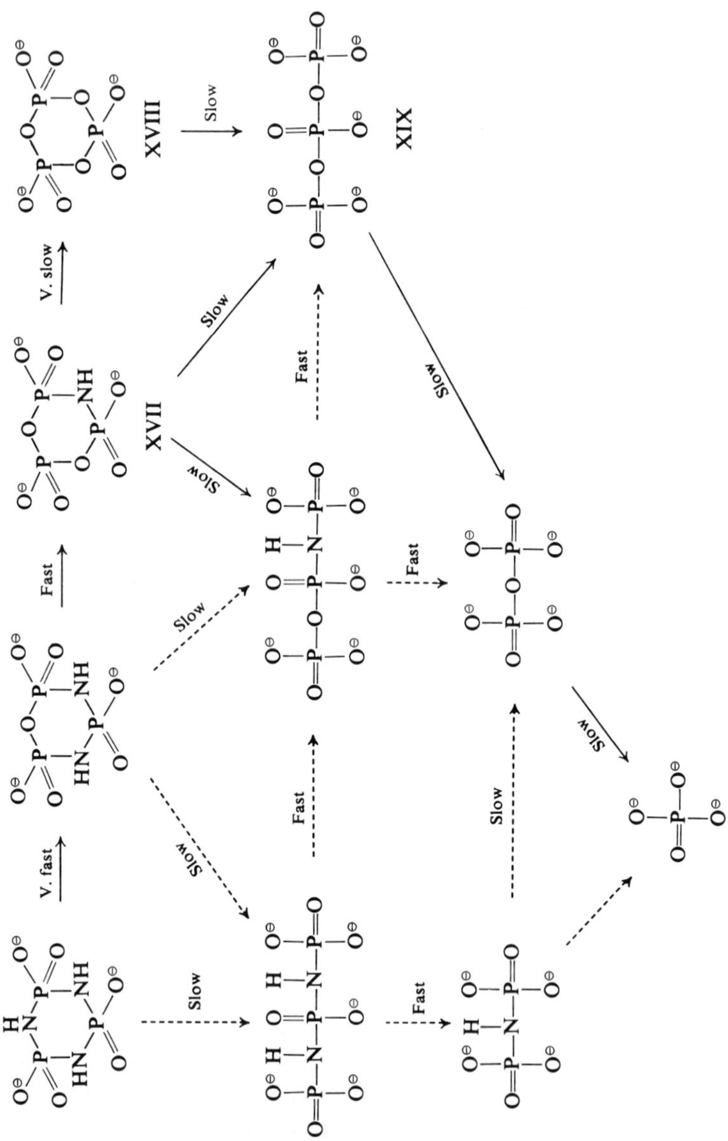

SCHEME I

raises the activation energy and lowers the hydrolysis rate.[27] For this reason, ring cleavage of XVII to give XIX is a faster reaction than conversion of XVII to trimetaphosphate (XVIII) followed by ring breakdown.

D. Hydrolysis of Organophosphazenes and Organophosphazanes

1. Hydrolysis of Alkyl- and Arylphosphazenes

Alkyl- and arylphosphazenes are generally stable to hydrolysis, especially in the solid state. Poly[bis(trifluoromethyl)phosphazene], $[NP(CF_3)_2]_n$, as a solid is unaffected by boiling sulfuric, nitric, and perchloric acids, but is partly hydrolyzed by aqueous 10% sodium hydroxide solution.[28] The phenylphosphazenes, $[NPPh_2]_{3\text{ and }4}$ are degraded to diphenylphosphinic acid, $Ph_2P(O)OH$, and ammonia, when heated with aqueous acid.[29]

2. Hydrolysis of Alkoxyphosphazenes

a. *Qualitative Observations*

Alkoxycyclophosphazenes of formula $[NP(OR)_2]_{3\text{ or }4}$ are stable to water when OR is OMe, OEt, OPr^i, OBu, or OCH_2Ph, but they are decomposed by hot, aqueous 10% hydrochloric acid.[30] Fluoroalkoxycyclophosphazenes, $[NP(OCH_2CF_3)_2]_{3\text{ and }4}$, are unaffected when the solid or liquid compounds are treated with boiling concentrated hydrochloric or nitric acids or with aqueous base.[31, 32] However, degradation does take place in hot methanolic[31] or ethanolic alkali[32] hydroxides in which the compounds are slightly soluble. The high polymers $[NP(OCH_2CF_3)_2]_n$, $[NP(OCH_2C_2F_5)_2]_n$, and $[NP(OCH_2C_3F_7)_2]_n$ are unaffected after months in contact with concentrated sodium hydroxide solution.[33]

b. *Kinetic and Mechanistic Studies*

Allcock and Walsh[34, 35] have investigated the alkaline hydrolysis reactions of a series of fluoroalkoxycyclophosphazenes, including $[NP(OCH_2CF_3)_2]_3$, $[NP(OCH_2CF_3)_2]_4$, $[NP(OCH_2C_2F_5)_2]_3$, $[NP(OCH_2C_3F_7)_2]_3$, and $[NP(OCH_2C_3F_7)_2]_4$. The reaction pathway was investigated by both product analysis and kinetic techniques. Hexakis(trifluoroethoxy)cyclotriphosphazene, XX (Scheme II), can be hydrolyzed in 50 vol.% aqueous methanolic 1 *N* sodium hydroxide solution at 75°C for 150 hours to yield the sodium salt of pentakis-(trifluoroethoxy)hydroxycyclotriphosphazene, XXI. This product is converted by acid to the phosphazadiene, XXII. When the reaction is allowed to continue for a total of 350 hours, the disodium salt of tetrakis(trifluoroethoxy)-

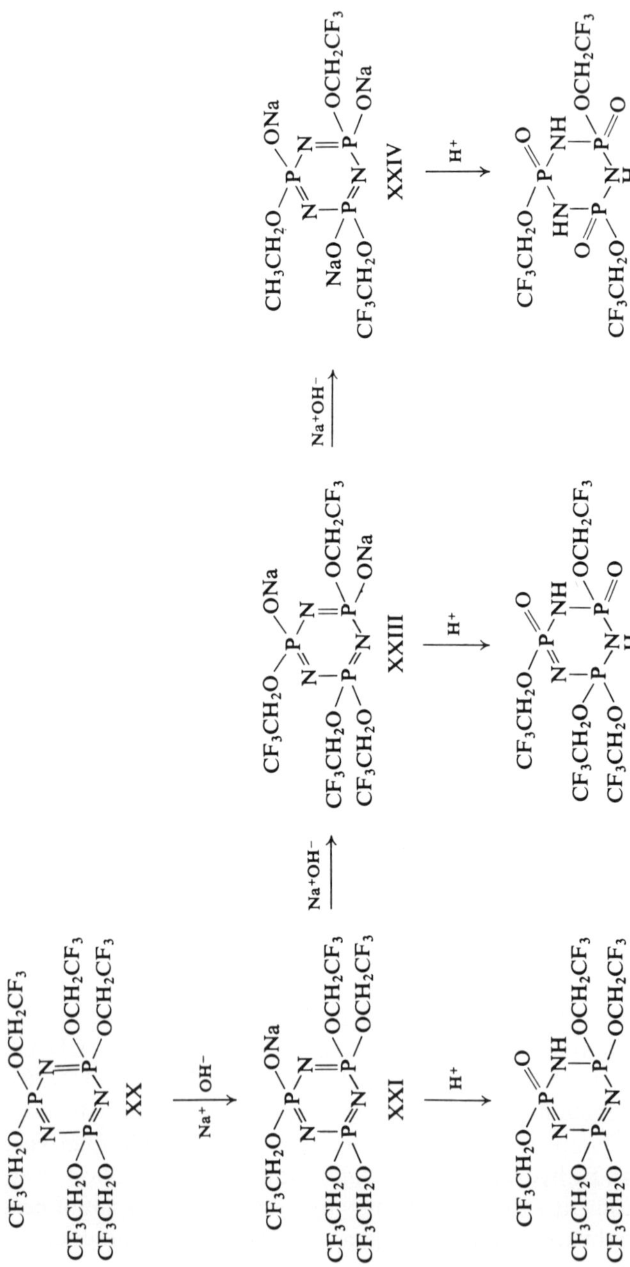

SCHEME II

dihydroxycyclotriphosphazene (XXIII) and the trisodium salt of tris(trifluoroethoxy)tris(hydroxy)cyclotriphosphazene (XXIV) are obtained. Both of these can be converted to the phosphazene (XXV) or phosphazane (XXVI) by treatment with acid. Substitution takes place by a non-geminal pathway. A similar initial mechanism is found when XX is hydrolyzed under milder conditions in a homogeneous system in 2×10^{-2} *M* sodium hydroxide in 25 vol.% aqueous diglyme. A quantitative yield of XXI is obtained after 24 hours at 80°C. In the same medium, $[NP(OCH_2CF_3)_2]_4$, $[NP(OCH_2C_3F_7)_2]_3$, and $[NP(OCH_2C_3F_7)_2]_4$ yield a monohydroxy sodium salt analogous to XXI. No further hydroxy-substituted products were detected.

Hydrolysis of $[NP(OCH_2CF_3)_2]_3$ and $[NP(OCH_2C_3F_7)_2]_4$ was also investigated in a solution of 2×10^{-2} *N* sodium hydroxide in 25 vol. % aqueous diglyme which contained 1% of $H_2{}^{18}O$. No ^{18}O was found in the displaced alcohols, and this indicates that the hydrolysis involves phosphorus–oxygen rather than oxygen–carbon bond cleavage.

The kinetics of removal of the first fluoroalkoxy group from phosphorus in 0.01 to 0.001 *N* sodium hydroxide in 25 vol. % aqueous diglyme showed that the reaction was first order in both phosphazene and hydroxide ion concentration, with the reactivity of the phosphazenes falling in the order: $[NP(OCH_2CF_3)_2]_4 > [NP(OCH_2C_2F_5)_2]_3 > [NP(OCH_2C_3F_7)_2]_4 \approx [NP(OCH_2CF_3)_2]_3 > [NP(OCH_2C_3F_7)_2]_3$. However, the second-order rate constants for all five compounds did not vary by more than a factor of 25, within the range of 9.15×10^{-1} to 4.11×10^{-2} M^{-1} sec^{-1}. The cyclic tetramers are more reactive than the appropriate trimers, and the trifluoroethoxy derivatives are slightly more reactive than the heptafluorobutoxy compounds. Steric factors apparently play only a minor role in affecting the hydrolysis rates (Table 5.1), but activation entropy effects are important. In particular, the faster hydrolysis

TABLE 5.1

ACTIVATION ENERGIES AND ENTROPIES FOR THE HYDROLYSIS OF FLUOROALKOXYCYCLOPHOSPHAZENES[a]

Phosphazene	*E* (kcal/mole)	*ΔS* (e.u.)
$[NP(OCH_2CF_3)_2]_3$	16.6	−11.7
$[NP(OCH_2C_2F_5)_2]_3$	17.5	−12.5
$[NP(OCH_2C_3F_7)_2]_3$	18.2	−8.3
$[NP(OCH_2CF_3)_2]_4$	20.5	−3.7
$[NP(OCH_2C_3F_7)_2]_4$	20.5	−4.9

[a] From ref. 35.

rates found for tetramers appears to be a consequence of the higher entropies of activation, which themselves may reflect the greater flexibility of the tetrameric transition state or a difference in solvation parameters.

3. Hydrolysis of Alkoxyphosphazanes

Alkoxyphosphazenes rearrange when heated to form alkoxyoxocyclophosphazanes (see Chapter 13). The ethoxy derivatives, XXVII and XXVIII, are formed by this reaction.[36] Both compounds can be hydrolyzed by 10%

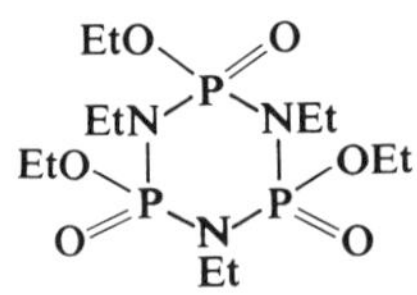

XXVII

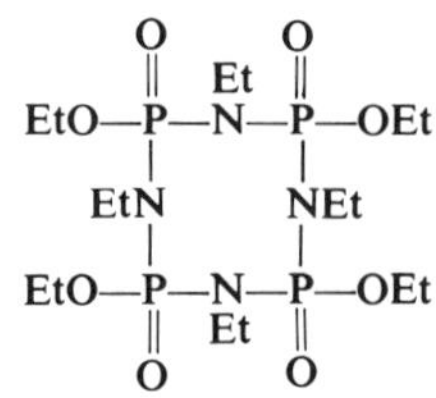

XXVIII

ethanolic hydrogen chloride at 160°C in sealed tubes. Treatment of the residue with potassium hydroxide yields ethylamine.[36] This degradation reaction has been used as a structural proof for phosphazanes.

4. Hydrolysis of Aryloxyphosphazenes

Aryloxycyclophosphazenes hydrolyze in basic media by a process which results in the cleavage of an aryloxy group from phosphorus and the replacement of it by an OH or ONa unit.[34, 35, 37] Four types of aryloxyphosphazenes have been examined by Allcock and Walsh[34, 35, 37] and these are depicted as XXIX–XXXII. In the case of XXIX, XXXI, and XXXII, the analogous cyclic tetramers have also been examined.

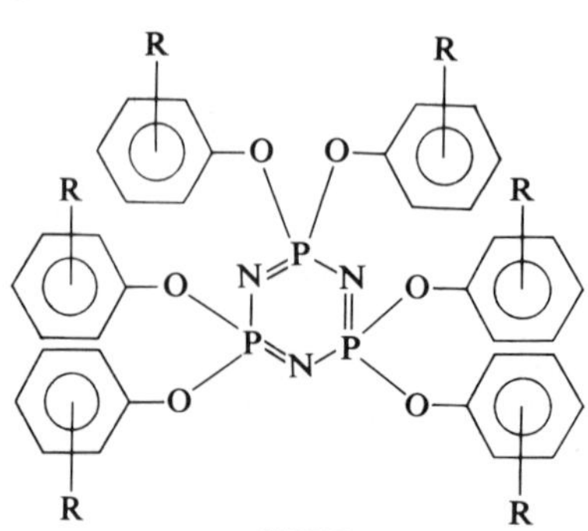

XXIX

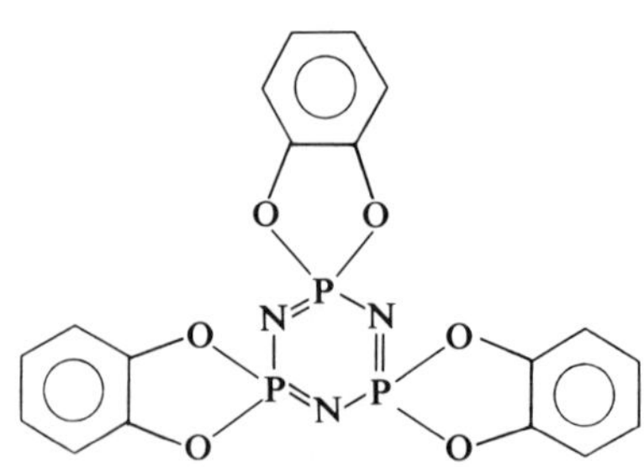

XXX

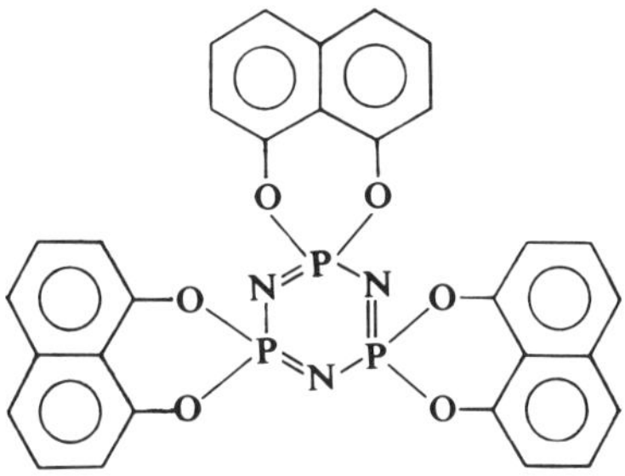

XXXI

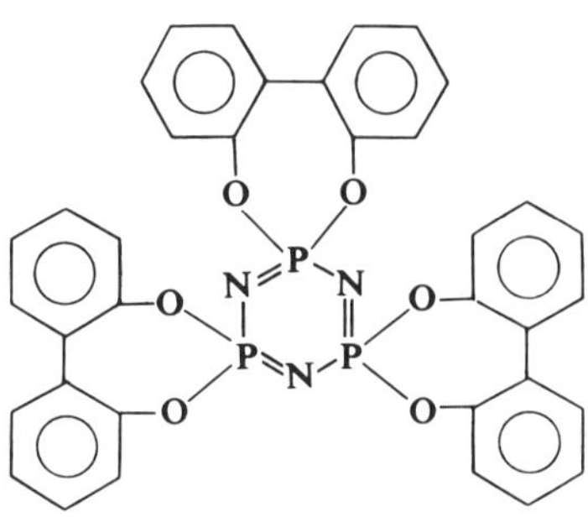

XXXII

Considering first derivatives of structure XXIX, it is found that the rate of hydrolytic cleavage of the first aryloxy group from phosphorus in 25 vol.% aqueous diglyme varies with the substituent R in the order R = p-NO_2 > m-NO_2 > o-NO_2 ≫ H > p-Me. The variation in specific rate constant over this series is quite large, from $5.16 \times 10^{-1}\ M^{-1}\ sec^{-1}$ to $6.93 \times 10^{-7}\ M^{-1}\ sec^{-1}$ for the p-NO_2 and p-Me derivatives, respectively, with the rate depending on the concentration of both phosphazene and hydroxide ion. Electron withdrawal in OC_6H_4R clearly enhances the rate of hydrolysis, and this is consistent with a mechanism in which hydroxide ion attacks phosphorus in an S_N2 rate-determining step (XXXIII). This mechanism is also consistent with the fact that

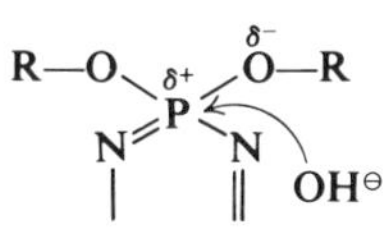

XXXIII

lowering of the polarity of the water–diglyme medium increases the hydrolysis rate. A pentacoordinate phosphorus transition state is further indicated by the very negative entropy values for this reaction, as illustrated in Table 5.2. The

TABLE 5.2

ACTIVATION ENERGIES AND ENTROPIES FOR HEXAKIS(ARYLOXYCYCLOTRIPHOSPHAZENES), $[NP(OC_6H_4R)_2]_3$[a]

R	E (kcal/mole)	S (e.u.)
p-NO_2	10.25	−30.5
m-NO_2	10.60	−32.0
o-NO_2	18.20	−15.7
p-Me	15.90	−40.5

[a] From Allcock and Walsh.[35]

unusually high activation energy found for the *o*-nitrophenoxy derivative is consistent with a steric hindrance effect by the *o*-nitro group, but the rapid hydrolysis rate is a function of the high entropy of activation. Octaphenoxycyclotriphosphazene, $[NP(OPh)_2]_4$, is hydrolyzed in basic media faster than the trimer, $[NP(OPh)_2]_3$, and this corresponds to the situation found in the fluoroalkoxyphosphazene series. It should be noted that the removal of a second aryloxy group from phosphorus takes place far less readily than removal of the first. For example, cleavage of the first aryloxy group from $[NP(OC_6H_4NO_2\text{-}p)_2]_3$ is 100% complete in 25 minutes at 80°C with 0.01 *N* sodium hydroxide in 25 vol. % aqueous diglyme, but removal of a second ligand is not observed during 48 hours of subsequent treatment.[37]

Strikingly different hydrolysis rates are observed when the spiroarylenedioxycyclophosphazenes, XXX, XXXI, and XXXII, are compared. Compound XXX, which contains a five-membered arylenedioxy exocyclic ring at phosphorus, hydrolyzes almost instantly in basic aqueous diglyme or aqueous dioxane. The hydrolysis rates can be followed easily only in neutral 25 vol. % aqueous diglyme or in 1×10^{-6} *N* base. This rapid hydrolysis rate is reminiscent of the hydrolysis behavior of organophosphates which contain a five-membered ring system.[38] It is believed that the hydrolysis pathway follows the scheme outlined below:

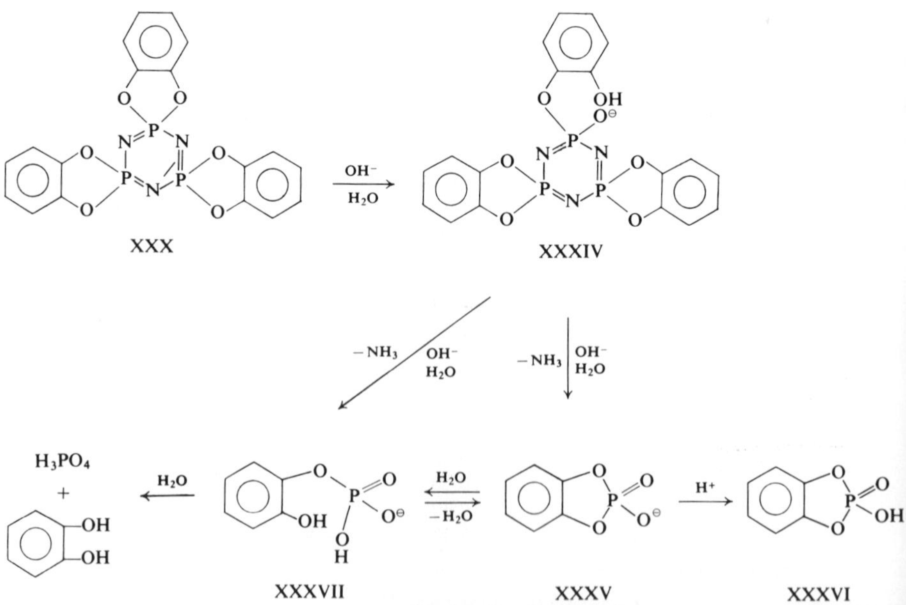

Attack by hydroxide ion on phosphorus leads to cleavage of a phosphorus–oxygen bond to yield a species such as XXXIV. Such a compound has not yet been identified as an intermediate. The first identifiable products formed in 2 vol. % aqueous diglyme are ammonia and the cyclic phosphate (XXXV), which can be converted by acid to the phosphoric acid ester (XXXVI). It is assumed that XXXV is formed by a subsequent ring-closure step in media which contain low concentrations of water. In solvents which contain a high proportion of water, XXXVII is formed. Products analogous to XXXV and XXXVI are isolated when tris(2,3-naphthyldioxy)cyclotriphosphazene (XXXVIII) is hydrolyzed in basic 2 vol. % aqueous diglyme. The reaction rates are comparable to those shown by XXX.

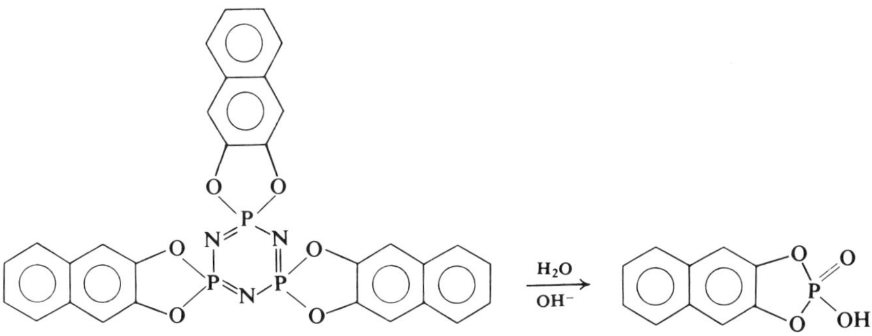

XXXVIII

By contrast, the spirophosphazenes XXXI and XXXII, which contain six- and seven-membered exocyclic arylenedioxy units at phosphorus, show no evidence of hydrolysis after 200 hours at 100°C in 25 vol. % aqueous diglyme containing 1×10^{-3} *N* sodium hydroxide. The remarkable rate enhancement which accompanies the presence of a five-membered ring at phosphorus represents a release of steric strain within that ring. The O—P—O bond angle in XXX is known to be 97°,[39] a significant decrease from the preferred angle of 102°–104° in most phosphazenes (see Chapter 3). Presumably, the hydroxide ion attacks along the axial position of a trigonal bipyramid,

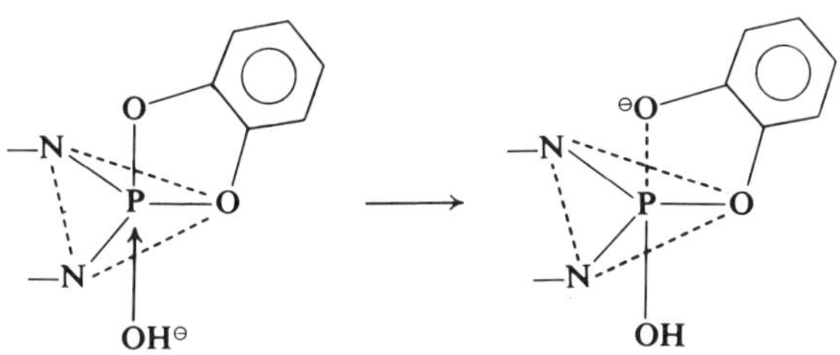

and axial P—O bond cleavage is followed by breakdown of the phosphazene ring and recyclization of the organophosphate residue. When a six- or seven-membered ring is present, or even if two independent phenoxy groups are attached to phosphorus, no steric inducement exists for the initial hydrolysis step to be accelerated.

5. Hydrolysis of Aminophosphazenes

Aminophosphazenes of formula, $[NP(NH_2)_3]_{3\text{ or }4}$, are decomposed hydrolytically when heated with water or aqueous sodium hydroxide solution.[40, 41] Ammonia is evolved initially by removal of the side groups, with concurrent formation of hydroxyoxophosphazanes.[19] Organoamino derivatives of formula $[NP(NHR)_2]_{3,4,\text{ or }n}$ or $[NP(NR_2)_2]_{3,4,\text{ or }n}$ are usually stable to water.

References

1. F. Seel and J. Langer, *Z. Anorg. Allg. Chem.* **295**, 316 (1958).
2. H. R. Allcock, R. L. Kugel, and E. G. Stroh, unpublished work (1969).
3. H. N. Stokes, *Amer. Chem. J.* **17**, 275 (1895).
4. H. N. Stokes, *Chem. Ber.* **28**, 437 (1895).
5. H. N. Stokes, *Amer. Chem. J.* **18**, 629 (1896).
6. H. N. Stokes, *Amer. Chem. J.* **18**, 780 (1896).
7. A. Besson and G. Rosset, *C. R. Acad. Sci.* **143**, 37 (1906).
8. R. Schenck and G. Römer, *Chem. Ber.* **57B**, 1343 (1924).
9. M. Yokoyama, H. Cho, and M. Sakuma, *Kogyo Kagaku Zasshi* **66**, 422 (1963).
10. B. I. Stepanov and G. I. Migachev, *Zavod. Lab.* **32**, 414 (1966).

10a. U. Einsele, *Melliand Textilber.* **3**, 299 (1969).

11. M. Becke-Goehring and G. Koch, *Chem. Ber.* **92**, 1188 (1959).
12. C. D. Schmulbach and V. R. Miller, *Inorg. Chem.* **5**, 162 (1966).
13. D. L. Herring and C. M. Douglas, *Inorg. Chem.* **4**, 1012 (1965).
14. B. W. Fitzsimmons, C. Hewlett, K. Hills, and R. A. Shaw, *J. Chem. Soc., A* p. 679 (1967).
15. C. D. Schmulbach and V. R. Miller, *Inorg. Chem.* **7**, 2191 (1968).
16. H. Bode and H. Bach, *Chem. Ber.* **75**, 212 (1942).
17. R. A. Shaw and C. Stratton, *Chem. Ind.* (*London*) p. 52 (1959).
18. K. G. Acock, R. A. Shaw, and F. B. G. Wells, *J. Chem. Soc., London* p. 121 (1964).
19. A. M. de Ficquelmont, *C. R. Acad. Sci.* **202**, 423 (1936).
20. F. G. R. Gimblett, *Trans. Faraday Soc.* **56**, 528 (1960).
21. J. A. Benckiser, Brit. Pat. 850,251 (1960) (G.m.b.H. Chemische Fabrik).
22. M. Becke, U.S. Pat. 3,179,489 (1965).
23. W. Haubold and M. Becke-Goehring, *Z. Anorg. Allg. Chem.* **352**, 113 (1967).
24. A. Narath, F. H. Lohman, and O. T. Quimby, *J. Amer. Chem. Soc.* **78**, 4493 (1956).
25. O. T. Quimby, A. Narath, and F. H. Lohman, *J. Amer. Chem. Soc.* **82**, 1099 (1960).
26. F. H. Pollard, G. Nickless, and R. W. Warrender, *J. Chromatogr.* **9**, 493 (1962).
27. F. H. Pollard, G. Nickless, and A. M. Bigwood, *J. Chromatogr.* **11**, 534 (1963).
28. G. Tesi and C. M. Douglas, *J. Amer. Chem. Soc.* **84**, 549 (1962).
29. R. A. Shaw and C. Stratton, *Chem. Ind.* (*London*) p. 52 (1959).
30. B. W. Fitzsimmons and R. A. Shaw, *Chem. Ind.* (*London*) p. 109 (1961); *J. Chem. Soc., London* p. 1735 (1964).

31. E. T. McBee, H. R. Allcock, R. Caputo, A. Kalmus, and C. W. Roberts, *U.S. Govt. Res. Rep.* **AD 209,669** (1959).
32. R. Ratz, H. Schroeder, H. Ulrich, E. Kober, and C. Grundmann, *J. Amer. Chem. Soc.* **84**, 551 (1962).
33. H. R. Allcock and G. Konopski, unpublished results (1968).
34. H. R. Allcock and E. J. Walsh, *J. Amer. Chem. Soc.* **91**, 3102 (1969).
35. H. R. Allcock and E. J. Walsh, unpublished results (1970).
36. B. W. Fitzsimmons, C. Hewlett, and R. A. Shaw, *J. Chem. Soc., London* p. 4459 (1964).
37. H. R. Allcock and E. J. Walsh, *Chem. Commun.* p. 580 (1970).
38. F. H. Westheimer, *Accounts Chem. Res.* **1**, 70 (1968).
39. L. A. Siegel and J. H. van den Hende, *J. Chem. Soc., A* p. 817 (1967).
40. D. B. Sowerby and L. F. Audrieth, *Chem. Ber.* **94**, 2670 (1961).
41. L. F. Audrieth and D. B. Sowerby, *Chem. Ind.* (*London*) p. 748 (1959).

Chapter 6

REACTIONS OF HALOPHOSPHAZENES WITH ALKOXIDES, ARYLOXIDES, AND THIOLATES

A. Introduction

The reactions to be discussed in this chapter are the interactions of halophosphazenes with a wide variety of alkoxides, aryloxides, or their mercapto analogues to yield organo-substituted phosphazenes. The overall general reaction scheme can be formulated by the following equations:

$$2n\ \mathrm{ROH} + (\mathrm{NPX_2})_n \rightarrow [\mathrm{NP(OR)_2}]_n + 2n\ \mathrm{HX}$$

$$2n\ \mathrm{RSH} + (\mathrm{NPX_2})_n \rightarrow [\mathrm{NP(SR)_2}]_n + 2n\ \mathrm{HX}$$

In practice, the nucleophile ROH or RSH can be varied to include almost any stable alcohol, phenol, mercaptan, or even diol or dithiol. Many reactions use sodium alcoholates, phenolates, or sodium thiolates instead of the free alcohol or phenol. In such cases, as shown below, sodium chloride is precipitated as a reaction product. When the free alcohol, phenol, or thiol is used, a

$$2n\ \mathrm{RONa} + (\mathrm{NPX_2})_n \rightarrow [\mathrm{NP(OR)_2}]_n + 2n\ \mathrm{NaX}$$

base such as triethylamine or sodium carbonate is usually employed to remove the hydrohalide. The halogen atom, X, can be fluorine, chlorine, or bromine, and there appear to be no limits to the degree of polymerization, n, of the halophosphazene used. A reaction solvent is nearly always employed.

It should be noted that nucleophilic substitution reactions of this type provide one of the easiest routes to the synthesis of organophosphazenes. The reactions are often clean processes with few side reactions, and the products usually are stable solids which are easily purified and characterized. Undoubtedly, this provides one reason for the impressive number of alkoxy, aryloxy, and organothiophosphazenes reported in the literature. However, another contributing factor is connected with the fact that certain aryloxy and fluoroalkoxycyclophosphazenes show very high thermal and hydrolytic stabilities and this has led to a detailed investigation of such derivatives for applications in the high-temperature materials and polymer fields.

B. Experimental Considerations

The following three examples illustrate the techniques that are commonly used to prepare aryloxy- and alkoxycyclophosphazenes.

1. Synthesis of Hexaphenoxycyclotriphosphazene[1-3]

The overall reaction scheme is

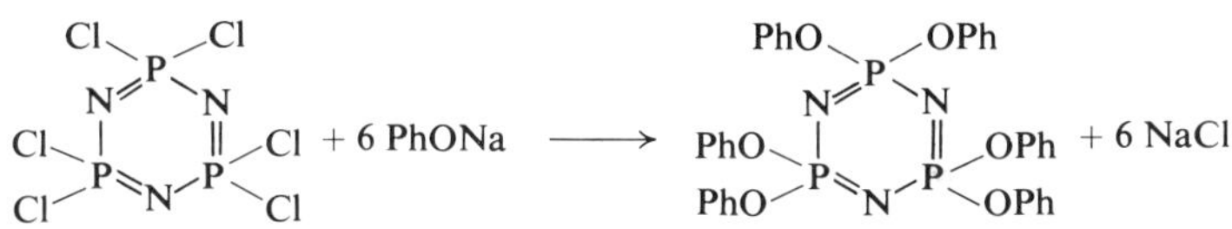

I II

A solution of sodium phenoxide in tetrahydrofuran (500 ml) is prepared from phenol (84.6 g, 0.91 mole) and sodium (20.7 g, 0.9 mole) in an atmosphere of dry argon. To this stirred solution is added a solution of hexachlorocyclotriphosphazene (I) (52.2 g, 0.15 mole) in tetrahydrofuran (500 ml). The mixture is then boiled at reflux for 8 hours and then allowed to stand at 25°C for 56 hours. Filtration of the mixture yields sodium chloride (55 g, 100% based on I), and evaporation of the filtrate, followed by precipitation of the residue from acetone into a large excess of water, yields 88.5 g of a white solid. This is then recrystallized twice from *n*-heptane and *n*-heptane–benzene mixtures, and is exhaustively dried in a vacuum oven at 60°C to yield hexaphenoxycyclotriphosphazene (II) (83.2 g, 93%), m.p. 112°–112.5°C. Characterization has been effected by C, H, N, and P microanalysis, by infrared spectroscopy, and by mass spectrometry. Compound II is, in fact, an excellent high-temperature mass spectrometry standard because of its high mass (MW = 693) and its high stability. The compound is a white, crystalline solid which is soluble in benzene or tetrahydrofuran. It is highly resistant to hydrolytic degradation.

2. Synthesis of Hexakis(trifluoroethoxy)cyclotriphosphazene[4-7]

The following equation illustrates the reaction scheme:

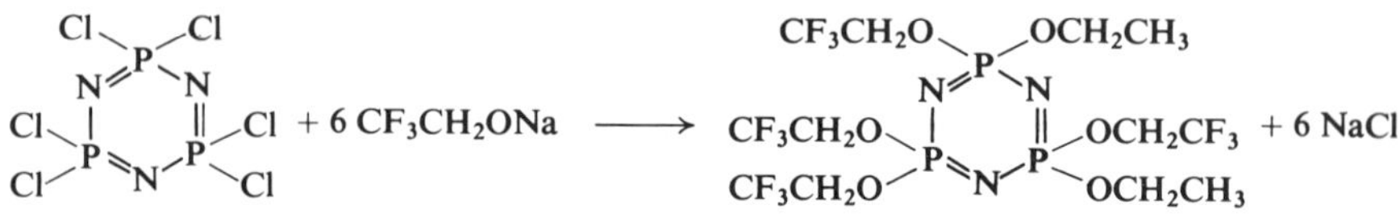

I III

Sodium (23.8 g, 1.05 moles) is added to a cooled solution of trifluoroethanol (103.3 g, 1.033 moles) in anhydrous ether (400 ml). To this stirred, cooled solution is added dropwise a solution of hexachlorocyclotriphosphazene (I) (59.6 g, 0.172 mole) in anhydrous ether (400 ml). The mixture is then stirred at 25°C for 8 hours, and the white precipitate of sodium chloride is filtered off. The filtrate is washed thoroughly with water to remove excess sodium trifluoroethoxide and the ethereal layer is distilled to remove excess trifluoroethanol and solvent. The residue is a viscous oil which solidifies on cooling. This is sublimed at 70°C/2 mm to give hexakis(trifluoroethoxy)-cyclotriphosphazene (III) (96.0 g, 77%), m.p. 49°C. It should be noted that omission of the washing step for removal of sodium trifluoroethoxide introduces the danger of an explosion during the subsequent distillation and sublimation steps.[7] The cyclophosphazene III is a white, waxy, crystalline substance, readily soluble in many organic solvents. It has an unusually high thermal stability and is quite stable to atmospheric moisture or neutral aqueous solvents.

3. Synthesis of Tris(*o*-phenylenedioxy)cyclotriphosphazene

The overall reaction scheme is as follows:

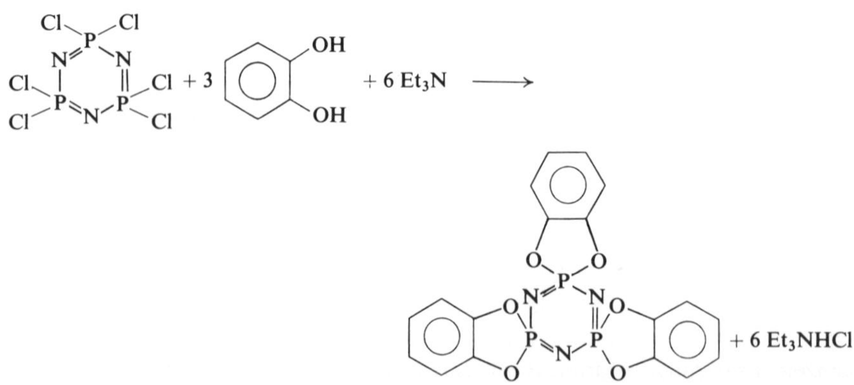

A solution of catechol (95 g, 0.864 mole) and triethylamine (175 g, 1.728 moles) in tetrahydrofuran (500 ml) is added dropwise to a stirred solution of hexachlorocyclotriphosphazene (100 g, 0.288 mole) in tetrahydrofuran (1000 ml) at 30°C. The mixture is then boiled at reflux for 2 hours and stirred at 30°C for 48 hours. The white precipitate is filtered off, dried (350 g), and then washed thoroughly with water to remove triethylamine hydrochloride. The powdery residue is then extracted with boiling benzene and the extracts are cooled to yield white crystals of tris(*o*-phenylenedioxy)cyclotriphosphazene (74 g, 50% yield). Subsequent recrystallizations from benzene followed by sublimation at 230°C/0.05 mm give material with a melting point of 244°–245°C. This compound readily forms channel clathrates when brought into contact with organic liquids (see Chapter 11).

4. Variations in the Experimental Technique

Two of the above examples use sodium salts of phenol or an alcohol as the nucleophilic reagent, but a number of modifications to this process are possible. For example, the reaction of sodium hydride with an alcohol has been used for preparation of the sodium alkoxide.[6,7] Direct reaction of a solution of an alcohol or phenol with the halophosphazene in the presence of a suspension of anhydrous sodium carbonate has also proved to be a convenient technique.[8,9]

$$6\ ROH + (NPCl_2)_3 + 6\ Na_2CO_3 \rightarrow [NH(OR)_2]_3 + 6\ NaCl + 6\ NaHCO_3$$

Sodium hydroxide pellets can be used when a phenol is allowed to react with a halocyclophosphazene in a high-boiling hydrophobic solvent, such as xylene.[10–12] Finally, a tertiary organic amine, such as triethylamine or pyridine, may be employed as a hydrogen chloride acceptor, as in the equation[2,8,9,13–19]

$$6\ ROH + (NPCl_2)_3 + 6\ Et_3N \rightarrow [NP(OR)_2]_3 + 6\ Et_3N \cdot HCl$$

Nearly any unreactive organic solvent may be used for these interactions, but aliphatic ethers, such as diethyl ether or tetrahydrofuran, are particularly useful. Although it is usual to perform substitution reactions of this type in standard ground glass equipment with exclusion of moisture by drying traps, it is often necessary to use a glass high-vacuum system when pyridine is used as a solvent.[15]

C. Scope of the Reaction

A wide variety of alkoxy-, aryloxy-, alkylthio-, and arylthiophosphazenes can be prepared by the techniques described above. For example, the following alkoxy groups have been attached to the phosphazene skeleton: CH_3O—,[2, 18, 20–25] C_2H_5O—,[2, 13, 18, 20, 24, 25] n-C_3H_7O—,[2, 20] i-C_3H_7O—,[2, 20] n-C_4H_9O—,[2, 18, 19, 22] $PhCH_2O$—,[2] $CH_2{=}CHCH_2O$—.[26] Fluoroalkoxy groups can be substituted at phosphorus with particular ease, and the following are some of the substituents that have been used: CF_3CH_2O—,[4–6, 24, 25, 27–32] $C_2F_5CH_2O$—,[4–6, 24, 25, 27, 29–31, 33] $C_3F_7CH_2O$—,[4–6, 24, 25, 27, 29–31, 33, 34] $H(CF_2)_6$-CH_2O—,[5, 6, 29, 35, 36] $H(CF_2)_{10}CH_2O$—,[5, 6] $H(CF_2)_2CH_2O$—,[27, 29, 30, 33, 35] $H(CF_2)_4CH_2O$—,[29, 30, 35, 36] $H(CF_2)_8CH_2O$—,[29, 35] and $(CF_3)_2CHO$—.[27]

Hexachlorocyclotriphosphazene also reacts with an equimolar amount of acetoxine in the presence of triethylamine to yield pentachloro-*N*-isopropylidenoximinocyclotriphosphazene.[36a] The product is a solid, m.p. 91°C.

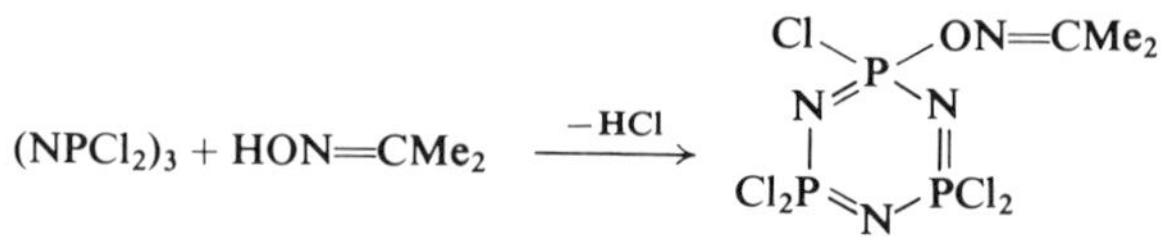

Aryloxy substitution has also been studied in detail and a wide range of phenols have been used as nucleophiles. The substituents include C_6H_5O—,[1–3, 24, 25, 35, 37–43] m-MeC_6H_4O—,[29] p-MeC_6H_4O—,[37] m-CF_3C_6-H_4O–,[10, 29] m-FC_6H_4O–,[29] m-ClC_6H_4O–,[29] p-ClC_6H_4O–,[29, 37] m-CF_3OC_6-H_4O–,[11] *o*, *m*, and p-$NO_2C_6H_4O$–,[7, 27, 37, 44] p-BrC_6H_4O–,[39] p-$MeOC_6H_4O$–.[2, 37]

Fewer mercapto derivatives have been reported than alkoxy or aryloxy phosphazenes, but those which are known cover a wide range of alkyl- and arylthio groups, as demonstrated by the following list: C_2H_5S–,[13, 19, 45, 46] C_3H_7S–,[19, 45, 46] iso-C_3H_7S–,[19, 45, 46] C_4H_9S–,[19, 45, 46] iso-C_4H_9S–,[45, 46] t-C_4-H_9S–,[46] $C_6H_{11}S$–,[45, 46] $C_8H_{17}S$–,[19] $C_{12}H_{25}S$–,[19] $C_{16}H_{33}S$–,[19] $PhCH_2S$–,[19, 45, 46] C_6H_5S–.[13, 19, 37, 45, 46]

The above syntheses lead to the formation of derivatives which possess two independent substituents at each phosphorus. However, it is also possible to perform substitution reactions using diols or dithiols in such a way that cyclized units are formed at the phosphorus atoms, as shown in the following equation[7, 8, 14–17]:

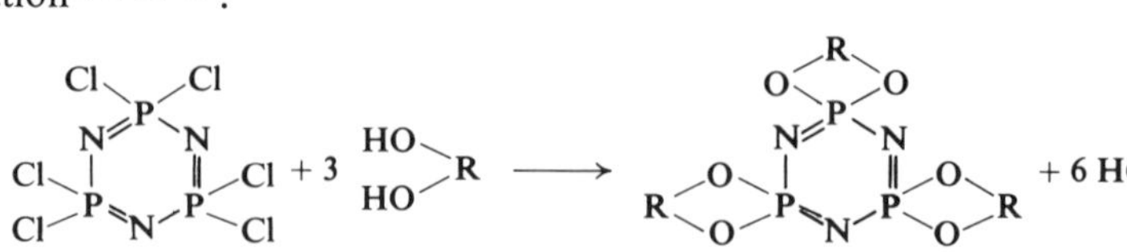

These derivatives are especially interesting because of their bearing on substitution mechanisms. They are considered in more detail in Section E5 of this chapter.

Finally, it is possible to carry out substitution reactions between alkoxides, aryloxides, or mercaptans and fluoro or bromophosphazenes. The reactions are not restricted to cyclic trimers or tetramers. Thus, although most of the foregoing examples have referred to reactions involving cyclic trimers or tetramers, $(NPX_2)_{3\text{ or }4}$, monophosphazenes react similarly,[43] as do medium-ring-sized cyclic compounds such as $(NPX_2)_5$, $(NPX_2)_6$, $(NPX_2)_7$, and $(NPX_2)_8$.[23] Even very high molecular weight chlorophosphazenes, $(NPCl_2)_{15,000}$, readily undergo complete replacement of halogen when treated with methoxide, ethoxide, trifluoroethoxide, phenoxide, or other related nucleophiles.[24, 25] Only for the cyclization processes with diols are there marked mechanistic differences between the substitution of cyclic trimers and the higher homologues.[7, 25]

D. General Properties of Alkoxy-, Aryloxy-, Alkylthio-, and Arylthiophosphazenes

In general, phosphazenes of structure $[NP(OR)_2]_n$ or $[NP(SR)_2]_n$ are among the most stable phosphorus–nitrogen derivatives known. However, the stability to heat and hydrolysis depends on the nature of the substituent OR or SR and, although this subject is considered in more detail in Chapters 5 and 12, some generalizations are appropriate at this point.

Most fully substituted alkoxy-, aryloxy-, alkylthio-, and arylthiophosphazenes are stable, white crystalline solids which are readily soluble in organic media. However, some of the alkoxy- or alkylthio-substituted species such as $[NP(OEt)_2]_3$,[2] $[NP(OBu^n)_2]_{3\text{ or }4}$, $[NP(OMe)_2]_{5\text{ or }7}$,[23] or $[NP(SBr^n)_2]_3$[46] are colorless liquids at room temperature. A comprehensive study of the effects of mixed substitution around the phosphazene ring by Kober, Lederle, and Ottmann[29, 30, 35] has shown that substitution of the trimeric or tetrameric ring by both fluoroalkoxy and aryloxy groups lowers the melting point to such a degree that the materials have useful properties as fire-resistant, low-temperature hydraulic fluids or lubricants. Alkoxy- or aryloxy-substituted high polymers are elastomers or flexible-film-forming materials. In some cases, the materials do not lose their flexible properties until cooled below $-80°C$[24, 25, 27] (see Chapter 16).

As mentioned earlier, hexaphenoxycyclotriphosphazene, $[NP(OPh)_2]_3$, is exceedingly stable to heat and hydrolysis, and this behavior is characteristic of many aryloxyphosphazenes. Alkoxyphosphazenes (except fluoroalkoxy derivatives) rearrange to cyclophosphazanes when heated, and they are also more sensitive to hydrolysis than aryloxy compounds. Most alkoxy derivatives are quite stable when stored at room temperature in a moist atmosphere, and fluoroalkoxyphosphazenes are exceptionally stable to heat and neutral hydrolysis. When five-membered cyclized groups are attached to phosphorus, hydrolysis takes place readily in solution,[47] but marked hydrolytic stability is displayed by species which contain six- or seven-membered ring systems attached to phosphorus in a spiro arrangement.[7] The trimer, $[NP(OMe)_2]_3$, is soluble in water.

Alkoxyphosphazenes have been developed as flame retardants for rayon. Godfrey and Schappel reported that *n*-propoxycyclophosphazenes and trimer-tetramer-linear oligomer mixtures are particularly effective flame retardants, especially when the phosphazene is encapsulated in discrete cavities within each fiber.[47a] Approximately 15 wt.% of the phosphazene is effective.

E. Influence of the Nucleophile on Substitution

The above information illustrates the wide versatility of this general substitution reaction but it indicates very little about the reaction mechanism. To obtain a deeper insight into this process it is necessary to examine the influence of a number of factors on the substitution. One of the most important of these is the influence of the nucleophile, $RO^{\ominus}$ or $RS^{\ominus}$. Useful information about the role of the nucleophile can be obtained by an examination of instances of incomplete substitution, evidence concerning the pattern of substitution, the effects of competition reactions, the influence of other substituents on the reaction, and the results of cyclization processes at phosphorus. Each of these factors will be considered in turn.

1. Degree of Halogen Replacement

Replacement of chlorine in chlorocyclophosphazenes by unbranched alkoxy groups takes place more readily than replacement by phenoxy or branched alkoxy groups. Thus, complete replacement of chlorine in $(NPCl_2)_3$, $(NPCl_2)_4$, or $(NPCl_2)_{\sim 15,000}$ occurs readily with methoxide, ethoxide, or *n*-propoxide,[2] but more drastic conditions are required to achieve complete substitution with isopropoxide[2] or phenoxide ions.[1, 2, 25, 37–42, 48] In fact, partly substituted phenoxychlorocyclophosphazenes, such as $N_3P_3Cl(OPh)_5$, $N_3P_3Cl_2(OPh)_4$, $N_3P_3Cl_3(OPh)_3$, $N_3P_3Cl_4(OPh)_2$, and $N_3P_3Cl_5(OPh)$ can be isolated from reactions which have not been forced to completion.[2, 38–41] The same is true for reactions between $(NPCl_2)_3$ and the *p*-bromophenoxide ion.[39]

However, partially substituted alkoxy derivatives, such as $N_3P_3Cl(OCH_2CF_3)_5$, $N_3P_3Cl_2(OCH_2CF_3)_4$, $N_3P_3Cl_3(OCH_2CF_3)_3$, or $N_3P_3Cl(OCH_2CF_3)_4$ can be prepared if the reagent stoichiometry is correct and if the reaction conditions are sufficiently mild.[4, 28] Similarly, butoxy derivatives of the types $N_3P_3Cl_5(OBu)$, $N_3P_3Cl_4(OBu)_2$, and $N_3P_3Cl_3(OBu)_3$ can be isolated from the reaction of $(NPCl_2)_3$ with butoxide ion at 20°C.[21]

A similar pattern can be discerned for the thioalkyl and thioaryl derivatives.[46] Thus, in boiling benzene, the reaction of $(NPCl_2)_3$ with EtS$^\ominus$, n-PrS$^\ominus$, n-BuS$^\ominus$, or $PhCH_2S^\ominus$ gives hexathioalkyl derivatives, but branched reagents, such as i-PrS$^\ominus$ or cyclohexyl-S$^\ominus$ yield only tetrasubstituted products. Drastic reaction conditions are required for the preparation of $[NP(SPh)_2]_3$ in boiling benzene. It should be noted that the reaction solvent exerts a profound effect on the reactivity of thioalkyl or thioaryl nucleophiles in this reaction.

The above results suggest that the steric dimensions of the nucleophile influence the degree of halogen replacement, with bulky side groups being the most difficult to introduce as substituents on the phosphazene skeleton.

2. Pattern of Halogen Replacement

Substitution at phosphorus in cyclo- or polyphosphazenes can take place either by geminal or non-geminal routes (IV or V). Furthermore, non-geminal

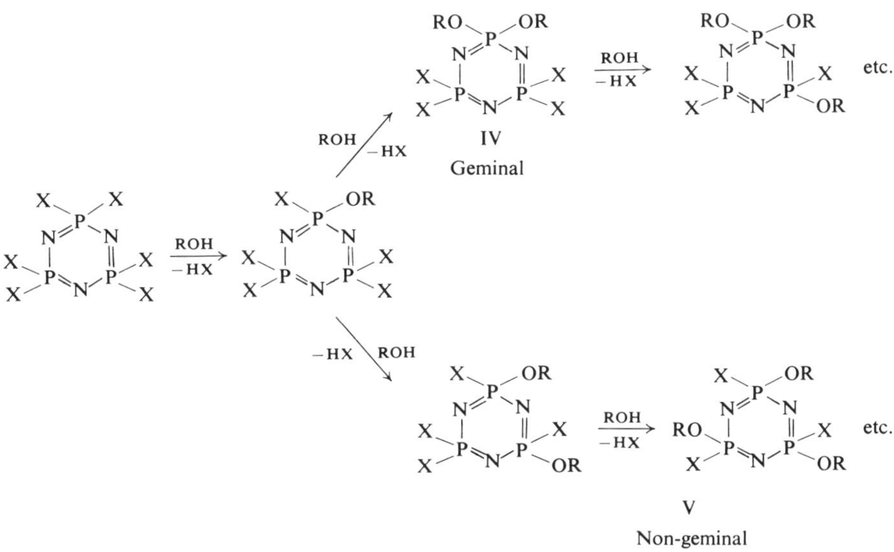

substitution can result in the formation of both *cis* (VI) and *trans* (VII) isomers.

VI VII

Examination of the pattern and stereochemistry of alkoxide, aryloxide, or mercaptide substitution has helped to clarify the mechanisms of these reactions.

The reaction of hexachlorocyclotriphosphazene, $(NPCl_2)_3$, with sodium phenoxide or sodium *p*-bromophenoxide proceeds non-geminally in benzene, acetone, or tetrahydrofuran.[38, 39, 40] In acetone solution, the *cis* substitution predominates slightly over the *trans*.[38] These conclusions are based on the work of Bezman[41] McBee,[38] Shaw,[39, 39a] and their co-workers, who identified the non-geminal products by conversion of the phenoxy chloro products to phenoxy dimethylamino[39, 41] or phenoxy amino derivatives,[38] followed by structural identification by melting points and proton NMR spectra. The reaction schemes studied are

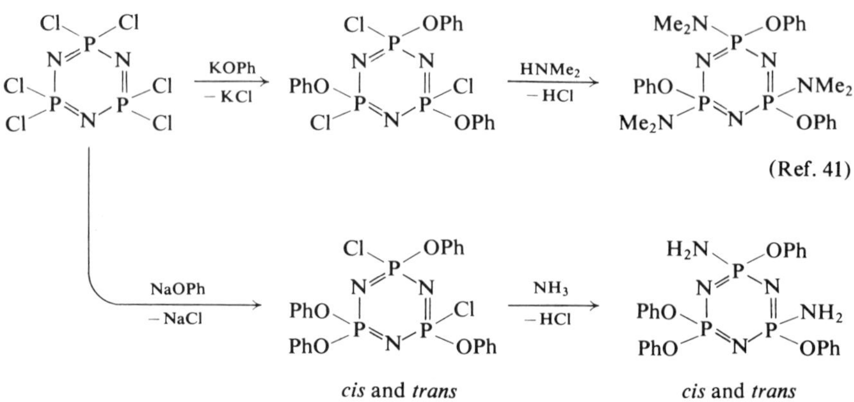

(Ref. 41)

(Ref. 38)

The partly substituted phenoxy chloro derivatives can be separated by column and thin-layer chromatography.[38, 39] It was also suggested that inversion of configuration probably does not occur during phenoxide attack.[38]

The situation is less clear when alkoxy nucleophiles are employed. For example, sodium trifluoroethoxide reacts with hexachlorocyclotriphosphazene to yield $N_3P_3Cl_3(OCH_2CF_3)_3$.[28] This material can then be ammonolyzed and three isomers of the compound $N_3P_3(NH_2)_3(OCH_2CF_3)_3$ are isolated. Two of these species are probably non-geminal *cis* and *trans* isomers and the third is a geminal form.[28] Sodium butylate reacts with hexachlorocyclotriphosphazene to give a dibutoxy derivative, $N_3P_3Cl_4(OBu)_2$.[21] It was suggested but not proved that this compound is non-geminal.

By contrast, it has been shown that the interaction of hexachlorocyclotriphosphazene with ethylthiolate or phenylthiolate ions leads to geminal substitution.[46] This information was derived from ^{31}P and ^{1}H NMR spectra of $N_3P_3Cl_4(SEt)_2$, $N_3P_3Cl_4(SPh)_2$, and $N_3P_3Cl_2(SEt)_4$. However, it should be noted that a number of these reactions are heterogeneous and markedly solvent-dependent, and some care must be exercised when interpreting the data.

3. Effect of Other Substituents

Additional information about the mechanism of chlorine replacement can be obtained from the influence of substituents already present on the ring. For example, it was reported by McBee, Okuhara, and Morton[38] that 1,1-diphenyl-3,3,5,5-tetrachlorocyclotriphosphazene (VIII) was phenoxylated and ammonolyzed to IX. The *cis* and *trans* non-geminal isomers of IX were isolated in

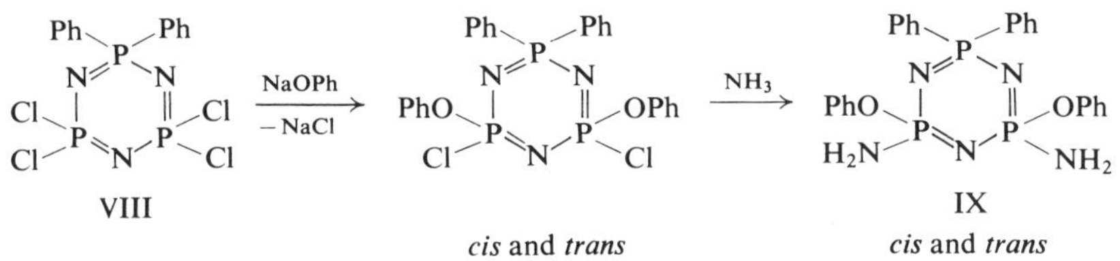

39.4% and 32.2% yields, respectively, and a geminal isomer was also recovered in 4.5% yield. Similar results were reported by Ford, Barr, Dickson, and Bezman[42] but, in this case, the bisdimethylamino and methylamino derivatives were identified by ^{31}P and ^{1}H NMR spectroscopy. Thus, phenoxylation is still predominantly non-geminal in the presence of adjacent geminal phenyl units.

Fitzsimmons, Hewlett, Hills, and Shaw[20] have reported the results of alcoholysis reactions of the geminal derivatives, $N_3P_3Ph_2Cl_4$ (VIII) and $N_3P_3Ph_4Cl_2$ using alkoxide ion in alcohol solvents. Methoxide, ethoxide, *n*-propoxide, and isopropoxide ions were used as nucleophiles. The results suggested that the ease of substitution decreases in the order $(NPCl_2)_3 > N_3P_3Ph_2Cl_4 > N_3P_3Ph_4Cl_2$. Phenyl groups thus retard the substitution reaction at adjacent phosphorus atoms. However, it should be noted that side products of types X or XI were also formed in this reaction.

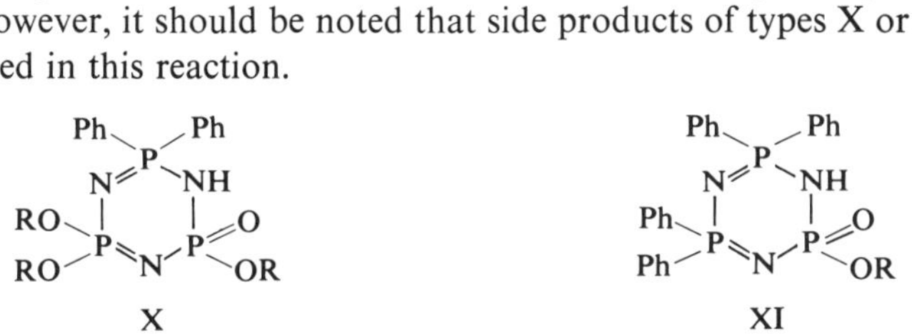

Non-geminal diamino groups apparently do not seriously interfere with the replacement of adjacent chlorine atoms by fluoroalkoxy units. For example, it has been reported that 1,3-diamino-1,3,5,5-tetrachlorocyclotriphosphazene (XII) reacts with the sodium salts of trifluoroethanol, 2,2,4,4-tetrafluoro-*n*-

XII

propanol, 2,2,4,4,4-pentafluoropropanol, 2,2,4,4,5,5,5-heptafluorobutanol, and 2,2,4,4,5,5-hexafluorobutanol to replace all the halogen atoms.[33] Similarly, 1,1,3,5-tetrakis(dimethylamino)-3,5-dichlorocyclotriphosphazene reacts with a wide variety of nucleophiles, such as $MeO^{\ominus}$, $EtO^{\ominus}$, n-$PrO^{\ominus}$, i-$PrO^{\ominus}$, n-$BuO^{\ominus}$, i-$BuO^{\ominus}$, s-$BuO^{\ominus}$, n-$C_5H_{11}O^{\ominus}$, i-$C_5H_{11}O^{\ominus}$, n-$C_6H_{13}O^{\ominus}$, n-$C_7H_{15}O^{\ominus}$, $PhCH_2O^{\ominus}$, and $CH_2{=}CH{-}CH_2^{\ominus}$ to replace the two non-geminal chlorine atoms.[49]

4. Competition Reactions

Very few examples have been reported in which two nucleophiles have been allowed to interact simultaneously with a halocyclophosphazene. However, equimolar amounts of the sodium salts of *m*-fluorophenol and trifluoroethanol were simultaneously allowed to react with hexachlorocyclotriphosphazene. The final product contained 3 to 4 times as many aryloxy as trifluoroethoxy groups, although the detailed molecular arrangement was not established.[29]

It has also been shown that the interaction of sodium phenoxide with hexachlorocyclotriphosphazene in butanol solvent really constitutes a competition reaction for the substrate between sodium phenoxide and sodium butoxide.[21] Both phenoxy and butoxy substituents were detected in the final product, although the relative proportions of each were not reported.

5. Cyclization Reactions at Phosphorus

When catechol is allowed to interact with hexachlorocyclotriphosphazene in the presence of a solvent and a suitable base, cyclization takes place at each phosphorus atom, to yield XIII.[14, 15] Triethylamine or anhydrous sodium

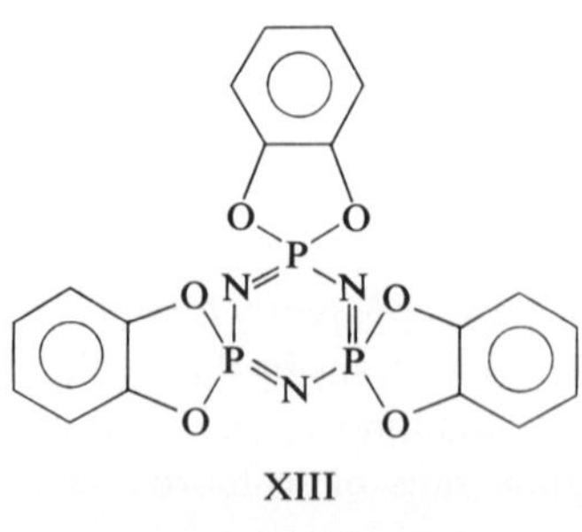

XIII

carbonate are satisfactory hydrogen chloride acceptors,[8, 14, 15] although pyridine can also be used.[15, 16]

Similar high-yield reactions take place between hexachlorocyclotriphosphazene and 2,3-dihydroxynaphthalene or toluene-3,4-dithiol to give XIV and XV.[8] It is worthwhile to note that the cyclic tetrameric analogue of XIII is

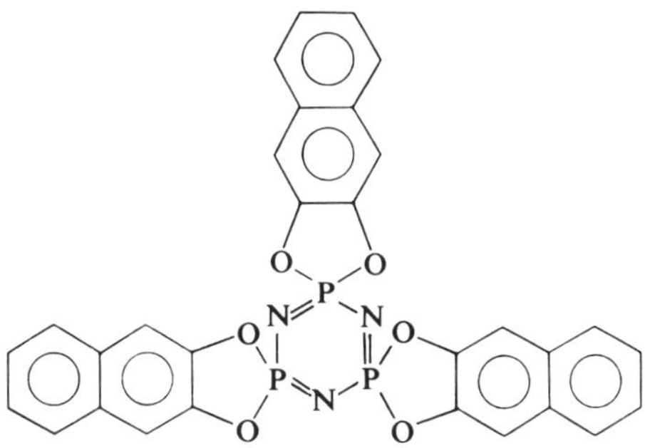

XIV

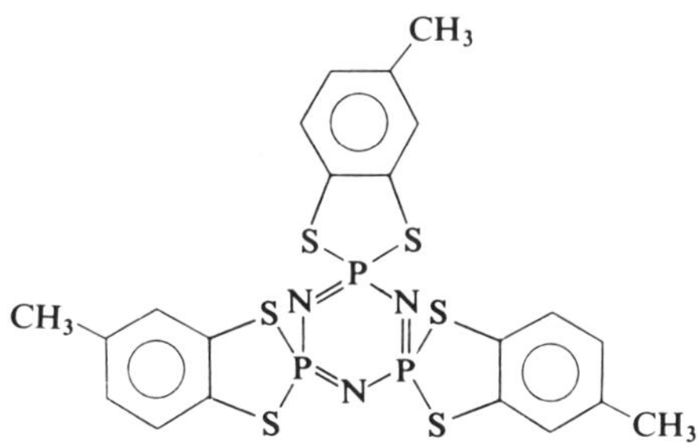

XV

extremely labile and difficult to isolate.[7] The principal reaction product when triethylamine is used as a base is a phosphorane (XVI) formed by phosphazene ring cleavage. The same phosphorane is also produced during formation of XIII, particularly when catechol is in excess.[14, 15]

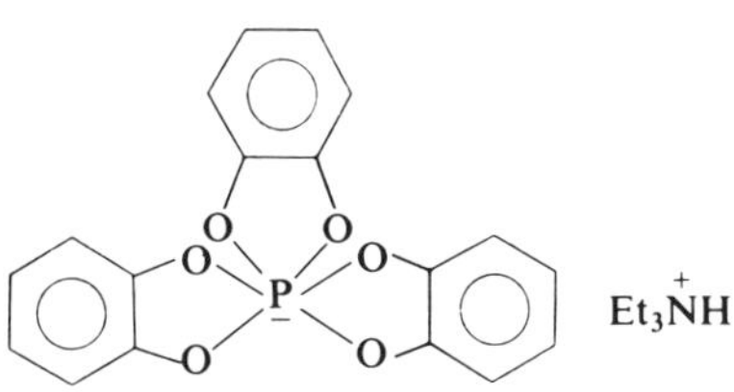

XVI

Six- and seven-membered spiro-aryloxy systems can also be introduced at phosphorus. For example, 1,8-dihydroxynaphthalene and 2,2′-dihydroxybiphenyl react with hexachlorocyclotriphosphazene to yield XVII and XVIII.[7]

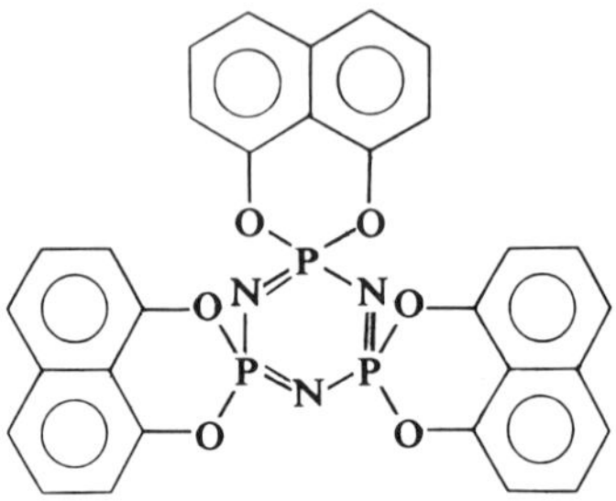

XVII

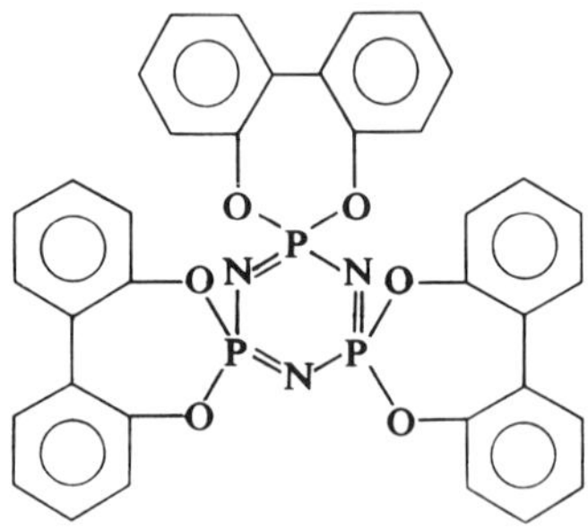

XVIII

Cyclotetraphosphazene analogues of these two compounds have also been prepared.[7]

Cyclic derivatives such as XIII–XVIII are particularly interesting because of the unexpected ease of cyclization. This is especially so for XVIII, where torsional motions of the Ph—Ph bond might be expected to reduce the probability of cyclization. However, manifestations of ring strain in XIII–XV become apparent during hydrolysis[50] (see Chapter 5) and during phosphazene degradations to form phosphoranes.[9, 51] For example, it has been shown that XIII, XIV, and XV form phosphorane XIX when treated with *o*-aminophenol. Compounds XVII and XVIII are inert under these conditions (see also Chapter 14).

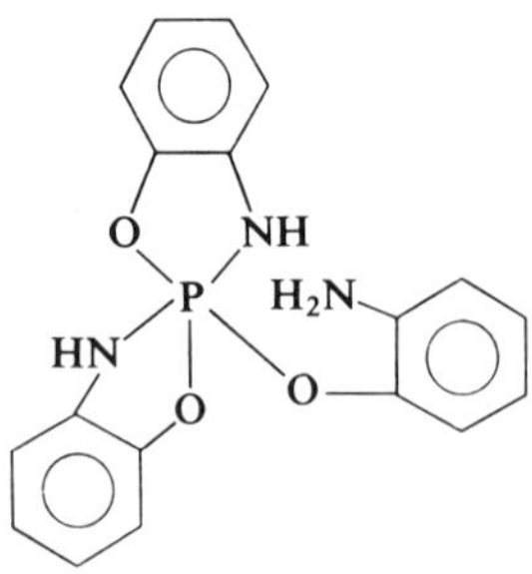

XIX

Cyclization at phosphorus is sterically favored by the presence of *ortho* functional groups on a rigid phenyl ring, but aliphatic diols might be expected to cyclize less readily because of the greater opportunities for cross-linking *between* rings with these ligands. However, it has been shown that spiro-type derivatives are formed when hexachlorocyclotriphosphazene reacts with the sodium or triethylamine salts of $HOCH_2(CF_2)_2CH_2OH$ or $HOCH_2(CF_2)_3CH_2OH$. Species such as XX are isolated.[5] Similarly, 2,2-bis(nitroxymethyl)-

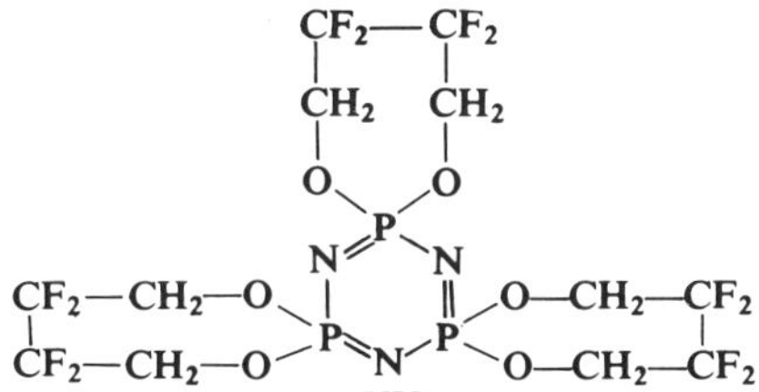

XX

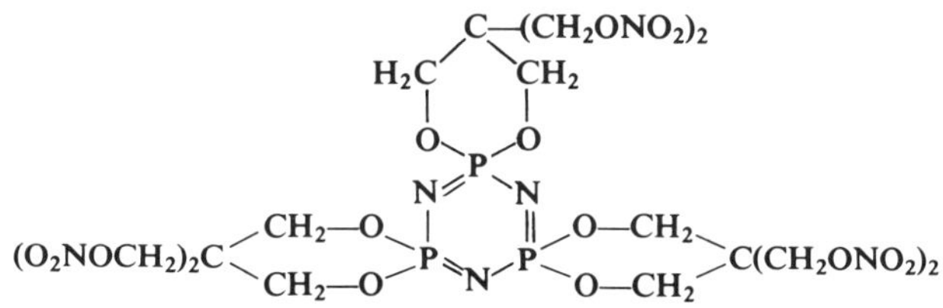

XXI

3-nitroxypropan-1-ol reacts with hexachlorocyclotriphosphazene in pyridine to form derivative XXI.[17] Experiments have also been reported which involve the interaction of hexachlorocyclotriphosphazene with ethylene glycol, propane-1,2-diol, butane-1,2-diol, and cyclohexane-1,2-diol in the presence of pyridine.[52, 53, 53a] Spirocyclic phosphazenes such as XXII are formed.

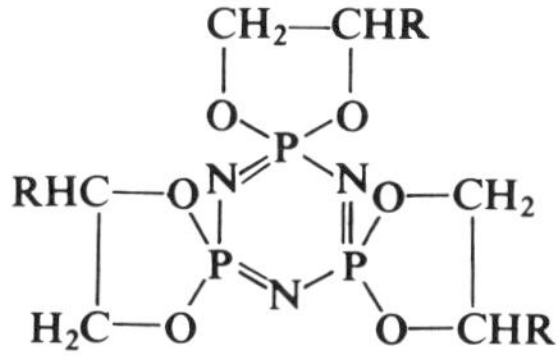

XXII

Analogous cyclic tetramers have also been described.[52] Pornin[52] reported that spiro compounds were also formed when propane-1,3-diol or butane-1,4-diol reacted with hexachlorocyclotriphosphazene in pyridine at −10°C, but Matuszko and Chang[53] observed that at higher temperatures (200°–210°C) the principal products were cyclic ethers formed by dehydration of the diol.

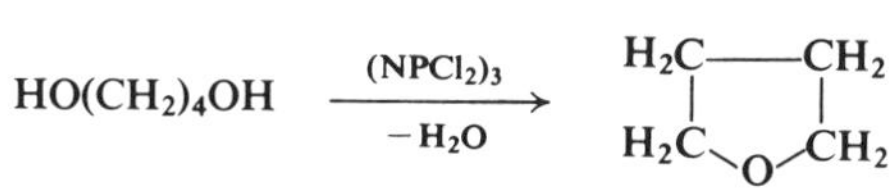

2-Aminoethanol has been reported to form spirocyclic derivatives with $(NPCl_2)_3$.[53a]

In a similar manner, thioamides such as dithiooxamide, and thiosemicarbazide apparently react with hexachlorocyclotriphosphazene in dry tetrachloroethane to yield the spiro derivatives shown below.[54] Thiourea reacts as

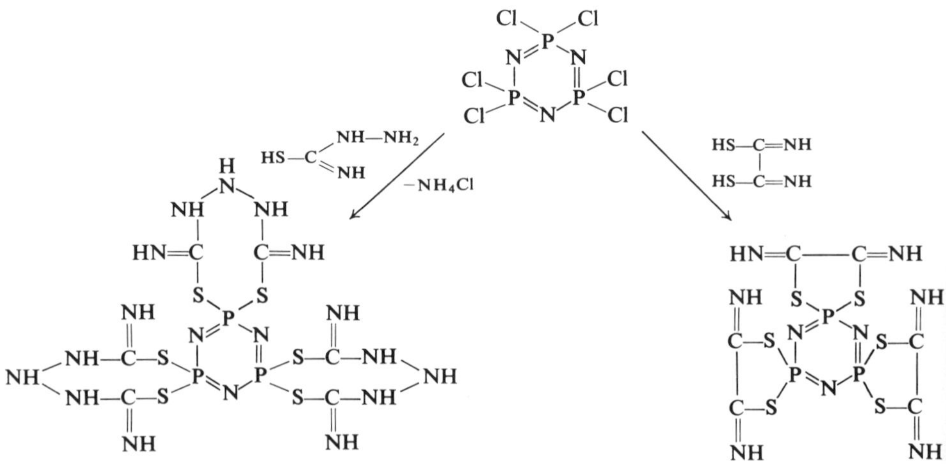

the species HS—C(NH_2)═NH to give hexa-substituted products of structure $[NP(SC(NH_2){=}NH)_2]_3$.[54]

It should be noted that spirocyclic phosphazenes are of interest as intermediates for polymer synthesis (see Chapters 15 and 16).

F. Influence of the Halogen in Halophosphazenes

Much of the work which has been reported for the alcoholysis, phenolysis, and mercaptanolysis of chlorophosphazenes is believed to be applicable also to fluorophosphazenes and bromophosphazenes, but very little experimental work has been reported for these latter halophosphazenes. However, Mao, Dresdner, and Young[36] showed that $(NPF_2)_3$ and $(NPF_2)_4$ react with the sodium salt of 1,1,7-trihydrodecafluoroheptanol in a fluorocarbon solvent to yield the appropriate fully-substituted cyclophosphazenes. Similarly, Paddock and co-workers[23] reported that species of structure $[NP(OMe)_2]_{5-8}$, $[NP(OCH_2CF_3)_2]_{5-8}$, and $[NP(OPh)_2]_{5-8}$ could be prepared from the appropriate fluorocyclophosphazenes. Niecke, Thamm, and Glemser have prepared mono-methoxy, ethoxy, phenoxy, and phenylthio-fluorophosphazenes from $(NPF_2)_3$.[57a]

G. Influence of Phosphazene Ring or Chain Size

In general, it appears that many of the reactions that have been described for hexachlorocyclotriphosphazene are equally applicable to cyclic tetramers,[2–6, 29, 30, 31, 35–37, 48, 52] higher cyclic oligomers,[24, 33, 55] and very high polymers.[24, 25] However, some noticeable exceptions are also found. For instance, although the cyclic trimer $(NPCl_2)_3$ reacts readily with catechol in the presence of base to yield a spirocyclophosphazene (XIII), the high polymer $(NPCl_2)_n$ degrades instead to phosphorane (XXVI) and ammonia.[15] The cyclic tetrameric spirophosphazene can perhaps be isolated under very critical conditions[52] but, here too, the predominant reaction is degradation to phosphorane.[15] Explanations for this behavior can be formulated in terms of the greater skeletal flexibility of the tetramer and higher homologues.[15]

Special problems are also involved when alkoxides or aryloxides interact with very high molecular weight poly(dichlorophosphazene).[24, 25] In the first place, longer reaction times and higher temperatures are required to effect complete substitution on a high molecular weight polymer. For example, total substitution of $(NPCl_2)_3$ by many alkoxides is complete in less than 12 hours in ether at 25°C[4] or in 4 hours in a boiling ether-toluene solution.[5] Complete substitution of the high polymer requires at least 16 hours reaction at 70°–80°C.[24, 25] When bulky substituents such as phenoxy are employed, more drastic conditions are required for total polymer substitution.[25] This results from the fact that polymer chains are often extensively coiled in solution, and considerable steric hindrance must be surmounted before a nucleophile can penetrate sufficiently close to the skeleton to effect substitution. A second problem encountered with the substitution of high polymers is that the temperature must not be so high that depolymerization to cyclic oligomers is induced (see Chapter 15).

H. Effect of the Solvent

A variety of anhydrous solvents have been used for substitution reactions of this type. These include diethyl ether,[4, 28, 46, 56] tetrahydrofuran,[8, 15, 25, 39, 42, 46] dioxane,[2, 25, 43] benzene,[2, 23, 25, 40–42, 46] toluene,[5, 30, 40, 42, 46, 48] xylene,[10, 11, 25, 44, 46, 48] acetone,[37, 38, 43] methyl ethyl ketone,[43] dimethylformamide,[43, 46] fluorocarbon-*t*-butylamine,[36] pyridine,[2, 16–18, 46, 52, 53] and excess of the alcohols or thiols being used as reagents for substitution.[18, 20, 21, 46]

A number of practical considerations must be taken into account when choosing a reaction solvent. Thus, anhydrous conditions must be maintained during substitution to avoid hydrolysis of the phosphorus–halogen bonds, and hydrophilic solvents such as tetrahydrofuran, dioxane, dimethylformamide, and pyridine must be exhaustively dried before use. The same is, of course, true for the alcohols or mercaptans used as reagents and solvents. Boiling xylene, benzene, or toluene is used as reaction medium when solid potassium or sodium hydroxide is employed as a base, since water formed during the reaction is rapidly removed by distillation. This can influence the reaction in curious ways. Thus, when phenoxylation of $N_3P_3Cl_4Ph_2$ takes place in benzene, the *cis*-non-geminal isomer of $N_3P_3Cl_2Ph_2(OPh)_2$ is formed. In tetrahydrofuran, from which water is not removed by azeotroping, both *cis* and *trans* isomers are formed.[42]

A second important solvent influence is connected with solvent polarity and with the solubility of sodium salts in the medium. For example, a preferred solvent is one in which the sodium alkoxide, aryloxide, or mercaptide is soluble, but in which sodium chloride is insoluble. Ethers often fall into this category.

The influence of the solvent on the reaction rate or mechanism is less easy to determine. It seems reasonable to assume that the more polar solvents should accelerate the rate of substitution by facilitating ionization of alkoxide, aryloxide, or mercaptide ions. Some evidence in favor of this view has been provided by the results of Carroll and Shaw for the alkylthiolysis and arylthiolysis of hexachlorocyclotriphosphazene.[45, 46] Thus, fully mercaptalated isopropylthio-substituted products were obtained when the solvent was dimethylformamide, four chlorine atoms were replaced in tetrahydrofuran, but only two were replaced in benzene. In general, the ease of substitution falls off with solvent changes in the order: dimethylformamide at 100°C $\gg$ boiling diglyme $>$ boiling tetrahydrofuran $>$ boiling benzene $>$ boiling diethyl ether. This order roughly parallels the expected decreasing solvation of $Na^{\oplus}$ in the reagent,[46] but it should be remembered that the reactions are heterogeneous in all of these media except dimethylformamide, and reagent solubility effects may thus be responsible for the differences. It is interesting to note that these solvent effects become manifest as an influence on the *number* of halogen atoms replaced and not simply as an effect on the reaction rate. For instance, in diethyl ether at room temperature, ethane-, propane-, butane-, or isobutanethiol bring about replacement of two chlorine atoms only, and further substitution does not normally occur if the reaction time or reagent concentration is increased. However, after 7 days in boiling ether, a tetrakis(ethylthio) derivative can be isolated. In boiling tetrahydrofuran, fully substituted ethane-, *n*-propane-, *n*-butane-, and benzylthiol derivatives were formed, but branched alkanethiolates (isopropyl or cyclohexyl) gave only tetrakis derivatives. Other nucleophilic substitutions, such as alcoholysis or aminolysis, do not show this striking sensitivity to different solvents.

Zhivukhin, Tolstoguzov, and Lukashevski[22] have examined the initial rate of the reaction between hexachlorocyclotriphosphazene and sodium methoxide in several different solvents using a potentiometric technique. The initial rate in dimethylformamide was found to be much more rapid than in acetone, dioxane, butanol, dioxane–acetone, or dioxane–methyl ethyl ketone mixtures. However, the almost instantaneous replacement of up to four chlorine atoms which occurred in dimethylformamide must be contrasted with the ultimate but slower replacement of all six halogen atoms in the other solvent systems.

I. Influence of the Base

In the majority of reactions, a base or sodium salt must be present before an alcohol, phenol, or thiol will interact cleanly with a chlorophosphazene. Only when reactions of this type are employed for the high-temperature curing of cross-linked polyphosphazene resins may a base be absent, but even here the presence of a hydrogen chloride acceptor markedly facilitates the process. In the absence of a base, alkanethiols or phenylthiol do not undergo substitution when heated with chloro- or alkoxycyclophosphazenes.[46] Benzyl alcohol reacts with $(NPCl_2)_3$ in dioxane in the absence of a base to yield a hydroxyphosphazane and benzyl chloride.[56a]

The use of sodium salts of alkoxides, aryloxides, or thiols has been mentioned throughout this chapter. Reactions involving such reagents are usually clean and rapid, and they yield sodium chloride as an easily removed side product. Sodium salts of alcohols, phenols, or thiols can be prepared by the addition of metallic sodium to a solution of the reagent in ether or tetrahydrofuran. When strongly acidic alcohols or phenols are employed, potassium or sodium hydroxides can be used *in situ* for salt formation. However, these methods are unsuitable when the alcohol or phenol undergoes side reactions with sodium or caustic alkali. Under these circumstances, milder conditions are employed, with the use of sodium carbonate, pyridine, or triethylamine as the hydrogen chloride acceptor. This is especially so for the formation of spirocyclic phosphazenes such as as XIII, XIV, XVII, XVIII, and XXI.

The designation of the base as a "hydrohalide acceptor" in these reactions is not strictly valid, since few reactions of this type take place at all in the absence of a base. It is more reasonable to view the base as a catalyst for facilitating ionization of the alcohol or phenol to alkoxide or aryloxide ion (XXIII).

$$ROH + Et_3N \rightleftharpoons RO^{\ominus} + \overset{\oplus}{Et_3NH}$$

XXIII

Sodium carbonate probably functions partly as a hydrohalide acceptor to remove hydrohalide liberated during a very slow direct phenolysis process

$$6\ ROH + (NPCl_2)_3 \rightarrow [NP(OR)_2]_3 + 6\ HCl$$

$$HCl + Na_2CO_3 \rightarrow NaCl + NaHCO_3$$

XXIV

(XXIV). However, it is also likely that the interaction of a phenol with sodium carbonate results in the formation of small amounts of sodium phenolate, which then react by the conventional route:

$$ROH + Na_2CO_3 \rightleftharpoons RONa + NaHCO_3$$

Reactions performed with pyridine as a base are especially sensitive to traces of moisture.[15] Furthermore, pyridine apparently forms an isolable crystalline complex with hexachlorocyclotriphosphazene.[52]

J. Influence of Temperature

The interaction of some nonfluorinated alcohols or alkoxides with chlorophosphazenes must be carried out at the lowest practicable temperature to avoid rearrangement of the alkoxycyclophosphazene to an oxophosphazane (XXV).[57] (see Chapter 13). Rearrangements of this kind occur slowly even at

$$[NP(OR)_2]_3 \xrightarrow{\text{Heat}} [N(R)\text{—}P(O)OR]_3$$

XXV

room temperature, when OR is methoxy or ethoxy.[7] For this reason, the reaction of sodium methoxide with hexachlorocyclotriphosphazene in benzene–methanol solution is carried out at 0°C.[2] The interaction of ethanol, pyridine, and hexachlorocyclotriphosphazenes is conducted at 0°–5°C.[2] When *n*-butanol replaces ethanol, the reaction takes place at room temperature.[2]

Mild reaction conditions are also preferred for the formation of some spirocyclophosphazenes, such as XIII,[14, 15] XXI,[17] or XXII[52] to avoid side reactions. For example, the reaction of $(NPCl_2)_{3\text{ or }4}$ with catechol in pyridine solution is conducted at –8° to 30°C.[15, 52]

However, more drastic temperature conditions can be used when the nucleophile is isopropoxy, benzyloxy, phenoxy, or fluoroalkoxy, and temperatures in the 30°–120°C range may be used. Aryloxy- and fluoroalkoxyphosphazenes do not rearrange thermally to oxophosphazanes. It should be noted that at 250°C fluoroalkoxy groups can be exchanged from one molecule to another.[57b]

When dimethylformamide is used as a solvent for the interaction of mercaptans or their sodium salts, the temperature must be kept below 100°C to avoid degradation reactions.[46]

K. Monophosphazenes and Related Compounds

All of the foregoing analysis has centered around the nucleophilic substitutions of halocyclophosphazenes or their linear high polymers. However, monophosphazenes are known to undergo similar reactions. For example, Zhmurova, Voitsekhovskaya, and Kirsanov have shown that *P*-trichloro-*N*-arylmonophosphazenes (phosphinimines) (XXVI) react with sodium phenoxide to yield the appropriate triphenoxy derivatives (XXVII).[43] Such tri-

$$\underset{\text{XXVI}}{ArN{=}PCl_3} + 3\ PhONa \rightarrow \underset{\text{XXVII}}{ArN{=}P(OPh)_3} + 3\ NaCl$$

phenoxymonophosphazenes are formed from both the monomeric trichloromonophosphazene, as shown above, or from the dimeric (cyclodiphosphazane) form. However, the triphenoxy derivatives exist only as monomers and not as cyclic dimers, possibly for steric reasons.

This reaction has been carried out for a number of *P*-trichloromonophosphazenes (XXVI) in which the group Ar is Ph; *o*-, *m*-, *p*-MeC_6H_4; 3,5-Me_2-C_6H_3; *o*-, *m*-, *p*-ClC_6H_4; 2,4-$Cl_2C_6H_3$; 3,5-$Cl_2C_6H_3$; 2,4,6-$Cl_3C_6H_2$; *o*-, *m*-, *p*-BrC_6H_4; 2,4-$Br_2C_6H_3$; 2,4,6-$Br_3C_6H_2$; *p*-$MeOC_6H_4$; *p*-$EtOC_6H_4$; *o*, *m*-, *p*-$NO_2C_6H_4$; 2,4-$(NO_2)_2C_6H_3$; 2,6-Cl_2-4-$NO_2C_6H_2$. The compounds in which Ar is 2,4,6-$Cl_3C_6H_2$; 2,4,6-$Br_3C_6H_2$; *p*-NO_2-C_6H_4; 2,4-$(NO_2)_2C_6H_4$; or 2,6-Cl_2-4-$NO_2C_6H_2$ are crystalline solids, but the rest are viscous liquids. Except in the case where Ar is phenyl, they cannot be distilled at reduced pressure without decomposition. All these materials are hydrolyzed rapidly by water or moist air to form diphenyl esters of arylamidophosphoric acids, by the reaction

$$ArN{=}P(OPh)_3 + H_2O \rightarrow ArNH{-}PO(OPh)_2 + PhOH$$

The compounds $SCl_2P{-}N{=}PCl_3$ and $SP(OPh)_2{-}N{=}PCl_3$ react with sodium phenoxide according to the schemes[58, 59]

$$S{=}PCl_2{-}N{=}PCl_3 + 5\ NaOPh \rightarrow S{=}P(OPh)_2{-}N{=}P(OPh)_3 + 5\ NaCl$$

$$S{=}P(OPh_2){-}N{=}PCl_3 + NaOPh \rightarrow S{=}P(OPh_2{-}N{=}P(OPh)_3 + 3\ NaCl$$

It has been shown that monophosphazenes of structure $Cl_3C{-}Cl_2C{-}N{=}PCl_3$ react with primary alcohols at 20°–25°C to yield compounds of structure $Cl_3C{-}Cl_2C{-}N{=}PCl_2(OAlk)$.[60] However, these materials are unstable and decompose with elimination of alkyl halide.

L. The Reaction Mechanism

The nature of the reaction mechanism during alcoholysis, phenolysis, or thiolysis is intimately connected with questions about the mechanisms of the hydrolysis, aminolysis, metathetical exchange, and skeletal degradation of halophosphazenes. These latter mechanisms are considered in Chapters 5, 7, 8, and 14. The available evidence about the mechanism of alcoholysis, phenolysis, and thiolysis is incomplete and fragmentary, but several pertinent facts can be discerned and these are listed below.

(a) The direct reaction between a halophosphazene and an alcohol, phenol, or thiol in the absence of a base is a slow reaction which often yields side products or decomposition species rather than substituted phosphazenes. However, when sodium alkoxides, aryloxides, or thiolates are used, or when a strong base is employed with the alcohol or phenol, the reaction is rapid. This suggests that alkoxide, aryloxide, or thiolate ions are the reactive species during "normal" substitution. This has been confirmed by the kinetic studies of Sorokin and Latov,[61] who showed that, when sodium ethoxide reacts with $(NPCl_2)_3$ or $N_3P_3Cl_4(OBu)_2$, the nucleophile is $EtO^{\ominus}$ and not EtONa. The ionization to ethoxide ion is enhanced in high dielectric constant solvents and this leads to increased substitution rates in such media.

(b) Both the degree of substitution and the pattern of halogen replacement are sensitive to the steric characteristics of the nucleophile. Thus, branched alkoxides, branched alkylthiolates, or phenoxides cause total substitution only with difficulty. Phenoxide ion substitutes non-geminally, possibly because of steric influences.

(c) The possibility also exists that the non-geminal replacement pattern observed with phenoxide and some alkoxides may result from electron supply from alkoxide or aryloxide to phosphorus, which lowers the reactivity of a Cl—P—OR unit below that of a Cl—P—Cl unit. In view of the results of polarographic experiments with alkoxy- and aryloxyphosphazenes,[62] it seems unlikely that resonance effects involving the phenyl groups are responsible for the electron supply. However, donation of the oxygen lone-pair electrons toward phosphorus is possible.

(d) Alkylthiolate ions substitute geminally, rather than non-geminally, in spite of the lower electronegativity of alkylthio substituents than chloro groups.[46] It has been suggested[46] that the high polarizability of the Cl—P—SR unit compared with that of the Cl—P—Cl grouping may be the reason for this behavior.

(e) Geminal *cyclization* to form spirocyclophosphazenes occurs readily when five-, six-, or seven-membered rings at phosphorus are involved.[8, 15] This implies that, when geminal substitution is *not* retarded by steric effects, there are no serious electronic influences which inhibit the geminal process. Some confirmation of this view is provided by the fact that when $(NPCl_2)_3$ reacts with catechol, 2,3-dihydroxynaphthalene, 2,2′-dihydroxybiphenyl, 1,8-dihydroxynaphthalene, or toluene-3,4-dithiol, no non-geminal, uncyclized species, such as XXVIII, are isolated.[8, 15]

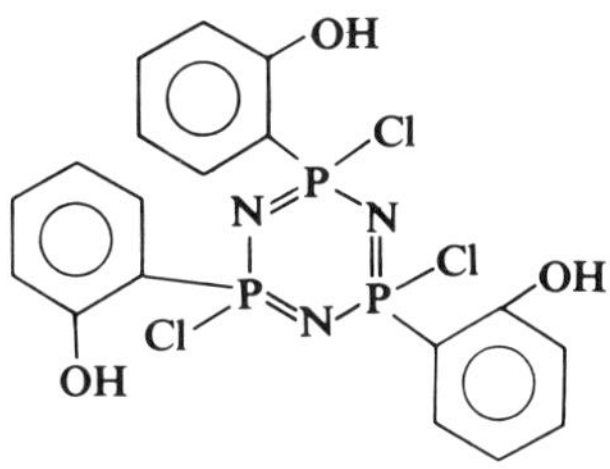

XXVIII

(f) When butoxide ion replaces chlorine in hexachlorocyclotriphosphazene, the reaction rate decreases in the order $(NPCl_2)_3 > N_3P_3Cl_5(OBu) > N_3P_3Cl_4(OBu)_2 > N_3P_3Cl_3(OBu)_3$, with the activation energy increasing in the order 10.0, 11.3, 14.5, and 17.2 kcal/mole, respectively.[61] If this rate decrease is the result of electron supply from butoxy groups to the ring, an S_N1-type mechanism is precluded since electron supply would enhance the ionization of P—Cl to $P^{\oplus}Cl^{\ominus}$ and, thus, increase the rate. Second-order kinetics (first order in both phosphazene and alkoxide) are, in fact, observed and this suggests an S_N2 type of mechanism.

(g) If the mechanism does proceed by an S_N2-type process, there are two possibilities for the approach of the nucleophile to phosphorus prior to attainment of the transition state. The first mechanism is one involving a side attack on phosphorus in the plane of the ring, as shown in XXIX. Either retention or

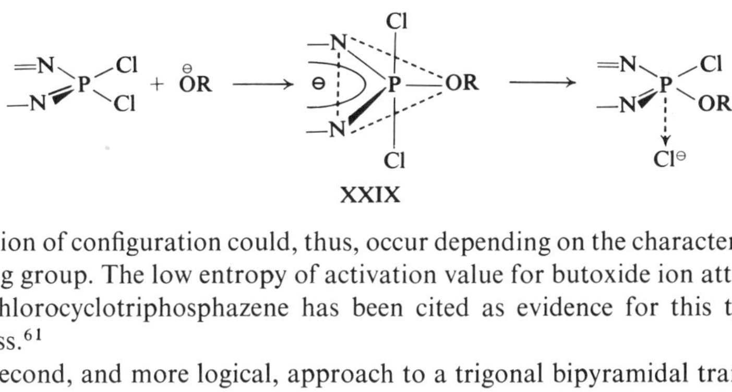

XXIX

inversion of configuration could, thus, occur depending on the character of the leaving group. The low entropy of activation value for butoxide ion attack on hexachlorocyclotriphosphazene has been cited as evidence for this type of process.[61]

A second, and more logical, approach to a trigonal bipyramidal transition state is via a backside attack, leading to inversion of configuration, as shown in XXX.

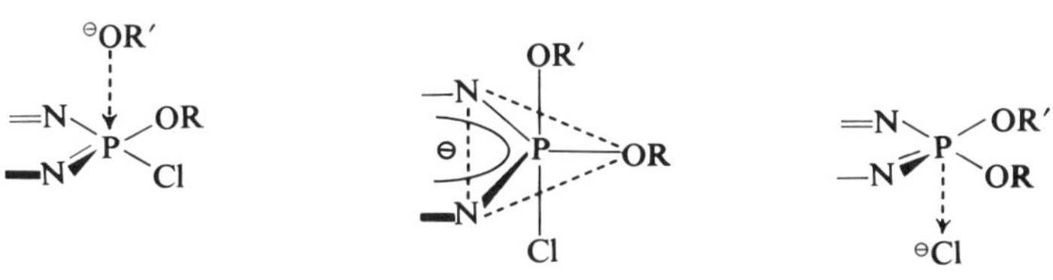

XXX

No unambiguous evidence for or against inversion of configuration has yet been produced. However, this type of mechanism seems reasonable on theoretical grounds since the attacking nucleophile could approach phosphorus along the z axis of an unfilled $d_z{}^2$ orbital. Additional evidence for this type of mechanism is provided by the results of reactions between nucleophiles such as HO^- or MeO^- and spirocyclic aryloxyphosphazenes. Thus, attack by RO^- on a unit such as XXXI cannot take place by any route which forces the two

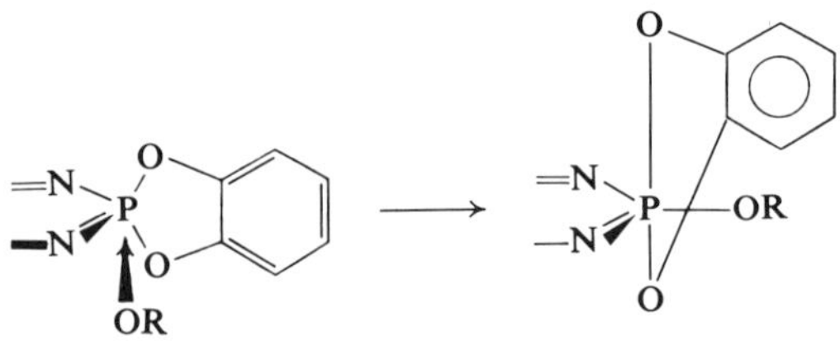

XXXI XXXII

P—O—Ph units to move to axial positions in the transition state (XXXII), and the side attack mechanism is, thus, prohibited. As discussed earlier in

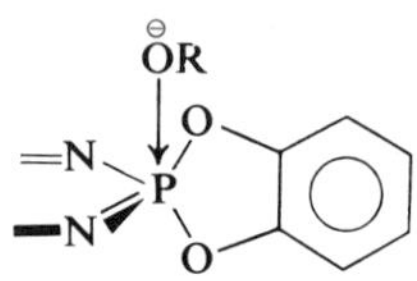

XXXIII

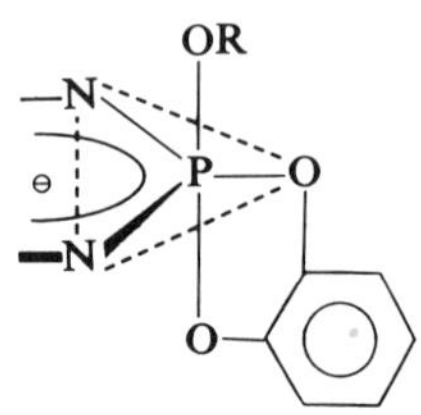

XXXIV

Chapter 4, nucleophilic attack by HO^- (and RO^-) on spiroaryloxyphosphazenes with five-membered rings at phosphorus, as in XXXIII, is much more rapid than attack on molecules with seven-membered rings or two independent substituents at phosphorus.[50] This is consistent with the ease with which the five-membered ring can assume the axial and equatorial positions (XXXIV), thus allowing the O—P—O bond angle to approximate to 90°–95°. Thus, the energy of the transition state would be lowered by the presence of a five-membered ring at phosphorus and the rate would be enhanced. In the absence of evidence to the contrary it must be assumed that attack by RO^- follows the same pathway as attack by HO^-, i.e., by an S_N2 backside attack. The main differences between substitution by HO^- and RO^- would be expected when the group, R, is bulky or highly polarizable.

REFERENCES

1. H. R. Allcock and R. J. Best, *Can. J. Chem.* **42**, 447 (1964).
2. B. W. Fitzsimmons and R. A. Shaw, *Chem. Ind.* (*London*) p. 109 (1961); *J. Chem. Soc. London* p. 1735 (1964).
3. M. Yokoyama and F. Yamada, *J. Chem. Soc. Jap., Ind. Chem. Sect.* **66**, 613 (1963).
4. Ė. T. McBee, H. R. Allcock, R. Caputo, A. Kalmus, and C. W. Roberts, *U.S. Govt. Astia Rep.* **AD 209,669** (1959).
5. R. Rätz, H. Schroeder, H. Ulrich, E. Kober, and C. Grundmann, *J. Amer. Chem. Soc.* **84**, 551 (1962).
6. R. F. Rätz and C. Grundmann, U.S. Pat. 2,876,247 (1959).
7. H. R. Allcock and E. J. Walsh, unpublished information (1969).
8. H. R. Allcock and R. L. Kugel, *Inorg. Chem.* **5**, 1016 (1966).
9. H. R. Allcock and R. L. Kugel, *J. Amer. Chem. Soc.* **91**, 5452 (1969).
10. G. M. Nichols, Ger. Pat. 1,182,660 (1964).
11. G. M. Nichols, U.S. Pat. 3,316,330 (1967); *U.S. Govt. Res. Rep.* **ASD-TDR 62–372** (1962).
12. E. Kober, H. Lederle, and G. Ottmann, *Inorg. Chem.* **5**, 2239 (1966).
13. B. W. Fitzsimmons and R. A. Shaw, *Inorg. Synt.* **8**, 77 (1966).
14. H. R. Allcock, *J. Amer. Chem. Soc.* **85**, 4050 (1963).
15. H. R. Allcock, *J. Amer. Chem. Soc.* **86**, 2591 (1964).
16. J. Parrod and R. Pornin, *C. R. Acad. Sci.* **258**, 3022 (1964).
17. M. S. Chang and A. J. Matuszko, *Chem. Ind.* (*London*) p. 410 (1962).
18. B. Dishon, *J. Amer. Chem. Soc.* **71**, 2251 (1949).
19. M. Yokoyama, *J. Chem. Soc. Jap., Pure Chem. Sect.* **81**, 158 (1960).
20. B. W. Fitzsimmons, C. Hewlett, K. Hills, and R. A. Shaw, *J. Chem. Soc., A* p. 679 (1967)
21. M. F. Sorokin and V. K. Latov, *Zh. Obshch. Khim.* **35**, 1471 (1965).
22. S. M. Zhivukhin, V. B. Tolstoguzov, and Z. Lukashevski, *Russ. J. Inorg. Chem.* **10**, 901 (1965).
23. G. Allen, D. J. Oldfield, N. L. Paddock, F. Rallo, J. Serregi, and S. M. Todd, *Chem. Ind.* (*London*) p. 1032 (1965).
24. H. R. Allcock and R. L. Kugel, *J. Amer. Chem. Soc.* **87**, 4216 (1965).
25. H. R. Allcock, R. L. Kugel, and K. J. Valan, *Inorg. Chem.* **5**, 1709 (1966).
26. C. Hamalainen, W. A. Reeves, and J. D. Guthrie, *Text. Res. J.* **26**, 145 (1956).
27. H. R. Allcock and G. Konopski, unpublished work (1969).
28. E. T. McBee, L. Brinkmann, and H. P. Braendlin, *U.S. Govt. Res. Rep.* **AD 254,982** (1960).
29. H. Lederle, E. Kober, and G. Ottmann, *J. Chem. Eng. Data* **11**, 221 (1966).
30. G. Ottmann, H. Lederle, and E. Kober, *Ind. Eng. Chem., Prod. Res. Develop.* **5**, 202 (1966).
31. E. Kober, H. Lederle, and G. F. Ottmann, *U.S. Govt. Res. Rep.* **AD 430,861** (1964).
32. M. V. Lenton, B. Lewis, and C. A. Pearce, *Chem. Ind.* (*London*) p. 1387 (1964).
33. M. V. Lenton and B. Lewis, *J. Chem. Soc., A* p. 665 (1966).
34. S. H. Rose, *J. Polym. Sci., Part B* **6**, 837 (1968).
35. E. Kober, H. Lederle, and G. Ottmann, *Amer. Soc. Lubric. Eng. Trans.* **7**, 389 (1964).
36. T. J. Mao, R. D. Dresdner, and J. A. Young, *J. Inorg. Nucl. Chem.* **24**, 53 (1962).

36a. M. R. Pitina, T. M. Ivanova, and N. I. Shvetsov-Shilovskii, *J. Gen. Chem. USSR* **37**, 1968 (1967).

37. E. T. McBee, P. Johncock, and H. P. Braendlin, *U.S. Govt. Res. Rep.* **AD 254,984** (1960).
38. E. T. McBee, K. Okuhara, and C. J. Morton, *Inorg. Chem.* **5**, 450 (1966).
39. D. Dell, B. W. Fitzsimmons, and R. A. Shaw, *J. Chem. Soc., London* p. 4070 (1965).

39a. D. Dell, B. W. Fitzsimmons, R. Keat, and R. A. Shaw, *J. Chem. Soc., A* p. 1680 (1966).
40. V. B. Tolstoguzov, V. V. Pisarenko, and V. V. Kireev, *Russ. J. Inorg. Chem.* **10**, 382 (1965).
41. C. T. Ford, F. E. Dickson, and I. I. Bezman, *Inorg. Chem.* **4**, 419 (1965).
42. C. T. Ford, J. M. Barr, F. E. Dickson, and I. I. Bezman, *Inorg. Chem.* **5**, 351 (1966).
43. I. N. Zhmurova, I. Yu. Voitsekhovskaya, and A. V. Kirsanov, *Zh. Obshch. Khim.* **31**, 3741 (1961).
44. E. Kober, H. Lederle, and G. Ottmann, *Inorg. Chem.* **5**, 2239 (1966).
45. A. P. Carroll and R. A. Shaw, *Chem. Ind. (London)* p. 1908 (1962).
46. A. P. Carroll and R. A. Shaw, *J. Chem. Soc., A* p. 914 (1966).
47. H. R. Allcock and E. J. Walsh, *J. Amer. Chem. Soc.* **91**, 3102 (1969).
47a. L. E. A. Godfrey and J. W. Schappel, *Ind. Eng. Chem., Prod. Res. Develop.*, **9**, 426 (1970).
48. E. Kober, H. Lederle, and G. T. Ottmann, *U.S. Navy Res. Rep.* **Nobs-90092** (1964).
49. M. R. Pitina and N. I. Shvetsov-Shilovskii, *J. Gen. Chem. USSR* **36**, 517 (1966).
50. H. R. Allcock and E. J. Walsh, *J. Amer. Chem. Soc.* **91**, 3102 (1969).
51. H. R. Allcock and R. L. Kugel, unpublished work (1969).
52. R. Pornin, *Bull. Soc. Chim. Fr.* [5] **258**, No. 9, 2861 (1966).
53. A. J. Matuszko and M. S. Chang, *J. Org. Chem.* **31**, 2004 (1966).
53a. A. Wende and D. Joel, *Z. Chem.* **3**, 467 (1963).
54. B. Yanik, V. Zheshutko, and T. Pel'char, *J. Gen. Chem. USSR* **36**, 1451 (1966).
55. F. Rallo, *Ric. Sci. Chim.* **8**, 1134 (1965).
56. M. R. Pitina, T. M. Ivanova, and N. I. Shvetsov-Shilovskii, *J. Gen. Chem. USSR* **37**, 1968 (1967).
56a. M. Kajiwara and H. Saito, *Kogyo Kagaku Zasshi*, **74**, 619 (1971).
57. B. W. Fitzsimmons, C. Hewlett, and R. A. Shaw, *J. Chem. Soc., London* p. 4459 (1964).
57a. E. N. Niecke, H. Thamm, and O. Glemser, *Z. Naturforsch.* **26B**, 366 (1971).
57b. V. N. Prons, M. P. Grinblat, A. L. Klebanskii, and G. A. Nikolaev, *J. Gen. Chem. USSR*, **40**, 2109 (1970).
58. E. Fluck, *Z. Anorg. Allg. Chem.* **320**, 64 (1963).
59. A. V. Kirsanov and I. N. Zhmurova, *Zh. Obshch. Khim.* **28**, 2478 (1958).
60. V. I. Shevchenko, A. A. Koval, and A. V. Kirsanov, *J. Gen. Chem. USSR* **38**, 541 (1968).
61. M. F. Sorokin and V. K. Latov, *Kinet. Katal.* **7**, 42 (1966); *Kinet. Catal. (USSR)* p. 35 (1966).
62. H. R. Allcock and W. J. Birdsall, *Inorg. Chem.* (1971).

Chapter 7

AMINOLYSIS OF HALOPHOSPHAZENES

A. Introduction

Halophosphazenes or halophosphazanes react with primary or secondary amines to eliminate hydrogen halide and form an amino-substituted phosphazene or phosphazane. Thus, ammonia or primary amines interact with hexachlorocyclotriphosphazene in the following manner:

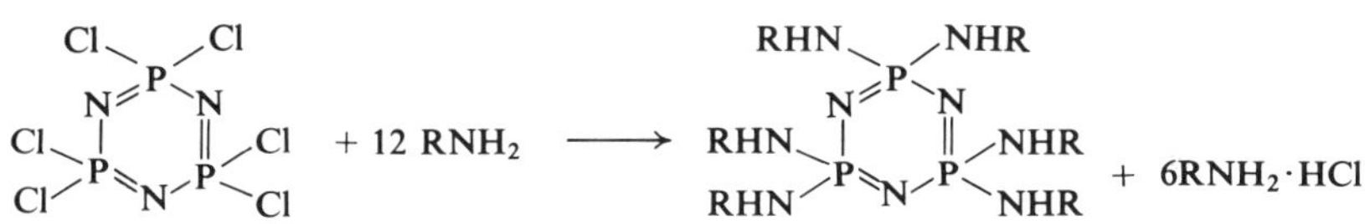

An excess of the amine often functions as the hydrohalide acceptor. Similarly, secondary amines, such as dimethylamine or piperidine, replace halogen atoms in halocyclophosphazenes to yield the appropriate aminophosphazenes by processes such as

$$(NPCl_2)_3 + 12\,R_2NH \rightarrow [NP(NR_2)_2]_3 + 6\,R_2NH \cdot HCl$$

Such reactions may be performed in a suitable solvent, with the process being facilitated by the insolubility of the amine hydrochloride in the medium, or they may be carried out in the liquid amine or even in a heterophase aqueous organic medium.

Aminolysis reactions have attracted considerable attention during recent years, and some mechanistic information is now available. The reactions of a wide variety of different amines have been studied and aminophosphazenes comprise (along with alkoxy- and aryloxyphosphazenes) one of the two most extensive classes of phosphorus–nitrogen compounds. Aminolysis reactions are not restricted to chlorophosphazenes nor to cyclic trimers. Fluoro- and bromophosphazenes undergo such reactions also, as do cyclic tetrameric, higher cyclic, and linear high polymeric halophosphazenes. Although aminophosphazenes have been investigated as intermediates in the synthesis of heat-resistant polymers, their pharmacological and biomedical properties have so far shown the most promise for useful applications.

The organization of this chapter will follow the format used for the alkoxide and aryloxide substitutions in Chapter 6. Thus, practical problems involving aminolysis reactions will be considered first, followed by an analysis of the influence of different amines, different halogen atoms, changes in ring or chain size, solvent, hydrohalide acceptor, and temperature. Monophosphazenes, cyclophosphazanes, and a summary of the mechanistic information will be covered at the end of the chapter.

B. Experimental Considerations

Two examples are given below describing the synthesis of a secondary alkylaminocyclophosphazene, and an aromatic primary aminophosphazene.

1. Synthesis of 1,3,5-Trichloro-1,3,5-tris(dimethylamino)cyclotriphosphazene, $[NPCl(NMe_2)]_3$

Hexachlorocyclotriphosphazene (76.0 g, 0.22 mole) is dissolved in anhydrous diethyl ether, and the solution is cooled to −78°C in a Dry Ice–methanol bath. Anhydrous dimethylamine (60 g, 1.33 moles) is then added slowly during 1.5 hours. The reaction mixture is stirred at 25°C for 18 hours and the dimethylamine hydrochloride precipitate is removed by filtration. Evaporation of diethyl ether from the filtrate yields 73 g (88 %) of a crude mixture of the 1,3,5-tris(dimethylamino)- and the 1,1,2-tris(dimethylamino)-trichlorocyclotriphosphazene derivatives. The non-geminal isomer is obtained by several fractional crystallizations from *n*-hexane to give material of m.p. 101°–104°C (46.8 g, 57 %).

2. Synthesis of Hexakis(phenylamino)cyclotriphosphazene, $[NP(NHPh)_2]_3$

Hexachlorocyclotriphosphazene (11.6 g, 0.033 mole) is dissolved in aniline (150 ml) and the solution is heated at 100°C for 15 hours. When cooled to 25°C, the mixture crystallizes, and it is then washed with water and with ethanol to remove excess aniline and aniline hydrochloride. The residue is dissolved in glacial acetic acid, from which it is precipitated by petroleum. Final purification is effected by solution in tetrahydrofuran, addition of ethanol, and slow crystallization of the product (21.6 g, 94.3 %). It is a white or pale pink solid, m.p. 271°–273°C.

C. Scope of the Aminolysis Reaction

Halophosphazenes have been allowed to react with a large number of different amines. Aliphatic primary amines that have been used include ammonia,[1–13] methylamine,[5, 7, 8, 11, 13, 14–26] ethylamine,[5, 13–17, 19–23, 27] *n*-propylamine,[15–17, 19, 20, 23, 28, 29] *n*-butylamine,[15–17, 19, 20, 28, 30–32] *n*-pentylamine,[15–17] *n*-hexylamine,[15] allylamine[31, 33] 2,2,2-trifluoroethylamine,[30, 31] isopropylamine,[19, 20, 34] isobutylamine,[19, 20, 32] *sec*-butylamine,[19, 20] *tert*-butylamine,[13, 19–21, 27, 31] cyclohexylamine,[13, 19, 21, 32, 35] and benzylamine.[21, 32] A number of aromatic primary amines have also been employed, including aniline,[5, 14, 18, 22, 36–42] *o*, *m*, and *p*-toluidines,[5, 14, 31, 42] *m*-trifluoromethylaniline,[43] and *m*- and *p*-chloroanilines.[31, 40]

For secondary amines, by far the most work has been reported for dimethylamine,[7, 8, 11, 13, 17, 18, 20–22, 24, 25, 44–52] although the reactions of halophosphazenes with diethylamine,[11, 17, 20–22, 53, 54] di-*n*-propylamine,[17] di-*n*-butylamine,[17] ethylenimine,[55–62] pyrrolidine,[61, 63–65] piperidine,[11, 13, 14, 21, 22, 37, 38, 47, 53, 66–68] morpholine,[14, 69–71] and *N*-methylaniline[14, 31, 41] have also been described in some detail.

Amino acid esters, such as the ethyl ester of glycine, react with chlorocyclophosphazenes[60, 66, 72] or polyphosphazenes.[23] The methyl ester of β-alanine,[66] the methyl esters of L-alanine, L-leucine, and L-phenylalanine, adenine, thymine, uracil, and guanine react similarly. Diamines such as hydrazine[73] or phenylhydrazine[38] react as monoamines, with two hydrazine or phenylhydrazine residues becoming attached to each phosphorus atom. However, *o*-phenylenediamine undergoes a cyclosubstitution at phosphorus,[5, 74] and so apparently does thiourea.[75] Reactions of halophosphazenes with *o*-aminophenol have also been reported,[76, 77] but the initial substitution is followed by degradation. Diamino compounds, such as pentamethylenediamine, hexamethylenediamine, octamethylenediamine, or 4,4′-diaminodicyclohexylmethane react in such a way that phosphazene rings are linked together to form cyclolinear polymers,[78] and similar results have been obtained for other aromatic diamines.[79] Halocyclophosphazenes also react with potassium pyrrole, to eliminate sodium chloride and yield the pyrrole-substituted phosphazenes.[30]

D. General Properties and Uses of Aminophosphazenes

Most aminocyclophosphazenes are stable, white, crystalline solids, They are usually soluble in, and can be recrystallized from, organic solvents, although the amino-substituted compounds, $[NP(NH_2)_2]_{3\text{ and }4}$, are soluble in water, rather than organic media. Amino-substituted high polymers are either glassy materials or transparent, flexible, film-forming plastics, similar in appearance to poly(ethyleneterephthalate) (Mylar) or poly(vinyl chloride–vinylidene chloride) (Saran). As might be expected, aminophosphazenes are basic and form hydrohalide salts. However, in aqueous media, aminophosphazenes as a class are less hydrolytically stable than alkoxy- or aryloxyphosphazenes, and hydrolysis to amine and phosphate may occur at a measurable rate in homogeneous media. At high temperatures also, aminophosphazenes show a tendency to decompose and condense by elimination of an amine and such processes limit the thermal stability of many derivatives.

However, certain aminophosphazenes have been shown to possess interesting pharmacological properties. For example, ethyleneiminocyclophosphazenes show antitumor activity,[80] and they have also been used as chemosterilant insecticides[81] (under the trade name of Apholate). Many aminophosphazenes are currently being tested for use in pharmacological and biomedical applications. The high polymers, in particular, are being examined for use as semipermeable membranes and possibly as body implantation plastics.[23] Hexakis(amino)cyclotriphosphazene has been studied for possible use as an ultra-high-capacity fertilizer.[81a]

E. Influence of the Amine on Substitution

The following discussion will consider three questions concerning the influence of the nucleophile. Firstly, how do different nucleophiles affect the ease and degree of halogen replacement? Secondly, which amines favor geminal replacement patterns and which favor non-geminal substitution? Thirdly, which amines give *cis* or *trans* products? The possibility of *cis–trans* isomerization during or following substitution will also be considered.

1. Substitution by Primary Amines

a. *Ammonia*

Ammonia reacts with hexachlorocyclotriphosphazene, $(NPCl_2)_3$, and with the tetramer $(NPCl_2)_4$ to give the fully ammonolyzed cyclophosphazenes, $[NP(NH_2)_2]_{3\text{ and }4}$. Complete halogen replacement can be effected either in liquid ammonia under pressure during 48 hours reaction or after 5 hours reaction at –40°C in a diethyl ether–liquid ammonia medium.[6] Both products are white, slightly hygroscopic compounds which decompose with loss of ammonia when heated. They are insoluble in organic solvents, but dissolve in water and undergo slow hydrolysis.

However, if an ethereal solution of $(NPCl_2)_3$ is allowed to react for 1 hour with aqueous ammonium hydroxide, or if ammonia gas is bubbled through a solution of $(NPCl_2)_3$ in ether or dioxane, a diaminotetrachloro derivative is formed:[4]

$$(NPCl_2)_3 + 4\,NH_3 \rightarrow N_3P_3Cl_4(NH_2)_2 + 2\,NH_4Cl$$

Similarly, when a stream of ammonia gas is passed through an ethereal solution of the tetramer, $(NPCl_2)_4$, ammonium chloride is precipitated and two partly substituted species, $N_4P_4Cl_6(NH_2)_2$ and $N_4P_4Cl_4(NH_2)_4$, are isolated.[4]

The pattern of halogen replacement in $(NPCl_2)_3$ has been examined in some detail[9, 10] and it has been shown that substitution by ammonia takes place by a geminal mechanism, as shown below. The structure of the diamino compound

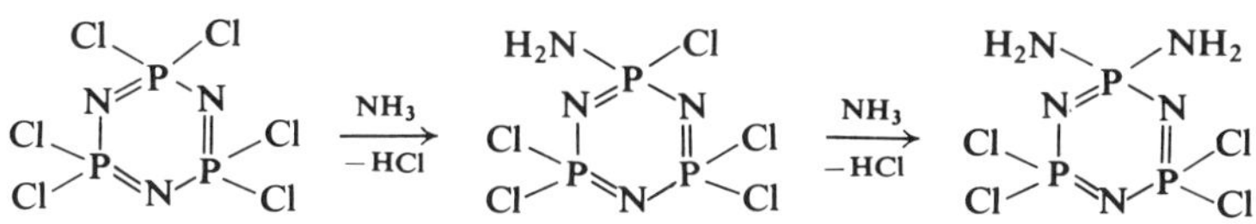

was elucidated by treatment with 2 moles of phosphorus pentachloride to convert both amino units to $N{=}PCl_3$ groupings. The ^{31}P NMR spectrum of this product was an AB_2C_2 system which was consistent with a geminal rather than a non-geminal arrangement of the ligands.[9, 10]

Simple inductive or conjugative arguments do not explain the geminal substitution pattern, since replacement of one chlorine by amino would be expected to enhance the electron density at that phosphorus and thereby retard nucleophilic attack at the same phosphorus. An explanation for this anomaly is given at the end of this chapter.

A similar geminal replacement pattern is evident when 1,1-diphenyl-3,3,5,5-tetrachlorocyclotriphosphazene (I) reacts with excess ammonia in chloroform at –78°C.[18] Subsequent treatment of this product (II) with an excess of dimethylamine yields the bis(dimethylamino) bisamino derivative (III). The

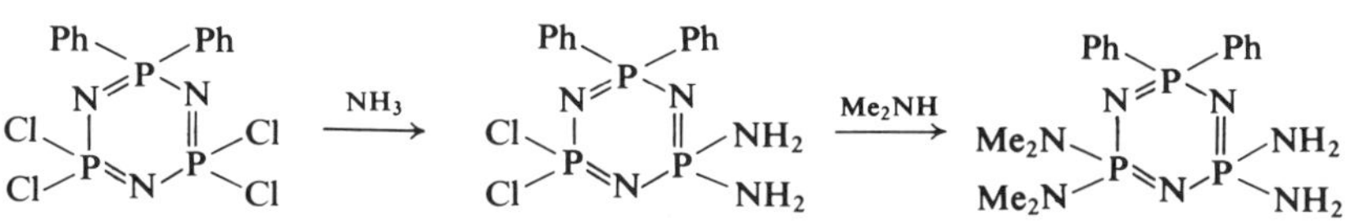

I II III

dimethylamino ^{1}H NMR spectrum of this product is a doublet, which confirms the geminal arrangement of ligands. Basicity evidence has also been used to confirm the geminal structure of II.[13] A tetrakisamino derivative is formed along with II when I reacts with excess of ammonia.

Octachlorocyclotetraphosphazene, $(NPCl_2)_4$, apparently reacts with ammonia by a non-geminal mechanism to give 1,5-diamino-1,3,3,5,7,7-hexachlorocyclotetraphosphazene.[81b]

It should be noted that the reaction of ammonia with linear high molecular weight poly(dichlorophosphazene) leads to the initial formation of cross-linked matrix polymers,[22] although if other substituents (such as diethylamino) are present, high polymers such as $[N—P(NEt_2)(NH_2)]_n$ are formed.[23]

b. *Unbranched Alkylamines*

In general, unbranched primary alkylamines readily replace all the halogen atoms in $(NPCl_2)_3$, $(NPCl_2)_4$, or high molecular weight $(NPCl_2)_n$. For example, $(NPCl_2)_3$ reacts rapidly with excess methylamine below room temperature to give the hexakis(methylamino) derivative $[NP(NHMe)_2]_3$.[19] *n*-Propylamine, *n*-butylamine, *n*-pentylamine, and *n*-hexylamine react similarly with $(NPCl_2)_3$ in boiling ether[19] or benzene.[15] Allylamine replaces all the halogen atoms in $(NPCl_2)_3$ within $5\frac{1}{2}$ hours at 20°–25°C in benzene solution. A hexakis(2,2,2-trifluoroethylamino)cyclotriphosphazene can be prepared at 35°C.[30] The reactions of unbranched primary alkylamines with octachlorocyclotetraphosphazene proceed as rapidly or even faster than with the trimer, and octakisamino derivatives are readily obtained even under mild conditions.[20, 33] The same is true for the reactions of high molecular weight poly(dichlorophosphazene) with methylamine, ethylamine, *n*-propylamine, or *n*-butylamine.[22, 23] This fact is particularly surprising in view of the steric hindrance expected from chain coiling in solution.

However, if the reaction conditions are sufficiently mild and particularly when the halophosphazene is in excess, partly substituted aminohalophosphazenes can be isolated. For example, the interaction of $(NPCl_2)_3$ in diethyl ether with a 30% aqueous dimethylamine solution for 10 minutes yielded a bis(methylamino)tetrachlorocyclotriphosphazene.[5] Similarly, monoamino pentachloro derivatives have been prepared by the reaction of 1:2 molar ratios of $(NPCl_2)_3$ to methylamine, ethylamine, *n*-propylamine, *n*-butylamine, or *n*-pentylamine in a diethyl ether–water medium:[16, 17, 24]

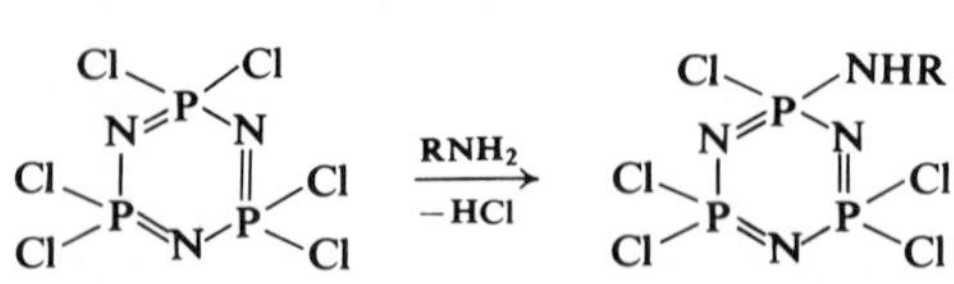

Methylamine and ethylamine both yield amino derivatives when allowed to react with the tetramer, $(NPCl_2)_4$.[14] Allylamine yields a monoallylamino compound, $N_3P_3Cl_5(NHC_3H_5)$, in ether solution, when triethylamine is used as a base.[33] A penta-*n*-butylamino monochloro derivative, $N_3P_3Cl(NHC_4H_9)_5$, has also been reported.[30] Methylamine gives a monosubstituted derivative with $(NPF_2)_3$.[25]

The question of the pattern of halogen replacement has been examined when methylamine or ethylamine is used as a base. ^{31}P NMR[8, 26] and dipole moment data[26] have been employed to show that methylamine reacts with $(NPCl_2)_3$ by both geminal and non-geminal routes. Thus, three isomers of formula $N_3P_3Cl_4(NHMe)_2$ can be isolated from the reaction of aqueous methylamine with ethereal $(NPCl_2)_3$. Lehr[26] showed that these were the geminal isomer (m.p. 128°C), the non-geminal *trans* (m.p. 103°C), and the non-geminal *cis* isomer (m.p. 149°C). Ethylamine yields a *non*-geminal *bis*ethylamino derivative $N_3P_3Cl_4(NHEt)_2$.[27]

Other work on this aspect has involved the aminolysis of di- and triphenylhalocyclotriphosphazenes[21, 28] and the results of these experiments are discussed in sections F and M of this chapter.

c. *Branched Primary Alkylamines*

Isopropylamine, isobutylamine, *sec*-butylamine, and cyclohexylamine can be induced to replace all the halogen atoms in $(NPCl_2)_3$ only in boiling benzene or at 140°–180°C under pressure.[19] With milder conditions in diethyl ether near room temperature, the tetrasubstituted derivatives are the preferred products.[19] *tert*-Butylamine yields no trimeric products containing more than four amino groups unless drastic reaction conditions are employed.[19, 82] Substitution of halogen by branched primary amines is easier when the tetramer, $(NPCl_2)_4$, is used.[20] Nevertheless, autoclave or sealed-tube techniques must be used, and milder conditions lead to partial replacement of halogen. For example, isopropylamine reacts with $(NPCl_2)_4$ in organic media to yield a bis(isopropylamino) hexachloro derivative,[14] and with $(NPCl_2)_3$ to yield mono-, di-, tetra-, and hexasubstituted products.[34]

The pattern of halogen replacement with branched primary amines is not clear-cut. Isopropylamine reacts with $(NPCl_2)_3$ to yield both geminal and non-geminal products.[34] This information was derived from basicity measurements, and from the replacement of residual chlorine atoms by dimethylamino units, followed by NMR examination. For example, the bis(isopropylamino)-tetrachlorocyclotriphosphazene reacts with excess dimethylamine to give the hydrochloride of *trans*-non-geminal $N_3P_3(NHPr^i)_2(NMe_2)_4$. However, the initial *cis* or *trans* configuration of the isopropylamino groups could not be inferred from this because of the possibility of isomerization during dimethylaminolysis.[34]

t-Butylamine interacts with $(NPCl_2)_3$ by a geminal reaction path.[82] Mono-, di-, tetra-, and hexasubstituted products are obtained, but, significantly, no trisubstituted derivatives are formed. Only one isomer was obtained for each of the derivatives prepared. The geminal structure of the disubstituted derivative was confirmed by replacement of the remaining halogen atoms by dimethylamino groups, followed by an ^1H NMR analysis of the product.[82] In any case, it is clear that the expected inductive electron supply from amino to phosphorus should retard attack at the same phosphorus, and it must be assumed that the factors favoring the geminal mechanism are sufficiently powerful to overcome this effect. Possible explanations are discussed later (Section M).

d. *Aromatic Primary Amines*

One of the earliest reported reactions of chlorocyclophosphazenes was the interaction of $(NPCl_2)_3$ or $(NPCl_2)_4$ with excess aniline to form hexa(phenylamino)cyclotriphosphazene, $[NP(NHPh)_2]_3$[5, 36–38], or the tetramer, $[NP(NHPh)_2]_4$.[37, 38] High polymeric $(NPCl_2)_n$ also undergoes total halogen replacement with aniline during 48 hours reaction at 80°C in benzene.[22] A number of substituted anilines will replace all the chlorine atoms in $(NPCl_2)_3$. These include *p*-toluidine,[5, 31] *p*-chloroaniline,[31] *p*-methoxyaniline,[31] and *m*-trifluoromethylaniline.[43] However, a tetrakis(phenylamino)dichlorocyclotriphosphazene is obtained from the reaction of $(NPCl_2)_3$ with aniline in benzene solution,[5] and a geminal bis(phenylamino)tetrachloro derivative can be obtained from 1:2 molar mixture of $(NPCl_2)_3$ and aniline in toluene at 80°C, with triethylamine used as a hydrohalide acceptor.[40] A series of bis(arylamino)-hexachlorocyclotetraphosphazenes are formed from $(NPCl_2)_4$ and aniline, *o*-, *m*-, or *p*-toluidine in an aqueous-organic heterophase mixture or in a boiling organic solvent.[14]

The pattern of halogen replacement appears to be geminal when aniline interacts with $(NPCl_2)_3$.[18, 40] Thus, the bis(anilino) derivative, $N_3P_3Cl_4(NHPh)_2$, shows an AB_2 ^{31}P NMR spectrum (after decoupling of the ^1H interactions), with a doublet arising from a PCl_2 unit and a triplet from the $P(NHPh)_2$ grouping. *m*-Chloroaniline also substitutes geminally.[40] Further evidence for the geminal pattern is provided by an analysis of the products formed by the reaction of 1,1-diphenyl-3,3,5,5-tetrachlorocyclotriphosphazene (IV) first with aniline and then with dimethylamine.[18] ^1H NMR spectra of the

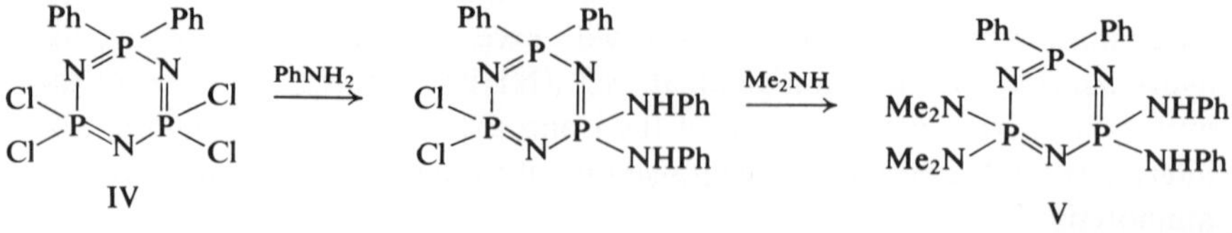

methyl protons in the final product (V) show one doublet, thus confirming the geminal arrangement.[18]

However, evidence has been presented that octachlorocyclotetraphosphazene, $(NPCl_2)_4$, yields a non-geminal bis(phenylamino) derivative[14] (VI).

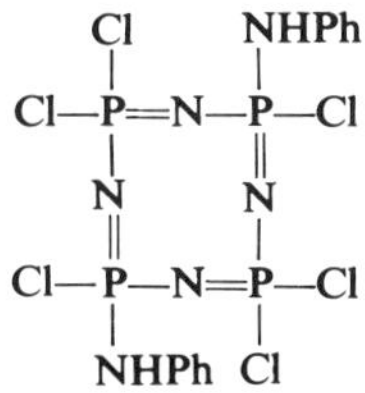

VI

Thus, a six-line ^{31}P spectrum was observed, which suggested a *trans*-ring structure, with each phosphorus atom being affected by two unlike phosphorus atoms.

e. *Amino Acid Esters*

Ethyl glycinate, NH_2CH_2COOEt, or the methyl ester of β-alanine, NH_2CH_2-CH_2COOEt, react with $(NPCl_2)_3$ to replace one, two, four, and six chlorine atoms.[72] No information is available on the pattern of substitution.

2. SUBSTITUTION BY SECONDARY AMINES

a. *Dimethylamine*

A considerable amount of research work has been devoted to studying the reactions of hexachlorocyclotriphosphazene with dimethylamine, and, for this reason, the characteristics of this reaction are more clearly understood than those of most other aminolysis processes. Hexachlorocyclotriphosphazene interacts with dimethylamine to yield mono-, di-, tri-, tetra-, and hexadimethylamino derivatives.[7, 8, 13, 17–19, 21, 24, 45, 46, 49, 51, 53, 83]

The mono(dimethylamino)pentachlorocyclotriphosphazene (VII), m.p.

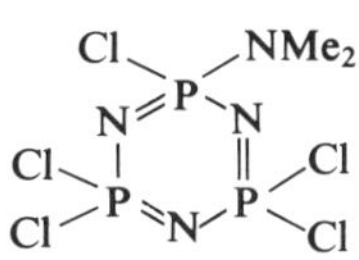

VII

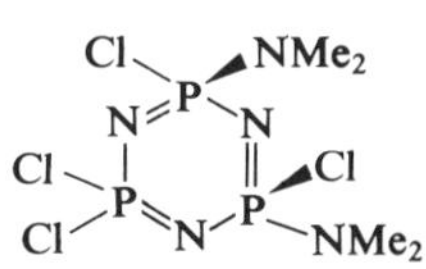

VIII

16°C, is formed when $(NPCl_2)_3$ is treated with two equivalents of dimethylamine in diethyl ether at −78°C,[46] or when an aqueous solution of dimethylamine is added to an ethereal or benzene solution of $(NPCl_2)_3$ at 10°–15°C.[17, 24, 49] Aqueous sodium hydroxide may be used as a hydrohalide acceptor in the heterophase reactions.[17, 24]

Bis(dimethylamino)tetrachlorocyclotriphosphazene is prepared by treatment of an ether or benzene solution of one equivalent of $(NPCl_2)_3$ with four equivalents of aqueous dimethylamine[8, 49] at 10°–16°C, or with four equivalents of amine at −78°C.[46] The heterophase reaction yields the *trans*-non-geminal isomer (VIII), one-third the amount of the *cis*-non-geminal form, and a trace of the geminal form,[49] whereas the low-temperature technique yields the *cis*- and *trans*-non-geminal isomers, with the latter predominating.[46]

Use of six equivalents of dimethylamine to one of $(NPCl_2)_3$ in ether at −78°C gives the *trans*-non-geminal tris (IX) and the geminal tris(dimethylamino) (X) derivatives in almost equal amounts.[46] Small amounts of the

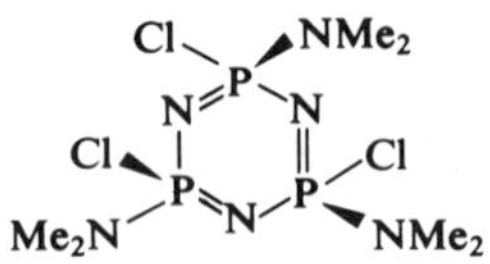

IX

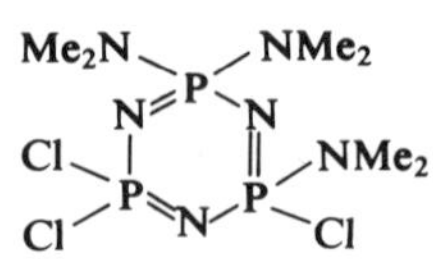

X

geminal-bis- and non-geminal-*cis*-tris are also formed under these conditions, and particularly at higher temperatures. A heterophase reaction yielded mainly the non-geminal *trans* species (IX).[49]

Only one tetrakis derivative, XI, is formed when excess dimethylamine reacts with $(NPCl_2)_3$ in boiling ether.[46]

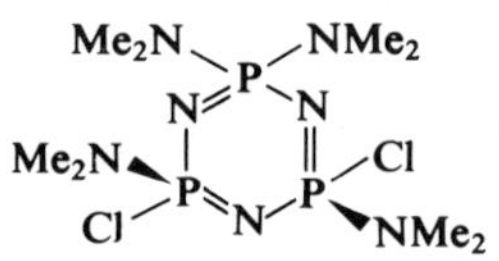

XI

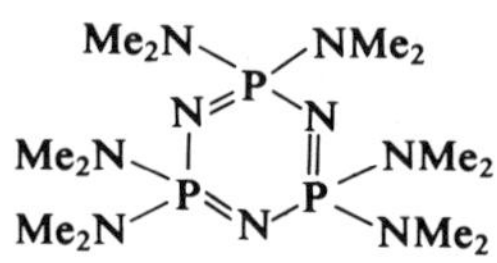

XII

No pentakis compound has yet been isolated, but hexakis(dimethylamino)-cyclotriphosphazene (XII) can be prepared from $(NPCl_2)_3$ and excess dimethylamine in boiling xylene.[46] It should be noted that most of the above structural assignments are based on proton NMR, basicity measurements, and dipole moments.

Thus, the evidence indicates that the reaction between dimethylamine and $(NPCl_2)_3$ mainly follows a non-geminal pattern, although geminal products are also formed, particularly during tris substitution. The non-geminal reaction path can be rationalized in terms of the inductive electron supply from dimethylamine to phosphorus, thus retarding subsequent nucleophilic attack on the same skeletal atom (XIII). However, it is also reasonable to suppose that bulky dimethylamino substituents may sterically retard the approach of the nucleophile.

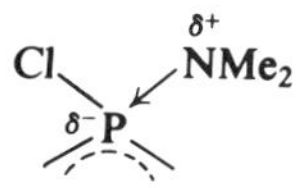

XIII

Dimethylamine also reacts with octachlorocyclotetraphosphazene, $(NPCl_2)_4$, and with the high polymer $(NPCl_2)_n$. For the interaction with $(NPCl_2)_4$ in ether solution, the reaction yields one mono-, one bis-, one tris-, three tetrakis-, two pentakis-, and two hexakis(dimethylamino) derivatives.[20] In another instance, five different bis(dimethylamino)hexachlorocyclotetraphosphazenes were reported.[52] The assignment of the exact structures of tetrameric isomers is a problem of some complexity, and little information is yet available.

Little difficulty is encountered in effecting complete replacement of halogen in the high polymer $(NPCl_2)_n$ by dimethylamino units in spite of the steric retardation which must be involved.[22, 23] It should also be noted that *fluoro*-cyclophosphazenes, such as $(NPF_2)_3$, undergo stepwise substitution with dimethylamine under vigorous conditions.[25]

b. *Higher Secondary Amines*

Diethylamine reacts with $(NPCl_2)_3$ to yield mono-, bis-, tris-, tetrakis-, and hexakis(diethylamino)derivatives.[17, 19] However, diethylamine and higher secondary amines, such as dibutylamines react much more slowly than dimethylamine.[19] Complete replacement of halogen in $(NPCl_2)_3$ by diethylamino occurs during 24 hours reaction at 150°C, but with di-*sec*-butylamine over 95% of the starting materials can be recovered after the most drastic reaction conditions have been employed.[19] This behavior is also found for reactions of the high polymer, $(NPCl_2)_n$, with diethylamine, where a maximum of 50% of the chlorine atoms can be substituted—one on each phosphorus.[23] However, it has been reported that the initial reaction of $(NPCl_2)_4$ with diethylamine is 10^2 to 10^3 times faster than with $(NPCl_2)_3$.[53] Clearly, there is a serious steric hindrance influence by the amine component, although the high electron-supplying powers of these ligands must also contribute. In any event, these factors favor the isolation of monosubstituted cyclic trimers, and such derivatives have been identified for reactions involving di-*n*-propylamine and di-*n*-butylamine.[17] It is also of interest to note that, although dimethylamine undergoes a conventional type of substitution with $(NPF_2)_3$, diethylamine cleaves the phosphazene ring, presumably because of its higher nucleophilicity.[25]

Interactions between chlorophosphazenes and very highly hindered secondary amines have also been reported. Thus, high molecular weight poly(dichlorophosphazene), $(NPCl_2)_n$, undergoes no substitution at all with diphenylamine, even after drastic reaction conditions have been employed.[22, 23] *N*-Methylaniline replaces only part of the chlorine atoms in $(NPCl_2)_3$, $(NPCl_2)_4$,[41] or $(NPCl_2)_n$.[23] When $(NPCl_2)_4$ is treated with a large excess of *N*-methylcyclohexylamine at 25°C for 24 hours, a 90% yield of a tetrasubstituted product is obtained.[35] Proton NMR assignments have been used to show that this product has a β *trans* configuration (XIV).[35] It was concluded that in

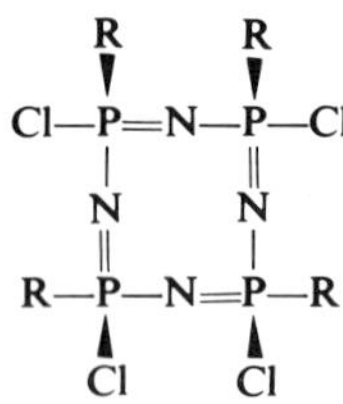

XIV

this reaction pairs of amine molecules react sequentially at the 1,5 and 3,8 positions, with each substitution taking place *trans* to the previous one.[35] The explanation for this unique mechanism must involve the steric hindrance of the amine and the nonplanarity of the tetrameric ring, but the details of this mechanism are not yet clear.

c. *Cyclic Secondary Amines*

Included in this section are the reactions of piperidine (XV), morpholine (XVI), pyrrolidine (XVII), and ethyleneimine (aziridine) (XVIII) with chlorophosphazenes.

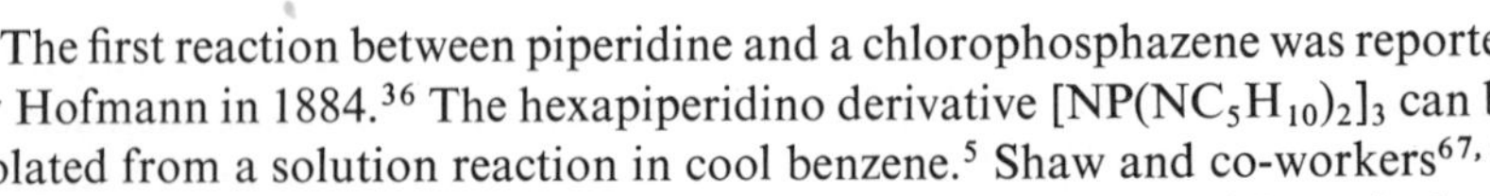

XV XVI XVII XVIII

The first reaction between piperidine and a chlorophosphazene was reported by Hofmann in 1884.[36] The hexapiperidino derivative $[NP(NC_5H_{10})_2]_3$ can be isolated from a solution reaction in cool benzene.[5] Shaw and co-workers[67, 68] have examined the reaction of piperidine with $(NPCl_2)_3$ and have isolated partly substituted derivatives. With a 2:1 mole ratio of piperidine to $(NPCl_2)_3$ in diethyl ether at −78°C a 42% yield of the monopiperidino derivative, $N_3P_3Cl_5(NC_5H_{10})$, is obtained. When the amine to phosphazene ratio is raised to 4:1 under the same conditions, a 1% yield of the non-geminal-*cis*-bis derivative (XIX) and a 34% yield of the non-geminal-*trans*-bis analogue (XX), are obtained. Two trispiperidino derivatives are known. The non-geminal-*trans*

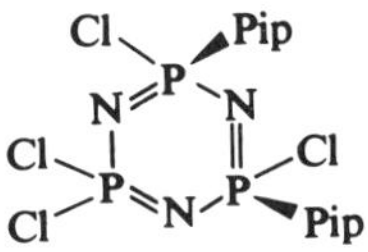

XIX

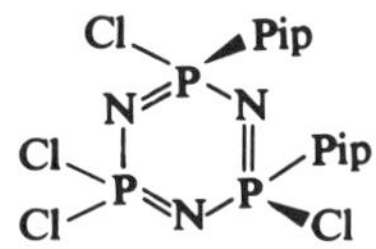

XX

(XXI) (24%) is formed from a 6:1 base to phosphazene mixture in ether at 0°C, and a geminal species (XXII) (33%) results from a similar reaction in benzene near 60°C. Only one tetrakis derivative was isolated (37%) when an 8:1

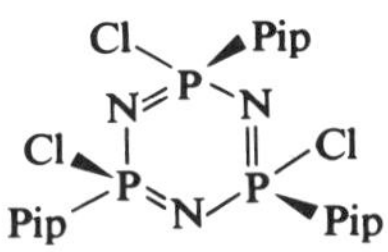

XXI

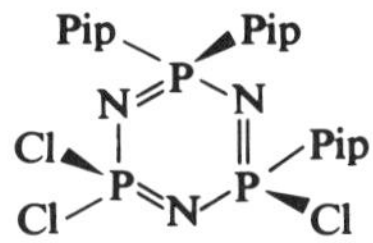

XXII

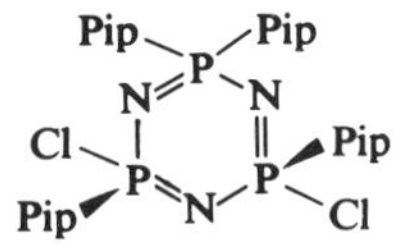

XXIII

molar mixture in ether was allowed to react at 0°C, and this was assigned a non-geminal-*cis* structure (XXIII). The hexakispiperidino compound, $[NP(NC_5H_{10})_2]_3$, is formed in 45% yield when a 12:1 molar mixture of piperidine to $(NPCl_2)_3$ reacts in benzene at 80°C. No pentakispiperidino derivative was isolated,[68] although a species corresponding to this structure has been mentioned in other work.[66]

Thus, the evidence favors a predominantly non-geminal mechanism, although some tendency toward a geminal pathway is evident at higher temperatures. A surprising fact is the tendency for the reaction to favor *trans*-bis, *trans*-tris, and geminal-tris derivatives. The reactions of piperidine with $(NPCl_2)_3$ thus closely parallel those of dimethylamine with the same phosphazene. An explanation for the non-geminal-*trans* preference of these reactions is presented at the end of this chapter. It is interesting to note that $(NPCl_2)_3$ reacts with piperidine in toluene about 10^3 times faster than with diethylamine, even though in water and nitrobenzene these amines are of similar basicity.[53] Furthermore, the monopiperidino pentachloro trimer reacts with piperidine 50 times slower than does $(NPCl_2)_3$, and the 1,3-dipiperidino compound reacts 500 times slower than $(NPCl_2)_3$.[67]

Octachlorocyclotetraphosphazene yields a dipiperidino derivative in anhydrous organic solvents when the amine to phosphazene molar ratio is 4:1.[14] No difficulty is experienced in replacing all the halogen atoms in $(NPCl_2)_n$ high polymer by piperidino units during 48 hours reaction in boiling benzene.[22]

Kropacheva and co-workers have examined the reactions of morpholine and pyrrolidine with $(NPCl_2)_3$. With morpholine in ether or benzene, with progressively greater concentrations of morpholine and increasing temperatures, it is possible to prepare one mono-, two bis-, two tris-, two tetrakis-, one pentakis-, and a hexakismorpholino derivative.[69–71] The structures of these species were examined by synthetic schemes based on the known geminal substitution behavior of ethyleneimine and the formation of mixed ethyleneimino-morpholine derivatives.[71] The two bismorpholino compounds were *cis*- and *trans*-non-geminal isomers in a ratio of approximately 6 to 1. The two tris-morpholino derivatives isolated were assigned geminal, and non-geminal-*trans* structures, the latter predominating in a 2 to 3 ratio. The two tetrakismorpholino derivatives are believed to be non-geminal *cis* and *trans* isomers.[71] Thus, a non-geminal substitution pattern appears to predominate for morpholine, with a possible preference for *cis* replacement.

With pyrrolidine as a reagent in ether or benzene, all degrees of substitution from one to six are found.[63–65] The two bispyrolidino derivatives are non-geminal *cis* and *trans* isomers, with the *cis* isomer predominating in a ratio of approximately 6 to 1.[64] Three trispyrolidino isomers are formed, these being the non-geminal *cis* and *trans* forms and the geminal form.[65] The tetrakis-pyrolidino derivative is isolated as two different isomers.

The reactions of ethyleneimine with $(NPCl_2)_3$ have been studied in some detail.[55–59, 61] All degrees of substitution of chlorine by ethyleneimine (from 1 to 6) have been reported.[56–59] Hexakis(ethyleneimino)cyclotriphosphazene (XXIV) has been prepared on a large scale as an insect chemosterilant.[58] The

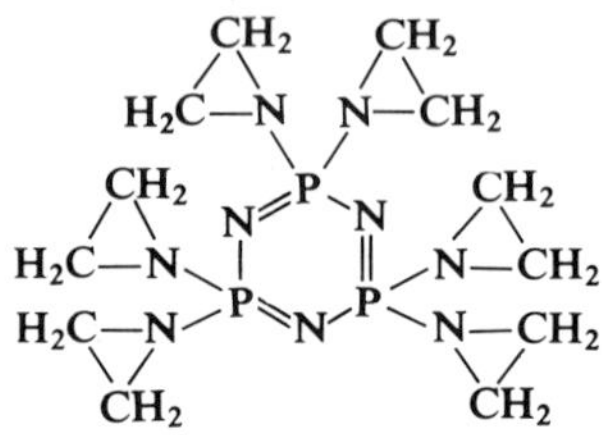

XXIV

reaction is conducted in a benzene–water heterophase system with aqueous alkali used as the hydrogen chloride acceptor.[58] However, the compound can be conveniently prepared in the laboratory with the use of chloroform or an aromatic hydrocarbon as a solvent and triethylamine or ammonia as a base.[58] Octakis(ethyleneimino)cyclotetraphosphazene, $[NP(NC_2H_4)_2]_4$, can be prepared by similar techniques; for instance, by the use of triethylamine as a base in benzene solution at 30°C for 24 hours.[58] Both materials are white crystalline solids.

The monoethyleneimino derivative $N_3P_3Cl_5(NC_2H_4)$ can be prepared from $(NPCl_2)_3$ and ethyleneimine in benzene at room temperature or below in the presence of triethylamine or dimethylamine as hydrogen chloride acceptors.[56, 57] Similar techniques are used to prepare the bis, tris, tetrakis, and pentakis derivatives.[56, 57] However, substitution beyond the tetrakis stage requires more forcing conditions.[57] ^{31}P and ^{1}H NMR results, and subsequent substitution of residual halogen by dimethylamine, demonstrate that the bis-, tris-, and tetrakisethyleneimino derivatives are geminal products.[56, 57] Thus, as in the case of ammonia and *t*-butylamine, replacement of one chlorine on phosphorus in $(NPCl_2)_3$ by ethyleneimino activates the same phosphorus to subsequent nucleophilic attack. Ethyleneimine is certainly a weaker base than most primary alkylamines, but this does not fully explain the striking preponderance of geminal attack.

3. Substitution by Diamines

Hydrazine reacts with $(NPCl_2)_3$ in ether solution to yield a hexahydrazide (XXV) and hydrazine hydrochloride.[73] The phosphazene hydrolyzes rapidly

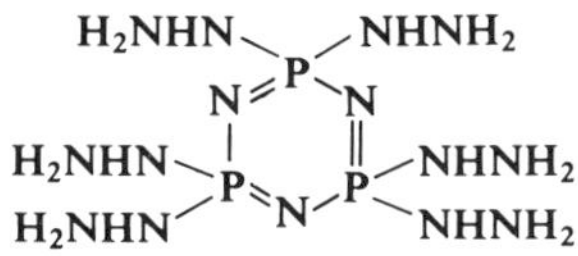

XXV

in acid solution. A reaction between $(NPCl_2)_3$ and phenylhydrazine has also been described which apparently yields a hexakis(phenylhydrazino)cyclotriphosphazene.[38]

o-Phenylenediamine reacts with $(NPCl_2)_3$ to yield mono-[5, 74] and tris(*o*-phenylenediamino)spirocyclotriphosphazenes[74] (XXVI and XXVII). A bis-

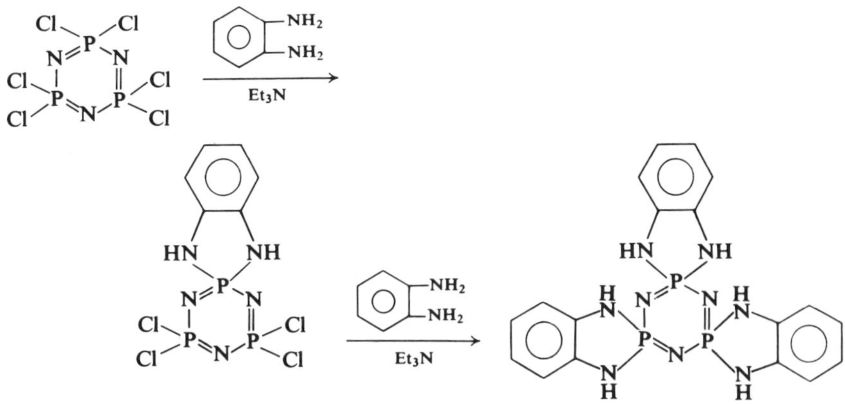

XXVI XXVII

o-phenylenediamino derivative was also isolated as a transient species. Compound XXVII is unusually stable at high temperatures.[74]

With *o*-phenylenediamine, the orientation of the amino groups strongly favors cyclization to the spirocyclophosphazene. With *meta* and *para*-aromatic diamines, and with aliphatic diamines, the predominant reaction usually involves the linkage of different cyclophosphazene rings[78, 79] (XXVIII). The

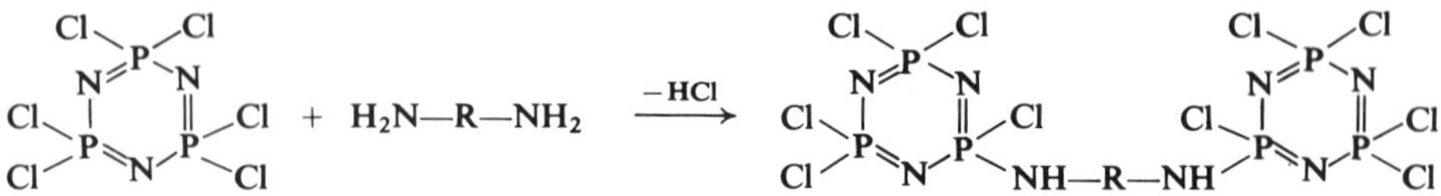

XXVIII

ultimate extension of this reaction leads to the formation of cyclomatrix polymers. This apparently occurs with *m*- and *p*-phenylenediamine, with 4,4′-diaminobiphenyl, with pentamethylenediamine, hexamethylenediamine, octamethylenediamine, and with 4,4′-diaminodicyclohexylmethane.[78, 79] Such materials will be discussed in Chapter 16.

However, ethylenediamine, 1,3-propylenediamine, and 1,4-butylenediamine react with $(NPCl_2)_3$ to form "ansa"-type compounds in which cyclization has occurred between two phosphorus atoms in the same ring (XXIX).[84, 85]

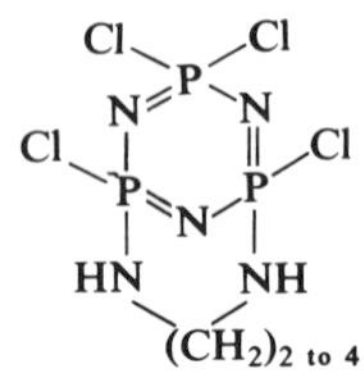

XXIX

Finally, it has been reported that thiourea reacts with $(NPCl_2)_3$ to form the hydrochloride of a spirocyclic derivative of formula $N_3P_3[(NH)_2CS]_3 \cdot 3\,HCl$.[75] The material was a hydrolytically unstable, viscous semisolid.

F. Influence of Other Substituents on the Aminolysis Reaction

A number of aminolysis reactions have been studied in which certain substitution sites on the ring have been blocked by unreactive, nonhalogeno groupings. The position at which an attacking amine molecule reacts will then depend on the location of the remaining phosphorus–halogen bonds and perhaps also on the electronic or steric characteristics of the blocking groups. Information derived from such reactions can also be used to elucidate the positions occupied by the blocking groups if this is unknown.

For example, treatment of the phenoxychlorocyclotriphosphazenes, $N_3P_3Cl_4(OPh)_2$, $N_3P_3Cl_3(OPh)_3$, and $N_3P_3Cl_2(OPh)_4$, with dimethylamine brought about replacement of all the chlorine atoms by dimethylamino. The isomers of these derivatives could then be separated and their structures determined by NMR techniques.[48] A danger inherent in this approach is that the attacking amine or amine hydrochloride may displace the blocking group or induce *cis–trans* isomerization reactions, and considerable care must be taken to eliminate this possibility before structural assignments are made.

The influence of geminal phenyl substituents on the reactivity of halogen atoms elsewhere on a phosphazene ring has been investigated.[21] It is found that the series $(NPCl_2)_3 > N_3P_3Ph_2Cl_4 > N_3P_3Ph_4Cl_2$ represents a sharp decrease in the reactivity of the phosphazene with a given amine. This can hardly represent a steric influence and explanations based on the electron-directing properties of phenyl groups have been put forward.[21] Hexaphenyldichlorocyclotetraphosphazene, $N_4P_4Ph_6Cl_2$, reacts with a variety of primary and secondary amines (including *N*-methylaniline) under relatively mild conditions to give diaminohexaphenyl derivatives.[11]

The influence of amino groups already present on the ring has also been studied. Thus, geminal-bis(ethyleneimino)tetrachlorocyclotriphosphazene (XXX) reacts with dimethylamine to yield a non-geminal bis(dimethylamino)

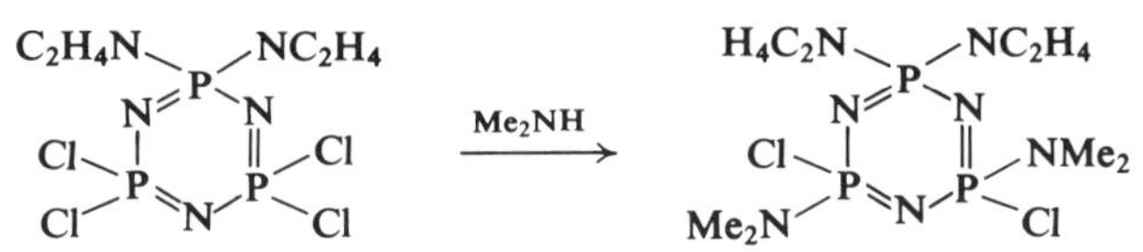

XXX XXXI

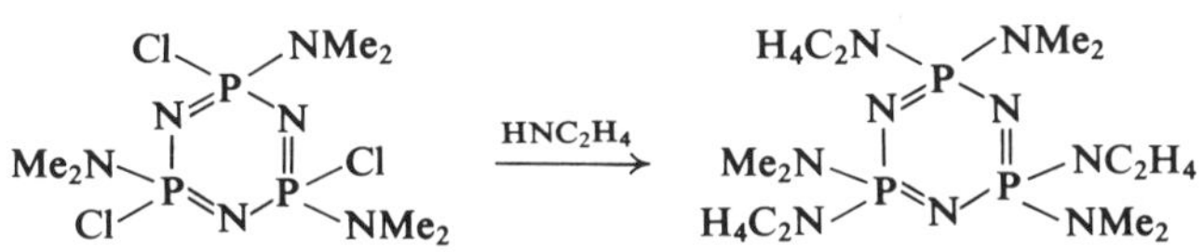

XXXII XXXIII

derivative (XXXI).[57] A non-geminal tris(ethyleneimino)tris(dimethylamino) derivative (XXXIII) can be prepared by the reaction of ethyleneimine with the non-geminal tris(dimethylamino)trichloro compound (XXXII), thus providing a route to such compounds not accessible by the direct reaction of ethyleneimine with $(NPCl_2)_3$.[57] Similar procedures have been employed to prepare non-geminal mixed ethyleneiminopyrrolidino derivatives[61] and a variety of other mixed aminophosphazenes with specific structures.[8]

G. *Cis–Trans* Isomerization during Substitution

The reactions of amines with aminochlorophosphazenes may involve simultaneous *cis-trans* isomerization processes. Gross positional changes of a substituent from one phosphorus atom to another apparently do not occur. Keat and Shaw[47] have shown that *cis-* or *trans*-non-geminal aminophosphazenes can be converted into mixtures which contain both forms by treatment with an amine hydrochloride, the latter being, of course, a product of most aminolysis reactions.

For example, *trans*-tris(dimethylamino)trichlorocyclotriphosphazene (XXXIV) can be converted to a mixture of the *cis* and *trans* isomers simply by heating with dimethylamine hydrochloride in chloroform solution. The same result is obtained if the pure *cis* isomer (XXXV) is used as a starting material.

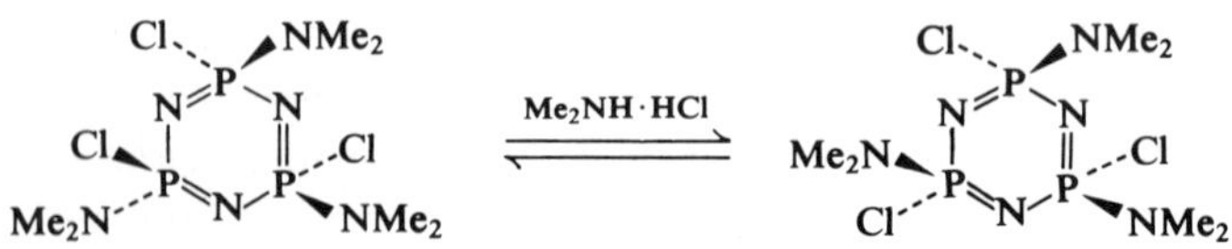

XXXIV XXXV

The same kind of isomerization occurs with piperidinochlorocyclophosphazenes, and with trimers and tetramers. Solvent effects are rather specific for this process since isomerization apparently does not occur in benzene or ether, but it does take place in chloroform, acetonitrile, or pyridine. A contributing factor is believed to be the poor solubility of some amine hydrochlorides in benzene or ether, but this is not the only factor involved. The structure of the amine hydrochloride is also important. Thus, in addition to dimethylammonium chloride and piperidium chloride, methyl- and tetramethylammonium chlorides, pyridinium chloride, tetra-*n*-butylammonium chloride, and even lithium chloride can promote inversions.[47] Foreign amino groups do not exchange for the original ligands, and a mechanism involving removal of the original amine thus appears unlikely. However, halogen exchange may occur between amine hydrochloride and chlorine bound to phosphorus. Thus, removal of chlorine from phosphorus in an ionization process (XXXVI) or formation of a pentacovalent intermediate (XXXVII) would lead to *cis–trans* scrambling of the amino ligands. These facts must be considered whenever the

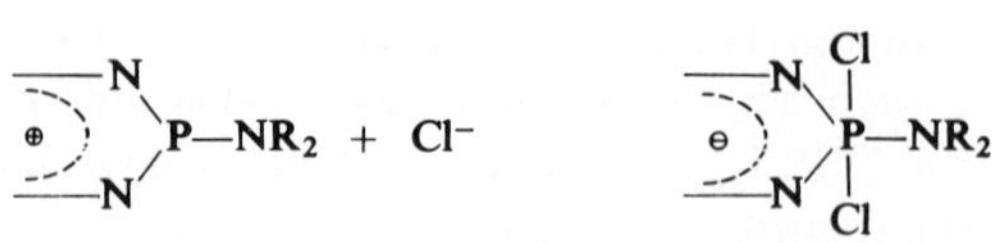

XXXVI XXXVII

cis or *trans* structure of an aminophosphazene has been determined by subsequent substitution steps. It is of interest that *cis–trans* isomerizations can also be induced by aluminum chloride, and this subject is covered in Chapter 13.

H. Influence of the Halogen Bound to Phosphorus

Bromophosphazenes and fluorophosphazenes react with primary and secondary amines to yield aminohalophosphazenes and fully substituted derivatives in a similar manner to chlorophosphazenes.[86–88] However, Moeller and Kokalis[29] have shown that the rate of aminolysis by *n*-propylamine in acetonitrile at 25°C decreases in the order $(NPBr_2)_3 > (NPCl_2)_3 > (NPF_2)_3$ and, for tetramers, in the order $(NPBr_2)_4 > (NPCl_2)_4 > (NPF_2)_4$. Several possible explanations have been suggested for this differing reactivity.[29] Firstly, the decreasing order of reactivity Br > Cl > F parallels the expected decreasing availability of phosphorus *d* orbitals during amine attack, since the strong electronegativity of fluorine should favor the strongest involvement of *d*-orbital electron density in skeletal d_π–p_π bonding. Secondly, the observed decrease in reactivity may parallel the decreasing polarizability of the halogen atoms. Thirdly, it seems reasonable to suppose that the most electronegative ligand should favor the strongest d_π–p_π skeletal binding which could confer greater rigidity on the ring. If the approach to the transition state for aminolysis requires some conformational mobility at phosphorus, the most flexible rings would react the fastest.

It was mentioned earlier that $(NPF_2)_3$ reacts with methylamine or dimethylamine to yield bis-, tris-, tetrakis-, pentakis-, and hexakisaminophosphazenes, but that diethylamine cleaves the phosphazene skeleton.[25]

I. Influence of Phosphazene Ring or Chain Size

Although the reactions of halocyclotetraphosphazenes have not been studied to the same degree as those of cyclic trimers, sufficient information is available to suggest that many of the same factors operate in both cases. For example, fully substituted tetramers are known with amino[6]; methylamino[20]; ethylamino[20]; *n*- and isopropylamino[20]; *n*-, *iso*-, *sec*-, and *t*-butylamino[20]; dimethylamino[20]; diethylamino[20]; allylamino[33]; and ethyleneimino groups.[58] Partly substituted products can also be prepared.[14, 20a, 52] Furthermore, nongeminal tetraphenyltetrachlorocyclotetraphosphazene, $(NPClPh)_4$[44], and hexaphenyldichlorocyclotetraphosphazene, $N_4P_4Ph_6Cl_2$,[11] react with dimethylamine to replace all the chlorine atoms.

However, cyclic tetramers react faster with *n*-propylamine than do cyclic trimers, irrespective of whether the halogen is fluorine, chlorine, or bromine,[29] and during replacement of the first chlorine by diethylamine, the tetramer, $(NPCl_2)_4$, reacts faster than the trimer, $(NPCl_2)_3$, by a factor of 10^2 to 10^3.[53] Explanations for this have included the greater ability of the tetrameric ring to accommodate a negative charge in the transition state,[53] and the known greater flexibility of the tetrameric ring, which perhaps facilitates the conformational change required during approach to the transition state.[29] The interaction of aniline with $(NPCl_2)_4$ leads to the formation of a 1,5-*trans*-dianilino derivative,[14] whereas geminal substitution occurs with the trimer.

Although the aminolysis reactions of higher cyclic species have not been described, there is every reason to believe that both fully and partly substituted cyclic pentamers, hexamers, heptamers, etc., can be prepared. It has, in fact, been demonstrated that the high molecular weight polymer, $(NPCl_2)_n$ ($n \approx 15{,}000$), yields fully substituted products in benzene or toluene with aniline, dimethylamine, ethylamine, and piperidine, ethyl glycinate, and numerous primary and secondary amines.[22, 23]

J. Monophosphazenes and Cyclophosphazanes

Some trichloromonophosphazenes react readily with aniline in boiling benzene to give almost quantitative yields of phenylamino-substituted derivatives. This reaction is summarized by the following equation[42]:

$$\mathrm{ArCON{=}PCl_3 + 6\,Ar'NH_2 \rightarrow ArCON{=}P(NHAr')_3 + 3\,ArNH_2\cdot HCl}$$

Amines such as aniline and *p*-toluidine can be used, and the group Ar may be phenyl, *p*-bromophenyl, or *m*- or *p*-nitrophenyl. The products are colorless, crystalline compounds which melt with decomposition. Unlike most amino-*cyclo*phosphazenes, the monomeric compounds hydrolyze readily even with atmospheric moisture.[42]

A similar aminolysis takes place when trichloromonophosphazenylsulfonalkyls interact with aniline in benzene solution.[89] These materials are

$$\mathrm{RSO_2N{=}PCl_3 + 6\,PhNH_2 \rightarrow RSO_2N{=}P(NHPh)_3 + PhNH_2\cdot HCl}$$

unchanged even after prolonged boiling with alcoholic alkali. A chlorodianilino derivative can also be formed under milder conditions.[89, 90] Similar reactions occur with trichlorophosphazenylsulfonaryls, $Cl_3P{=}NSO_2Ar$.[91]

Ethyleneiminomonophosphazenes can also be prepared from trichloromonophosphazenes of structure $Cl_3P{=}N{-}P(O)(OAr)Alk$ [aryl-*P*-alkyl-*N*-

(trichlorophosphoranylidene)phosphonamidates] according to the following equation[62]:

$$\mathrm{Alk(O)(OAr)N{=}PCl_3 + 3\,C_2H_4NH \xrightarrow[-HCl]{Et_3N} Alk(O)(OAr)N{=}P(NC_2H_4)_3}$$

These materials are relatively unstable.

A number of monophosphazenes are known to exist as cyclic dimers (cyclodiphosphazanes). *P*-Trichloro-*N*-alkylmonophosphazenes, $Cl_3P{=}NAlk$, for example, exist as cyclic dimers.[32] It might have been anticipated that the interaction of such compounds with aliphatic primary amines would yield either the monomeric or dimeric trisamino derivatives. Instead, the amine cleaves the phosphazane ring as shown in the following equation[32]:

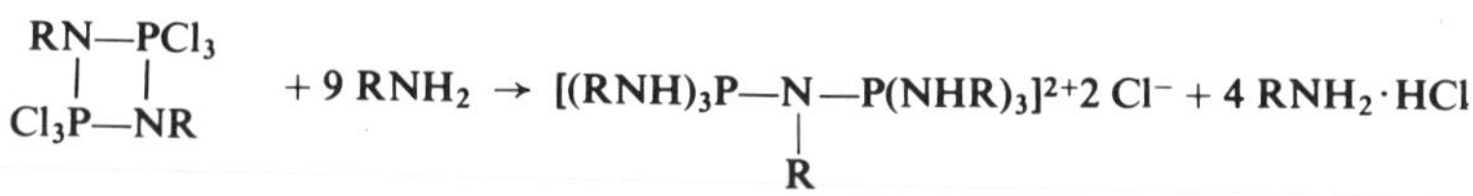

However, amino-substituted cyclodiphosphazanes have been reported. For example, phosphorus pentachloride reacts with an excess of primary amine, in the manner shown[91-94]:

The group R can be phenyl, *o*- or *p*-methoxyphenyl, or an aliphatic unit. A different group of aminocyclodiphosphazanes are apparently formed when thiophosphoryl chloride reacts wth primary aromatic amine hydrochlorides.[95]

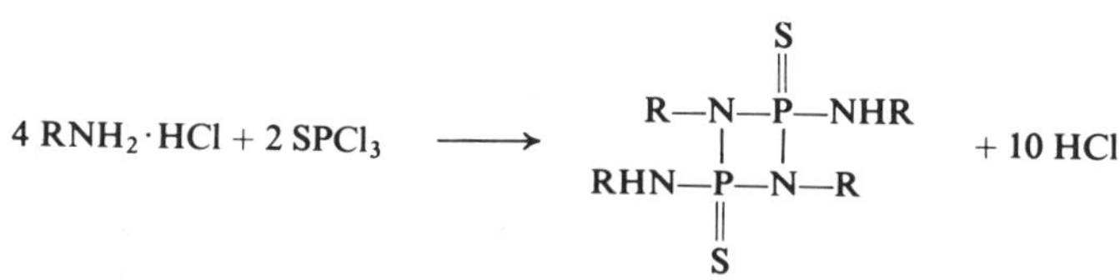

K. Effect of Solvent and Temperature on Aminolysis

Although a great deal of work has been reported on the aminolysis of cyclophosphazenes, very little information is available about the influence of the reaction solvent on reaction rate, on the pattern of halogen replacement, or on the substitution mechanism. Solvents which have been used for aminolysis reactions include the liquid amine, chloroform, hydrocarbons such as benzene, toluene, or xylene, acetonitrile, ethers such as diethyl ether or tetrahydrofuran, or heterophase systems such as benzene–water. Ideally, the solvent is chosen so that the amine hydrochloride formed during the reaction is insoluble in the medium, but the aminophosphazene remains in solution. Often, the solvent is chosen on the basis of its boiling point or freezing temperature. Examples do exist where the pattern of chlorine replacement apparently varies with different solvents,[46, 49] but these differences may result from different reaction temperatures. However, solvent effects appear to exert a strong influence on *cis–trans* isomerization reactions, as discussed earlier.

Higher reaction temperatures have the effect of accelerating substitution and increasing the number of halogen atoms replaced by amino residues. Thus, low temperatures often provide the best conditions for the formation of mono- or bisamino derivatives, whereas high-temperature conditions in boiling solvents or even within sealed tubes favor complete halogen replacement, particularly for the bulkiest substituents. Higher reaction temperatures in some instances reduce the selectivity of geminal or non-geminal substitution patterns.

L. Influence of the Hydrohalide Acceptor

It will be clear from the data presented in earlier sections that, in the majority of cases, an excess of the reacting amine functions as a hydrohalide acceptor. However, in a few cases, and particularly when the nucleophile is a weak base, triethylamine may be employed to remove hydrogen chloride. In organic–water heterophase systems, aqueous alkali hydroxide may also be used. In principle, there is no reason why solid, anhydrous sodium carbonate should not be used to remove hydrogen chloride formed in an organic solvent. This technique has been used successfully for phenolysis reactions.[74, 96] A further modification of preparative technique involves the use of an alkali metal salt of the amine. For example, potassium pyrrol can be prepared from potassium and pyrrole.[30] The potassium salt reacts with $(NPCl_2)_3$ or $(NPCl_2)_4$ in benzene to replace all the halogen atoms with concurrent precipitation of potassium chloride.[30]

$$(NPCl_2)_3 + 6\,KNC_4H_4 \rightarrow [NP(NC_4H_4)_2]_3 + 6\,KCl$$

M. The Reaction Mechanism

1. Possible Mechanisms

The two simplest interpretations of the reaction mechanism are an S_N2, bimolecular type of reaction (Scheme A), or one involving an S_N1-type ionization of halogen prior to attack by the amine (Scheme B). The synchro-

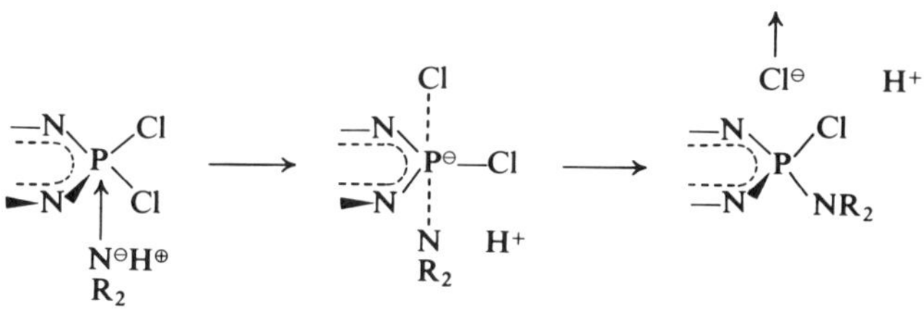

SCHEME A

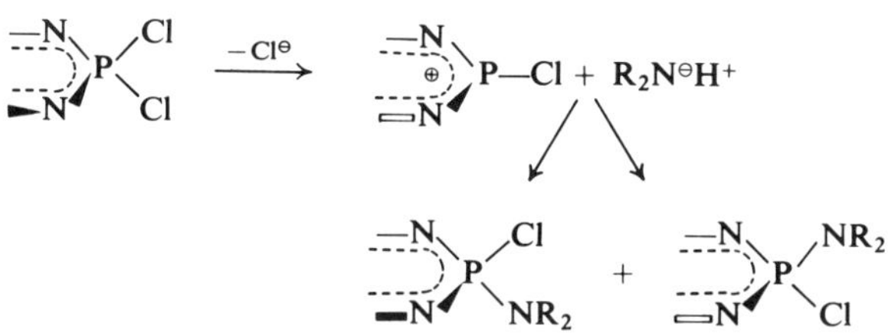

SCHEME B

nous bimolecular type of mechanism is well known in the chemistry of tetracoordinate phosphorus.[97] The mechanism is characterized by the formation of a trigonal bipyramidal transition state from which the displaced anion leaves axially from the opposite side of the complex to the attacking group. Inversion of configuration thus occurs, and second-order or pseudo-first-order kinetics may be expected. On the other hand, the S_N1-type process shown in Scheme B involves the prior dissociation of chloride ion from phosphorus followed by rapid amine attack on the cation thus formed. Such a mechanism would lead to scrambling of the conformation at phosphorus and should be characterized by first-order kinetics. In actual fact, these mechanisms represent the two extremes of the spectrum. Nevertheless, an analysis of the experimental data should provide a clue to the reaction path.

First, let us consider the evidence based on the pattern of halogen replacement. If aminolysis of a halocyclophosphazene follows a simple S_N2-type mechanism, replacement of one halogen atom on phosphorus by amino should decrease the tendency of a second amino molecule to attack at the same phosphorus atom. Both the inductive electron supply from amino to phosphorus, and π effects involving donation of the nitrogen lone-pair electrons into phosphorus d_{z^2} orbitals would raise the electron density at phosphorus and retard nucleophilic attack (XXXVIII). Steric hindrance by bulky amino groups would also be expected to inhibit the approach of the nucleophile to that phosphorus atom. Thus, non-geminal substitution patterns should be observed for most amines and particularly for strongly electron supplying or bulky amino units. Only the weakest amines with the smallest dimensions would be expected to

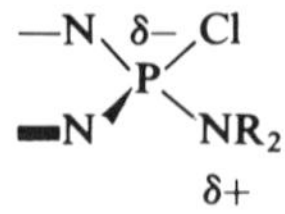

XXXVIII

yield geminally substituted products. However, if an S_N1-type mechanism operates, it could be argued that ionization of a halogen atom from phosphorus might be facilitated by the presence of an electron-supplying substituent at the same phosphorus atom. In such an event, geminal substitution patterns should predominate and these should be relatively insensitive to steric influences in the amine.

The following is a summary of the experimental facts reported for aminolysis reactions: (a) Geminal substitution pathways are observed when ammonia, *t*-butylamine, aniline, or ethyleneimine react with $(NPCl_2)_3$. (b) Non-geminal mechanisms predominate when dimethylamine, piperidine, morpholine, pyrrolidine, *N*-methylaniline, and probably diethylamine react with chlorophosphazenes. (c) Methylamine and isopropylamine give both geminal and non-geminal products. Ethylamine appears to be anomalous in yielding non-geminal products with $(NPCl_2)_3$, and aniline yields a non-geminal product with $(NPCl_2)_4$. (d) With piperidine and dimethylamine there is a curious preponderance of non-geminal *trans* substitution over *cis*.

It is clear that the simple mechanistic alternatives outlined above are inadequate to account for the experimental facts. Thus, although non-geminal reaction patterns are, indeed, observed with the relatively bulky, strongly electron supplying amines such as dimethylamine or piperidine, and it could be argued that ethyleneimine substitutes geminally because of its small size and weak basicity, one of the bulkiest and most strongly electron supplying amines—*t*-butylamine—substitutes geminally. In fact, primary amines appear to show a special preference for geminal substitution irrespective of their nucleophilicity or size.

Two explanations have been suggested for this phenomenon, specifically, for geminal substitution by ammonia and *t*-butylamine. Moeller[9] has suggested that when ammonia reacts with $(NPCl_2)_3$, the initial monoamino derivative (which is never isolated) gives up a proton to ammonia, and that the resultant imino intermediate then reacts rapidly with ammonia to form the geminal diamino products, as shown.

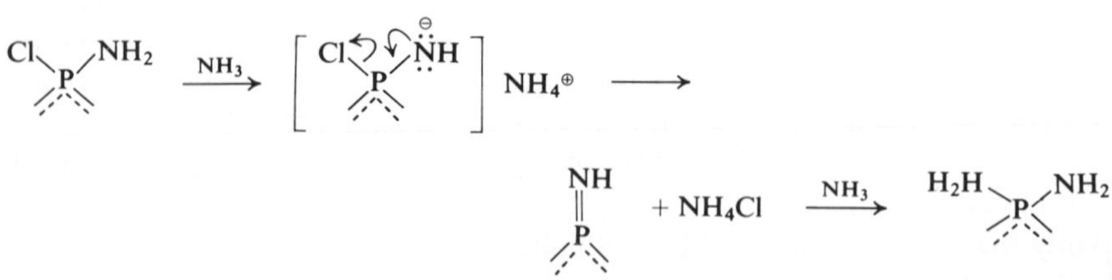

Shaw[82] has proposed a similar mechanism for geminal substitution by *t*-butylamine:

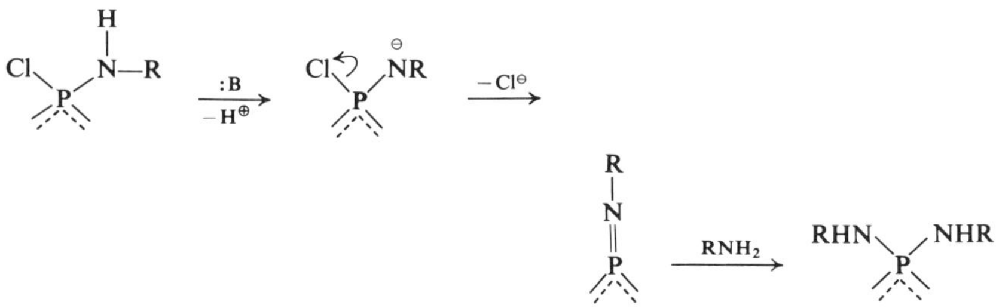

Such a process might give rise to S_N1 characteristics. An alternative possibility is that coordination occurs between an attacking nucleophile and an amino group already present, as shown in structure XXXIX, thus forcing the nucleo-

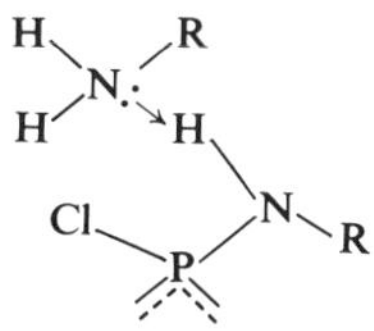

XXXIX

phile to attack geminally.[82]

A second anomaly is connected with the fact that the non-geminal pathway, which exists during replacement of two or three chlorine atoms in $(NPCl_2)_3$ by dimethylamine or piperidine, yields predominantly the *trans* products.[46, 68] A possible explanation for this has been proposed by Shaw, and referred to as a "cis effect".[68] Molecular models suggest that this *trans*-substitution pattern cannot be accounted for by steric effects. Instead, it is suggested that electrons can be transferred from an amino group preferentially to a *cis*-non-geminal chlorine atom, thus imparting a partial negative charge to the *cis* halogen, as shown in XL and XLI. This effect is presumed to be greatest when an electron supplier (the amino group) and an electron acceptor (the halogen) can contribute to a concerted process. Thus, if aminolysis is an S_N2-type process, the attacking nucleophile would be expected to approach from the side of the molecule away from the most negatively charged chlorine atoms, and therefore give

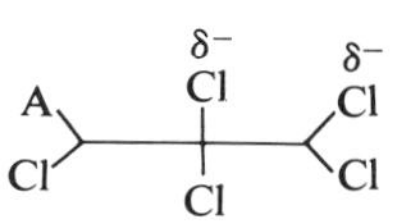

XL

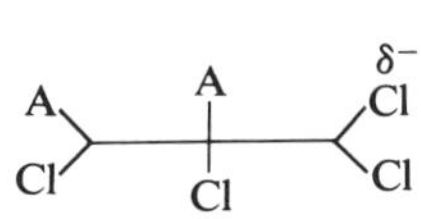

XLI

rise to *trans*-diamino and -triamino products. The relative proportions of many of the partly substituted piperidino and dimethylamino cyclic trimers can be explained by such arguments.

2. Factors which Influence the Degree of Substitution

It is clear from the foregoing evidence that the ease of halogen replacement and the number of chlorine atoms replaced under a given set of circumstances depends on the structure of the amine. For example, the bulkier amines (particularly *t*-butylamine) require more drastic conditions to effect total halogen replacement than do unbranched reagents. Diphenylamine does not displace the halogen from the high polymer, $(NPCl_2)_n$, whereas diethylamine replaces only one half the chlorine atoms in the polymer even with the most drastic conditions. There is a further point which bears on the mechanism, that considerable difficulty is encountered in preparing pentakisaminomonochlorocyclotriphosphazenes. Tetrakisamino and hexakisamino derivatives are usually easier to prepare. Two explanations for these observations can be formulated; one based on strong electron supply from amino substituents to the ring, and the other based on steric effects.

Basicity measurements have shown that alkylamino substituents (particularly dimethylamino groups) strongly supply electrons to the skeletal nitrogen atoms, and that the more halogens replaced by amino, the stronger is this effect.[13, 46] Since greater electron density at the skeletal atoms will retard the approach of a nucleophile, a decreasing reactivity would be expected as the degree of amine substitution increases. Phenyl-substituted phosphazenes are weaker bases than amino-substituted derivatives, but they are stronger bases than $(NPCl_2)_3$.[13] This provides an explanation for the decreasing ease of aminolysis in the order $(NPCl_2)_3 > N_3P_3Ph_2Cl_4 > N_3P_3Ph_4Cl_2$. However, the increasing electron supply to the ring which occurs as additional amino units replace chlorines in $(NPCl_2)_3$ should favor the ionization of the remaining halogen atoms in an S_N1-type process. This provides an explanation for the fact that monochloro pentakisamino trimers are so difficult to isolate. Presumably, the remaining chlorine is replaced particularly rapidly by an S_N1 process, more rapidly in fact than the fourth chlorine.

Steric effects must also play a powerful role in many aminolysis reactions. This is strikingly illustrated by the reactions of primary amines with high molecular weight poly(dichlorophosphazene).[22, 23] Ammonia and methylamine not only replace halogen on one chain, but also cross-link neighboring chains (XLIII). Ethylamine, propylamine, butylamine, and aniline, are sufficiently hindered that the bis-substitution route (XLII) is overwhelmingly favored. The inability of diphenylamine to react with $(NPCl_2)_n$, and the non-geminal 50%

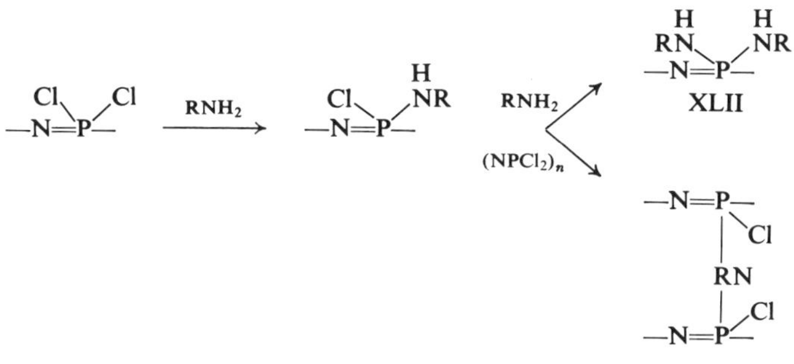

XLII

XLIII

substitution by diethylamine are further evidence of powerful steric influences. This in itself is evidence in favor of a bimolecular mechanism.

3. Kinetic Studies

Additional evidence about the aminolysis mechanism has been obtained from kinetic studies. The general observations that have emerged are as follows: (a) Cyclic tetramers react with amines faster than do cyclic trimers. (b) The rate of aminolysis with *n*-propylamine varies with halogen in the order: Br > Cl > F. (c) Aminolysis rates with piperidine decrease in the order: $(NPCl_2)_3$ > *gem*—$N_3P_3Ph_2Cl_4$ > *gem*—$N_3P_3Ph_4Cl_2$.

A kinetic study of the reaction between *n*-propylamine and the fluoro, chloro, and bromo cyclic trimers and tetramers, $(NPX_2)_{3 \text{ and } 4}$ in acetonitrile at 25°C showed that the rate-determining step was second order in each case.[29] Mixed second- and third-order kinetics were recorded for the reaction of the first chlorine in $(NPCl_2)_3$ with piperidine in toluene at 0°C.[53] The reaction was also found to be catalyzed by tri-*n*-butylamine. These results provide some evidence for a bimolecular process with a pentacoordinate transition state, at least during the initial step of the reaction. However, kinetic analysis of the competition reaction between aniline and ethanol for $(NPCl_2)_3$[39] suggested appreciable bond breaking (i.e., S_N1) character to the reaction, even though second-order kinetics were observed. It is clear that a more complete understanding of this mechanism must await more extensive kinetic studies on different systems.

References

1. H. N. Stokes, *Amer. Chem. J.* **17**, 275 (1895); *Chem. Ber.* **28**, 437 (1895).
2. H. Moureu and A. M. de Ficquelmont, *C. R. Acad. Sci.* **198**, 1417 (1934); **213**, 306 (1941).
3. H. Moureu and P. Rocquet, *C. R. Acad. Sci.* **198**, 1691 (1934); *Bull. Soc. Chim. Fr.* [5] **3**, 821 (1936).
4. A. M. de Ficquelmont, *C. R. Acad. Sci.* **199**, 1045 (1935); *Ann. Chim. (Paris)* [11] **12**, 169 (1939).

5. H. Bode, K. Bütow, and G. Lienau, *Chem. Ber.* **81**, 547 (1948).
6. L. F. Audrieth and D. B. Sowerby, *Chem. Ind.* (*London*) p. 748 (1959); *Chem. Ber.* **94,** 2670 (1961).
7. M. Becke-Goehring and K. John, *Angew. Chem.* **70**, 657 (1958).
8. M. Becke-Goehring, K. John, and E. Fluck, *Z. Anorg. Allg. Chem.* **302**, 103 (1959).
9. G. R. Feistel and T. Moeller, *J. Inorg. Nucl. Chem.* **20**, 2731 (1967).
10. W. Lehr, *Z. Anorg. Allg. Chem.* **350**, 18 (1967).
11. A. J. Bilbo, C. M. Grieve, D. L. Herring, and D. E. Salzbrunn, *Inorg. Chem.* **7**, 2670 (1968).
12. M. V. Lenton and B. Lewis, *J. Chem. Soc., A* p. 665 (1966).
13. D. Feakins, S. N. Nabi, R. A. Shaw, and P. Watson, *J. Chem. Soc., A* p. 10 (1968).
14. K. John, T. Moeller, and L. F. Audrieth, *J. Amer. Chem. Soc.* **82**, 5616 (1960).
15. S. G. Kokalis, K. John, T. Moeller, and L. F. Audrieth, *J. Inorg. Nucl. Chem.* **24**, 191 (1962).
16. T. Moeller and S. Lanoux, *J. Inorg. Nucl. Chem.* **25**, 229 (1963).
17. T. Moeller and S. Lanoux, *Inorg. Chem.* **2**, 1061 (1963).
18. V. B. Desai, R. A. Shaw, and B. C. Smith, *J. Chem. Soc., A* p. 1977 (1969); p. 2023 (1970).
19. S. K. Ray and R. A. Shaw, *J. Chem. Soc., London* p. 872 (1961).
20. S. K. Ray, R. A. Shaw, and B. C. Smith, *J. Chem. Soc., London* p. 3236 (1963).
20a. G. Mattogno and A. Monaci, *Ric. Sci.* **8**, 1139 (1965).
21. K. Hills and R. A. Shaw, *J. Chem. Soc., London* p. 130 (1964).
22. H. R. Allcock and R. L. Kugel, *Inorg. Chem.* **5**, 1716 (1966).
23. H. R. Allcock and D. P. Mack, unpublished work (1969).
24. J. M. E. Goldschmidt and J. Weiss, *J. Inorg. Nucl. Chem.* **26**, 2023 (1964).
25. O. Glemser, E. Niecke, and H. W. Roesky, *Chem. Commun.* p. 282 (1969).
26. W. Lehr, *Z. Anorg. Allg. Chem.* **352**, 27 (1967).
27. R. Keat and R. A. Shaw, *Angew. Chem., Int. Ed. Engl.* **7**, 212 (1968).
28. T. Moeller and P. Nannelli, *Inorg. Chem.* **1**, 721 (1962).
29. T. Moeller and S. G. Kokalis, *J. Inorg. Nucl. Chem.* **25**, 1397 (1963).
30. E. T. McBee, P. Johncock, S. E. French, and H. P. Braendlin, *U.S. Govt. Res. Rep.* **AD 254,985** (1960).
31. S. E. French, *Diss. Abstr.* **24**, 4993 (1963).
32. B. S. Drach, I. N. Zhmurova, and A. V. Kirsanov, *J. Gen. Chem. USSR* **37**, 2402 (1967).
33. H. R. Allcock, P. S. Forgione, and K. J. Valan, *J. Org. Chem.* **30**, 947 (1965).
34. S. K. Das, R. Keat, R. A. Shaw, and B. C. Smith, *J. Chem. Soc., A* p. 1677 (1966).
35. A. J. Berlin, B. Grushkin, and L. R. Moffett, *Inorg. Chem.* **7**, 589 (1968).
36. A. W. Hofmann, *Chem. Ber.* **17**, 1905 (1884).
37. R. Schenck and G. Römer, *Chem. Ber.* **57B**, 1343 (1924).
38. R. Schenck, *Chem. Ber.* **60B**, 160 (1927).
39. J. V. Bailey and R. E. Parker, *Chem. Ind.* (*London*) p. 1823 (1962).
40. H. Lederle, G. Ottmann, and E. Kober, *Inorg. Chem.* **5**, 1818 (1966).
41. G. Ottmann, H. Lederle, and E. Kober, *Ind. Eng. Chem., Prod. Res. Develop.* **5**, 202 (1966).
42. G. I. Derkach, E. S. Gubnitskaya, and A. V. Kirsanov, *J. Gen. Chem. USSR* **31**, 3500 (1961).
43. E. T. McBee, H. R. Allcock, R. Caputo, A. Kalmus, and C. W. Roberts, *U.S. Govt. Res. Rep.* **AD 209,666** (1958).
44. J. H. Smalley, F. E. Dickson, and I. I. Bezman, *Inorg. Chem.* **3**, 1780 (1964).
45. C. T. Ford, F. E. Dickson, and I. I. Bezman, *Inorg. Chem.* **3**, 177 (1964).
46. R. Keat and R. A. Shaw, *J. Chem. Soc., London* p. 2215 (1965).

47. R. Keat and R. A. Shaw, *J. Chem. Soc., London* p. 4067 (1965).
48. D. Dell, B. W. Fitzsimmons, R. Keat, and R. A. Shaw, *J. Chem. Soc., A* p. 1680 (1966).
49. H. Koopman, F. J. Spruit, F. van Deursen, and J. Bakker, *Rec. Trav. Chim. Pays-Bas* **84**, 341 (1965).
50. N. I. Shvetsov, K. A. Nuridzhanyan, A. Ya. Yakubovich, and F. F. Sukhov, *J. Gen. Chem. USSR* **33**, 3874 (1963).
51. R. Stahlberg and E. Steger, *J. Inorg. Nucl. Chem.* **28**, 684 (1966).
52. W. Lehr, *Naturwissenschaften* **56**, 214 (1969).
53. B. Capon, K. Hills, and R. A. Shaw, *J. Chem. Soc., London* p. 4059 (1965).
54. S. P. Joffre, U.S. Pat. 3,311,622 (1967) (to Shulton Inc.).
55. E. T. McBee, L. Brinkmann, and H. P. Braendlin, *U.S. Govt. Res. Rep.* **AD 254,983** (1960).
56. Y. Kobayashi, L. A. Chasin, and L. B. Clapp, *Inorg. Chem.* **2**, 212 (1963).
57. G. Ottmann, H. Agahigian, H. Hooks, G. D. Vickers, E. Kober, and R. Rätz, *Inorg. Chem.* **3**, 753 (1964).
58. R. Rätz, E. Kober, C. Grundmann, and G. Ottmann, *Inorg. Chem.* **3**, 757 (1964).
59. G. Ottmann, R. P. Alexander, and E. Kober, U.S. Pat. 3,157,638 (1964) (to Olin Mathieson Chem. Corp.).
60. A. A. Kropacheva and N. M. Kashnikova, *J. Gen. Chem. USSR* **35**, 2219 (1965).
61. A. A. Kropacheva and N. M. Kashnikova, *J. Gen. Chem. USSR* **38**, 135 (1968).
62. L. D. Protsenko, V. A. Shokol, G. A. Golik, and S. N. Koval, *J. Gen. Chem. USSR* **39**, 370 (1969).
63. A. A. Kropacheva and N. M. Kashnikova, *J. Gen. Chem. USSR* **32**, 645 (1962).
64. A. A. Kropacheva and N. M. Kashnikova, *J. Gen. Chem. USSR* **33**, 1036 (1963).
65. A. A. Kropacheva and N. M. Kashnikova, *J. Gen. Chem. USSR* **35**, 1978 (1965).
66. A. A. Kropacheva, L. E. Mukhina, N. M. Kashnikova, and V. A. Parshina, *J. Gen. Chem. USSR* **31**, 957 (1961).
67. B. Capon, K. Hills, and R. A. Shaw, *Proc. Chem. Soc., London* p. 390 (1962).
68. R. Keat and R. A. Shaw, *J. Chem. Soc., A* p. 908 (1966).
69. A. A. Kropacheva and L. E. Mukhina, *J. Gen. Chem. USSR* **32**, 512 (1962).
70. A. A. Kropacheva and L. E. Mukhina, *J. Gen. Chem. USSR* **33**, 699 (1963).
71. L. E. Mukhina and A. A. Kropacheva, *J. Gen. Chem. USSR* **38**, 314 (1966).
72. A. A. Kropacheva, N. M. Kashnikova, and V. A. Parshina, *J. Gen. Chem. USSR* **34**, 532 (1964).
73. R. J. A. Otto and L. F. Audrieth, *J. Amer. Chem. Soc.* **80**, 3575 (1958).
74. H. R. Allcock and R. L. Kugel, *Inorg. Chem.* **5**, 1016 (1966).
75. A. V. Babaeva and G. V. Derbisher, *Russ. J. Inorg. Chem.* **10**, 156 (1965).
76. H. R. Allcock and R. L. Kugel, *Chem. Commun.* p. 1606 (1968).
77. H. R. Allcock and R. L. Kugel, *J. Amer. Chem. Soc.* **91**, 5452 (1969).
78. M. Becke-Goehring and D. Neubauer, Ger. Pat. 1,143,027 (1963) (to Olin Mathieson Chem. Corp.).
79. M. Yokoyama and S. Konya, *J. Chem. Soc. Jap., Ind. Chem. Sect.* **68**, 2444 (1965).
80. V. A. Chernov, V. B. Lytkina, S. I. Sergievskaya, A. A. Kropacheva, V. A. Parshina, and L. E. Sventsitkaya, *Farmakol, Toksikol. (Moscow)* **22**, 365 (1959); *Chem. Abstr.* **54**, 7900 (1960).
81. *Chem. Week* Dec. 14, p. 83 (1963).
81a. Z. T. Wakefield, B. B. Luff, and J. J. Kohler, *J. Chem. Eng. Data* **15**, 314 (1970).
81b. W. Lehr and J. Pietschmann, *Chem.-Ztg.* **94**, 362 (1970).
82. S. K. Das, R. Keat, R. A. Shaw, and B. C. Smith, *J. Chem. Soc., London* p. 5032 (1965).
83. C. T. Ford, F. E. Dickson, and I. I. Bezman, *Inorg. Chem.* **6**, 1594 (1967).

84. M. Becke-Goehring and B. Boppel, *Z. Anorg. Allg. Chem.* **322**, 239 (1963).
85. B. Cardillo, G. Mattogno, A. Melera, and F. Tarli, *Atti Accad. Naz. Lincei, Cl. Sci. Fis., Natur., Rend.* **35**, 328 (1963); **67**, 194 (1964).
86. H. W. Roesky and W. Grosse-Böwing, *Inorg. Nucl. Chem. Lett.* **6**, 781 (1970).
87. O. Glemser, E. Niecke, and H. Thamm, *Z. Naturforsch. B* **25**, 754 (1970).
88. T. Chivers and R. Hedgeland, *Inorg. Nucl. Chem. Letters* **7**, 767 (1971).
89. A. V. Kirsanov and N. L. Egorova, *J. Gen. Chem. USSR* **28**, 1636 (1958).
90. A. V. Kirsanov and E. A. Abrazhanova, *Zh. Obshch. Khim.* **23**, 1370 (1953).
91. A. Michaelis and G. Schroeter, *Chem. Ber.* **27**, 490 (1894).
92. H. W. Grimmel, A. Guenther, and J. F. Morgan, *J. Amer. Chem. Soc.* **68**, 539 (1946).
93. G. S. Harris, *Proc. Chem. Soc., London* p. 118 (1957).
94. R. R. Holmes and J. A. Forstner, *Inorg. Chem.* **2**, 380 (1963).
95. A. Michaelis and W. Kärsten, *Chem. Ber.* **28**, 1237 (1895).
96. H. R. Allcock and E. J. Walsh, unpublished results (1968).
97. R. F. Hudson, "Structure and Mechanism in Organo-phosphorus Chemistry." Academic Press, New York, 1966.

Chapter 8

METATHETICAL REACTIONS

A. Introduction

Metathetical replacement reactions are those in which an exchange of ligands occurs between two or more species. In phosphazene chemistry this may involve the replacement of halogen at phosphorus by a second ligand originally present as an ion in a simple salt, or it can also take place when an organic anion displaces an organic ligand from phosphorus. Both reaction types can be illustrated by the process

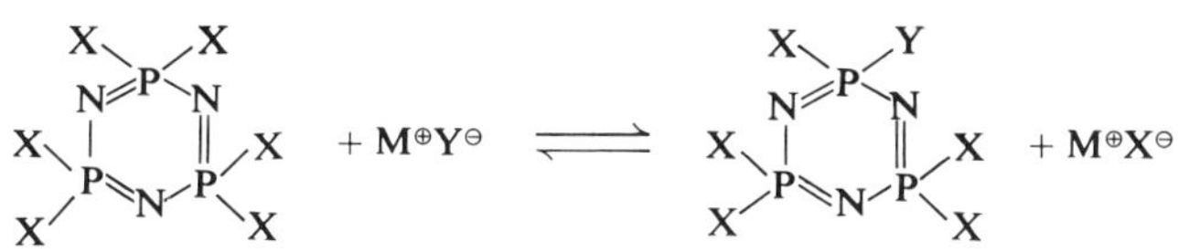

where X can be an inorganic or an organic unit. The idea of an equilibrium is implicit in the concept of metathetical exchange, although in practice it is not uncommon to select the solvent and cation, $M^{\oplus}$, to ensure the insolubility of $M^{\oplus}X^{\ominus}$ in order to enhance the yield of products.

It will be apparent that such interchanges closely resemble nucleophilic substitutions of the type discussed in Chapters 6 and 7. However, many metathetical exchanges involve the exchange of halogen atoms by other halogen or pseudohalogen units and the concept of ionic scrambling of ligands becomes more important. In addition, although many such reactions apparently take place by nucleophilic attack by $Y^{\ominus}$ on phosphorus, the prospect of S_N1 type reactions must be considered seriously. In the sections which follow, the exchange of halogen by other halogens, by cyano, by isothiocyano, and by azido will be considered in turn, and some metathetical-type exchanges between organic ligands will then be described.

B. Exchange of Chlorine by Chlorine

The fundamental question of the nature of the exchange reaction between chloride ion and various chlorophosphazenes has been studied by Sowerby.[1] The system examined was the interaction of tetraethylammonium chloride, which contained radioactive chlorine, with $(NPCl_2)_3$, $(NPCl_2)_4$, $(NPCl_2)_5$, and $(NPCl_2)_6$ in acetonitrile solution. Since tetraethylammonium chloride is a strong electrolyte in acetonitrile, chloride ion is considered to be the attacking species. The exchange rates were found to be first order in chloride ion and in phosphazene, and the overall second-order kinetics were taken as evidence for a bimolecular mechanism with a pentacoordinate phosphorus intermediate. It is interesting to note that the exchange rate with $(NPCl_2)_3$ was several orders of magnitude slower than with phosphoryl chloride, a result which may reflect the influence of the cyclic π system.

For the different chlorophosphazenes, the rates varied in the order: $(NPCl_2)_4 > (NPCl_2)_5 > (NPCl_2)_6 > (NPCl_2)_3$. The slower exchange rate of the trimer compared to the tetramer is understandable in view of the greater skeletal flexibility in the latter case. This facilitates the molecular distortions required during formation of the transition state. However, the fact that the pentamer and hexamer are less reactive than the tetramer is surprising in view of their expected even greater flexibility, and this suggests a supplementary influence by conformational or electron density factors.

No evidence was obtained for kinetically different chlorine atoms attached to phosphorus for any homologue studied. This is as expected for the planar cyclic trimer and pentamer. However, the tetramer is known to be puckered in the crystalline state, and the halogen exchange results imply that rapid conformational averaging occurs in solution.

C. Replacement of Chlorine by Fluorine

A variety of reagents can be used to effect the replacement of chlorine in chlorophosphazenes by fluorine.[2-17] These include potassium fluorosulfite or potassium fluoride in sulfur dioxide,[2-9] lead difluoride,[10-12] silver monofluoride,[10, 11, 13] sodium fluoride,[6, 14-16] or antimony trifluoride.[7, 17] The reactions can be carried out in the solid phase, in solution, or in a heterophase system, and partly or fully fluorinated derivatives can be obtained.

1. Influence of Different Fluorinating Reagents

Different fluorinating agents have markedly different characteristics in promoting metathetical exchange. Thus, lead(II) fluoride reacts with $(NPCl_2)_3$ at

300°C to give *tetrameric* as well as trimeric chlorofluoro or fluoro derivatives such as $N_3P_3Cl_4F_2$, $N_3P_3Cl_2F_4$, $(NPF_2)_3$, or $(NPF_2)_4$.[6, 10–12] Although this was the method used by Schmitz-DuMont[10–12] and co-workers for the preparation of the first fluorophosphazenes, it has now been superseded by more convenient techniques.

The use of potassium fluorosulfite (potassium fluorosulfinate), KSO_2F, was first developed by Seel and Langer,[2, 3] and this method has been used for the synthesis of a wide range of additional fluorophosphazene derivatives.[4, 5] Thus, a nitrobenzene solution of $(NPCl_2)_3$ or $(NPCl_2)_4$ with potassium fluorosulfite is allowed to react at 120°–125°C according to the following equation:

$$(NPCl_2)_n + 2n\,KSO_2F \rightarrow (NPF_2)_n + 2n\,KCl + 2n\,SO_2$$

Potassium fluorosulfite can be prepared by shaking anhydrous potassium fluoride with liquid sulfur dioxide in an autoclave for 12–16 days.[4] The fluorophosphazenes, $(NPF_2)_3$ and $(NPF_2)_4$, are isolated in 64.5% and 75% yields, respectively. The trimer is a colorless solid, m.p. 27.1°C, and the tetramer melts at 30.4°C. Both are relatively stable under ambient conditions, but they can be hydrolyzed to hydroxyphosphazenes.[3] Potassium fluorosulfite can also be used to prepare fluorochlorocyclophosphazenes in a solid-phase reaction. Thus, $N_3P_3FCl_5$, $N_3P_3F_2Cl_4$, $N_3P_3F_3Cl_3$, $N_3P_3F_4Cl_2$, and $N_3P_3F_5Cl$ can be prepared from $(NPCl_2)_3$,[4, 5] and $N_4P_4FCl_7$, $N_4P_4F_2Cl_6$, $N_4P_4F_3Cl_5$, $N_4P_4F_4Cl_4$, $N_4P_4F_5Cl_3$, $N_4P_4F_6Cl_2$, and $N_4P_4F_7Cl$ can be obtained from $(NPCl_2)_4$.[5] For tetramers in particular, mixed fluorochloro derivatives can only be prepared satisfactorily in the absence of a solvent.[6]

A variation of this preparative route involves the use of a mixture of potassium fluoride and liquid sulfur dioxide in an autoclave at ~100°C for 22 days.[8, 9] Use of this technique has permitted a broad polymeric series of fluorophosphazenes to be prepared from the appropriate chlorophosphazenes. The products include $(NPF_2)_{3-17}$. Fluorination takes place without change in ring size.

Silver monofluoride has also been used to synthesize $(NPF_2)_3$ and $(NPF_2)_4$ from the appropriate chlorophosphazenes.[13] Both fully and partially fluorinated compounds, such as $N_3P_3F_4Cl_2$, are obtained in the absence of a solvent. Conversion of $N_3P_3F_4Cl_2$ to $(NPF_2)_3$ can be effected by treatment with silver monofluoride or silver difluoride, even though the latter reagent degrades $(NPCl_2)_3$. Solution reactions in acetonitrile may also be used, but these yield partially substituted products only. An advantage of silver monofluoride as a reagent, compared to heavy metal difluoride, is that skeletal bond cleavage apparently does not occur. However, an appreciable excess of silver monofluoride is required since the first product formed, AgF·AgCl, is ineffective for further fluorination.[13]

Sodium fluoride is a particularly convenient reagent for fluorination processes.[6, 14–16] Thus, $(NPF_2)_3$ and $(NPF_2)_4$ can be prepared from the corresponding chlorides in 65–79% yield by treatment with sodium fluoride in nitrobenzene.[14, 16] Partly substituted compounds such as $N_3P_3F_4Cl_2$ and $N_3P_3F_5Cl$ can also be obtained by this route.[6] The reaction is strongly catalyzed by small quantities of water or aqueous hydrogen fluoride. Use of nitromethane as a solvent instead of nitrobenzene gives good yields of the lower fluorinated derivatives such as $N_3P_3FCl_5$, $N_3P_3F_2Cl_4$, and $N_3P_3F_3Cl_3$.[6] Acetonitrile is also an excellent medium for this reaction,[15, 18] and the following typical procedure makes use of this solvent.[19]

Acetonitrile (600 ml) is distilled from calcium hydride into a flask which contains dried sodium fluoride (168 g) and hexachlorocylcotriphosphazene (174 g, 0.5 mole), and the mixture is stirred and heated at reflux temperature (~80°C) for 1 hour. After 12 hours of additional reaction at 25°C, the mixture is fractionally distilled through a 1-ft, helix-packed column to remove all products which boil below 80°C. The distillate is washed twice with 200 ml portions of water to remove acetonitrile, and the water-insoluble portion is fractionally distilled to yield hexafluorocyclotriphosphazene, b.p. 40°–51°C (99 g, 79%).

Antimony trifluoride in boiling tetrachloroethane has been used to replace chlorine in dimethylaminochlorocyclophosphazenes by fluorine.[7, 17]

2. Replacement Pattern during Fluorination

The replacement of chlorine in $(NPCl_2)_3$ by fluorine using potassium fluorosulfite or potassium fluoride in sulfur dioxide proceeds by a geminal reaction pathway as shown in the sequences[4–6]

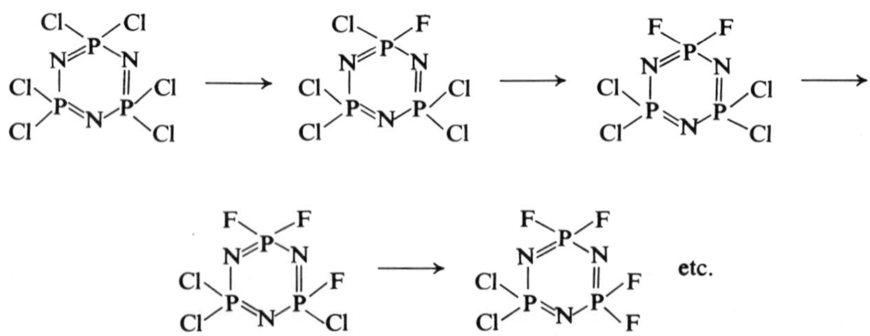

A geminal mechanism is also found for replacement of chlorine in the cyclic tetramer,[4, 6] with the added observation that the third and fourth atoms introduced are at the phosphorus atom closest to the site of original substitution.

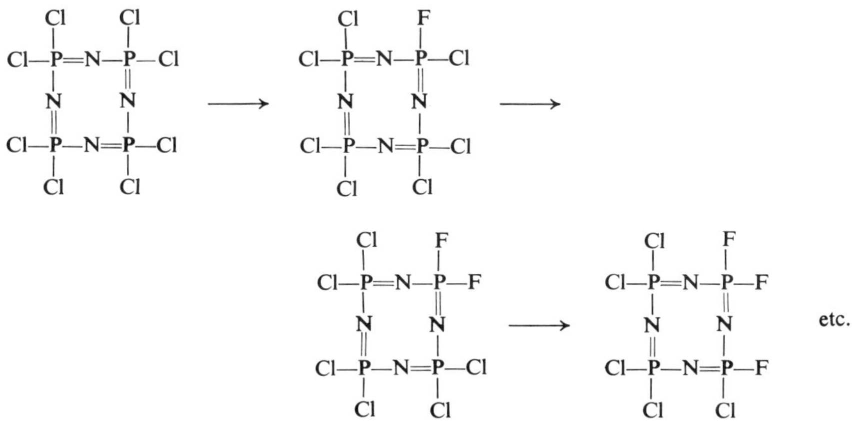

The preferential fluorination of a PFCl group rather than a PCl_2 group is consistent with the belief that electron withdrawal by fluorine enhances the susceptibility of the phosphorus atom to nucleophilic attack. Similarly, a PF_2 group would activate the adjacent phosphorus atoms to substitution rather than the remote PCl_2 group. These effects can be most readily explained in terms of an inductive electron withdrawal by fluorine from phosphorus. However, it has been observed that replacement of a second chlorine by fluorine is faster than expected in the tetramer than in the trimer and this has been used as evidence that π inductive effects are responsible.[6]

Although geminal fluorophosphazenes are produced by direct substitution with potassium fluorosulfite, an alternative sequence exists for the preparation of non-geminal fluoro derivatives.[7] Thus, treatment of *cis*- or *trans*-1,3-bis-(dimethylamino)tetrachlorocyclotriphosphazene (I) with antimony trifluoride

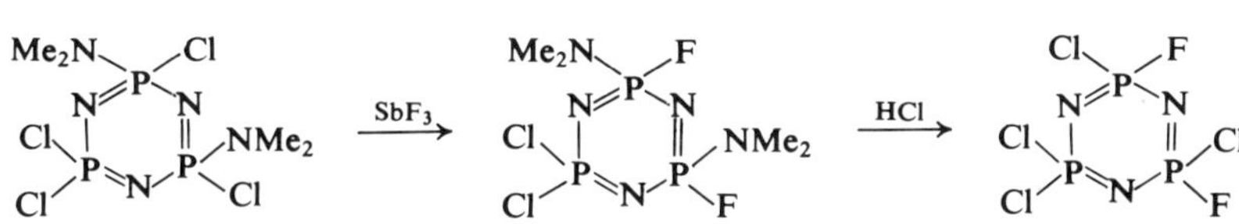

I II III

replaces the chlorine atoms geminal to the amino groups by fluorine to give the *cis* and *trans* isomers of II. Subsequent treatment of II in nitrobenzene with hydrogen chloride yielded the non-geminal difluoro derivative III. Potassium fluorosulfite reacts with both isomers of I to give the *cis* and *trans* isomers of the 5,5-difluoro derivative (IV) which can be converted by hydrogen chloride to the geminal difluorotetrachlorocyclotriphosphazene (V).[7] The different substitution patterns with potassium fluorosulfite and antimony trifluoride are ascribed

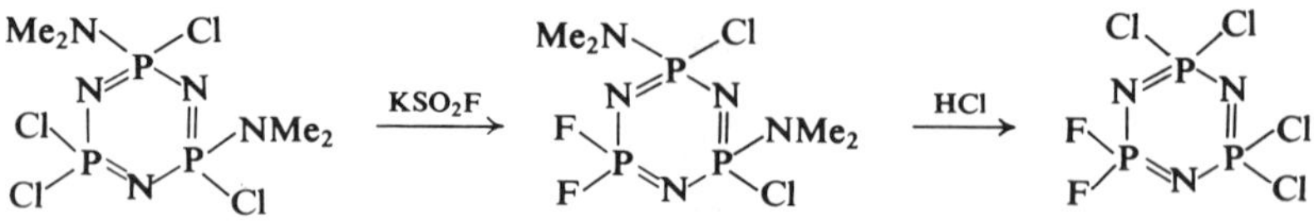

I IV V

to coordination between the latter reagent and the most basic ring nitrogen atoms. Similar differences between the reactions of antimony trifluoride and potassium fluorosulfite can be detected in the cyclic tetrameric series.

D. Replacement of Bromine by Fluorine and Chlorine

Metathetical halogen exchange reactions are not restricted to replacement of chlorine by fluorine. Hexabromocyclotriphosphazene reacts with silver monofluoride in benzene to yield $N_3P_3FBr_5$, $N_3P_3F_2Br_4$, $N_3P_3F_3Br_3$, $N_3P_3F_4Br_2$, $N_3P_3F_5Br$, and $(NPF_2)_3$.[20] With mercury dichloride, hexabromocyclotriphosphazene forms $N_3P_3Cl_3Br_3$ and (with a large excess of $HgCl_2$) $(NPCl_2)_3$.[21]

E. Replacement of Chloro by Cyano Groups

Attempts to replace chloro groups in $(NPCl_2)_3$ by the use of potassium cyanide in tetrahydrofuran or acetonitrile yielded a product which was tentatively assigned the formula $N_3P_3Cl_4(CN)_2$.[22] The use of trimethylcyanosilane as a reagent apparently resulted in the formation of a monocyano derivative.[22]

F. Replacement of Chloro by Isothiocyano Groups

Chlorophosphazenes can be converted to isothiocyano derivatives by treatment with ammonium or alkali metal thiocyanates in acetone, nitriles, tertiary amines, tertiary amides, acetic acid, or liquid sulfur dioxide.[23–27] The general equation for this reaction is

$$(NPCl_2)_n + 2n\,KSCN \rightarrow [NP(NCS)_2]_n + 2n\,KCl \qquad (n = 3 \text{ or } 4)$$

Complete replacement of chloro by isothiocyano is facilitated by the insolubility of potassium chloride in the medium. One of the main difficulties inherent in this reaction is the isolation and purification of the isothiocyanophosphazenes, which appear to be unstable in the impure state.[24, 27] However,

85% yields of $[NP(NCS)_2]_3$ have been reported,[24] the pure product being a white crystalline material, m.p. 41°–42°C.

A bis(isothiocyano)tetrachlorocyclotriphosphazene, $N_3P_3(NCS)_2Cl_4$, can be prepared by the interaction of $(NPCl_2)_3$ with ammonium thiocyanate in acetone at 0°C.[26] The tetramer, $N_4P_4(NCS)_2Cl_6$, was prepared in acetonitrile at 0°C.[26] Slow addition of ammonium thiocyanate in acetonitrile to a solution of 1,1-diphenyltetrachlorocyclotriphosphazene, $N_3P_3Ph_2Cl_4$, yielded the bis-(isothiocyano) derivative, $N_3P_3Ph_2(NCS)_2Cl_2$, m.p. 199°C.[26] Fluoroalkoxychlorocyclotriphosphazenes also undergo metathetical replacements with potassium thiocyanate in acetone. Thus, 1,1,3,3-tetrakis(heptafluorobutoxy)-5,5-dichlorocyclotriphosphazene (VI) yields the bis(isothiocyano) derivative (VII).[25] So far as is known, the isothiocyano- rather than the thiocyano-

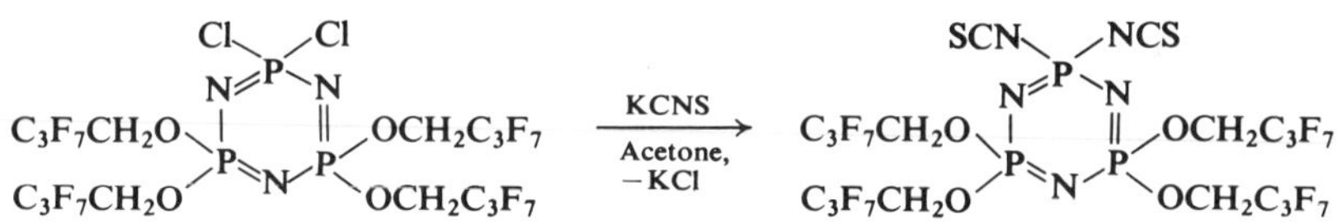

VI VII

phosphazene is obtained from all of these reactions.

Attempts have been made to prepare high polymeric isothiocyanophosphazenes by the interaction of poly(dichlorophosphazene) with alkali metal or ammonium thiocyanates in tetrahydrofuran.[27] The structure of the product

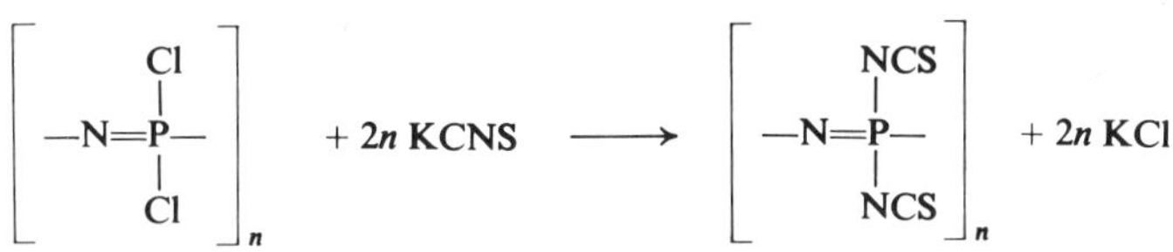

was almost indistinguishable from that of a polymer prepared by thermal polymerization of the trimer $[NP(NCS)_2]_3$.

A metathetical exchange takes place between hexabromocyclotriphosphazene and silver cyanate, but the product is not stable.[28]

G. Replacement of Chloro by Azido Groups

A metathetical exchange occurs between $(NPCl_2)_3$ and sodium azide in acetone.[29] The hexaazide, VIII, is a colorless oil, which is soluble in organic solvents, and stable to aqueous alkali. It is a violent detonator when heated or subjected to shock.

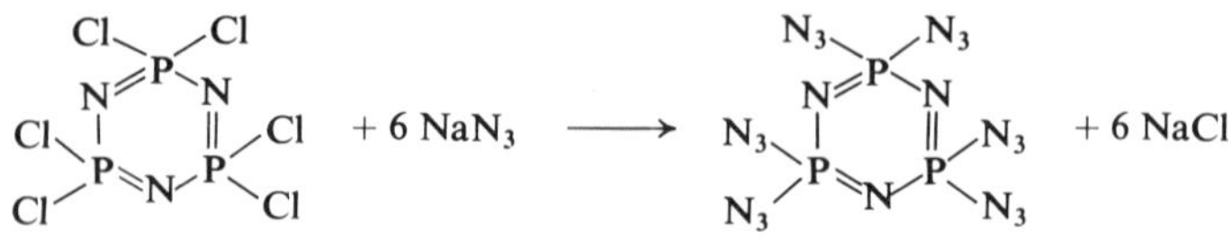

VIII

1,3-Dichlorohexaphenylcyclotetraphosphazene (IX) reacts with lithium azide in boiling acetonitrile to give the *cis*- and *trans*-bis(azido) derivatives (X).[30] Equilibration reactions between *cis*-X, and *trans*-X occurred in the

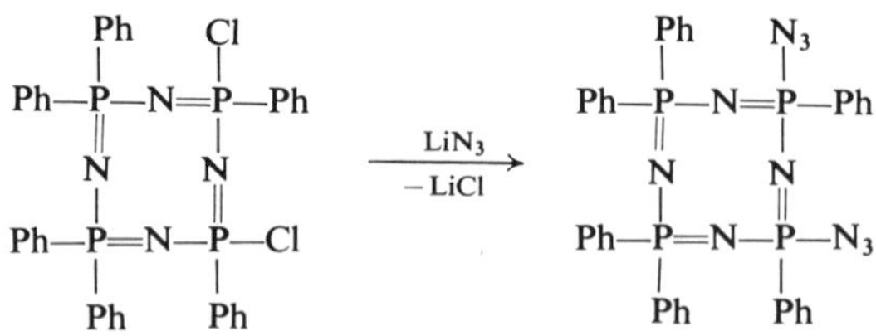

IX X

presence of lithium azide in boiling acetonitrile, the equilibrium mixture consisting of 28% *cis*-X and 72% *trans*-X. Both isomers eliminated nitrogen smoothly above 200°C by a first-order process, apparently to yield cyclolinear polymers.

H. Exchange between Organic Ligands

The concept of an exchange process between similar ligands has also been extended to the replacement of an organic ligand at phosphorus by a related anion.[31] Thus, sodium trifluoroethoxide reacts with hexakis(*p*-nitrophenoxy)-cyclotriphosphazene (XI) in tetrahydrofuran at 65°C to provide a 70% yield of hexakis(trifluoroethoxy)cyclotriphosphazene (XII). Cyclotriphosphazenes which contain both ligands have been isolated from this reaction at 25°–65°C. Sodium phenoxide also reacts with XI to yield mixed phenoxy-*p*-nitrophenoxy derivatives, and anilide ion interacts with XI to yield anilinotrifluoroethoxy-cyclotriphosphazenes.

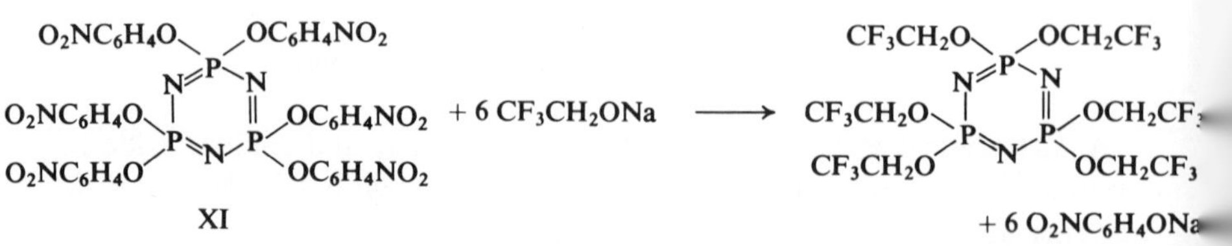

XI XII

TABLE 8.1

EXCHANGE OF LIGANDS WITH VARIOUS ORGANOPHOSPHAZENES[a]

	$O_2NC_6H_5O^-$	NH_2^-	$CF_3CH_2O^-$	$C_6H_5O^-$	$C_2H_5O^-$	CH_3MgI[c]	C_6H_5MgI[c]	PhLi[c]	CH_3Li[c]	$C_6H_5NH^-$
$[NP(OC_6H_4NO_2\text{-}p)_2]_3$ (I)[b]	*d*	+	+	+	+	+	+	+	+	+
$[NP(OCH_2CF_3)_2]_4$	–	–	*d*	+	+	+	+	+	+	+
$[NP(OCH_2CF_3)_2]_3$ (II)	–	–	*d*	–	+	–	+	+	+	+
$[NP(OC_6H_5)_2]_3$	–	–	–	*d*	–	–	–	–	+	+
$[NP(NHC_6H_5)_2]_3$	–	–	–	–	–	–	–	–	–	*d*

[a] From Allcock, Kugel, and Walsh.[31] A + sign indicates that ligand exchange occurs.

[b] The *o*-nitrophenoxy derivative behaves in a similar manner to the *p*-nitro compound.

[c] With Grignard and organolithium reagents, ring cleavage accompanies ligand exchange.

[d] Although exchange of identical ligands was not followed, evidence for the related halogen–halogen process suggests that side-group exchange occurs in these cases also.

Two series of exchange reactions have been examined; those which involve cyclophosphazenes with two independent ligands at each phosphorus, such as XI or XII, and those where each phosphorus atom forms part of an exocyclic ring. For the first group, oxy anions are found to be better nucleophiles toward phosphorus than uncharged primary amines, and substantial differences are observed between the effects of different nucleophiles and different leaving groups. Thus *p*-nitrophenoxide ion is displaced from XI by phenoxide, ethoxide, trifluoroethoxide, amide, or anilide ions but not by *n*-propylamine, aniline, dimethylamine, or *o*-aminophenol. However, trifluoroethoxide ion is displaced from XII by sodium anilide, but not by sodium amide. These and other exchange reactions are illustrated in Table 8.1, where + indicates that exchange occurs with a particular nucleophile within 24 hours at 65°C in tetrahydrofuran and – indicates that no exchange is detected. For the phosphazenes shown, the sequence of leaving-group displacement roughly parallels the electron-withdrawing power of the displaced ligands.

Ligand displacement reactions of this type have been used to cross-link high polymers, $[NP(OR)_2]_n$, with the dianions of diols, $O^-—R—O^-$, and to prepare copolymers of alkoxyphosphazene high polymers.[32]

Organometallic reagents such as methylmagnesium iodide, phenylmagnesium bromide, methyllithium, or phenyllithium displace ligands from $[NP(OC_6H_4NO_2\text{-}p\text{-})_2]_3$ (XI), $[NP(OC_6H_4NO_2\text{-}o)_2]_3$, $[NP(OCH_2CF_3)_2]_3$ (XII), $[NP(OCH_2CF_3)_2]_4$ or $[NP(OC_6H_5)_2]_3$, as illustrated in Table 8.1, but substitution is accompanied by rapid phosphazene ring cleavage to give linear derivatives. Both lithium and Grignard reagents which contain the same organic grouping apparently give the same ring-cleavage product. It should also be noted that high polymers participate in these reactions in a similar manner to cyclic trimers and tetramers. For example, methyllithium displaces piperidine from $[NP(NC_5H_{10})_2]_n$.

When a five-membered exocyclic ring exists at phosphorus, ring opening and ligand exchange occur readily with a wider variety of reagents. Thus, trifluoroethoxide ion, *n*-propylamine, diethylamine, and piperidine readily replace one aryloxy bond from each phosphorus in tris (*o*-phenylenedioxy)cyclotriphosphazene, $[NP(O_2C_6H_4)]_3$, as shown in the sequence:

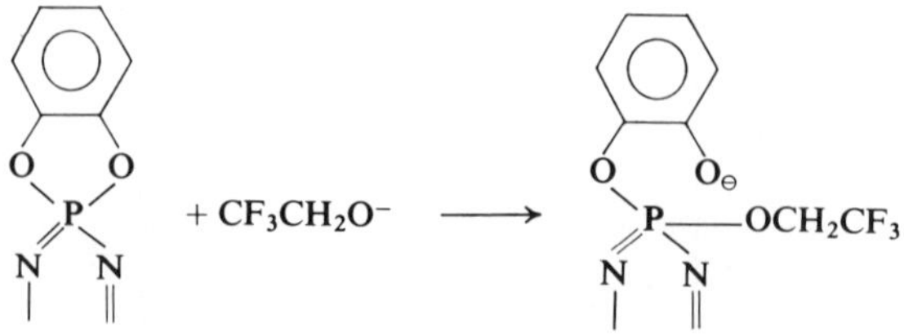

Similar exchanges are observed with tris(*o*-phenylenediamino)cyclotriphosphazene, $[NP(NH)_2C_6H_4]_3$, but not with tris(2,2′-dioxybiphenyl)cyclotriphosphazene, $[NP(O_2C_{12}H_8)]_3$, which contains a seven-membered exocyclic ring. Thus, the leaving-group ability of a substituent is markedly enhanced by the presence of a five-membered exocyclic ring.

In view of the fact that organic groups undergo ligand-exchange reactions in phosphazene systems, considerable care must be taken during the examination of *halogen* displacement sequences by mixtures of nucleophiles or consecutive reactions of the phosphazene with different nucleophiles.

References

1. D. B. Sowerby, *J. Chem. Soc., London* p. 1396 (1965).
2. F. Seel and J. Langer, *Angew. Chem.* **68**, 461 (1956).
3. F. Seel and J. Langer, *Z. Anorg. Allg. Chem.* **295**, 317 (1958).
4. A. C. Chapman, D. H. Paine, H. T. Searle, D. R. Smith, and R. F. M. White, *J. Chem. Soc., London* p. 1768 (1961).
5. G. Allen, M. Barnard, J. Emsley, N. L. Paddock, and R. F. M. White, *Chem. Ind. (London)* p. 952 (1963).
6. J. Emsley and N. L. Paddock, *J. Chem. Soc., A* p. 2590 (1968).
7. B. Green and D. B. Sowerby, *Chem. Commun.* p. 628 (1969); *J. Chem. Soc., A*, p. 987 (1970).
8. A. C. Chapman, N. L. Paddock, D. H. Paine, H. T. Searle, and D. R. Smith, *J. Chem. Soc., London* p. 3608 (1960).
9. C. P. Haber and R. K. Uenishi, *Chem. & Eng. Data Ser.* **3**, 323 (1958).
10. O. Schmitz-DuMont and H. Külkens, *Z. Anorg. Allg. Chem.* **238**, 189 (1938).
11. O. Schmitz-DuMont and A. Braschos, *Z. Anorg. Allg. Chem.* **243**, 113 (1939).
12. O. Schmitz-DuMont and M. Walther, *Z. Anorg. Allg. Chem.* **298**, 193 (1959).
13. R. Rätz and C. Grundmann, *J. Inorg. Nucl. Chem.* **16**, 60 (1960).
14. T. Moeller, K. John, and F. Tsang, *Chem. Ind. (London)* p. 347 (1961).
15. R. Schmutzler, *Inorg. Syn.* **9**, 75 (1967).
16. T. Moeller and F. Tsang, *Inorg. Syn.* **9**, 78 (1967).
17. B. Green and D. B. Sowerby, *Inorg. Nucl. Chem. Lett.* **5**, 989 (1969).
18. C. W. Tullock and D. D. Coffman, *J. Org. Chem.* **25**, 2016 (1960).
19. H. R. Allcock, W. J. Birdsall, and R. L. Kugel, unpublished results (1969) (based on refs. 14–16).
20. E. Steger and D. Klemm, *J. Inorg. Nucl. Chem.* **29**, 1812 (1967).
21. G. E. Coxon and D. B. Sowerby, *Inorg. Chim. Acta* **1**, 381 (1967).
22. D. L. Herring, *U.S. Govt. Res. Rep.* **AD 276–920** (1962).
23. R. J. A. Otto and L. F. Audrieth, *J. Amer. Chem. Soc.* **80**, 5894 (1958).
24. G. Tesi, R. J. A. Otto, F. G. Sherif, and L. F. Audrieth, *J. Amer. Chem. Soc.* **82**, 528 (1960).
25. R. Rätz, U.S. Pat. 3,074,989 (1963) (to Olin Mathieson Corp.).
26. L. F. Audrieth, T. Moeller, G. Tesi, A. F. Vandi, and K. John, WADC Tech. Rep. 58-51 (III) (1960).
27. H. R. Allcock, G. Konopski, and R. W. Tomlinson, unpublished results (1968).
28. E. Steger and G. Bachmann, *Z. Chem.* **10**, 306 (1970).
29. C. Grundmann and R. Rätz, *Z. Naturforsch. B* **10**, 116 (1955).
30. C. M. Sharts, A. J. Bilbo, and D. R. Gentry, *Inorg. Chem.* **5**, 2140 (1966).
31. H. R. Allcock, R. L. Kugel, and E. J. Walsh, *Chem. Commun.* p. 1283 (1970).
32. H. R. Allcock and G. Y. Moore, unpublished work (1970).

Chapter 9

REACTIONS OF PHOSPHAZENES WITH ORGANOMETALLIC REAGENTS

A. Introduction

The synthesis of phosphazenes which contain alkyl or aryl groups bonded directly to phosphorus can be accomplished by three routes: (a) by direct synthesis (see Chapter 4, Section A,1), (b) by Friedel–Crafts reactions (Chapter 10), or (c) by the interaction of organometallic compounds with halophosphazenes. This chapter deals with the latter route. Until recently, relatively few syntheses had been reported which utilized organometallic reagents. Nevertheless, many reactions can be envisaged in which halophosphazenes might interact with organolithium or Grignard reagents to yield organophosphazenes by processes such as

$$(NPX_2)_n + 2n\ RLi \rightarrow (NPR_2)_n + 2n\ LiX$$

$$(NPX_2)_n + 2n\ RMgX \rightarrow (NPR_2)_n + 2n\ MgX_2$$

The advantages of this synthetic route over direct synthesis or Friedel–Crafts reactions include the prospect of selective partial substitution, attachment of two or more different types of substituents to the ring, and (compared with Friedel–Crafts processes) more rapid substitution. Selective geminal or non-geminal reaction pathways can also be anticipated. A possible disadvantage inherent in the use of organometallic reagents is the prospect that attack on the skeleton may precede or follow replacement of halogen, thus leading to low yields of organophosphazene products. In the following sections, the greatest emphasis will be placed on the interaction of organometallic reagents with halophosphazenes.

B. Interaction of Halophosphazenes with Organolithium Reagents

Early work on the reactions of hexachlorocyclotriphosphazene with phenyllithium,[1,2] benzyllithium,[2] or methyllithium,[2] showed that decomposition products were formed rather than organocyclotriphosphazenes, and attack by the organolithium reagents on the skeleton was suspected. Furthermore, the formation of polymeric materials in certain instances[2,3] raised the possibility that substitution is accompanied by ring-opening polymerization.

However, fluorocyclophosphazenes react with organolithium reagents to yield organocyclophosphazenes in moderate to good yields. Thus, hexafluorocyclotriphosphazene, $(NPF_2)_3$, reacts with phenyllithium,[4–6] *m*- and *p*-tolyllithium,[4] or *n*-butyllithium[4,7] to yield partly substituted cyclophosphazenes. With phenyllithium, mono-, tri-, tetra-, and pentaphenyl derivatives were isolated.[4] Mono-*m*- and *p*-tolylpentafluorocyclotriphosphazenes have also been isolated.[4]

In a typical reaction, the organolithium reagent is allowed to interact with $(NPF_2)_3$ in boiling diethyl ether in an atmosphere of dry nitrogen for several hours.[4] The ether is then removed by distillation and vacuum distillation of the residue yields the product or products. Final purification is effected by recrystallization from heptane or pentane.

The mechanism of the reaction between phenyllithium and $(NPF_2)_3$ has been investigated by Moeller, Allen, and Tsang[5,6] and the results provide an interesting contrast with other nucleophilic substitutions. Thus, the degree of substitution is controlled by the mole ratio of the reactants, and the formation of geminal, non-geminal, *cis*-, or *trans* isomers is a rather random process. For example,[5] a 2:1 molar mixture of phenyllithium and $(NPF_2)_3$ yields the geminal bis derivative I, and the *cis* and *trans* non-geminal bisphenyl derivatives II and III. These were characterized by ^{31}P NMR, infrared, and dipole moment techniques. Derivative I is the minor product, the ratio of I:II:III being

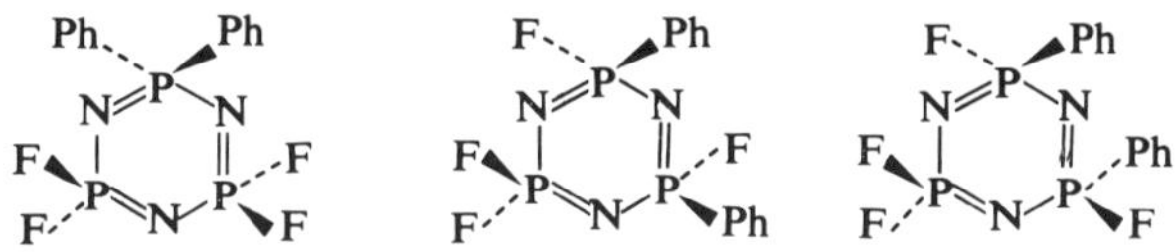

I II *cis* III *trans*

0.2:3:1. This situation contrasts with Friedel–Crafts arylation methods which favor geminal substitution.

It has been suggested that the exocyclic group which has the better ability to form π bonds to phosphorus labilizes the phosphorus–halogen bond which lies *cis* to it.[5] Thus, phenyl groups might be expected to exhibit less π bonding to phosphorus than fluorine, and attack by a second phenyl anion would take place *cis* to the phenyl group on the initially substituted phosphorus. This would result from a transfer of electrons from the fluorine of the F—P—Ph unit to the non-geminal *cis* fluorine atom, as shown in IV. Thus, attack by $Ph^{\ominus}$

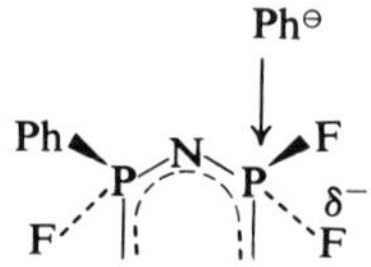

IV

would be from the side away from the *cis*-$F^{\delta-}$ unit, and an S_N2 process with inversion would preferentially yield the *cis*-diphenyl derivative. This contrasts with disubstitution by dimethylamine on $(NPCl_2)_3$, where the first dimethylamino substituent presumably forms a stronger π bond with phosphorus than does chlorine, and non-geminal disubstitution would then give the observed *trans* products. Some evidence in favor of the *cis* effect in *cis*-1,3-$N_3P_3Ph_2F_4$ has been obtained from phosphorus–fluorine coupling constant data.[5]

Methyllithium reacts with $(NPF_2)_3$ to form a geminal $N_3P_3F_4Me_2$ derivative, but further reaction results in addition of the reagent to the phosphazene ring.[6a] The reaction between methyllithium and $(NPF_2)_4$ also proceeds geminally, with the third methyl group being introduced at the 5-position. The fully methylated derivative, $(NPMe_2)_4$, is formed in 70% yield by a more extensive reaction. The pentamer, $(NPMe_2)_5$, can be obtained from $(NPF_2)_5$, but the yield is low because of the competitive addition process.[6a]

n-Butyllithium is less reactive with fluorophosphazenes than are aryllithium reagents.[7] Thus, monosubstitution is the predominant reaction. The reaction with the trimer $(NPF_2)_3$ using a 1:1 molar ratio gave the monosubstituted product, V, in only 11–14% yield.[7] When the phosphazene to *n*-butyllithium ratio was 1:2, only a small amount of dibutyl compound was formed, and when the ratio was 1:10, degradation products only were obtained. The tetrameric fluoride, $(NPF_2)_4$, is normally considered to be more reactive than the trimer and, with a 1:1 ratio of phosphazene to *n*-butyllithium, 30–33% yields of the

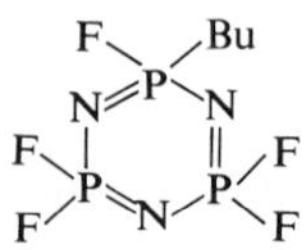

V

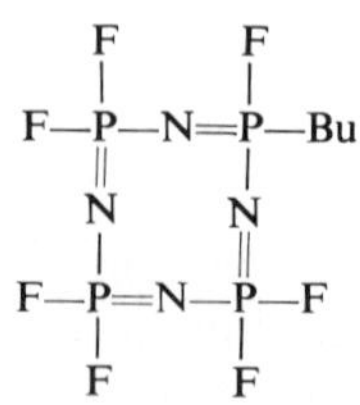

VI

monosubstituted tetramer (VI) are indeed obtained. Further reaction of VI with *n*-butyllithium yields a dibutyl product, but additional butyl groups apparently can be introduced only with great difficulty.[7] Thus, the future synthesis of fully butylated products by this route does not appear promising.

C. The Reactions of Halophosphazenes with Grignard Reagents

The interactions of chlorophosphazenes with phenylmagnesium bromide to yield phenylphosphazenes have been reported by a number of investigators.[1, 2, 8–14] For example, hexachlorocyclotriphosphazene reacts with phenylmagnesium bromide to yield hexaphenylcyclotriphosphazene.[1, 8, 10, 14]

$$(NPCl_2)_3 + 6\ PhMgBr \rightarrow (NPPh_2)_3 + 6\ MgClBr$$

However, the yields of the hexaphenylphosphazene are very low (~5%) and strong evidence exists that this product arises from a secondary process rather than a primary substitution mechanism.[14] Thus, Biddlestone and Shaw[14] have shown that the interaction of $(NPCl_2)_3$ with phenylmagnesium bromide is characterized by the following behavior: (1) no partially substituted cyclotriphosphazenes are ever isolated; only hexaphenyl or unreacted hexachloro cyclic trimers are detected; (2) the principal reaction products are *open-chain* phenylphosphazenes, with structures such as VII, or branched analogues; (3)

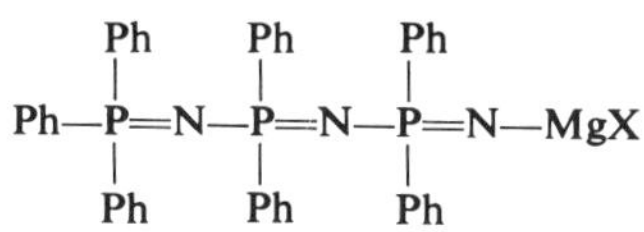

VII

these can be converted to species with end groups such as VIII after treatment with cold, aqueous ammonium chloride; (4) species with end groups of

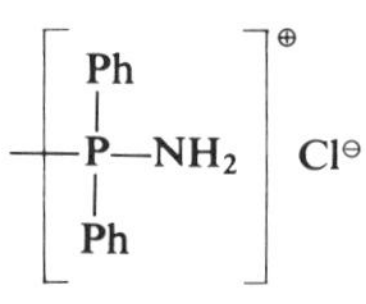

VIII

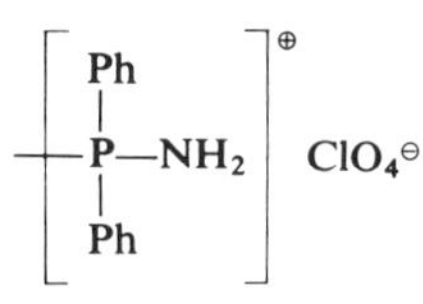

IX

structure IX can be prepared by treatment of VIII with silver perchlorate, and these products correspond to derivatives reported by earlier workers[1]; (5) treatment of hexaphenylcyclotriphosphazene with phenylmagnesium bromide does not result in skeletal cleavage.[14, 15]

These experimental facts are consistent with the following mechanism.[14] Hexachlorocyclotriphosphazene reacts with phenylmagnesium bromide immediately following or coincident with introduction of the first phenyl

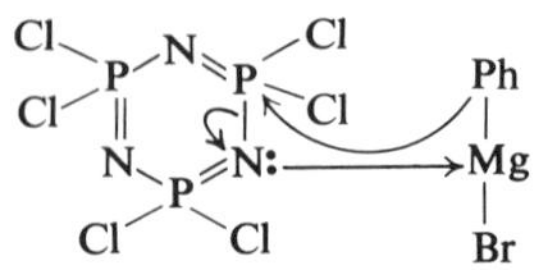

$\longrightarrow$ $PhPCl_2{=}N{-}PCl_2{=}N{-}PCl_2{=}N{-}MgBr$

X

group. The process is assumed to be a slow step. It is followed by rapid replacement of the chlorine atoms in the linear species by phenyl groups. The rapidity

$$\underset{\text{X}}{PhPCl_2{=}N{-}PCl_2{=}N{-}PCl_2{=}N{-}MgBr} \xrightarrow{\text{PhMgBr}} \underset{\text{XI}}{Ph_3P{=}N{-}PPh_2{=}N{-}PPh_2{=}N{-}MgBr}$$

of this process is probably facilitated by the flexibility of the linear chlorophosphazene. Reaction of XI with Brønsted acids would then yield derivatives of type VIII. Smaller linear fragments than X could presumably be formed by Grignard-induced chain-cleavage reactions. Finally, hexaphenylcyclotriphosphazene (XIII) may be formed by cyclization of a linear hexaphenyl intermediate such as XII. The formation of linear derivatives by the

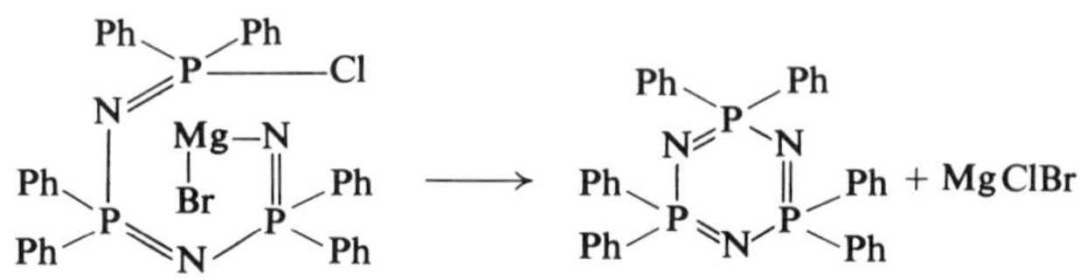

XII XIII

Grignard cleavage of XIII or partially phenylated cyclic trimers appears to be ruled out by the available experimental evidence.

Similar ring-cleavage processes appear possible when hexachlorocyclotriphosphazene is treated with benzyl,[15] methyl,[15] ethyl,[15] *n*-propyl,[15] and *n*-amyl[2] Grignard reagents.

The reaction between octachlorocyclotetraphosphazene $(NPCl_2)_4$, and phenylmagnesium bromide has also been investigated.[16] Two principal reaction products are formed in boiling diethyl ether. These are 1,1,5,5-tetraphenyl-3,3,7,7-tetrachlorocyclotetraphosphazene (XIV) and the cyclic trimer XV. A tetraphenyltetrachlorocyclotetraphosphazene, $N_4P_4Ph_4Cl_4$, and the octaphenyl derivative, $[NPPh_2]_4$, are also formed, together with open-chain species. Compound XV is isolated in 79% yield. The yield of XV rises to 93% when PhMgBr reacts with $(NPCl_2)_4$ in 1,4-dioxane. In toluene solution, phenyl-

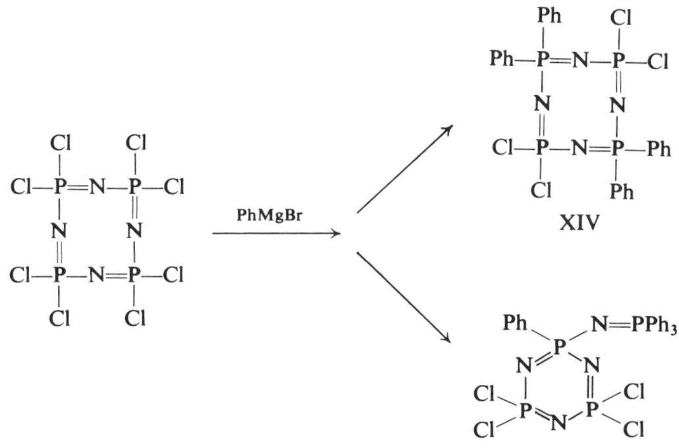

XIV, XV

magnesium bromide reacts with $(NPCl_2)_4$ to provide higher yields of acyclic derivatives and $N_4P_4Ph_4Cl_4$. The reaction mechanism for Grignard attack on $(NPCl_2)_4$ resembles that proposed for the reaction of the trimer. Monophenylation is believed to accompany ring cleavage to yield a species such as $PhPCl_2{=}N{-}PCl_2{=}N{-}PCl_2{=}N{-}PCl_2{=}N{-}MgBr$. Halogen replacement by phenyl groups and recyclization then occur.

Trimer XV has also been synthesized by Feldt and Moeller[12] by Grignard substitution with the chlorophosphazene precursor, XVI. It is particularly

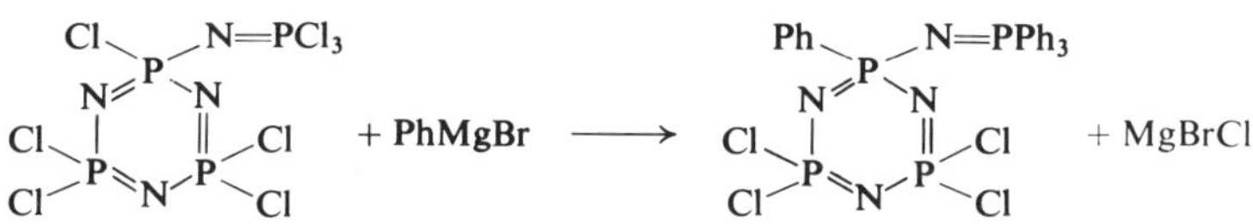

XVI XV

interesting that the reaction of the acyclic PCl_3 unit with phenylmagnesium bromide proceeds most rapidly, and that the only ring-bonded halogen subsequently replaced is the one bonded to the same phosphorus atom as the triphenylphosphazo grouping. When two trichlorophosphazo groups are present at the same phosphorus, these six halogen atoms are the only ones replaced by phenyl. Thus, these results confirm the higher reactivity of halogen atoms in acyclic species and reinforce the ring-cleavage mechanism proposed for reactions of $(NPCl_2)_3$.

Reactions have been reported in which linear, high polymeric chlorophosphazenes, $(NPCl_2)_n$, react with Grignard reagents to yield linear organophosphazene high polymers, but the products were ill-defined.[17] Chloromonophosphazenes react with Grignard reagents to yield phenylmonophosphazenes, as illustrated by the equations[18,19]:

$$\mathrm{ArC(O)N{=}PCl_3 + 3\ PhMgBr \rightarrow ArC(O)N{=}PPh_3 + 3\ MgBrCl}$$

$$\mathrm{R_2NS(O_2)N{=}PCl_3 + 3\ PhMgBr \rightarrow R_2NS(O_2)N{=}PPh_3 + 3\ MgBrCl}$$

$$\mathrm{Cl_3P{=}NS(O_2)N{=}PCl_3 + 6\ PhMgBr \rightarrow Ph_3P{=}NS(O_2)N{=}PPh_3 + 6\ MgBrCl}$$

The presence of dimethylamino groups on a phosphazene ring apparently facilitates the Grignard induced substitution of halogen atoms.[11] Thus, 1,3,5-tris(dimethylamino)-1,3,5-trichlorocyclotriphosphazene (XVII), reacts with

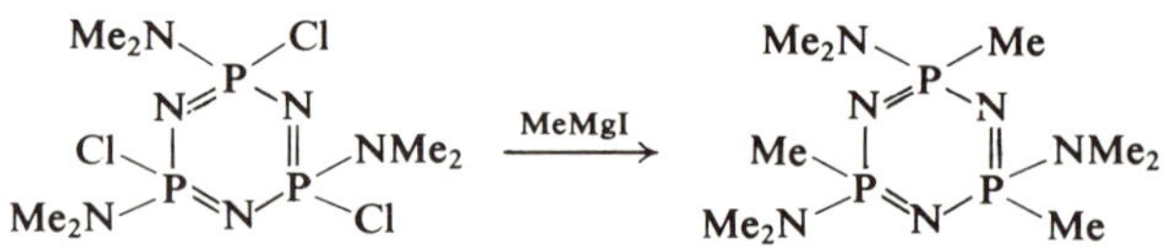

XVII XVIII

methylmagnesium iodide to replace the chlorine atoms by methyl groups (XVIII). Similar reactions are reported for phenyl- and vinylmagnesium bromides.

Although phenylmagnesium bromide reacts with hexachlorocyclotriphosphazene to yield linear and monocyclic phenylphosphazenes, the use of diphenylmagnesium in boiling dioxane results in the formation of a bicyclophosphazene (XIX).[13]

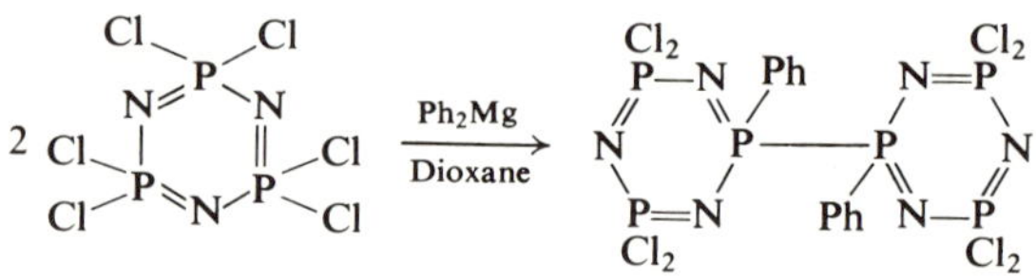

XIX

The reactions between Grignard reagents and hexabromocyclotriphosphazene have also been investigated,[15] but no organocyclophosphazenes were isolated, and ring-cleavage reactions appeared to predominate.

However, hexafluorocyclotriphosphazene reacts with phenylmagnesium bromide in boiling tetrahydrofuran to yield the mono- and geminal-diphenylcyclotriphosphazenes XX and XXI by a stepwise process[20]. Thus, there is a

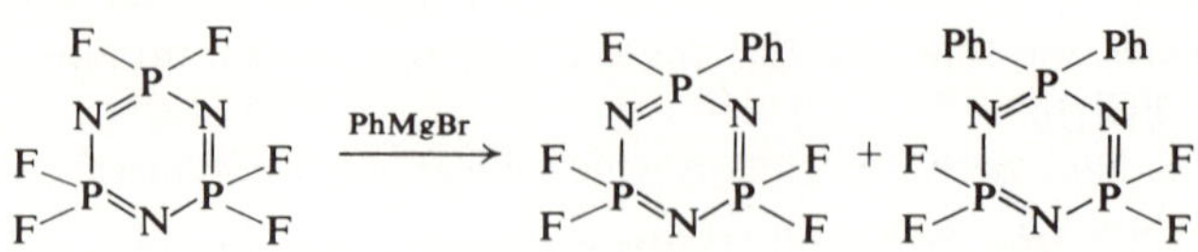

XX XXI

marked contrast between the behavior of $(NPCl_2)_3$ and $(NPF_2)_3$ which may be connected with the greater polarity of the P—F bond and the enhanced stability of the skeletal bonds when fluorine ligands are present.

D. Reactions of Organophosphazenes with Organometallic Reagents

Although hexaphenylcyclotriphosphazene, $(NPPh_2)_3$, is relatively stable to phenylmagnesium bromide in tetrahydrofuran at 70°C,[15] other organophosphazenes are attacked by Grignard or organolithium reagents with ring cleavage and ligand exchange.[21] For example, methylmagnesium iodide, phenylmagnesium iodide, and phenyllithium react with $[NP(OC_6H_4NO_2\text{-}p)_2]_3$ or $[NP(OCH_2CF_3)_2]_{3\text{ or }4}$ to displace the original ligands and cleave the phosphazene ring. Methyllithium behaves similarly with these compounds and with $[NP(OPh)_2]_3$.

References

1. H. Bode and H. Bach, *Chem. Ber.* **75B**, 215 (1942).
2. E. T. McBee, H. R. Allcock, R. Caputo, A. Kalmus, C. W. Roberts, and L. Brinkmann, *U.S. Govt. Res. Rep.* **AD 209,668** (1959).
3. R. A. Shaw, *U.S. Govt. Res. Rep.* **AD 272,009** (1961).
4. T. Moeller and F. Y. Tsang, *Chem. Ind. (London)* p. 361 (1962).
5. C. W. Allen and T. Moeller, *Inorg. Chem.* **7**, 2178 (1968): *Inorg. Synth.* **12,** 293 (1970).
6. C. W. Allen, F. Y. Tsang, and T. Moeller, *Inorg. Chem.* **7**, 2183 (1968).

6a. N. L. Paddock, T. N. Ranganathan, and S. M. Todd, *Can. J. Chem.* **49**, 164 (1971).

7. T. Moeller, A. Failli, and F. Y. Tsang, *Inorg. Nucl. Chem. Lett.* **1**, 49 (1965).
8. G. Rosset, *C. R. Acad. Sci.* **180**, 750 (1925).
9. H. Bode and R. Thamer, *Chem. Ber.* **76**, 121 (1943).
10. R. Ramain, Y. Runavot, and P. Scheebeli, *J. Chim. Phys. Physicochim. Biol.* **56**, 659 (1959).
11. G. Tesi and P. J. Slota, *Proc. Chem. Soc., London* p. 404 (1960).
12. M. K. Feldt and T. Moeller, *J. Inorg. Nucl. Chem.* **30**, 2351 (1968).
13. M. Biddlestone and R. A. Shaw, *Chem. Commun.* p. 407 (1968).
14. M. Biddlestone and R. A. Shaw, *J. Chem. Soc., A* p. 178 (1968).
15. H. R. Allcock and J. E. Brown, unpublished results (1968); J. E. Brown, M.S. Thesis, The Pennsylvania State University (1969).
16. R. A. Shaw and M. Biddlestone, *Chem. Commun.* p. 205 (1965): *J. Chem. Soc., London* p.1750 (1970).
17. C. F. Liu and R. L. Evans, U.S. Pat. 3,169,933 (1965).
18. G. I. Derkach, E. S. Gubnitskaya, and A. V. Kirsanov, *J. Gen. Chem. USSR* **31**, 3434 (1961).
19. A. F. Vandi and T. Moeller, *Chem. Ind. (London)* p. 221 (1962).
20. C. W. Allen, *Chem. Commun.* p. 152 (1970).
21. H. R. Allcock, R. L. Kugel, and E. J. Walsh, *Chem. Commun.* p. 1283 (1970).

Chapter 10

FRIEDEL–CRAFTS SUBSTITUTIONS

A. Scope and Reaction Characteristics

1. Conditions for Phenylation

The synthesis of 1,1-diphenyl-3,3,5,5-tetrachlorocyclotriphosphazene (II) by the interaction of hexachlorocyclotriphosphazene (I) with benzene in the presence of aluminum chloride was first reported by Bode and Bach in 1942.[1]

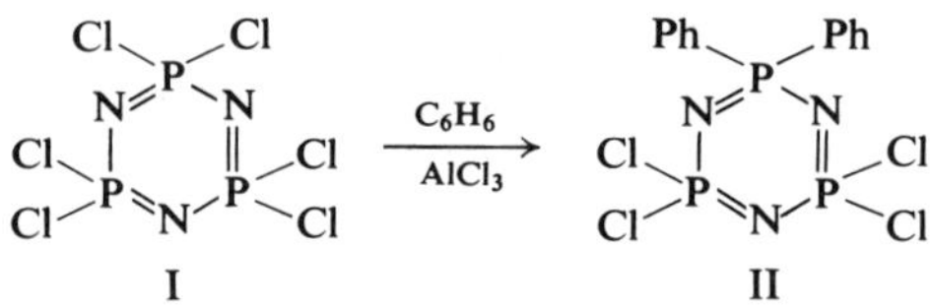

The geminal structure of II was proved first by hydrolysis to diphenylphosphinic acid, $Ph_2P(O)OH$,[1] and later by nuclear magnetic resonance techniques.[2]

Shaw and his co-workers have shown that the introduction of more than two phenyl groups requires very drastic reaction conditions.[3,4] For example, the diphenyl derivative, II, is formed in 25–40% yields when a 1:2 molar mixture of I and aluminum chloride is allowed to react in boiling benzene for several days. Greater proportions of aluminum chloride do not increase the yield of II appreciably for a given reaction time,[3,4] but evidence exists that the addition of aluminum powder or triethylamine accelerates the rate of arylation.[5] However, 6 *weeks* reaction is required to obtain a 41% yield of the geminal tetraphenyl compound (III) together with a 16% yield of II. The tetraphenyl derivative (III) could be prepared in slightly improved yield (46%) when II was

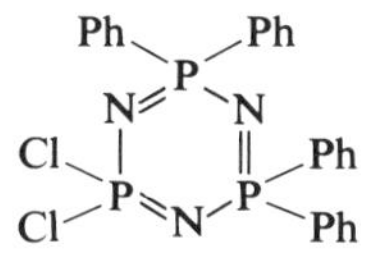

III

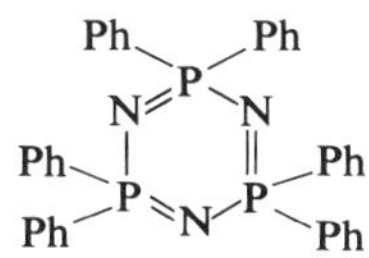

IV

used as a starting material for a 6-week Friedel–Crafts reaction, and a 6% yield of hexaphenylcyclotriphosphazene (IV) was also obtained. To prepare IV in higher yields (20%) the reaction of II with benzene and aluminum chloride must be conducted at 150°C for 48 hours in a stainless steel autoclave. The structures of III and IV have been confirmed by hydrolysis to diphenylphosphinic acid and phosphoric acid (for III) and diphenylphosphinic acid alone (for IV).

2. Introduction of Other Ligands

Tolylcyclophosphazenes can be prepared by the reaction between $(NPCl_2)_3$ (I) or $N_3P_3Cl_4Ph_2$ (II) and toluene in the presence of aluminum chloride, but drastic reaction conditions are required.[4] For example, bis-tolyl derivatives of formula $N_3P_3(C_6H_4Me)_2Cl_4$ were prepared from I after 80 hours reaction in boiling toluene. Bis- and tetrakis(xylyl) derivatives can also be prepared,[4] but mesitylene, 1,2,4,5-tetramethylbenzene, anisole, biphenyl, thiophene, and furan are apparently unreactive.

However, chlorobenzene does react with $(NPCl_2)_3$ under Friedel–Crafts conditions to yield *p*-chlorophenylcyclophosphazenes. Thus, after 42 hours reaction in boiling chlorobenzene, a monosubstituted derivative, $N_3P_3(C_6H_4\text{Cl-}p)Cl_5$, and two isomeric bis-aryl derivatives, $N_3P_3(C_6H_4\text{Cl-}p)_2Cl_4$, were formed. After 11 days reaction, a tetrakis- and the hexakis(*p*-chlorophenyl) derivatives were formed. The reactivity of chlorobenzene in this reaction has been ascribed to the relatively high reaction temperature.[4]

Alkylation of chlorophosphazenes by Friedel–Crafts techniques has not been described.

3. Friedel–Crafts Reactions with Cyclotetraphosphazenes

Octachlorocyclotetraphosphazene, $(NPCl_2)_4$, undergoes arylation reactions in boiling benzene in the presence of aluminum chloride and triethylamine,[6] but only very low yields of identifiable products are obtained after 48 hours. One of the products (obtained in 0.6% yield) is V, apparently formed by a ring contraction mechanism. Another product (also isolated in 0.6% yield following treatment with dimethylamine) is a dichlorodimethylaminopentaphenylcyclotetraphosphazene.

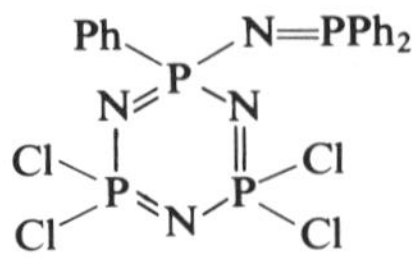

V

It seems clear that $(NPCl_2)_4$ is a less suitable substrate for Friedel–Crafts reaction than is the trimer. The mechanistic reasons for this are not yet clear.

4. Influence of Other Substrates

The influence of geminal aryl groups in retarding the ease of further arylation has already been mentioned. However, the influence of non-geminal phenyl groups has also been investigated.[7] 1,3,5-Trichloro-1,3,5-triphenylcyclotriphosphazene (VI) was treated batchwise with a 30–35 molar excess of

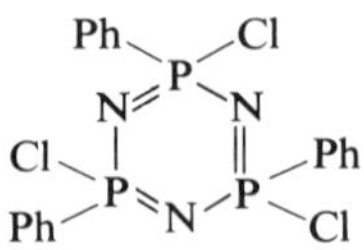

VI

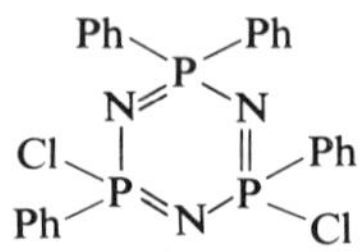

VII

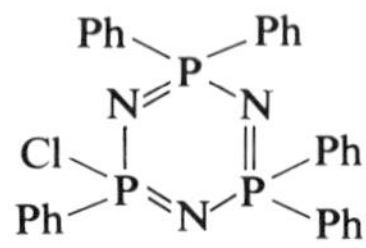

VIII

aluminum chloride in boiling benzene during 3 to 6 hours, to give good yields of the *cis* and *trans* isomers of the tetraphenyl derivative, VII. However, when only a 15- to 20-fold excess of aluminum chloride to VI was used, and the catalyst was added in one portion, 70–90% yields of the pentaphenyl compound (VIII) were obtained after 2 to 5 days of reaction. Ferric chloride and stannic chloride do not catalyze the arylation of VI, although they do induce the *cis–trans* isomerization of VI.

Phenylpentafluorocyclotriphosphazene (IX) can also be arylated in the presence of aluminum chloride, boiling benzene, and triethylamine during 24 hours to yield the geminal diphenyl derivative (X)[8]. Similarly, 1,3-diphenyl-

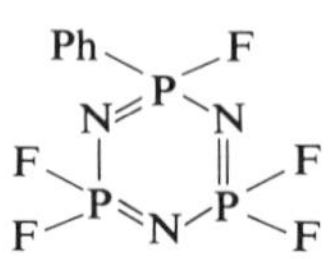

IX

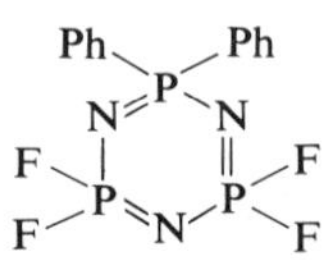

X

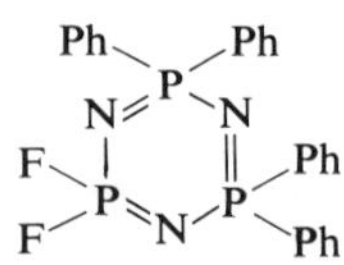

XI

1,3,5,5-tetrafluorocyclotriphosphazene is arylated to the geminal tetraphenyl compound XI. The geminal structure of XI was deduced from ^{31}P and ^{19}F NMR data.

The influence of dimethylamino groups on Friedel–Crafts arylation has also been examined.[9] Arylation of 1,3-bis(dimethylamino)-1,3,5,5-tetrachlorocyclotriphosphazene (XII) by benzene and aluminum trichloride yielded the non-geminal diphenyl derivative (XIII). Proton magnetic resonance spectra of

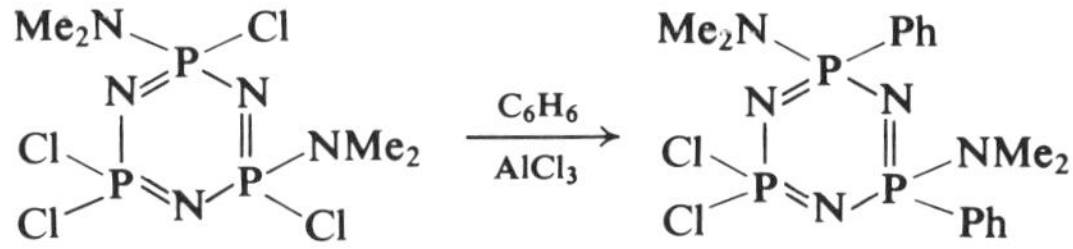

XII XIII

the product of the reaction of XIII with dimethylamine suggested that the phenyl groups were non-geminal and *cis*. Furthermore, acid hydrolysis of XIII gave phenylphosphonic acid, thus confirming the non-geminal structure. Dimethylamino groups apparently direct arylation geminally to themselves, whereas the PCl_2 unit remains unactivated. An explanation for this behavior is considered below.

B. The Mechanisms of Friedel–Crafts Arylation

A valid mechanism for this process must explain (a) the general absence of mono-, tris-, or pentakisaryl derivatives during arylation of $(NPCl_2)_3$, (b) the geminal pattern of substitution, (c) the striking rate-retarding effect which geminal diphenyl or tetraphenyl groups have on subsequent arylation, and (d) the fact that dimethylamino groups direct aryl groups to the same phosphorus atom and not to a remote PCl_2 unit. The relationship between this mechanism and the Lewis acid-catalyzed *cis–trans* isomerization of arylphosphazenes[6, 10] (see Chapter 13) must also be considered.

It seems clear that a phenyl group in the unit, Ph—P—Cl, is more effective in facilitating heterolytic cleavage of the P—Cl bond, than is a chlorine atom in the grouping Cl—P—Cl. This indicates that electron supply from one ligand (in this case phenyl) favors separation of chloride with the electrons of the P—Cl bond (XIV). Further confirmation of this effect is provided by the fact

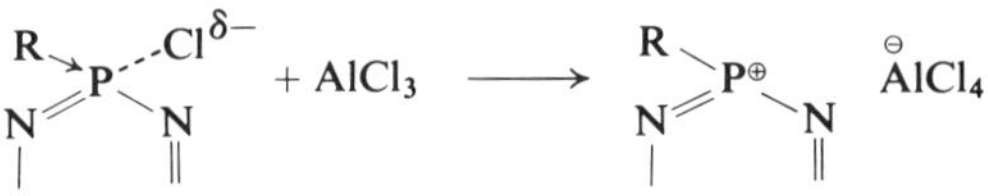

XIV XV

that an electron-supplying dimethylamino group directs a phenyl group to the position geminal to it, rather to a remote PCl_2 site.[9] Similarly, tolyl groups substitute geminally. On the other hand, the electron-withdrawal effects inherent in a *p*-chlorophenyl unit may explain why a mono(*p*-chlorophenyl)-pentachlorocyclotriphosphazene can be isolated.[4]

The aluminum chloride catalyst is essential for this reaction since, if the catalyst is omitted, even for high-temperature reactions, $(NPCl_2)_3$ is recovered quantitatively from benzene solution. It seems reasonable to suppose that the function of the catalyst, as in other Friedel–Crafts reactions is to facilitate removal of chloride ion from phosphorus to produce an electrophilic site at phosphorus, as shown in structure XV. Two alternative interpretations of this step have been suggested by different investigators. In one, the isolation of a complex of structure $N_3P_3Cl_6 \cdot 2AlCl_6$, and the reaction stoichiometry, were taken as evidence that a doubly ionized phosphazenium structure (XVI) is a reactive intermediate.[1, 2] The other suggestion is that a simple phosphazenium-

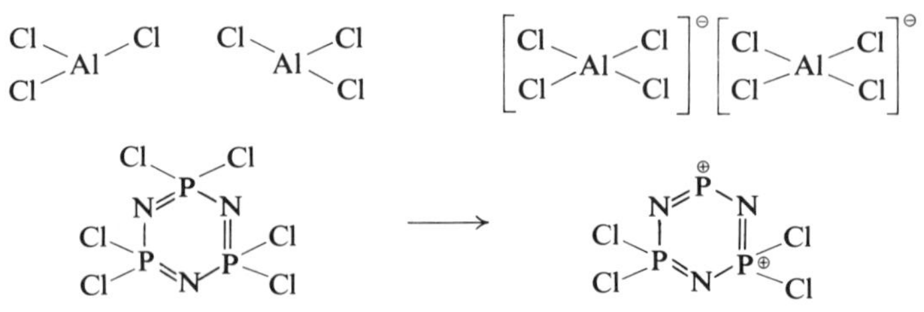

XVI

type species is formed by removal of one chloride ion (XVII).[4] This species

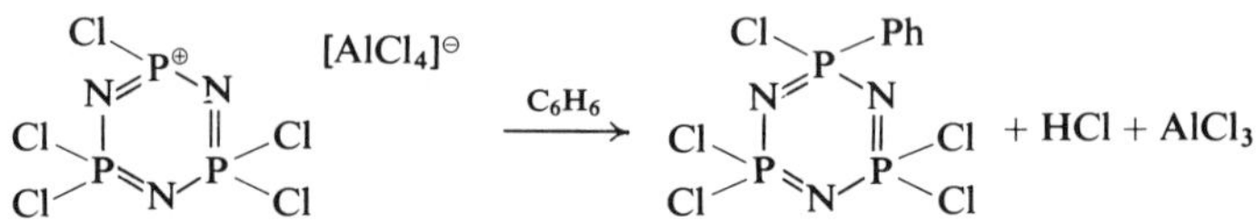

XVII

then undergoes electrophilic attack on an aromatic molecule, and hydrogen chloride and aluminum chloride are liberated. The existence of phosphazenium-type intermediates such as XVII is reasonable in view of the evidence for similar species as intermediates in chlorophosphazene polymerization reactions.[11] Furthermore, phosphazenium-type intermediates probably account for the *cis–trans* isomerization of phenyl and tertiary amino phosphazenes in the presence of aluminum chloride.[7, 10] However, these substitution reactions are almost certainly much more complex than implied by this mechanism, for it is never possible to account for all the chlorophosphazene consumed.

The question of the reasons for the slowness of the reaction can be answered in terms of the size of cation XVII and the electronic characteristics of the phosphazene ring.[4] Thus, attack by species such as XVII on benzene could be progressively retarded as bulky phenyl groups are added to the structure. However, a more likely explanation is that the electrophilicity of XVII is reduced by dissipation of the positive charge due to the influence of the nitrogen lone-pair electrons, as shown in XVIII. Furthermore, evidence exists that the lone-pair

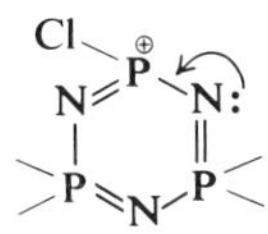

XVIII

electrons on nitrogen are more available for salt formation when phenyl groups replace chloro groups on phosphorus.[12] Thus, as Friedel–Crafts substitution progresses, the greater availability of the lone pairs may progressively reduce the electrophilicity of the attacking species.

REFERENCES

1. H. Bode and H. Bach, *Chem. Ber.* **75**, 215 (1942).
2. M. Becke-Goehring and K. John, *Z. Anorg. Allg. Chem.* **304**, 126 (1960).
3. R. A. Shaw and F. B. G. Wells, *Chem. Ind.* (*London*) p. 1189 (1959).
4. K. G. Acock, R. A. Shaw, and F. B. G. Wells, *J. Chem. Soc., London* p. 121 (1964).
5. K. Okuhara, *Nagoya Kogyo Gijutsu Skikensho Hokuko* **16**, 280 (1967).
6. V. B. Desai, R. A. Shaw, and B. C. Smith, *Angew. Chem., Int. Ed. Engl.* **7**, 887 (1968).
7. B. Grushkin, M. G. Sanchez, M. V. Ernest, J. L. McClanahan, G. E. Ashby, and R. G. Rice, *Inorg. Chem.* **4**, 1538 (1965).
8. C. W. Allen, F. Y. Tsang, and T. Moeller, *Inorg. Chem.* **7**, 2183 (1968).
9. I. I. Bezman and C. T. Ford, *Chem. Ind.* (*London*) p. 163 (1963).
10. R. Keat, R. A. Shaw, and C. Stratton, *J. Chem. Soc., London* p. 2223 (1965).
11. H. R. Allcock and R. J. Best, *Can. J. Chem.* **42**, 447 (1964).
12. D. Feakins, W. A. Last, and R. A. Shaw, *Chem. Ind.* (*London*) p. 510 (1962).

Chapter 11

COMPLEX AND ADDUCT FORMATION

A. Introduction

Phosphazenes form a wide range of coordination complexes, saltlike adducts, and molecular inclusion adducts. This behavior can be attributed to five characteristics of phosphazenes; (1) the fact that the skeletal nitrogen atoms can donate their lone-pair electrons to acceptor molecules to form coordination complexes, (2) the ability of halophosphazenes to release a halogen anion to a Lewis acid to form an anion-cation complex, (3) the ability of phosphazenes to accept a cation to form "onium"-type salts, (4) the availability of the π electrons for coordination with metals, and (5) the rigidity of the cyclotriphosphazene ring, which facilitates the specific crystalline molecular-packing arrangements required for clathration.

In the following sections, those complexes formed between phosphorus–nitrogen compounds and inorganic halides or organometallic compounds will be considered first, followed by "onium"-type complexes, complexes in solution, and crystalline inclusion clathrates.

B. Complexes between Phosphazenes and Inorganic Halides or Organometallic Compounds

1. Complexes with Metal Halides

Two types of coordination appear possible between halophosphazenes and metal halides. The first involves coordination or ion transfer at the phosphorus dihalide unit, and the second requires coordination between the skeletal nitrogen atoms and the metal atom of the metal halide acceptor molecule. Apparently both types are found.

It is believed that halophosphazenes, such as $(NPCl_2)_3$ (I), have the ability to release a halide ion to a Lewis acid to form anion–cation complexes such as II. Complexes such as II are believed to be the intermediates formed during the

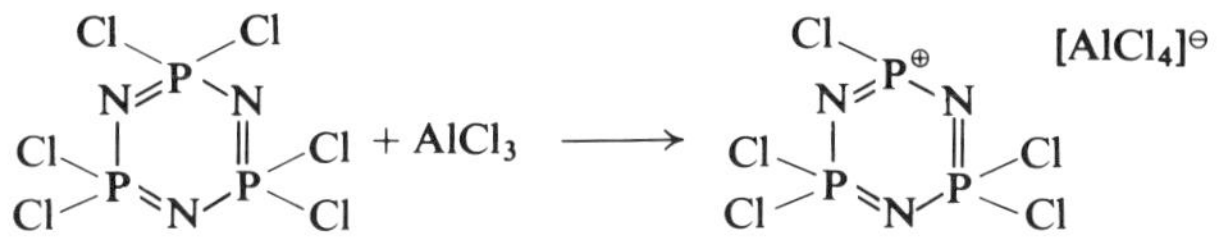

I II

Friedel–Crafts arylation of phosphazenes (see Chapter 10), and complexes of formula $N_3P_3Cl_6 \cdot 2\,AlCl_3$ have been isolated and analyzed.[1] The structure of this latter complex has not yet been unambiguously elucidated.

Fluorophosphazenes form related complexes with antimony pentafluoride[2] but the structure of these products appears to be different from the type shown in II. Thus $(NPF_2)_{3-6}$ ring systems form 1:2 complexes of formula $(NPF_2)_n \cdot 2\,SbF_5$ when condensed with antimony pentafluoride and then allowed to warm to 25°C. The adducts are white crystalline compounds which can be sublimed in vacuum below 110°C. A weak 1:1 complex can also be formed between $(NPF_2)_5$ and arsenic pentafluoride, but phosphorus pentafluoride or boron trifluoride form no complexes at 25°C. Three types of structure for the antimony pentafluoride adducts have been considered. These are (1) an anion–cation complex similar to structure II, (2) a coordination adduct between skeletal nitrogen and the metal fluoride, or (3) a halogen-bridged structure.

The first alternative is considered to be unlikely.[2] The adducts are more volatile than would be expected for salts. Furthermore, hydrolysis in moist ether regenerates $(NPF_2)_5$ from $(NPF_2)_5 \cdot 2SbF_5$ rather than yielding a hydroxyphosphazene. The same complex, when heated with potassium chloride, again gives $(NPF_2)_5$ rather than $N_5P_5F_9Cl$ or $N_5P_5F_8Cl_2$. Mass spectra of $(NPF_2)_3 \cdot 2SbF_5$ and $(NPF_2)_5 \cdot 2SbF_5$ show parent ion peaks for $(NPF_2)_3^{\oplus}$ and $(NPF_2)_5^{\oplus}$ rather than $N_3P_3F_5^{\oplus}$ or $N_5P_5F_9^{\oplus}$. The second possibility, that skeletal nitrogen is the coordination site, is also unlikely. Fluorophosphazenes should be too weakly basic to react in this manner because of the strong inductive electron withdrawal from nitrogen by the fluorine atoms. Furthermore, $(NPCl_2)_5$, which is more basic than the fluorphosphazenes, forms no adducts with antimony pentafluoride. The remaining alternative is that coordination takes place through fluorine, but that ionization of the fluorine does not occur.[2] Infrared spectra provide some support for this view.

However, more evidence is available on the structures of complexes formed between $(NPCl_2)_3$ or $(NPBr_2)_3$ with aluminum tribromide in carbon disulfide,[3] and these do not correspond to structures such as II. Coxon and Sowerby have shown that aluminum tribromide complexes can be prepared relatively easily.[3] For example, equimolar amounts of aluminum tribromide and $(NPBr_2)_3$ react exothermically in anhydrous carbon disulfide to give white crystals of a 1:1 adduct in 86% yield. A 2:1 adduct is also formed in 73% yield when a twofold excess of aluminum tribromide is employed. Only a 1:1 complex can be formed between aluminum tribromide and $(NPCl_2)_3$. The complexes hydrolyze on exposure to the atmosphere. When heated at 180°C in vacuum they apparently form linear, end-capped molecules such as $Br(NPCl_2)_3AlBr_2$.

Aluminum tribromide appears to be unique in this respect since no adducts are formed between $(NPBr_2)_3$ and boron tribromide, methyl iodide, stannic chloride, or titanium tetrachloride in carbon disulfide, or with antimony pentachloride in carbon tetrachloride. Furthermore, $(NPCl_2)_3$ forms no solid addition complexes with stannic chloride in carbon disulfide or with antimony pentachloride in carbon tetrachloride.

Infrared spectral evidence suggests that coordination is through the ring nitrogen atoms to form σ complexes, as shown in III. The synthetic evidence

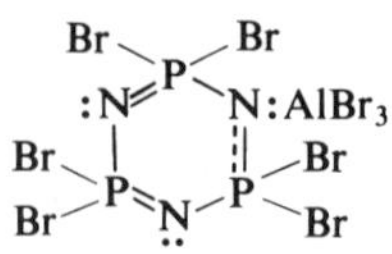

III

also favors this formulation since the more electronegative chloro ligands in $(NPCl_2)_3$ presumably compete more effectively than bromo ligands for the lone-pair electrons on nitrogen, and this can be used to rationalize the absence of 2:1 complexes with $(NPCl_2)_3$. Furthermore, stannic chloride and titanium tetrachloride are presumably weaker Lewis acids in this situation and cannot form coordination adducts.

If this interpretation is correct, electron-supplying ligands at phosphorus should enhance the availability of the nitrogen lone pair for complex formation. This is, in fact, observed when hexamethylcyclotriphosphazene (IV) interacts with stannic chloride or titanium tetrachloride in dichloromethane to form stable, crystalline, 1:1 adducts.[4] Infrared evidence was used here also to confirm the σ-bonded structure, although no distinction was made between a five-coordinate metal structure (V) or a six-coordinate metal involving two nitrogen atoms.[4] Significantly, hexaphenylcyclotriphosphazene, $(NPPh_2)_3$, 1,1-di-

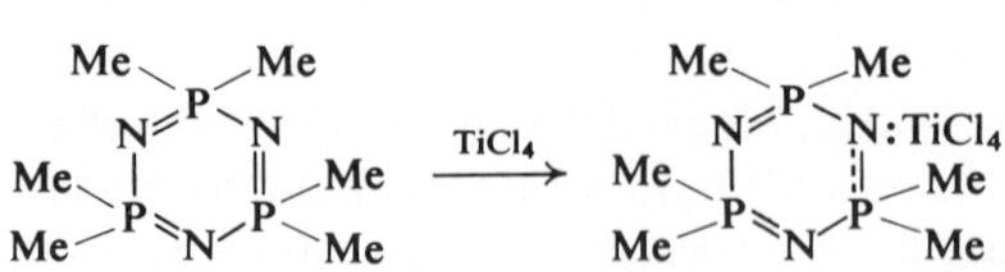

IV V

phenyl-3,3,5,5-tetrachlorocyclotriphosphazene, $N_3P_3Ph_2Cl_4$, and hexachlorocyclotriphosphazene, $(NPCl_2)_3$, do not form complexes with these ligands.[4] Apparently phenyl groups as well as chloro groups lower the availability of the nitrogen lone-pair electrons.

It has also been reported that $(NPCl_2)_3$ reacts with stannic chloride in the presence of pyridine to form a three-dimensional polymeric complex of structure VI.[5] Other metal halide complexes are formed when the adduct of

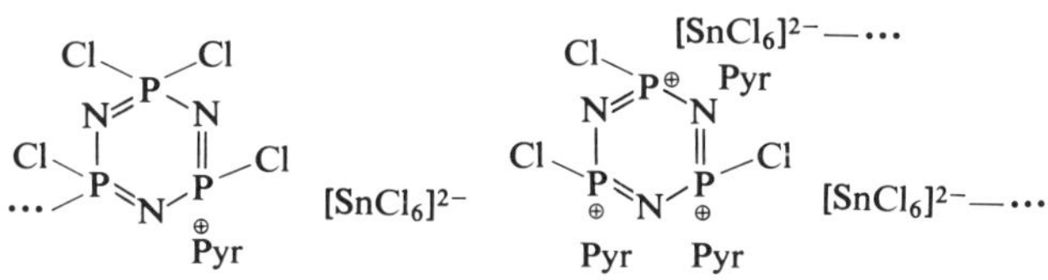

VI

pyridine with $(NPCl_2)_3$ reacts with aluminum chloride, cuprous bromide, cuprous chloride, ferric chloride, or titanium tetrachloride.[6] These crystalline complexes are believed to have structures similar to VI except that direct pyridine to phosphorus bonding may exist.[6]

Octamethylcyclotetraphosphazene, $(NPMe_2)_4$, forms a yellow crystalline complex when treated with anhydrous copper(II) chloride in methyl ethyl ketone.[7] An X-ray single-crystal structure determination of this complex has shown that the molecule is a σ complex, apparently with a hydrogen atom attached to a skeletal nitrogen, as shown in VII.[8] A related, pale blue, crystal-

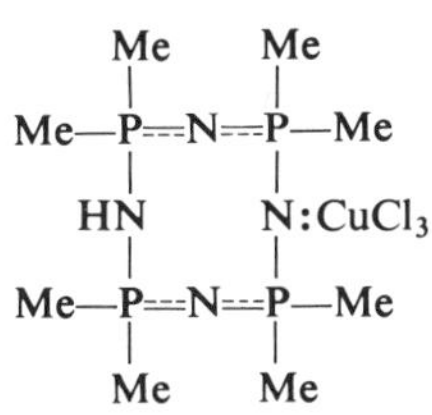

VII

line complex is formed when anhydrous cobalt(II) chloride reacts with $(NPMe_2)_4$ in methyl ethyl ketone.[7] An X-ray structure analysis of this complex[9] revealed that two protonated phosphazene rings were associated with one cobalt atom, the formula being $[(NPMe_2)_4H^+]_2[CoCl_4]^{2-}$. Thus, the association is ionic and with no evidence of cobalt–nitrogen bonding being found. Cyclophosphazonium cations of this type are discussed in a later section.

By contrast, the cyclic hexamer $[NP(NMe_2)_2]_6$ reacts with copper(II) chloride to form the complex $[N_6P_6(NMe_2)_{12}CuCl]^{\oplus}CuCl_2^{\ominus}$, in which the coordinated copper atom lies inside the puckered phosphazene ring in such a way that four of the six ring nitrogen atoms occupy the basal sites of a distorted square-pyramidal arrangement.[10] The chlorine atom occupies the apical position.

Evidence of complexation has also been obtained when $[NP(NHPr^n)_2]_3$ is treated with Co^{2+}, Ni^{2+}, or Cu^{2+} in acetonitrile, when $[NP(NHBu^n)_2]_3$ is treated with Co^{2+} ions in acetonitrile, or when iodine in carbon tetrachloride interacts with $[NP(NHMe)_2]_3$, $[NP(NHPr^n)_2]_3$, or $[NP(NMe_2)_2]_3$.[11]

Complexes of the type $(NPCl_2)_3 \cdot 3SO_3$,[12] $(NPCl_2)_3 \cdot NO_2$,[1] and $[NP(NHR)_2]_3 \cdot BF_3$[13] have also been reported, with the skeletal nitrogen atoms providing possible coordination sites.

Hexachlorocyclotriphosphazene, $(NPCl_2)_3$, reacts with sodium hexachloroplatinate(IV) in acetone to yield a brown solid with the composition $N_3P_3Cl_2(PtCl_4)_2$.[14] The same reagent interacts with hexakis(isothiocyano)-cyclotriphosphazene, $[NP(NCS)_2]_3$, in acetone to yield a compound with the composition $N_3P_3(NCS)_2(PtCl_4)_2$. The structure of these species is not known.

Evidence has been presented[15] that hexaethoxycyclotriphosphazene, $[NP(OEt)_2]_3$, reacts with titanium or zirconium tetrachlorides to form adducts of formula $N_3P_3(OEt)_4O_2TiCl_2$ or $N_3P_3(OEt)_4O_2ZnCl_2$. The metal halide components are believed to be associated with the oxygen ligands of a cyclophosphazane structure.

A variety of white or colored complexes are formed when hexakis(*S*-thiosemicarbazido)cyclotriphosphazene interacts in aqueous media with Ag^+, Hg^{2+}, Hg^{2+}, Cu^+, Cu^{2+}, Cd^{2+}, Zn^{2+}, Pb^{2+}, Bi^{3+}, Ni^{2+}, Co^{2+}, and Fe^{2+}.[16, 17] This is the basis of a quantitative potentiometric determination of Ag^+ and Cu^{2+} ions. The structures of the products have not been elucidated.

2. Reactions of Phosphazenes and Phosphazanes with Metal Carbonyls

Molybdenum hexacarbonyl reacts almost quantitatively with octamethylcyclotetraphosphazene, $(NPMe_2)_4$, to yield a yellow solid with the composition $(NPMe_2)_4Mo(CO)_4$.[18] The infrared spectrum of this compound is consistent with a C_{2v} structure in which the coordination to the molybdenum octahedron involves two nitrogen atoms at opposite ends of the molecule. The methiodide salt of $(NPMe_2)_4$, $(N_4P_4Me_9)^{\oplus}I^{\ominus}$, displaces only one molecule of carbon monoxide from molybdenum or chromium hexacarbonyls[18] to yield complexes of formula $[N_4P_4Me_9]^{\oplus}[Mo(CO)_5I]^{\ominus}$ or $[N_4P_4Me_9]^{\oplus}[Cr(CO)_5I]^{\ominus}$.

Chromium hexacarbonyl reacts with the cyclodiphosphazane VIII in benzene to yield a light red compound with the formula $(MeN{-}PCl_2)_2Cr(CO)_4$.[19]

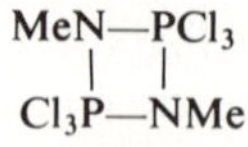

VIII

Presumably, coordination is through the nitrogen atoms by donation of the lone pairs to chromium, perhaps in a chelating arrangement.

Triphenylmonophosphazene (triphenylphosphinimine), $Ph_3P{=}NH$, reacts with molybdenum or tungsten hexacarbonyls at 84°C to yield yellow or red crystals of $(Ph_3P{=}NH)_2Mo(CO)_4$ or $(PH_3P{=}NH)_2W(CO)_4$.[20] At higher temperatures (above 110°C), a molybdenum tricarbonyl adduct, $[(PH_3P{=}NH)_2Mo(CO)_3]_2$, is formed both from the starting materials and from the tetracarbonyl adduct. The same product can also be prepared by treatment of $Ph_3P{=}NH$ with cycloheptatrienylmolybdenum tricarbonyl or bis(dialkylcyanamido)molybdenum tetracarbonyl.[20] Structure IX has been proposed for this complex.

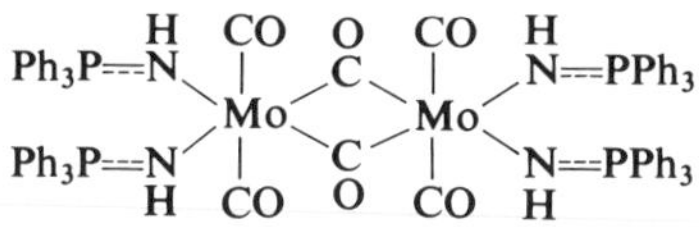

IX

It is also known that bis(diphenylphosphino)alkylamines, $(Ph_3P)_2NR$, displace two molecules of carbon monoxide from molybdenum hexacarbonyl to yield complexes of the type $(Ph_3P)_2NRMo(CO)_4$, with the structure shown in X.[21, 22]

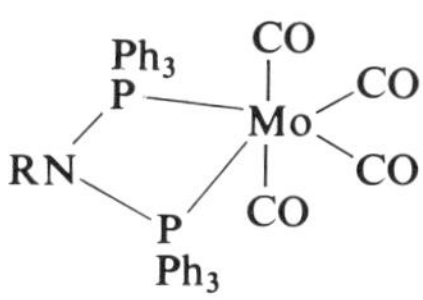

X

3. Other Phosphorus-Nitrogen Complexes

Beryllium acetylacetonate reacts with the linear phosphazene $O{=}PPh_2{-}N{=}PPh_2OH$ to yield the chelate coordination compound XI[23] and treatment of XI with more of the phosphazene at a higher temperature (175°C) yields the diphosphazenyl chelate, XII. Similar chelates were formed by the phos-

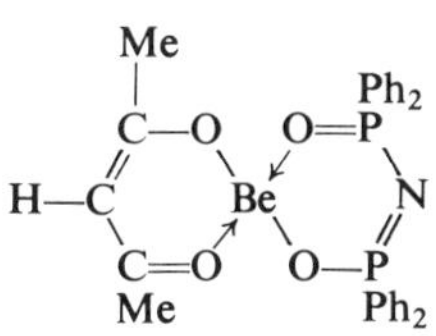

XI

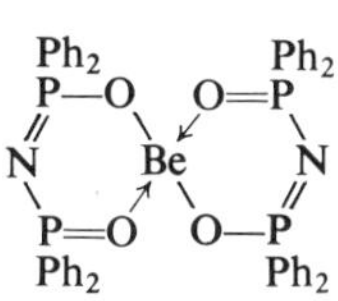

XII

phazene, $O{=}PPh_2{-}(N{=}PPh_2)_3OH$, but in these cases there is some question as to whether chelation involves the terminal oxygen atoms of the phosphazene ligands or the nitrogen atoms adjacent to each terminal phosphorus.[23]

It has also been reported that aluminum and gallium trialkyls form 1:1 coordination complexes with *N*-trialkylsilyl triorganophosphinimes.[24] Coordination is through the nitrogen atoms, as shown in XIII.

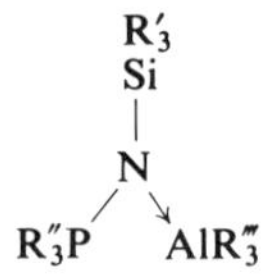

XIII

C. "Onium"-Type Complexes

Allen, Dyson, and Paddock[25] have shown that methylcyclophosphazenes such as $(NPMe_2)_3$ and $(NPMe_2)_4$ react smoothly and completely with methyl or ethyl iodide to form *N*-alkylphosphazonium iodides, with the cations having structures such as XIV and XV. These salts are insoluble in nonpolar solvents

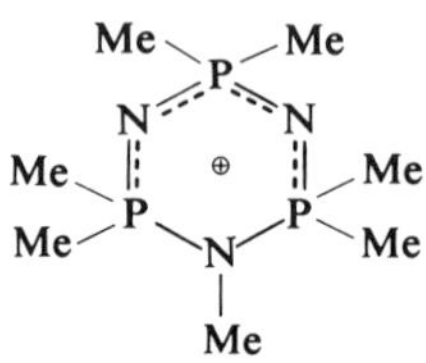

XIV

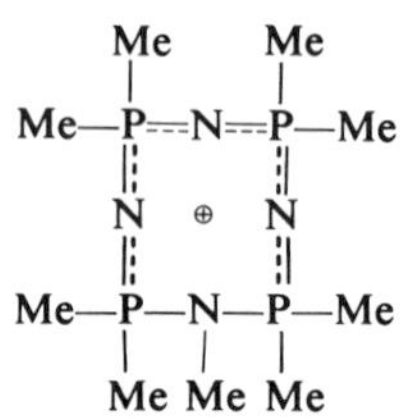

XV

but they are soluble and stable in water. The iodide salts can be converted to chlorides, insoluble iodomercurates, $[N_3P_3Me_7]^+[HgI_3]^-$ or $[N_4P_4Me_9]^+$-$[HgI_3]^-$, and insoluble tetraphenyl borates. Isothiocyanozincate salts can also be prepared and recrystallized from water.

It was mentioned in an earlier section that $(NPMe_2)_4$ reacts with $CoCl_2$ to yield a cyclophosphazonium ion complex of structure $[(NPMe_2)_4H^+]_2$-$[CoCl_4]$.[2–9, 26] One of the cyclophosphazonium rings occupies a saddle and the other a tub conformation.

The formation of onium ions by the addition of cations to cyclophosphazene side groups has also been reported by Rapko and Feistel.[27] Thus, hexakis-(dimethylamino)cyclotriphosphazene, XVI, reacts with trimethyloxonium tetrafluoroborate to yield the dication, XVII. The dication is a white solid which is stable in air and soluble in polar organic solvents. Monoalkylation also occurs with $N_3P_3(NMe_2)_6$, $N_3P_3Cl_2(NMe_2)_4$, $N_3P_3Cl_3(NMe_2)_3$, $N_3P_3Cl_4$-

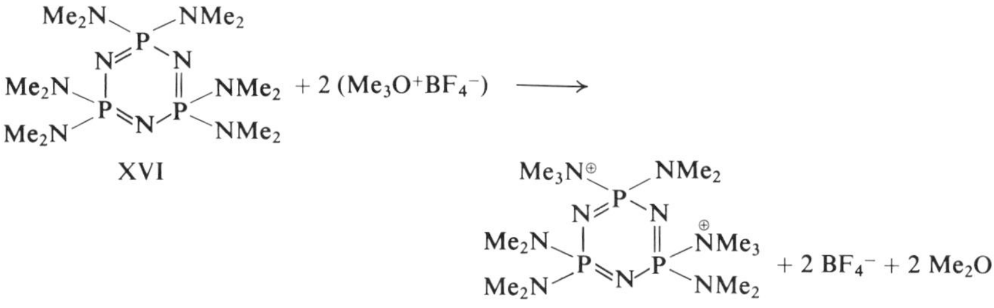

XVI

XVII

$(NMe_2)_2$, and $N_3P_3Cl_5(NMe_2)$. In each case, the proton NMR spectra indicate that the exocyclic nitrogen atoms are the methylation sites. However, the alkylation of 1,1-dichloro-3,3,5,5-tetrakis(isopropylamino)cyclotriphosphazene, $N_3P_3Cl_2(NHPr^i)_4$, with trimethyloxonium fluoroborate results in methylation of a ring nitrogen atom. A similar ring alkylation takes place with hexaphenylcyclotriphosphazene, $(NPPh_2)_3$.[27] It is interesting that methylation of dimethylaminocyclotriphosphazenes occurs at the *exocyclic* nitrogen atoms, whereas protonation is believed to take place at the *ring* nitrogen atoms.

It has also been suggested[28] that onium ions such as XVIII are formed when pyridine interacts with chlorophosphazenes and that species of this type are

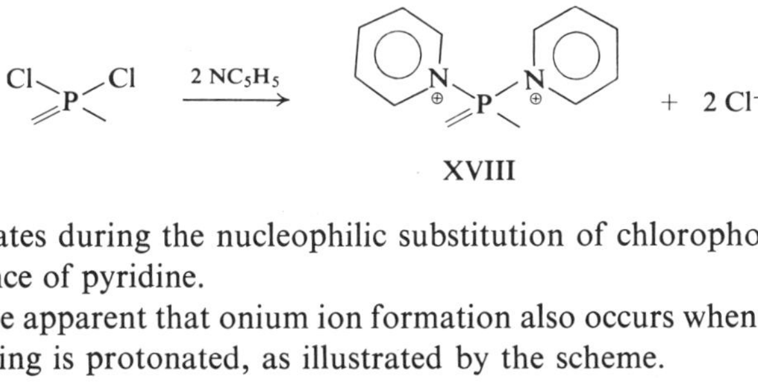

XVIII

intermediates during the nucleophilic substitution of chlorophosphazenes in the presence of pyridine.

It will be apparent that onium ion formation also occurs when a cyclophosphazene ring is protonated, as illustrated by the scheme.

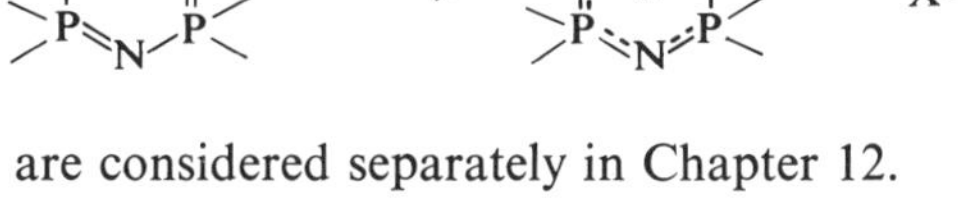

These reactions are considered separately in Chapter 12.

D. π and Charge-Transfer Complexes

The prospect that cyclophosphazenes might form π complexes with electron donors or acceptors in solution has been examined with the use of ultraviolet spectroscopy, but no evidence for π-complex formation was obtained.[29] For example, solutions of $(NPCl_2)_3$ and tetracyanoethylene, $(NPCl_2)_3$ and hexamethylbenzene, or $(NPF_2)_3$ and hexamethylbenzene in cyclohexane showed no spectral shifts at all. The spectra were superimposable upon those of the individual components.[29] However, spectral shifts were observed when $[NP(NMe_2)_2]_{3\text{ or }4}$ was added to solutions of iodine in nonpolar solvents.[11, 13]

It has been reported that solid molecular addition compounds are formed between $[NP(NMe_2)_2]_3$ and tetracyanoethylene, $[NP(NMe_2)_2]_3$ and $(NPCl_2)_3$, $[NP(NHPr^i)_2]_3$ and $(NPCl_2)_3$, $N_3P_3Cl_4(NHBu^t)_2$ and $N_3P_3Cl_2(NHBu^t)_4$, or $[NP(NMe_2)_2]_4$ and $(NPCl_2)_4$.[30, 31] Phase diagrams were used as proof of adduct formation. The same technique was used to demonstrate complex formation in the solid state between hexamethylbenzene and $(NPCl_2)_{3-6}$ or $[NP(NMe_2)_2]_{4\text{ and }6}$. In the case of the chlorophosphazenes, the even-numbered systems formed both 1:1 and 2:1 complexes, whereas the odd-numbered systems formed only 2:1 adducts.[32] Both chloro- and dimethylaminophosphazenes are assumed to function as π acceptors in this system. However, it has been shown that the infrared spectrum of a fused 1:1 molar solid mixture of hexamethylbenzene and $(NPCl_2)_3$ is identical to a superimposition of the spectra of the individual components.[29] Evidence for complex formation between $(NPCl_2)_3$ or $(NPCl_2)_4$ and naphthalene in cyclohexane has been deduced from cryoscopic data.[33]

Proton NMR evidence has been used to provide evidence for π interactions in solution.[34] Thus, mono-, bis-, tris-, and tetrakis(dimethylamino)chlorocyclotriphosphazenes show upfield shifts in benzene compared to measurements for carbon tetrachloride solutions. Similar effects are noticed for mono-, bis-, and tris(dimethylamino)bromocyclotriphosphazenes, but no shift was observed with hexakis(dimethylamino)cyclotriphosphazene. The shifts were more pronounced with *cis*- rather than *trans*-dimethylamino compounds and when a dimethylamino group was flanked by a dimethylamino and a chloro group rather than by two chlorine atoms. Thus, the suggestion has been made[34] that the benzene molecules approach the cyclophosphazene in such a way that dimethylamino rather than chloro groups are encountered, possibly because electron density has been transferred from exocyclic nitrogen to skeletal phosphorus.

E. Crystalline Inclusion Clathrates

1. Spirocyclophosphazenes

Certain cyclotriphosphazenes form an important series of molecular inclusion clathrates.[35–43] The most thoroughly studied of these are the clathrates formed by tris(*o*-phenylenedioxy)cyclotriphosphazene (XIX), a derivative which is readily synthesized from $(NPCl_2)_3$ and catechol in the presence of a base.[44–46] When purified by sublimation, XIX is a white crystalline solid, m.p. 244°–245°C. However, when sublimed crystals of XIX are brought into

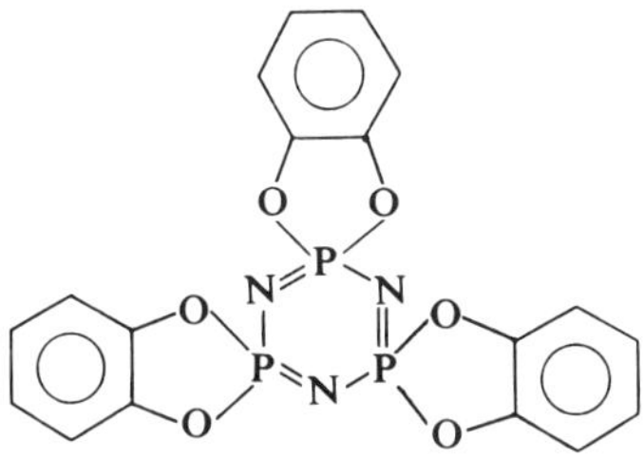

XIX

contact with organic liquids or vapors, the crystals swell and disintegrate, and analysis shows that the organic component has been incorporated into the crystallites.[35] This process is accompanied by a crystal structure change from triclinic or monoclinic for the pure compound to hexagonal for the adduct, as indicated by X-ray diffraction data. A similar process takes place when pure XIX is recrystallized from organic media.

In general, the dried adducts are stable at room temperature, but volatile guest molecules with small molecular dimensions, such as carbon disulfide or methanol, can escape from the lattice over a period of several weeks. However, recrystallized adducts with other guest components are stable at room temperature for years or when held in vacuum at 0.1 mm at 25°C for several days. The included components can be driven off with more drastic treatment, and prolonged heating at 100° to 150°C at 0.1 mm pressure removes included benzene. Rapid removal of the guest component can be effected above 170°C, and loss of the guest is accompanied by reversion of the crystal structure to the triclinic form. Thus, absorption and release of the guest can be made part of a recycling process.

Adducts can be prepared between XIX and a wide variety of organic compounds. Representative examples are shown in Table 11.1.[35] The fact that adduct formation is not a coordination or π-complex phenomenon is indicated by the wide range of different guest molecules which form stable adducts. Thus, aliphatic and aromatic hydrocarbons, halocarbons, esters, ethers, ketones, nitriles, carbon disulfide, and ethanol all function as guest components. Within wide limits, the molecular dimensions of the guest molecules do not limit adduct formation. Thus, compounds as small as carbon disulfide, acrylonitrile, or ethanol, or as bulky as decalin, norbornadiene, or isooctane are included. However, the dimensions of the guest compounds do influence the ratio of host to guest, as illustrated in Table 11.1, with larger quantities of the

TABLE 11.1

INCLUSION COMPOUNDS OF TRIS(*o*-PHENYLENEDIOXY)CYCLOTRIPHOSPHAZENE (XIX)[a]

Included compound	Moles guest / Moles XIX[b]	Melting range (°C)	Included compound	Moles guest / Moles XIX[b]	Melting range (°C)
Tetralin	0.59	220–230	Tetrahydrofuran	0.35	233–244
Norbornadiene	0.51	236–275	Chloroform	0.34–0.42	224–245
Isooctane	0.48–0.50	234–245	—	(0.48–0.66)	240–244
trans-Decalin	0.47–0.66	237–245	Methyl methacrylate	0.33	230–267
o-Xylene	0.46	235–259	Diethyl ether	0.31	226–243
o-, *m*-, and *p*-Xylenes	(0.50)	241–256	Benzene	0.20–0.25	222–245
Cumene	0.45	228–248	—	(0.44–0.50)	244–255
Cyclohexane	0.41–0.42	233–245	—	—	—
	(0.50)	—	Acetone	0.20	231–250
n-Heptane	0.38–0.5	247–253	Carbon disulfide	0.20	226–239
Styrene	0.38	239–256	Acrylonitrile	0.19	230–257
	(0.50)	244–257	Ethanol	0.11	231–243
Ethyl acetate	0.37	233–243	(Methanol)	0.03	236–243
Carbon tetrachloride	0.37	233–244	—	—	—

[a] From Allcock and Siegel.[35]

[b] Guest to host molar ratios were determined by weight increase or (in parentheses) by mass spectrometry or microanalysis.

bulkier guest molecules being retained by the host crystal lattice, and this suggests that the guest molecules are retained mechanically.

Adduct formation by XIX can also be used to separate mixtures of liquids by preferential absorption of one guest component into the adduct. Typical results are shown in Table 11.2.[35] In each case, the noncyclic component is

TABLE 11.2

INCLUSION COMPOUNDS OF XIX FORMED FROM MIXED LIQUIDS[a]

Organic liquid (guest)	Molar ratio of guest compounds in mixture	Molar ratio of guest compounds in adduct[b]
Heptane–cyclohexane	0.74	100
Hexane–benzene	0.73	20
Hexane–cyclohexane	0.91	33
Carbon tetrachloride–benzene	0.91	20

[a] From Allcock and Siegel.[35]
[b] By mass spectrometry.

incorporated into the lattice with the almost total exclusion of the cyclic component. Presumably, during spontaneous clathration the linear molecules can penetrate the lattice more easily. It is also possible to displace one included compound by treatment of the adduct with an excess of a second liquid or vapor. Thus, included benzene can be totally displaced by xylene or carbon disulfide.

The mechanism of clathration has been elucidated by X-ray single-crystal investigations of benzene, xylene, chloroform, or bromobenzene adducts of XIX[35, 37] and by X-ray powder examination of the adducts listed in Tables 11.1 and 11.2.[35] The powder data confirmed the hexagonal crystal structure for every adduct, and all the powder patterns were strikingly similar for different guest molecules, with the lattice dimensions varying slightly with the dimensions of the guest molecules. Although pure crystals of XIX give monoclinic or triclinic-type X-ray patterns, a crystal transformation to hexagonal occurs above 170°C, the same temperature at which the adduct crystals release the guest. Thus, it seems clear that clathration is a phenomenon associated with the energetics of molecular packing in the crystalline state.

Single-crystal X-ray work on the adducts revealed a space group of $P6_3$, with alternating layers of paddlewheel-shaped host molecules arranged in the *a–a* plane, as shown in Fig. 11.1. The bulky *o*-phenylenedioxy side groups are almost perpendicular to the layer plane with the vertical separation between successive layers along the *c* axis being nearly 5 Å. The structure is oriented in the unit cell as shown in Fig. 11.2,[37] and the molecular arrangement creates distinct channels of 4.5 Å diameter between the side groups, as shown in Fig. 11.3. It is in these channels that the guest compounds are retained. Presumably, in the absence of crystal dislocations, the channels would be continuous throughout the crystal, and this explains the fact that one guest component can be displaced by another.

The process of clathration can be visualized as follows. In the pure crystalline

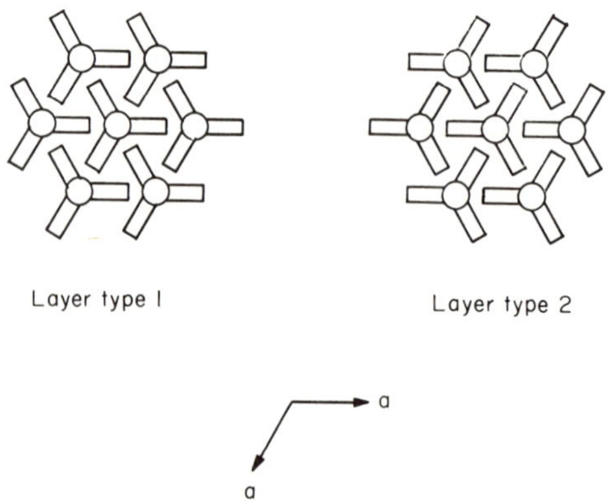

FIG. 11.1. Molecular orientation in hexagonal adduct crystals of tris(*o*-phenylenedioxy)-cyclotriphosphazene (XIX). The structure consists of alternating layers of types 1 and 2, where the circles represent the phosphazene rings and the radial arms depict the *o*-phenylene-dioxy side groups. The layers are separated by nearly 5 Å. From Allcock and Siegel.[35]

state, XIX occupies a large, eight-molecule, triclinic unit cell. Crystal packing within this cell is assumed to be inefficient because of the unusual shape of the molecules. However, in the presence of guest molecules, a facile rearrangement to the lower-energy hexagonal clathrate cell occurs, with the guest components randomly "cushioning" the side groups of XIX. The highly exothermic nature of this crystal transition has been demonstrated by differential thermal analysis.[35] When the temperature of the adduct is raised to 170°C, the thermal motions present in the hexagonal structure can stabilize that form in the absence of guest molecules, and the latter are, therefore, released at this temperature. Thus, the clathration phenomenon is connected with the bulkiness of the side group and with the deformability of the lattice.

This interpretation of the phenomenon implies that related cyclophosphazenes with bulky side groups should behave in a similar way and this is, indeed, found to be the case. Tris(2,3-dioxynaphthyl)cyclotriphosphazene (XX)

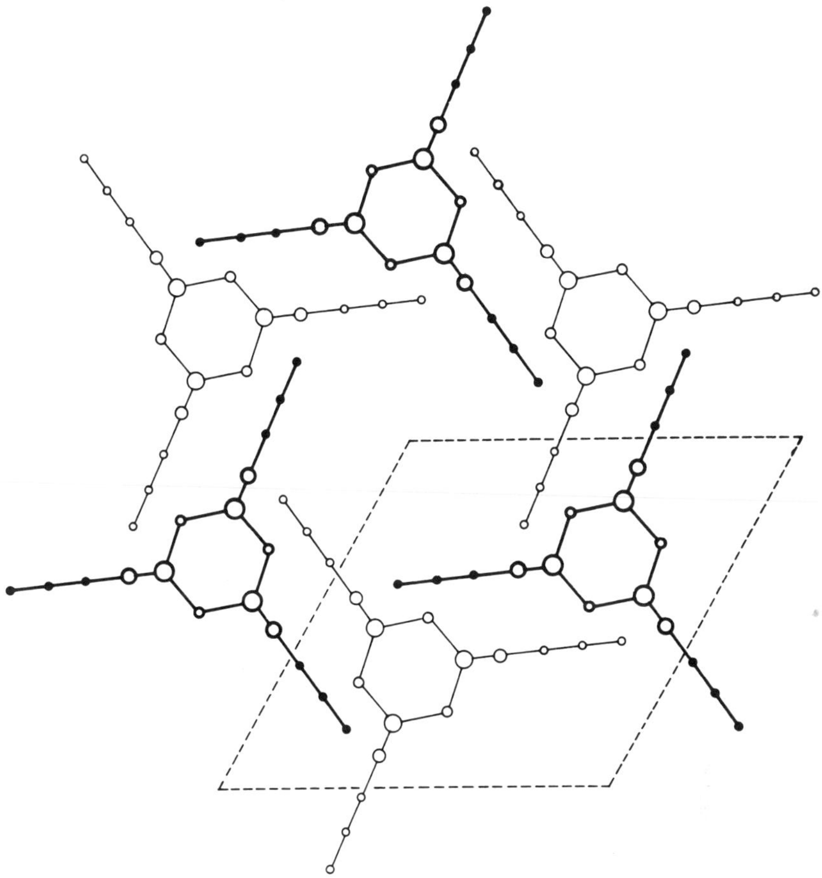

FIG. 11.2. Idealized *c*-axis projection of molecules of XIX in the clathrate adduct structure. From Siegel and van den Hende.[37]

crystallizes from benzene as hexagonal crystals which retain appreciable amounts of solvent.[36] Single-crystal X-ray studies suggested a molecular arrangement in the crystal similar to that found for XIX. Thus, the separation between the molecular layers (*c*-axis distance) in XX is very similar to that found for XIX, but the *a*-axis cell parameter is appreciably larger (15.75 Å) than for XIX, a fact which suggests an increased radial diameter of the channels in the naphthyl derivative. Furthermore, the ratio of benzene to phosphazene in the lattice is approximately 2:1, more than four times the amount of benzene than is retained by XIX. The fact that the channels in XX are wider than those in XIX is also indicated by the fact that crystals of the adduct lose benzene

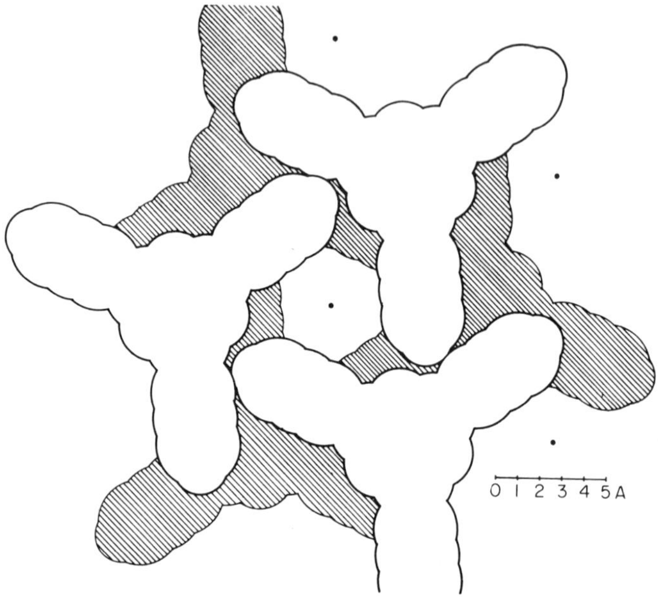

FIG. 11.3. Van der Waals boundaries of tris(*o*-phenylenedioxy)cyclotriphosphazene in the hexagonal, adduct crystal form, showing the location of the channels in which included guest components are retained. Courtesy of L. A. Siegel.

slowly on standing at 25°C in the atmosphere and that relatively mild conditions are required to remove all the guest component under vacuum, with the concurrent formation of a lower symmetry crystalline modification. This is also consistent with the observation that, although spontaneous benzene absorption (by mixing of the components) occurs, and the hexagonal modification can be identified by X-ray powder patterns, the included benzene is released and the crystal structure reverts to the original form as the solid dries in the atmos-

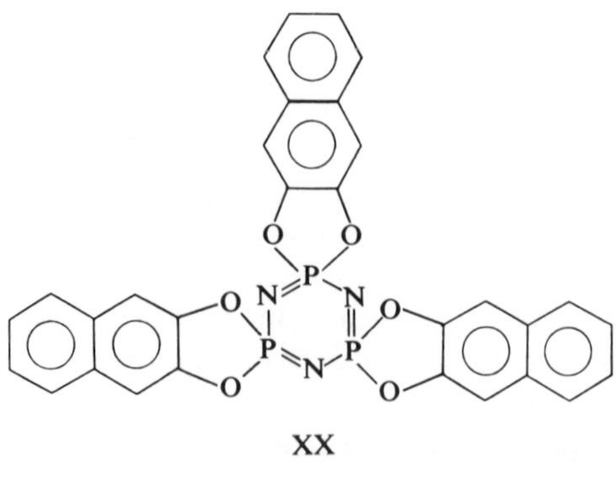

XX

phere. Bulkier guest molecules such as *o*- or *p*-xylene do not escape from the lattice at 25°C, and permanent crystalline adducts can be obtained with these compounds.[38]

Tris(*o*-phenylenediamino)cyclotriphosphazene (XXI) forms a different

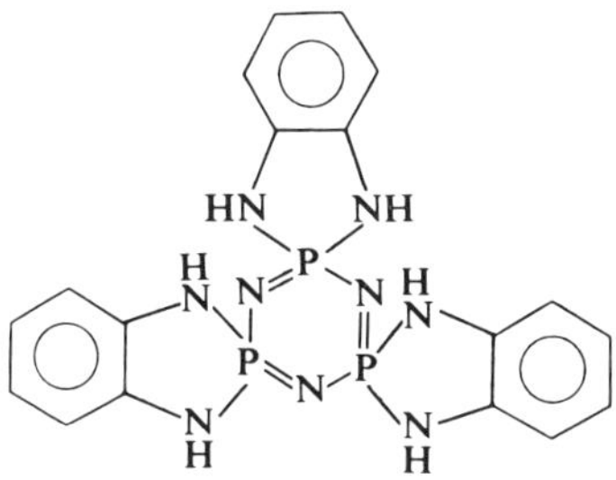

XXI

type of adduct with methyl ethyl ketone and other carbonyl compounds.[36] Crystals of XXI obtained from methyl ethyl ketone solution have orthogonal rather than hexagonal symmetry, but prolonged heating at 80°C/0.1 mm removes the solvent and induces a crystal structure change. The process can be recycled indefinitely. In the adduct, the molar ratio of XXI to ketone is 1:1.8. It seems likely that hydrogen-bonding effects may participate in formation of these adducts.

2. Other Cyclophosphazenes

It has also been reported that other cyclophosphazenes retain solvents after recrystallization. For example, 1,1-dichloro-3,3,5,5-tetraphenylcyclotriphosphazene (XXII) retains acetonitrile in a 1:1 ratio when crystallized from that

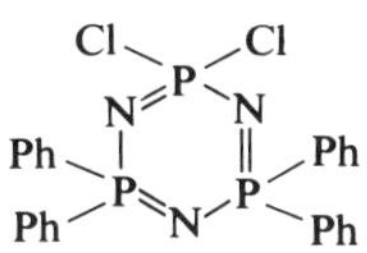

XXII

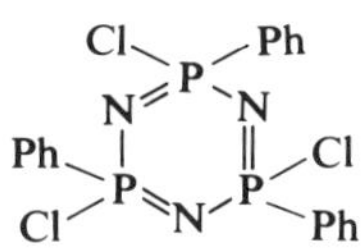

XXIII

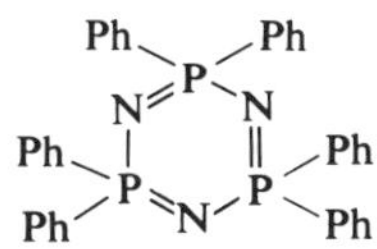

XXIV

solvent,[40,43] *cis*-1,3,5-trichloro-1,3,5-triphenylcyclotriphosphazene (XXIII) retains benzene,[41] and hexaphenylcyclotriphosphazene (XXIV) retains appreciable quantities of *sym*-tetrachloroethane when recrystallized from that solvent.[39,42] These adducts are almost certainly clathrates comparable to those formed by XIX and XX.[42,43,47] The adduct formed between XXIV and *sym*-tetrachloroethylene contains the components in a 1:3 molar ratio. Vapor pressure–composition curves show that loss of the guest compound in vacuum is not stepwise, and thermodynamic data reveal that the binding energy per mole for tetrachloroethane in the adduct is only about 8.7 kcal.[42] This latter value eliminates the possibility of electron donor–acceptor complexation and confirms that the guest molecules are mechanically retained. The retention is, in fact, so weak that mere grinding of the crystals is sufficient to release the guest component.[42, 47] X-Ray powder patterns suggest that pure XXIV exists as a triclinic modification, whereas crystals of the adduct may be hexagonal.[42] No clathration could be detected for crystals recrystallized from benzene, *o*- or *p*-xylene, or from acetonitrile even when X-ray patterns were taken of the powder in contact with the potential guest liquid.[47] Tetrachloroethane apparently is not clathrated by XXII, $(NPPh_2)_4$, the *trans* isomer of XXIII, or *gem*-$N_3P_3Cl_4Ph_2$.[42]

Phosphazenes which contain only one bulky side unit are also apparently capable of forming clathrates. For example, compound XXV crystallizes with

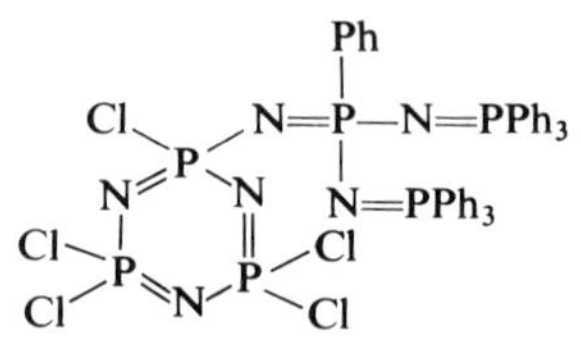

XXV

the inclusion of benzene, diethyl ether, or *n*-hexane solvent, but nitromethane or ethanol are not retained.[48] Presumably the packing arrangement of this phosphazene in the crystalline state leads to the formation of cavities or channels.

3. Practical Applications of Phosphazene Clathrates

The unusual clathration behavior described earlier for tris(*o*-phenylenedioxy)cyclotriphosphazene (XIX) has important practical ramifications for the separation of organic compounds in the vapor or liquid states. In particular, the ease of recycling and regeneration of the pure host compound, and the ability of the host to selectively absorb guest molecules on the basis of subtle structural differences, has widespread implications in petroleum technology.[49] As a result, several patents[50–52] have appeared which describe the technological development of hydrocarbon separation processes based on the selective absorption and substrate regeneration steps described earlier.[35, 36]

In particular, processes have been claimed for the separation of saturated from unsaturated hydrocarbons,[50] monoolefins from diolefins,[50] hydrocarbons with different degrees of branching,[51] hydrocarbons with different chain lengths,[51] the xylene isomers, ethylbenzene, styrene, and the C_8 and C_9

alkenyl aromatic hydrocarbons.[52] As an illustration of the efficiency of the process, it is possible to prepare *m*-xylene containing only 4% impurities in one step from a 25% *p*-xylene, 25% ethylbenzene, 50% *m*-xylene mixture.[52] The host substrate (XIX) is deposited from xylene solution onto diatomaceous earth, ground firebrick, or thermosetting resins, and various guest-mixture sorption and purging cycles are conducted.

It appears that additional research being conducted with other phosphazene clathrates may stimulate technology in this general area.

References

1. H. Bode and H. Bach, *Chem. Ber.* **75B**, 215 (1942).
2. T. Chivers and N. L. Paddock, *J. Chem. Soc., A* p. 1687 (1969).
3. G. E. Coxon and D. B. Sowerby, *J. Chem. Soc., A*, p. 3012 (1969).
4. M. F. Lappert and G. Srivastava, *J. Chem. Soc., A* p. 210 (1966).
5. S. M. Zhivukhin and V. V. Kireev, *Russ. J. Inorg. Chem.* **9**, 1439 (1966).
6. G. I. Migachev and B. I. Stepanov, *Russ. J. Inorg. Chem.* **11**, 929 (1966).
7. J. Dyson, Ph.D. Thesis, University of Manchester (1964) (quoted by Trotter and Whitlow[8]).
8. J. Trotter and S. H. Whitlow, *J. Chem. Soc., A* p. 455 (1970).
9. J. Trotter and S. H. Whitlow, *J. Chem. Soc., A* p. 460 (1970).
10. W. C. Marsh, N. L. Paddock, C. J. Stewart, and J. Trotter, *Chem. Commun.* p. 1190 (1970).
11. T. Moeller and S. G. Kokalis, *J. Inorg. Nucl. Chem.* **25**, 875 (1963).
12. M. Becke-Goehring, H. Hohenschutz, and R. Appel, *Z. Naturforsch. B* **9**, 678 (1954).
13. S. K. Das, R. A. Shaw, B. C. Smith, W. A. Last, and F. B. G. Wells, *Chem. Ind. (London)* p. 866 (1963).
14. G. V. Derbisher and A. V. Babaeva, *Russ. J. Inorg. Chem.* **10**, 1194 (1965).
15. Yu. A. Buslaev, B. V. Rumyantseva, S. P. Petrosants, and V. V. Levin, Z. G. Mironova, *Zh. Neorg. Khim.* **14**, 3245 (1969).
16. B. Janik, W. Ryeszutko, and T. Pel'char, *J. Anal. Chem. USSR* **22**, 930 (1967).
17. B. Janik, W. Ryeszutko, and T. Pel'char, *Zh. Obshch. Khim.* **8**, 1444 (1966).
18. J. Dyson and N. L. Paddock, *Chem. Commun.* p. 191 (1966).
19. H.-G. Horn and M. Becke-Goehring, *Z. Anorg. Allg. Chem.* **367**, 165 (1969).
20. H. Bock and H. Dieck, *Z. Naturforsch. B* **21**, 739 (1966).
21. D. S. Payne and A. P. Walker, *J. Chem. Soc., C* p. 498 (1966).
22. D. S. Payne, J. Mokuolu, and J. C. Speakman, *Chem. Commun.* p. 599 (1965).
23. K. L. Paciorek and R. H. Kratzer, *Inorg. Chem.* **5**, 538 (1966).
24. H. Schmidbaur and W. Wolfsberger, *Chem. Ber.* **100**, 1000 (1967).
25. G. Allen, J. Dyson, and N. L. Paddock, *Chem. Ind. (London)* p. 1832 (1964).
26. J. Trotter, S. H. Whitlow, and N. L. Paddock, *Chem. Commun.* p. 695 (1969).
27. J. N. Rapko and G. R. Feistel, *Chem. Commun.* p. 474 (1968); *Inorg. Chem.* **9**, 1401 (1970).
28. B. I. Stepanov and G. I. Migachev, *J. Gen. Chem. USSR* **36**, 1454 (1966).
29. H. R. Allcock and P. C. Ostrowski, unpublished results (1968).
30. S. K. Das, R. A. Shaw, and B. C. Smith, *Chem. Commun.* p. 176 (1965).
31. S. K. Das, D. Feakins, W. A. Last, S. N. Nabi, S. K. Ray, R. A. Shaw, and B. C. Smith, *J. Chem. Soc., A* p. 616 (1970).
32. S. K. Das, R. A. Shaw, B. C. Smith, and C. P. Thakur, *Chem. Commun.* p. 33 (1966).

33. P. A. D. DeMaine and R. D. Srivastava, *U.S. Govt. Res. Rep.* **AD 452-033** (1964).
34. R. Keat and R. A. Shaw, *J. Chem. Soc., A* p. 703 (1968).
35. H. R. Allcock and L. A. Siegel, *J. Amer. Chem. Soc.* **86**, 5140 (1964).
36. H. R. Allcock and R. L. Kugel, *Inorg. Chem.* **5**, 1016 (1966).
37. L. A. Siegel and J. H. van den Hende, *J. Chem. Soc., A* p. 817 (1967).
38. H. R. Allcock and M. T. Stein, unpublished work (1969).
39. H. H. Sisler, H. S. Ahuja, and N. L. Smith, *Inorg. Chem.* **1**, 84 (1962).
40. C. D. Schmulbach and C. Derderian, *J. Inorg. Nucl. Chem.* **35**, 1395 (1963).
41. B. Grushkin, M. G. Sanchez, and R. G. Rice, *Inorg. Chem.* **3**, 623 (1964).
42. R. D. Whitaker, A. J. Barreiro, P. A. Furman, W. C. Guida, and E. S. Stallings, *J. Inorg. Nucl. Chem.* **30**, 2921 (1968).
43. R. D. Whitaker and W. C. Guida, *J. Inorg. Nucl. Chem.* **31**, 875 (1969).
44. H. R. Allcock, *J. Amer. Chem. Soc.* **85**, 4050 (1963).
45. H. R. Allcock, *J. Amer. Chem. Soc.* **86**, 2591 (1964).
46. H. R. Allcock and E. J. Walsh, *Inorg. Chem.* **10**, 1643 (1971).
47. H. R. Allcock and E. G. Stroh, unpublished work (1970).
48. M. Biddlestone, R. A. Shaw, and D. Taylor, *Chem. Commun.* p. 320 (1969).
49. H. R. Allcock and L. A. Siegel, U.S. Pat. 3,356,768 (1967) (to American Cyanamid Company).
50. A. Goldup and M. T. Westaway, U.S. Pat. 3,472,762 (1969) (to British Petroleum Company).
51. A. Goldup and M. T. Westaway, U.S. Pat. 3,499,944 (1970) (to British Petroleum Company).
52. J. N. Haresnape, U.S. Pat. 3,504,047 (1970) (to British Petroleum Company).

Chapter 12

PHOSPHAZENES AS BRØNSTED-LOWRY BASES

A. Introduction

The ability of phosphazenes to form adducts with hydrohalides has been reported by a number of investigators. In 1938, Schmitz-DuMont and Kulkens[1] described an adduct of formula $(NPF_2)_3 \cdot 2HF \cdot 2H_2O$, and 10 years later, Bode, Bütow, and Lienau[2] reported the existence of the adducts, $(NPCl_2)_3 \cdot HClO_4$, $(NPCl_2)_3 \cdot 2HClO_4$, $[NP(NHPh)_2]_3 \cdot HClO_4$, $[NP(NHPh)_2]_3 \cdot HCl$, $N_3P_3Cl_4(NHMe)_2 \cdot HClO_4$, and $N_3P_3Cl_4(NHC_2H_4NH_2)_2 \cdot 2HCl$, formed from the appropriate phosphazene and the acid. It later became apparent that weak acids could also interact with phosphazenes in the same way, when Sowerby and Audrieth[3] described complexes of formula $[NP(NH_2)_2]_3 \cdot HOOCCH_3$ and $[NP(NH_2)_2]_4 \cdot HOOCCH_3$. As new hydrohalide–phosphazene adducts were reported it became clear that the following questions required answers: (1) Are the adducts clathrates, π complexes, or salts? (2) If they are salts, is the site of protonation the side group or a ring nitrogen atom? (3) What effects are associated with the presence of different substituent groups attached to phosphorus? (4) What is the influence of different size ring systems on adduct formation? Initial evidence on the first question was provided by Moeller and Kokalis,[4] and answers to the remaining questions have resulted from the extensive studies of Shaw, Feakins, Last, and their collaborators.[5-14]

B. Preparation of Hydrohalide Adducts

Aminophosphazenes are usually prepared by the interaction of a primary or secondary amine with a chlorophosphazene (see Chapter 7). Normally an excess of the amine functions as a hydrohalide acceptor.

$$(NPCl_2)_3 + 12\,RNH_2 \rightarrow [NP(NHR)_2]_3 + 6\,RNH_2 \cdot HCl$$

In practice, the aminophosphazene may be a stronger base than the parent amine and a hydrohalide adduct of the aminophosphazene may be formed. For example, interaction of *n*-propylamine with $(NPCl_2)_3$, followed by recrystallization of the product from *n*-heptane, yields the monohydrochloride, $[NP(NHPr^n)_2]_3 \cdot HCl$.[4] Reaction of *n*-propylamine with $(NPBr_2)_3$ gives the hydrobromide salt, and interaction of *n*-butylamine with $(NPCl_2)_3$ yields the salt, $[NP(NHBu^n)_2]_3 \cdot HCl$.[4] These salts have definite melting points, and they can be converted to the neutral aminophosphazene by washing with water, dimethylformamide, or liquid ammonia.

Hydrohalide salt formation is particularly noticeable when primary or secondary amines react with high molecular weight poly(dichlorophosphazene) to yield poly(diaminophosphazenes), $[NP(NHR_2]_n$, $[NP(NR_2)_2]_n$, or $[NP(NR_2)(NHR)]_n$.[15] Total removal of hydrogen chloride from the high polymers can be accomplished only by prolonged washing with water or treatment with a stronger amine or sodium carbonate.

Aminophosphazene hydrohalide salts can also be prepared by bubbling dry hydrogen chloride gas through a solution of the aminophosphazene in benzene, diethyl ether, or ethyl acetate.[16]

C. Structure of the Adducts

Phosphazene–hydrohalide adducts of the types $[NP(NHPr^n)_2]_3 \cdot HCl$, $[NP(NHPr^n)_2]_3 \cdot HBr$, or $[NP(NHBu^n)_2]_3 \cdot HCl$, can be recrystallized repeatedly from hot heptane without change in composition.[4] Thus, the hydrogen halide must be chemically bonded to the phosphazene rather than physically trapped within the crystal lattice. However, the hydrogen halide can be displaced from the complex by dimethylformamide, water, or liquid ammonia,[4] and this suggests that the complexes are acid–base adducts which can be disrupted by stronger bases than the phosphazene. Furthermore, the adducts react instantly with silver perchlorate to precipitate silver halide, and this indicates the presence of ionic halide ions in the original complex.[4] Thus, the structures can be likened to those of amine hydrochlorides.

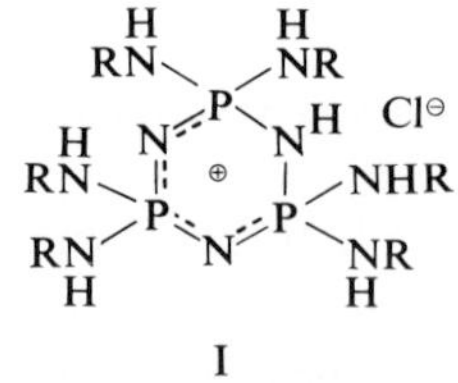

I

Protonation of the phosphazenes apparently occurs at the ring nitrogen atoms to give structures such as I. Evidence for this is derived from the striking P—N skeletal infrared spectral shifts which accompany protonation,[4] from the broadening and shift of the proton NMR spectrum,[4] and from the results of X-ray crystallographic studies on $N_3P_3Cl_2(NHPr)_4 \cdot HCl$.[17] This interpretation is also confirmed by the fact that protonation can occur when no basic side substituents are present.[7, 9, 13] It has also been pointed out that the differences in pK_a values for mono- and diprotonation can be adequately explained only in terms of protonation of two ring nitrogen atoms in cyclic tetramers.[8] Although the second protonation site in cyclic trimers is less obvious, ring protonation is believed to operate here also.[8] Further evidence, to be discussed later, is based on the influence of different substituents on the base strength of the phosphazene.

D. Degree of Protonation

Mono-, di-, and trihydrohalide salts have been isolated and identified. Hydrogen fluoride forms a dihydrohalide salt with $(NPF_2)_3$[1], as does perchloric acid with $(NPCl_2)_4$,[2] and hydrogen chloride with $[NP(NHPr^n)_2]_3$ or $[NP(NHBu^n)_2]_3$.[15] A trihydrochloride is formed from $[NP(NHC_5H_{11})_2]_3$ when hydrogen chloride is bubbled through a benzene solution of the aminophosphazene.[16] pK_a Measurements for many cyclophosphazenes (see Tables 12.1–12.7) demonstrate the fact that introduction of one proton into the phosphazene unit lowers the base strength by a considerable amount. Thus, formation of a dihydrohalide salt is much more difficult. Introduction of a second proton commonly requires the presence of a strong electron-supplying substituent attached to phosphorus. Presumably, trihydrohalide salts are formed only when this influence is present to an extreme degree. However, it is important to note that, when the presence of different substituents around the ring causes different nitrogen atoms to have different basicities, some protonation of the weaker basic centers is expected. Thus, although the starred nitrogen atom of II is the most basic center, some protonation of the other nitrogen atoms can also be expected.[13]

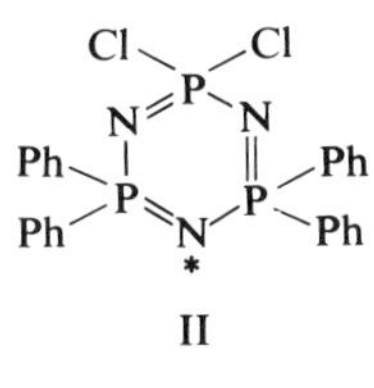

II

TABLE 12.1

pK_a VALUES FOR CYCLOPHOSPHAZENES IN WATER AT 25°C[a]

Compound	pK_a	pK_{a_2}	References
$(NPEt_2)_3$	5.85	—	7, 9
$(NPEt_2)_4$	6.45	—	7, 9
$[NP(NH_2)_2]_3$	7.70	—	8
$[NP(NH_2)_2]_4$	7.50	5.15	8
$[NP(NHMe)_2]_3$	7.80	—	8
$[NP(NHMe)_2]_4$	7.95	5.20	8
$[NP(NHEt)_2]_3$	8.65	—	8
$[NP(NHEt)_2]_4$	8.70[b]	6.10[b]	8
$[NP(NMe_2)_2]_3$	7.45	—	8

[a] From Shaw *et al.*[7,8]
[b] Measured in 3% methanol in water.

TABLE 12.2

pK_a' VALUES FOR HALO-, ALKYL-, AND ARYLCYCLOPHOSPHAZENES IN NITROBENZENE AT 25°C[a]

Compound	pK_{a_1}'	pK_{a_2}'	References
$(NPCl_2)_3$	<−6.0	—	7
$(NPCl_2)_4$	<−6.0	—	7
$(NPEt_2)_3$	6.40	—	7, 9
$(NPEt_2)_4$	7.60	0.20	7, 9
$[NP(CF_3)_2]_3$	<−6.0	—	7, 9
$[NP(CF_3)_2]_4$	<−6.0	—	7
$(NPPh_2)_3$	1.50	—	7, 9
$(NPPh_2)_4$	2.20	−5.80	7, 9
$[NP(C_6H_4Cl\text{-}p)_2]_3$	−1.4	—	7, 9
$N_3P_3Cl_2Ph_4$-*gem*	−3.90	—	7
$N_3P_3Cl_4Ph_2$-*gem*	<−6.0	—	7, 13
$N_3P_3Cl_3Ph_3$-*non-gem*	<−6.0	—	13
$N_3P_3Cl_2(C_6H_4Cl\text{-}p)_4$-*gem*	−5.9	—	13

[a] From Shaw *et al.*[7,13]

TABLE 12.3

pK_a′ VALUES FOR ALKOXY-, ARYLOXY-, AND MERCAPTOCYCLOTRIPHOSPHAZENES IN NITROBENZENE AT 25°C[a,b]

Compound	pK_{a_1}′	References
$[NP(OMe)_2]_3$	−1.9	9
$[NP(OMe)_2]_4$	−1.0	9
$[NP(OEt)_2]_3$	0.2	7, 9
$[NP(OEt)_2]_4$	0.6	7, 9
$[NP(OPr^i)_2]_3$	1.4	9
$[NP(OPr^i)_2]_4$	2.1	9
$[NP(OBu^n)_2]_3$	0.1	9
$[NP(OBu^n)_2]_4$	0.7	9
$[NP(OCH_2Ph)_2]_3$	−2.1	7, 9
$[NP(OCH_2Ph)_2]_4$	−1.6	9
$[NP(OCH_2CF_3)_2]_3$	<−6.0	9
$[NP(OCH_2CF_3)_2]_4$	<−6.0	9
$[NP(OPh)_2]_3$	−5.8	9
$[NP(OPh)_2]_4$	−6.0	9
$[NP(OC_6H_4OMe\text{-}p)_2]_3$	−5.4	9
$[NP(OC_6H_4OMe\text{-}p)_2]_4$	−5.2	9
$[NP(OC_{10}H_7\text{-}\beta)_2]_3$	<−6.0	9
$[NP(OC_{10}H_7\text{-}\beta)_2]_4$	<−6.0	9
$[NP(SEt)_2]_3$	−2.8	9
$[NP(SPr^n)_2]_3$	−2.6	9
$[NP(SC_6H_{11})_2]_3$	−2.2	9
$[NP(SCH_2Ph)_2]_3$	−4.2	9
$[NP(SPh)_2]_3$	−4.8	9

[a] From Shaw *et al.*[7,9,13]
[b] No pK_{a_2}′ values are available.

TABLE 12.4

pK_a′ VALUES FOR AMINOCYCLOPHOSPHAZENES IN NITROBENZENE AT 25°C[a]

Compound	pK_{a_1}′	pK_{a_2}′	References
$[NP(NHMe)_2]_3$	8.8 ± 0.6	−2.0 ± 0.4	8
$[NP(NHMe)_2]_4$	8.2	3.4	8
$[NP(NHEt)_2]_3$	8.2	−1.3	8
$[NP(NHEt)_2]_4$	8.1	3.8	8
$[NP(NHPr^n)_2]_3$	7.9	−1.3	8
$[NP(NHPr^n)_2]_4$	8.3	3.9	8
$[NP(NHPr^i)_2]_3$	8.4	−1.5	8
$[NP(NHPr^i)_2]_4$	8.1	2.2	8
$[NP(NHBu^n)_2]_3$	7.9	−1.8	8

TABLE 12.4 (*Continued*)

Compound	pK_{a_1}'	pK_{a_2}'	References
$[NP(NHBu^n)_2]_4$	7.6	3.1	8
$[NP(NHBu^i)_2]_4$	8.0	3.2	8
$[NP(NHBu^t)_2]_3$	8.0	−1.7	8
$[NP(NHBu^t)_2]_4$	8.8	0.4	8
$[NP(NHC_6H_{11})_2]_3$	7.9	−1.4	8
$[NP(NHC_6H_4Me\text{-}p)_2]_3$	3.0	—	7
$[NP(NHC_6H_4Me\text{-}p)_2]_4$	0.9	−3.0	7
$[NP(NMe_2)_2]_3$	7.6	−3.3	8
$[NP(NMe_2)]_4$	8.3	0.6	8
$[NP(NEt)_2]_3$	8.5	−3.9	8
$[NP(NEt)_2]_4$	8.3	−0.9	8
$[NP(NC_5H_{10})_2]_3$	8.4	−3.2 ± 0.4	8
$[NP(NC_5H_{10})_2]_4$	8.4	0.5	8
$[NP(NMePh)_2]_3$	3.5	—	14

[a] From Shaw *et al.*[7,8,11,14]

TABLE 12.5

pK_a' VALUES FOR AMINOCHLOROCYCLOTRIPHOSPHAZENES IN NITROBENZENE AT 25°C[a,b]

Compound	pK_{a_1}'	References
$N_3P_3Cl_5NMe_2$[c]	<−6.0	11
$N_3P_3Cl_4(NMe_2)_2$[c]	<−6.0	11
$N_3P_3Cl_3(NMe_2)_3$-*cis*[d]	−5.5	11
$N_3P_3Cl_3(NC_5H_{10})_3$-*cis*[d]	−5.1	11
$N_3P_3Cl_3(NEt_2)_3$-*cis*[d] and *trans*[d]	−5.2	11
$N_3P_3Cl_3(NHEt)_3$-*cis*[d] and *trans*[d]	−4.9	11
$N_3P_3Cl_3(NMe_2)_3$-*trans*[d]	−5.4	11
$N_3P_3Cl_3(NC_5H_{10})_3$-*trans*[d]	−5.3	11
$N_3P_3Cl_3(NMe_2)_3$-*gem*	−4.4	11
$N_3P_3Cl_3(NC_5H_{10})_3$-*gem*	−3.9	11
$N_3P_3Cl_2(NMe_2)_4$-*cis*[d]	−1.4	11
$N_3P_3Cl_2(NC_5H_{10})_4$-*cis*[d]	−0.9	11
$N_3P_3Cl_2(NHEt)_4$-*gem*	3.2	11
$N_3P_3Cl_2(NHPr^i)_4$-*gem*	3.4	11
$N_3P_3Cl_2(NHBu^t)_4$-*gem*	3.5	11

[a] From Shaw *et al.*[11,14]
[b] No pK_{a_2}' values are available.
[c] Compounds of formula $N_3P_3Cl_5R$ and $N_3P_3Cl_4R_2$, where R is NEt_2, NC_5H_{10}, NHEt, $NHPr^i$, $NHBu^t$, and NHPh, all have pK_{a_1}' values below −6.0 units.
[d] Non-geminal.

TABLE 12.6

pK_a′ VALUES FOR AMINO DERIVATIVES OF GEMINAL DIPHENYL AND GEMINAL TETRAPHENYLCYCLOTRIPHOSPHAZENES IN NITROBENZENE AT 25°C[a]

Compound	pK_{a_1}′	pK_{a_2}′	References
$N_3P_3Ph_2(NHMe)_4$	7.5	−4.5	12, 14
$N_3P_3Ph_2(NHEt)_4$	7.4	−4.3	12, 14
$N_3P_3Ph_2(NHC_6H_{11})_4$	7.6	−4.0	12, 14
$N_3P_3Ph_2(NHBu^t)_4$	7.2	—	12
$N_3P_3Ph_2(NHCH_2Ph)_4$	6.6	—	14
$N_3P_3Ph_2(NMe_2)_4$	6.2	−5.1	12, 14
$N_3P_3Ph_2(NEt_2)_4$	6.8	—	12
$N_3P_3Ph_2(NC_5H_{10})_4$	6.7	−5.4	12, 14
$N_3P_3Ph_2(NHPh)$	1.7	—	14
$N_3P_3Ph_2Cl_2(NHMe)_2$-*gem*	−0.4	—	12
$N_3P_3Ph_2Cl_2(NHEt)_2$-*gem*	−0.5	—	12
$N_3P_3Ph_2Cl_2(NHC_6H_{11})_2$-*gem*	0.4	—	12
$N_3P_3Ph_2Cl_2(NHBu^t)_2$-*gem*	0.2	—	12
$N_3P_3Ph_2Cl_2(NC_5H_{10})_2$-*gem*	−1.0	—	12
$N_3P_3Ph_2Cl_2(NH_2)_2$-*gem*	0.0	—	12
$N_3P_3Ph_2Cl_2(NHMe)_2$-*non-gem*	−2.8	—	12
$N_3P_3Ph_2Cl_2(NMe_2)_2$-*non-gem*	−4.0	—	12
$N_3P_3Ph_2Cl_2(NEt_2)_2$-*non-gem*	−3.2	—	12
$N_3P_3Ph_2Cl_2(NC_5H_{10})_2$-*non-gem*	−3.6	—	12
$N_3P_3Ph_2Cl_3(NHC_6H_{11})$	−5.5	—	12
$N_3P_3Ph_2Cl_3(NC_5H_{10})$	<-6.0	—	12
$N_3P_3Ph_4(NHMe)_2$	4.7	—	14
$N_3P_3Ph_4(NHC_6H_{11})_2$	5.0	—	12, 14
$N_3P_3Ph_4(NHBu^t)_2$	4.7	—	12, 14
$N_3P_3Ph_4(NHCH_2Ph)_2$	3.8	—	14
$N_3P_3Ph_4(NMe_2)_2$	3.8	—	14
$N_3P_3Ph_4(NEt_2)_2$	4.0	—	12, 14
$N_3P_3Ph_4(NC_5H_{10})_2$	4.3	—	12, 14
$N_3P_3Ph_4(NHPh)_2$	1.6	—	14
$N_3P_3Ph_4Cl(NHMe)$	−0.75	—	12
$N_3P_3Ph_4Cl(NEt_2)$	−1.2	—	12

[a] From Shaw *et al.*[12,14]

TABLE 12.7

pK_a' VALUES FOR MISCELLANEOUS MIXED-LIGAND CYCLOTRIPHOSPHAZENES IN NITROBENZENE AT 25°C[a, b]

Compound		pK_{a_1}'	pK_{a_2}'	References
$N_3P_3Ph_3(NH_2)_3$-*non-gem-cis*		~5.1	~−3.8	14
$N_3P_3Ph_3(NHMe)_3$-*non-gem-cis*		5.7	−5.2	14
$N_3P_3Ph_3(NHEt)_3$-*non-gem-cis*		6.0	−4.8	14
$N_3P_3Ph_3(NHPr^n)_3$-*non-gem-cis*		6.1	—	14
$N_3P_3Ph_3(NHBu^n)_3$-*non-gem-cis*		6.0	—	14
$N_3P_3Ph_3(NMe_2)_3$-*non-gem-cis*		(4.1)	(−5.3)	14
$N_3P_3Ph_3(NC_5H_{10})_3$-*non-gem-cis*		(5.0)	—	14
$N_3P_3Ph_3(NH_2)_3$-*non-gem-trans*		5.6	−4.5	14
$N_3P_3Ph_3(NHMe)_3$-*non-gem-trans*		5.8	−5.2	14
$N_3P_3Ph_3(NHEt)_3$-*non-gem-trans*		6.0	−4.7	14
$N_3P_3Ph_3(NHPr^n)_3$-*non-gem-trans*		6.1	—	14
$N_3P_3Ph_3(NHBu^n)_3$-*non-gem-trans*		6.0	—	14
$N_3P_3Ph_3(NMe_2)_3$-*non-gem-trans*		(4.1)	(−5.3)	14
$N_3P_3Ph_3(NC_5H_{10})_3$-*non-gem-trans*		(5.0)	—	14
$N_3P_3(NMe_2)_4(NHPr^i)_2$	1,1,3,5:3,5	7.6	−2.3	14
$N_3P_3(NMe_2)_2(NHPr^i)_4$	1,1:3,3,5,5	7.8	−1.5	14
$N_3P_3(NMe_2)_2(NHPr^i)_4$	1,3:1,3,5,5	7.7	−1.7	14
$N_3P_3(NMe_2)_5(NHPh)$	1,1,3,3,5:5	5.5	−3.7	14
$N_3P_3Cl_2(NMe_2)_2(NHPr^i)_2$	1,1:3,5:3,5	2.5	—	14
$N_3P_3Cl_2(NMe_2)_3(NHPh)$	1,1:3,3,5:5	0.5	—	14
$N_3P_3PhCl_2(NMe_2)_3$	1:3,3:1,5,5	−0.1	—	14
$N_3P_3PhCl_3(NMe_2)_2$	1:3,3,5:1,5	−5.0	—	14
$N_3P_3PhCl_3(NEt_2)_2$	1:3,3,5:1,5	−4.5	—	14
$N_3P_3PhCl_3(NC_5H_{10})_2$	1:3,3,5:1,5	−4.8	—	14
$N_3P_3PhCl_4(NC_5H_{10})$	1:3,3,5,5:1	<-6.0	—	14
$N_3P_3Ph_2Cl(NC_5H_{10})_3$	1,3:5:1,3,5	1.9	—	14
$N_3P_3Ph_2Cl_2(NHEt)_2$-*cis*	1,3:5,5:1,3	0.7	—	14
$N_3P_3Ph_2Cl_2(NHEt)_2$-*trans*	1,3:5,5:1,3	0.6	—	14
$N_3P_3Ph_2Cl_2(NMe_2)_2$	1,3:5,5:1,3	−0.9	—	14
$N_3P_3Ph_2Cl_2(NC_5H_{10})_2$	1,3:5,5:1,3	0.8	—	14
$N_3P_3Ph_2Cl_2(NHPh)_2$	1,1:3,3:5,5	−2.7	—	14
$N_3P_3Ph_2Cl_3(NMe_2)$	1,1:3,3,5:5	−5.8	—	14
$N_3P_3Ph_2Cl_3(NHPh)$	1,1:3,3,5:5	<-6.0	—	14
$N_3P_3Ph_2Cl_2(NMe_2)(NHPh)$	1,1:3,3:5:5	~−2.0	—	14
$N_3P_3(NMe_2)OPh_5$		3.0	—	13
$N_3P_3(NMe_2)_3(OPh)_3$-*non-gem*		1.05	—	13
$N_3P_3(NMe_2)_3Me_3$-*non-gem*		7.3	—	13
$N_3P_3(NMe_2)_4(SPh)_2$-*non-gem*		2.8	—	13
$N_3P_3(NMe_2)_4(OPh)_2$-*non-gem*		3.3	—	13
$N_3P_3(NMe_2)_4(SEt)_2$-*non-gem*		3.8	—	13

TABLE 12.7 (*Continued*)

Compound	pK_{a_1}'	pK_{a_2}'	References
$N_3P_3(NMe_2)_4(SPh)_2$-*gem*	4.0	—	13
$N_3P_3(NMe_2)_4(SEt)_2$-*gem*	4.7	—	13
$N_3P_3(NMe_2)_5OPh$	5.6	—	13
$N_3P_3Ph_2Cl_4$-*gem*	<−6.0	—	13
$N_3P_3Ph_3Cl_3$-*non-gem*	<−6.0	—	13
$N_3P_3Ph_4Cl_2$-*gem*	−3.9	—	13
$N_3P_3Ph_4(OMe)_2$-*gem*	0.4	—	13
$N_3P_3Cl(OMe)_5$	−4.2	—	13
$N_3P_3(OPh)_5(OMe)$	~−5.6	—	13
$N_3P_3Ph_2(C_6H_4Me)_4$-*gem*	2.5	—	13
$N_3P_3Cl_2(C_6H_4Cl$-*p*$)_4$-*gem*	−5.9	—	13
$N_3P_3Cl_2(SEt)_4$-*gem*	<−6.0	—	13
$N_3P_3(NHEt)_4(OCH_2CF_3)_2$-*gem*	3.7	—	13

[a] From Shaw *et al.*[7–14]

[b] Parentheses indicate that the exact isomeric structure (cis or trans) is not known.

E. pK_a Measurements of Phosphazenes in Nitrobenzene and Water

pK_a Measurements are normally made using aqueous solutions. However, many phosphazenes are insoluble in water and some are even highly insoluble in aqueous-organic media. Accordingly, Feakins, Last, and Shaw have developed a technique for the potentiometric titration of bases with perchloric acid in nitrobenzene solution.[18] Most of the basicity values recorded in this chapter were obtained by these authors and their co-workers for nitrobenzene solutions, and these are designated pK_a' values to distinguish them from the pK_a values obtained for aqueous solutions. When diprotonation can be detected potentiometrically, pK_{a_1}' and pK_{a_2}' values are quoted for the addition of the first and second protons, respectively.

A few phosphazenes are soluble in water, and pK_a values have been measured in aqueous media. These values are shown in Table 12.1. Unfortunately, because of the sparsity of data for measurements in aqueous media, it is difficult to interpret these results in terms of substituent effects.

F. Influence of Different Substituents

The basicity of a skeletal nitrogen atom in a cyclophosphazene depends on the availability of the nitrogen lone-pair electrons. The availability of these electrons may be reduced by inductive electron-withdrawing substituents attached to the phosphorus atoms or by factors which draw the lone pair into a π' system. Strongly electron-withdrawing substituents at phosphorus favor the latter influence also. Thus, irrespective of the mechanism of migration of the lone-pair electrons, electron-withdrawing substituents should reduce the basicity of skeletal nitrogen, and electron-supplying substituents should enhance it. That this interpretation is essentially correct is illustrated by the pK_a data shown in Tables 12.2–12.7. Phosphazenes containing amino substituents are the strongest bases, whereas derivatives which contain halogeno groups are among the weakest. In the following sub-sections, these substituent effects will be reviewed in more detail, and the influence of mixed substituents will also be discussed.

1. Halogeno Groups

The base strengths of $(NPCl_2)_3$ and $(NPCl_2)_4$ (Table 12.2) are among the lowest recorded for phosphazenes, with pK_{a_1}' values of less than −6.0.[7] This is consistent with the expected strong electron-withdrawing effect of chlorine, which reduces the availability of the nitrogen lone-pair electrons by both inductive and π'-type effects. Basicity values for fluoro- and bromocyclophosphazenes are not available.

2. Alkyl and Aryl Groups

As shown in Table 12.2, the ethyl derivatives, $(NPEt_2)_3$ and $(NPEt_2)_4$, are stronger bases than trifluoromethyl- or phenylcyclophosphazenes. The striking reduction in base strength which occurs when ethyl groups are replaced by trifluoromethyl groups is consistent with the inductive electron-withdrawing properties of the latter group. Phenylphosphazenes, such as $(NPPh_2)_3$ or $(NPPh_2)_4$, are intermediate in base strength between ethyl and trifluoromethyl derivatives. However, it is interesting to note that replacement of phenyl groups by *p*-chlorophenyl units, as in III, has the effect of reducing the base strength

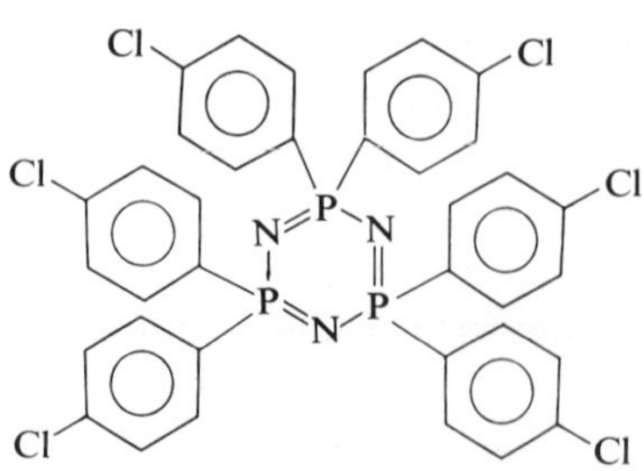

III

from 1.50 pK_a' units to −1.4. This must reflect the electron-withdrawing behavior of the chlorine atoms, even though they are far removed from the site of protonation. It is also noteworthy that replacement of two or three chlorine atoms in $(NPCl_2)_3$ by phenyl groups has no discernible effect on the observed pK_a' values, and the introduction of four phenyl units raises the base strength only slightly. This illustrates the powerful electron-withdrawing influence of chlorine in these compounds.

3. Alkoxy, Aryloxy, and Mercapto Groups

Cyclophosphazene basicity decreases in $[NP(OR)_2]_{3\ or\ 4}$ occur with changes in alkoxy or aryloxy side groups in the order: $OPr^i > OEt \approx OBu^n > OMe > OCH_2PH > OC_6H_4OMe$-$p > OPh > OCH_2CF_3$, $OC_{10}H_7$-β. In the mercapto series, $[NP(SR)_2]_3$, the order is $SC_6H_{11} > SPr^n > SEt > SCH_2Ph > SPh$ (Table 12.3). These orders also generally reflect the inductive influences of the side groups. This is especially obvious when ethoxy (pK_a' = 0.2–0.6) is compared with trifluorethoxy (pK_a' = <−6.0). Thus, for alkoxy and thioalkoxy groups, the relative basicities of the phosphazenes appear to be determined predominantly by inductive effects, and no multiple-bonding influences need be evoked. On the other hand, aryloxy substituents could participate through "resonance-type" effects. The differences between alkoxy and thioalkoxy substituents are consistent with a relatively weaker electron supply by the sulfur-containing grouping. However, $[NP(OPh)_2]_3$ is a weaker base than $[NP(SPh)_2]_3$, a result which has been ascribed to the different extent of π bonding between phenyl groups and the oxygen or sulfur lone-pair orbitals.[9] (However, this is opposite to the effect expected on the basis of the relative basicities of aryloxides and arylthiolates.)

4. Amino Groups

The basicities of amino-substituted cyclotri- and cyclotetraphosphazenes are considerably higher than those of alkoxy-, aryloxy-, or mercapto-phosphazenes, and they are comparable to, or greater than, the values for ethylphosphazenes (see Table 12.4). In fact, aminophosphazenes are slightly more basic than the parent amines for monoprotonation, as illustrated in Table 12.8. However, this order is reversed in aqueous solution.[8] Comparisons of the pK_a' values for the different aminophosphazenes shown in Table 12.4 reveal no trends which can be rationalized in terms of the electronic or steric properties of the exocyclic groups. Thus, although some correlation can be found between the base strengths of amino-, methylamino-, and ethylamino-phosphazenes and the expected electron-supplying characteristics of the ligands in aqueous media, no comparable correlation is possible for nitrobenzene solutions. Protonation of one ring nitrogen invariably yields a species with a lower base strength than the parent aminocyclophosphazene (Table 12.4). This pK_a' lowering is generally greater for cyclic trimers than for cyclic tetramers. Thus, for the addition of the first proton, the hexakis(amino)cyclotriphosphazenes are all bases comparable in strength to the parent amines, but the lower basicity of the monoprotonated species indicates that the protonation sites are not independent of each other. This is consistent either with the existence of two different types of protonation sites (a ring nitrogen and a side-group nitrogen) or, more likely, with protonation of two electronically interdependent ring nitrogen sites.

TABLE 12.8

COMPARISON OF BASICITIES (pK_a') OF AMINOPHOSPHAZENES AND AMINES[a]

R =	NHMe	NHEt	NHPr^i	NMe_2	NEt_2
HR	6.9	7.1	6.5	7.5	7.1
$(NPR_2)_3$	8.8	8.2	8.4	7.6	8.5
$(NPR_2)_4$	8.2	8.1	8.1	8.3	8.3

[a] From Shaw *et al.*[8]

It is somewhat surprising that different amino substituents do not exert a more selective influence on the base strength for monoprotonation. It has been suggested[8] that substituent steric effects will disguise the expected electronic influences by affecting the shapes of the rings, and that an electronic "saturation effect" serves to level the pK_{a_1}' values. A solvation effect by nitrobenzene molecules may also contribute to the basicity leveling process.

An expected increase in base strength occurs when chlorine atoms in $(NPCl_2)_3$ are replaced by amino substituents[11] (Table 12.5). The nitrogen atoms adjacent to an amino-substituted phosphorus should be the most basic sites. Rationalizations for this can be made both in terms of inductive influences and three-center π-bond electronic transmission effects. Thus, in compounds IV–VIII, the starred nitrogen atoms are the expected sites of protonation.[11] These considerations can be used to establish the geminal or non-geminal structures of aminocyclotriphosphazenes. Thus, the geminal tetrakis(ethylamino) derivative has a higher base strength than the non-geminal analogue (Table 12.5), and the same is true of the appropriate tetrakis(isopropylamino) compounds. Similarly, the geminal tris(dimethylamino) derivative is a stronger base than the non-geminal species. The geminal isomer is invariably the stronger base.[11]

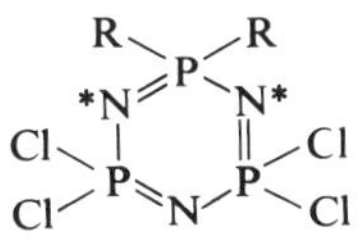

IV

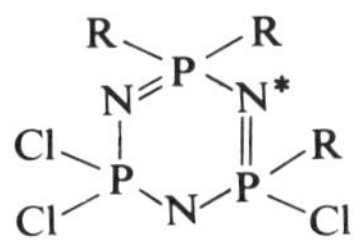

V

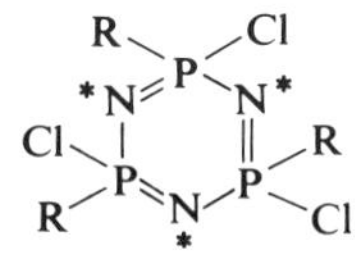

VI

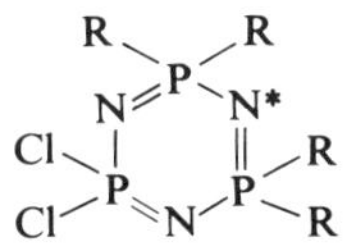

VII

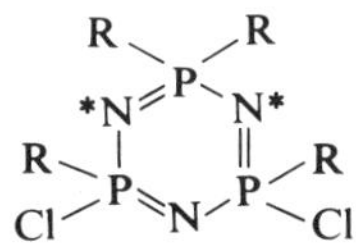

VIII

5. Other Mixed Substituents

pK_a' Values for a wide variety of mixed substituent cyclotriphosphazenes are shown in Tables 12.5 and 12.7. In the preceding sections it was noted that specific substituents raise the base strength of the skeletal nitrogen atoms while other substituents lower it. An analysis of the pK_a' values for a large range of mixed substituents allows for these effects to be placed on a more quantitative basis.[12–14] Consider first the geminal di- and tetraphenylcyclotriphosphazenes listed in Table 12.6. It is known that replacement of chlorine by phenyl and amino results in base-strengthening of the phosphazene ring, and that replacement of phenyl by amino results in further base strengthening. Both of these changes are expected to raise the electron density more at the adjacent (α) nitrogen atoms than at the distant (γ) nitrogen. If it is assumed that substituent effects are largely additive, it is possible to (a) distinguish between different positional isomers and (b) deduce the site of protonation for a wide range of mixed substituent phosphazenes. This can be illustrated by the following examples. The base strength data listed in Table 12.6 show that, for geminal diphenyl and tetraphenyl derivatives, the "saturation effect" observed for fully aminolyzed phosphazenes is not so obvious in this series. Thus, substituent effects can be studied with more reliability.

An assessment of the influence of various substituents on basicity can be made in the following way.[12] Compared to $(NPCl_2)_3$, the influence of each substituent in IX can be depicted in terms of an α or γ influence by phenyl

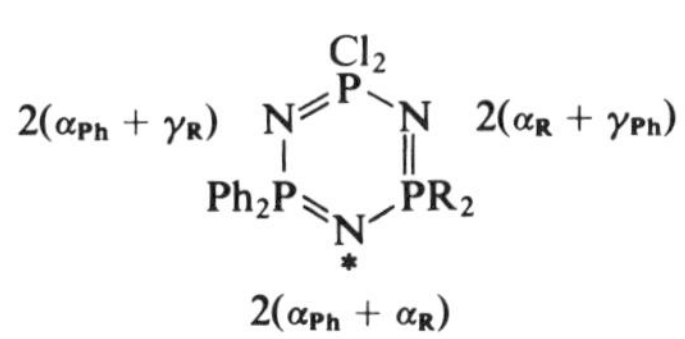

IX

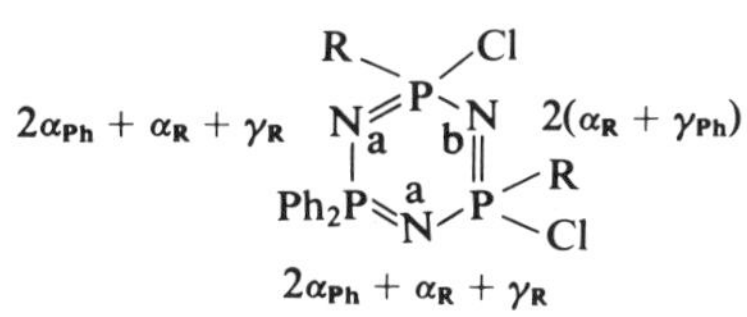

X

TABLE 12.9

SUBSTITUENT CONSTANTS FOR DIFFERENT LIGANDS[a]

	NH_2	NHMe	NHEt	$NHPr^n$	$NHPr^i$
α_R	6.0	5.8	5.8	5.8	5.8
γ_R	2.7	3.1	3.6	3.7	3.2
	$NHBu^n$	$NHBu^t$	NHC_6H_{11}	$NHCH_2Ph$	NHPh
α_R	5.8	5.9	6.0	5.4	4.4
γ_R	3.6	3.4	3.4	—	—
	NHC_6H_4Me-*p*	NHNHPh	NMe_2	NEt_2	NC_5H_{10}
α_R	4.6	4.8	5.6	5.5	5.6
γ_R	2.2	—	2.8	3.1	2.8
	C_6H_4Me	Ph	C_6H_4Cl-*p*	SEt	OMe
α_R	4.5	4.2	3.7	3.6	3.6
γ_R	—	2.3	2.0	1.8	1.8
	SPh	OPh	OCH_2CF_3		
α_R	3.0	3.1	—		
γ_R	1.5	1.3	0.3		

[a] From Shaw *et al.*[13,14]

(α_{Ph} or γ_{Ph}) and an α or γ influence by a second substituent group α_R or γ_R). Since substituent effects are assumed to be additive, each ring nitrogen atom in IX and X can be assigned the terms shown. If the influence of α_{Ph} or α_R is greater than that of γ_{Ph} or γ_R, then the geminal isomer should be the stronger base as, in fact, is found experimentally (Table 12.6). Actual values for the substituent constants, α_{Ph}, γ_{Ph}, α_R, γ_R, etc., can be deduced from the basicity differences between closely related cyclophosphazenes.[12-14]. Typical values are shown in Table 12.9.

The use of substituent constants for the calculation of pK_a' values can be illustrated by reference to structures XI and XII. Suppose that the pK_a' value of XII is unknown, but that the value for XI has been determined (actual value, $pK_a' = 1.5$). Calculation of the expected pK_a' value of XII can be performed in the following manner. First, estimate the most basic site or sites on both rings

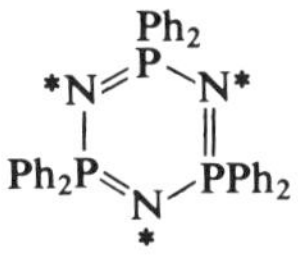

XI

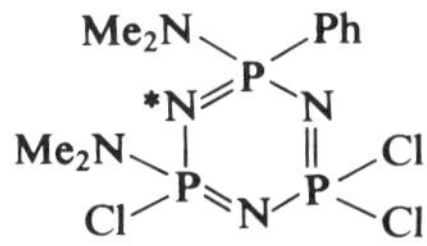

XII

and depict these by stars. Use of the substituent constant table facilitates this step. Second, estimate which of the two phosphazenes is the more basic, and write down a term for the pK_a' difference.

$$\Delta pK_a' = pK_a'_{XII} - pK_a'_{XI}$$

Third, assign substituent contributions for $pK_a'_{XII}$ and $pK_a'_{XI}$ for the most basic (starred) sites on each ring.

$$pK_a'_{XII} = 2\alpha_{NMe_2} + \alpha_{Ph}$$

$$pK_a'_{XI} = 2(2\alpha_{Ph} + \gamma_{Ph}) + 0.5$$

The 0.5 term is a statistical factor employed to take into account the fact that there are three equivalent basic sites. The presence of two equivalent sites requires the use of a statistical term of 0.3.

$$\therefore \Delta pK_a' = 2\alpha_{NMe_2} + \alpha_{Ph} - 4\alpha_{Ph} - 2\gamma_{Ph} - 0.5$$

Substitute values for the α and γ terms from Table 12.7.

$$\Delta pK_a' = 2(5.6) + 4.2 - 4(4.2) - 2(2.3) - 0.5$$
$$\therefore \Delta pK_a' = -6.5$$

Now, $pK_a'_{XI}$ is known to be 1.5, and $-6.5 = pK_a'_{XII} - 1.5$

$$\therefore pK_a'_{XII} = -5.0$$

The value compares well with the measured pK_a' value for XII of -5.0 (Table 12.7). Thus, the pK_a' value for any cyclotriphosphazene should be calculable once values have been obtained for closely related species, and comparison of calculated and observed values can be used to distinguish between different positional isomers. The method does not distinguish between *cis* and *trans* isomers, however. Use of these techniques has provided a pK_a' value of -20 ± 1 calculated for the parent compound, $(NPCl_2)_3$.[12]

It should also be noted than an alternative set of substituent constants can be calculated for diprotonation, and arguments have been developed which allow the site of the second protonation to be predicted.[14]

G. Theoretical Implications of Basicity Studies

The results discussed in this chapter raise a number of questions. For example, why does protonation take place at a ring nitrogen atom rather than at the nitrogen of an amino side group? Which molecular influences account for the effects of different substituent groups on the basicity of the ring? Do the basicity results imply anything about the electronic structure of the phosphazene ring and, in particular, about the differences between cyclic trimers and tetramers?

1. The Position of Protonation

Evidence was presented earlier that, even when amino ligands are attached to phosphorus, the ring nitrogen atoms are the protonation sites. This must mean that electron density is transferred from the ligand to the skeletal nitrogen atoms by an inductive effect or through multiple bonding between exocyclic nitrogen and phosphorus. In theory at least, interaction between the lone-pair orbital of nitrogen (or oxygen) and a phosphorus *d* orbital is permitted on symmetry grounds. The greater basicity of an α-nitrogen compared to a γ-nitrogen can be explained either in terms of the expected decrease in inductive influence at points distant from the ligand, or in terms of three-center islands of π character if π effects are assumed.

2. Influence of Different Substituent Groups

First, an explanation is required of the "saturation effect" observed when every ligand site at phosphorus is occupied by an amine group. This effect apparently swamps out the expected electronic differences between different ligands. It has been suggested[8] that the electron supply from the amino ligands to the ring is so strong that π bonding within the ring is weakened. Thus, beyond a certain point of saturation, the ring can accept no more electron density without losing stability. In these circumstances, the charge density at the ring may be almost independent of the nature of the amino ligands.

Secondly, it is of interest to compare the substituent effects of various groups in cyclophosphazenes with their influence in substituted pyridines. In α- and γ-substituted pyridines (XIII and XIV), the basicity is determined by both

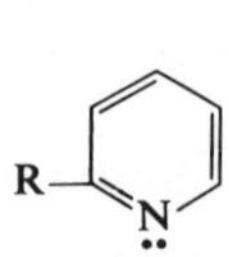

XIII

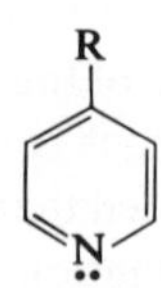

XIV

inductive and multiple-bonding influences by the side group.[13] Thus, methoxy and thioalkoxy groups are base-weakening in the α position when compared to phenyl, but they are base-strengthening with respect to phenyl in the γ position. This is because the inductive effect decreases from α to γ, but the mesomeric electron supply from methoxy or thioalkoxy is comparable from both α and γ positions. By contrast, in cyclotriphosphazenes, methoxy and thioethoxy are base-weakening compared to phenyl in both the α and γ positions, and this suggests that multiple-bonding effects are transmitted only weakly from the γ position in phosphazenes.[13] This is further confirmed by the fact that a reasonable correlation exists between the pK_a values of α-phenyl-, α-ethyl-, α-dimethylamino-, and α aminopyridines and the pK_a values of the appropriate phosphazenes, but no correlation exists when pK_a values for β- or γ-substituted pyridines are used.[9]

It appears, therefore, that in phosphazenes the strongest influence by both inductive and multiple-bonding effects is to the α position, the γ position being affected mainly by an attenuated inductive effect.

3. Influence of Ring Size

Comparisons of the pK_a' values of a number of analogous cyclic trimers and tetramers suggest that some tetramers show a slightly higher base strength than the trimer (Tables 12.1–12.4), although this is by no means a general rule. In the following series,[8, 9] the base strength of the tetramer exceeds that of the trimer: $(NPR_2)_{3\text{ and }4}$ where R = Et, Ph, OMe, OEt, OPr^i, OBu^n, OCH_2Ph, OC_6H_4OMe-*p*, NMe_2, NHBu^t, and NHPr^n. However, the reverse is true when R = OPh, NHMe, NHEt, NHPr^i, NHBu^n, NHC_6H_4Me-*p*, and NEt_2. Thus, there is no general correlation between side-group influence and higher basicity in the trimer or tetramer, except perhaps that the stronger electron-supplying substituents may marginally favor a higher base strength in the trimer.

This absence of a marked general basicity difference between cyclic trimers and tetramers is suprising in view of theoretical predictions that have been made. Thus, because the P—N—P bond angle in a cyclic tetramer is expected to be generally wider than in a trimer, the lone-pair hybrid orbitals at nitrogen are expected to contain more *p* character in tetramers. This could either raise or lower the base strength of the tetramer depending on whether π or π' interactions predominate.[9] In practice, the experimental results throw no light on the different bonding effects in trimers and tetramers. If anything, they favor the view that changes in ring size have very little influence on π bonding, and this is consistent with the existence of three-center islands of d_π–p_π character.

References

1. O. Schmitz-DuMont and H. Kulkens, *Z. Anorg. Allg. Chem.* **238**, 189 (1938).
2. H. Bode, K. Bütow, and G. Lienau, *Chem. Ber.* **81**, 547 (1948).
3. D. B. Sowerby and L. F. Audrieth, *Chem. Ber.* **94**, 2670 (1961).
4. T. Moeller and S. G. Kokalis, *J. Inorg. Nucl. Chem.* **25**, 875 (1963).
5. S. K. Ray and R. A. Shaw, *Chem. Ind.* (*London*) p. 1173 (1961).
6. D. Feakins, W. A. Last, and R. A. Shaw, *Chem. Ind.* (*London*) p. 510 (1962).
7. D. Feakins, W. A. Last, N. Neemuchwala, and R. A. Shaw, *Chem. Ind.* (*London*) p. 164 (1963).
8. D. Feakins, W. A. Last, and R. A. Shaw, *J. Chem. Soc., London* p. 4464 (1964).
9. D. Feakins, W. A. Last, N. Neemuchwala, and R. A. Shaw, *J. Chem. Soc., London* p. 2804 (1965).
10. S. K. Das, R. Keat, R. A. Shaw, and B. C. Smith, *J. Chem. Soc., A* p. 1677 (1966).
11. D. Feakins, W. A. Last, S. N. Nabi, and R. A. Shaw, *J. Chem. Soc., A* p. 1831 (1966).
12. D. Feakins, S. N. Nabi, R. A. Shaw, and P. Watson, *J. Chem. Soc., A* p. 10 (1968).
13. D. Feakins, W. A. Last, S. N. Nabi, R. A. Shaw, and P. Watson, *J. Chem. Soc., A* p. 196 (1969).
14. D. Feakins, R. A. Shaw, P. Watson, and S. N. Nabi, *J. Chem. Soc., A* p. 2468 (1969).
15. H. R. Allcock and D. P. Mack, unpublished observations (1969).
16. K. Denny and S. Lanoux, *J. Inorg. Nucl. Chem.* **31**, 1531 (1969).
17. N. V. Mani and A. J. Wagner, *Chem. Commun.* p. 658 (1968).
18. D. Feakins, W. A. Last, and R. A. Shaw, *J. Chem. Soc., London* p. 2387 (1964).

Chapter 13

REARRANGEMENTS

A. Introduction

Under certain conditions, phosphazenes can be induced to undergo ligand migration reactions. The known rearrangements fall into two categories—those in which a ligand component migrates from phosphorus to skeletal nitrogen to form a phosphazane, and those where *cis–trans* isomerization takes place at one phosphorus atom. An example from the first category is when an alkoxycyclophosphazene (I) undergoes thermal rearrangement to an *N*-alkylcyclophosphazane (II).[1–4] A related migration has been reported for linear

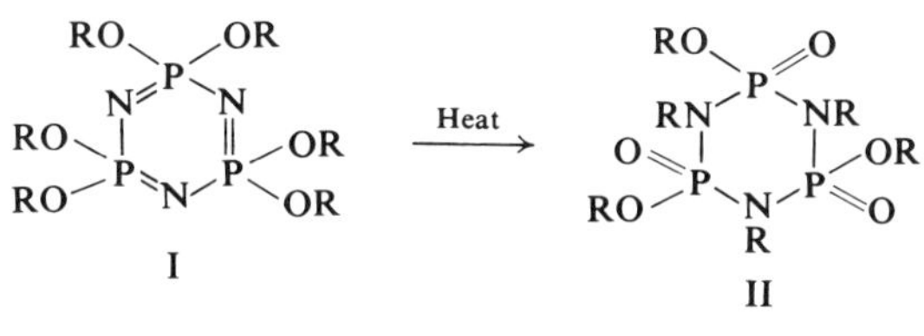

phosphazenes.[5, 6] The degradation reactions which follow the interactions of chlorophosphazenes with carboxylic acids are also believed to result from ligand migrations of this type.[7]

In the second category are the *cis–trans* isomerization reactions of phenylbromocyclotriphosphazenes in the absence of a catalyst,[8] the hydrohalide and ammonium halide-catalyzed *cis–trans* isomerization of aminocyclophosphazenes,[9–11] and the aluminum chloride induced *cis–trans* isomerization of amino- or phenyl-substituted cyclophosphazenes.[12, 13] Each of these rearrangements will be considered in turn.

B. Alkoxyphosphazene–Oxophosphazane Rearrangements

1. Scope and Experimental Facts

Hexakis(ethoxy)cyclotriphosphazene (III) can be converted to 1,3,5-triethoxy-2,4,6-triethyl-1,3,5-trioxocyclotriphosphazane (IV) in 29% yield

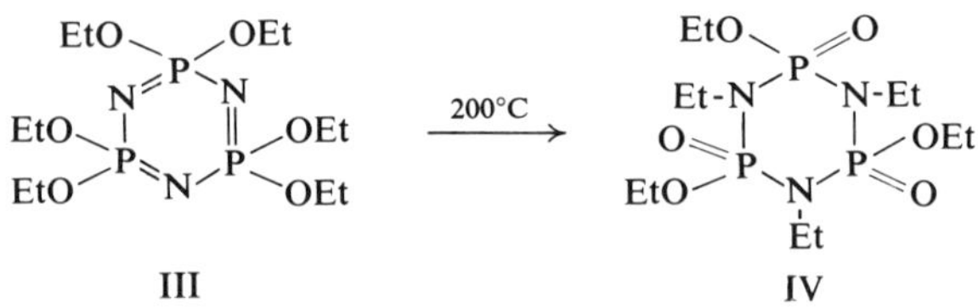

simply by heating at 200°C for 1 hour.[2] The rearrangement results in the conversion of a colorless liquid (III) to a white solid (IV), m.p. 74°–75°C. Octakis-(ethoxy)cyclotetraphosphazene (V) gives the appropriate phosphazane (VI) in 6% yield after 4 hours heating at 200°C. Similar rearrangements occur with

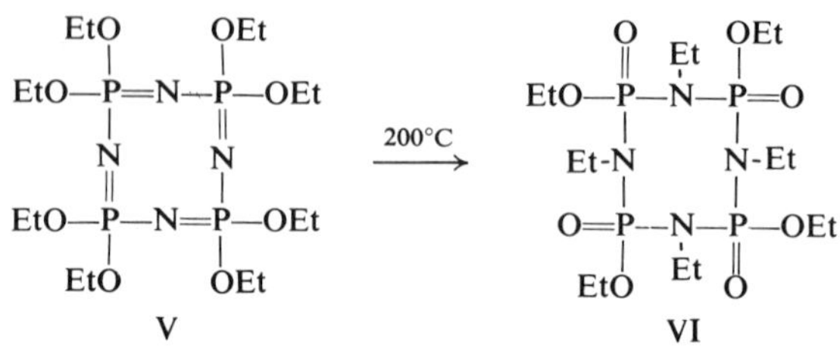

$[NP(OMe)_2]_3$ during distillation, and with $[NP(OCH_2Ph)_2]_3$ at 160°C.

The ethylcyclophosphazane, IV, was identified by comparison with an authentic specimen prepared by an alternative route, and by the absence of the characteristic cyclotriphosphazene 1225 cm^{-1} infrared band and the presence of 1250 cm^{-1} (P=O) and 700 cm^{-1} (P—N) bands. Further structural confirmation was obtained by acid degradation. Thus, IV was degraded by 10% ethanolic hydrogen chloride at 160°C to ethylamine and ethylammonium chloride, whereas III degrades under similar circumstances to ammonia or ammonium chloride.[1, 2] Use of similar techniques confirmed the structure of the cyclotetraphosphazane (VI).

Alkyl halides are catalysts for this rearrangement, permitting lower reaction temperatures to be used, and higher yields of cyclophosphazane to be isolated. Table 13.1 illustrates the reaction conditions which have been employed. In practice, relatively large amounts of alkyl halide are employed. For example, a 14-fold molar excess of ethyl iodide was used for the rearrangement of III.[2]

However, other related alkoxy- and aryloxycyclophosphazenes do not rearrange, even when drastic reaction conditions are used. For example, the phenoxycyclophosphazenes, $[NP(OPh)_2]_3$ and $[NP(OPh)_2]_4$, are stable for long periods of time at 300°C. The same is true of trifluoroethoxycyclophosphazenes, such as $[NP(OCH_2F_3)_2]_3$ and $[NP(OCH_2CF_3)_2]_4$. An intermediate example is provided by the benzyloxycyclophosphazenes, $[NP(OCH_2Ph)_2]_3$ and $[NP(OCH_2Ph)_2]_4$, where the trimer rearranges under relatively mild conditions, but the tetramer appears to be resistant to rearrangement at 175°C or below, whereas higher temperatures (200°C) result in decomposition.[2] Apparently, the methoxy derivatives, $[NP(OMe)_2]_5$, $[NP(OMe)_2]_6$, and $[NP(OMe)_2]_7$, also decompose at high temperatures, and no evidence for phosphazane formation has been recorded.[3]

TABLE 13.1

REARRANGEMENT OF ALKOXYPHOSPHAZENES[a]

Phosphazene	Conditions (catalyst absent) (°C)	Catalyst	Conditions (°C)
$[NP(OMe)_2]_3$	—	MeI	170°/3.5 hr
$[NP(OMe)_2]_4$	—	MeI	170°/3 hr
$[NP(OEt)_2]_3$	200°/1 hr	EtI	170/3 hr
$[NP(OEt)_2]_4$	200°/4 hr	EtI	170°/4.7 hr
$[NP(OPr^n)_2]_3$	—	Pr^nI	170°/6 hr
$[NP(OPr^i)_2]_3$	—	Pr^iI	175°/0.7 hr
$[NP(OCH_2Ph)_2]_3$	160°/2.7 hr	—	—

[a] From Fitzsimmons, Hewlett, and Shaw.[2]

2. Reaction Mechanism

The rearrangement mechanism is believed to involve attack of a ring nitrogen atom on the α-carbon atom of an alkoxy group, as depicted in VII and VIII. Although the mechanism could proceed by either an intermolecular or an intramolecular mechanism, the latter is preferred in the absence of conflicting data.[2]

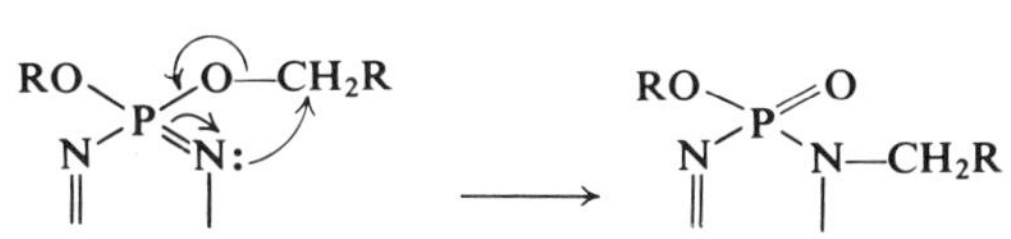

VII VIII

Evidence exists from other systems that the α-carbon atoms in phosphorus esters are susceptible to nucleophilic attack. For example, amidate IX decomposes thermally to ethylaniline and diethylaniline[14] and monophosphazene X undergoes ethyl group migration.[15]

$$\underset{\mathbf{IX}}{(EtO)_2P(O)NHPh} \xrightarrow{260°C} \underset{85\%}{PhNEt_2} + \underset{5\%}{PhNHEt}$$

$$\underset{\mathbf{X}}{(EtO)_3P{=}NPh} \xrightarrow[75°C]{EtI} (EtO)_2P(O){-}NEtPh$$

In terms of this mechanism, it is possible to understand why the thermal rearrangement does not occur with trifluoroethoxy and phenoxycyclotriphosphazenes. Since the skeletal nitrogen atoms are the most basic centers (see Chapter 12), the ease of rearrangement will depend on the basicity of these centers. The following order of decreasing pK_a' values roughly parallels the decreasing rearrangement tendency in $[NP(OR)_2]_3$: R = Pr^i, 1.4; Et, 0.2; Me, −1.9; CH_2Ph, −2.1; Ph, −5.8; OCH_2CF_3, <−6.0. Thus, electron withdrawal from phosphorus and nitrogen by trifluoroethoxy or phenoxy groups reduces the base strength of skeletal nitrogen and inhibits nucleophilic attack by that atom. This is in spite of the fact that a trifluoromethyl group in a trifluoroethoxy unit would be expected to activate the α-carbon atom to nucleophilic attack. Of course, in phenoxyphosphazenes, attack by nitrogen on an aromatic carbon atom would be particularly difficult.

The alkyl halide-catalyzed reaction has been studied in some detail,[4] and the following facts have been obtained. When the rearrangement is conducted with alkyl halides which contain an alkyl group different from those present in the phosphazene side groups, oxophosphazanes can be prepared which contain mixed alkyl groups. For example, isopropyl iodide reacts with hexakis(ethoxy)-cyclotriphosphazene to yield an oxophosphazane (XI) which contains isopropyl groups attached to nitrogen. 4-Nitrobenzyl bromide reacts similarly.

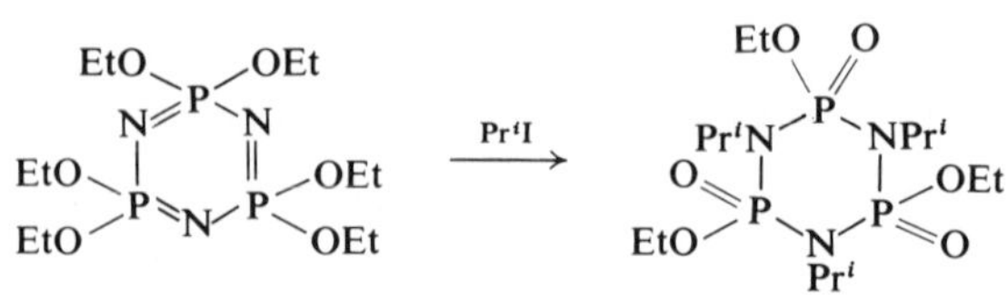

XI

However, the interaction of hexakis(methoxy)cyclotriphosphazene with benzyl bromide yielded a product which contained only two benzyl groups (XII). Evidence that these rearrangements proceed by a stepwise mechanism

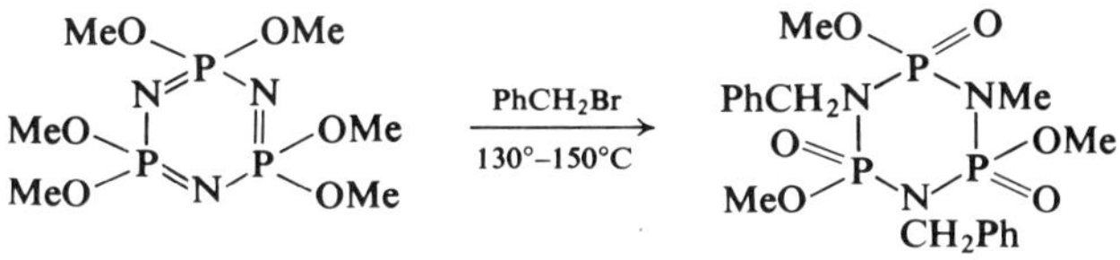

XII

was obtained by the isolation of a partially rearranged product (XIII) from the reaction of hexakis(ethoxy)cyclotriphosphazene with excess ethyl iodide. This ultimately rearranges fully to the oxophosphazane.

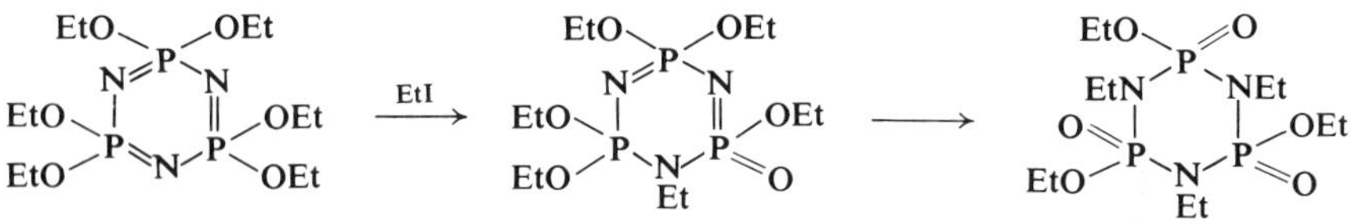

XIII

It is interesting to note that the thermal rearrangement of alkoxyphosphazenes closely resembles the rearrangement of alkoxy-*s*-triazines. For example, trimethyl cyanurate (XIV) rearranges to *N*-trimethyl-isocyanurate (XV) when boiled at 265°C.[16]

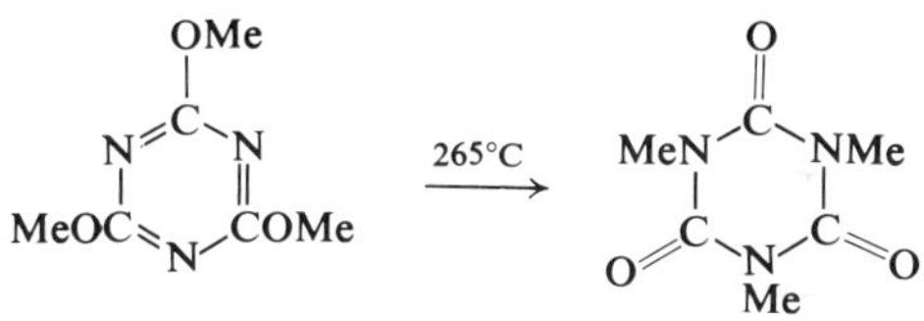

XIV XV

3. Rearrangement of Linear Phosphazenes

A related thermal rearrangement, which involves migration of alkyl groups to a terminal oxygen atom of a monophosphazene, is depicted for the conversion of XVI to XVII.[5, 6] The reaction has been studied for systems where

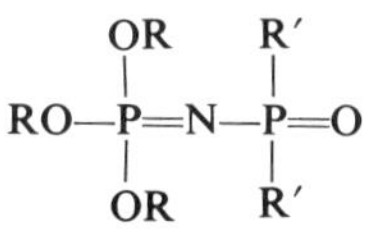

XVI

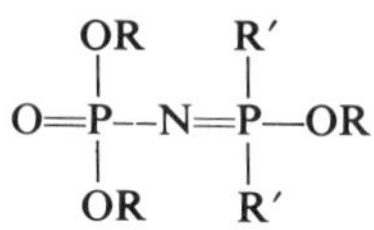

XVII

R = Me, R′ = Et; R and R′ = Et; R = Pr^n, R′ = Et, and R = Et, R′ = Ph. For example, in the case where both R and R′ are ethyl, the rearrangement was accomplished in high yield by heating at 130°C for 10 hours. The rate of the rearrangement decreases as the size of the R group increases, but it increases when R′ = ethyl is replaced by phenyl. An intermolecular reaction has been proposed.[6]

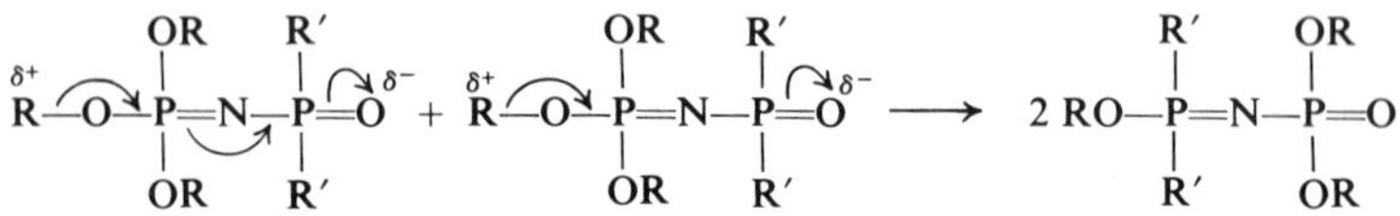

4. Rearrangement of Phosphazene Carboxylate Derivatives

Hexachlorocyclotriphosphazene reacts with the sodium salts of carboxylic acids to yield sodium chloride, a nitrile, an acid anhydride, and a metaphosphimate precursor[7] (see Chapter 14). This reaction is believed to proceed via a phosphazene–phosphazane rearrangement, as depicted in the following sequence:

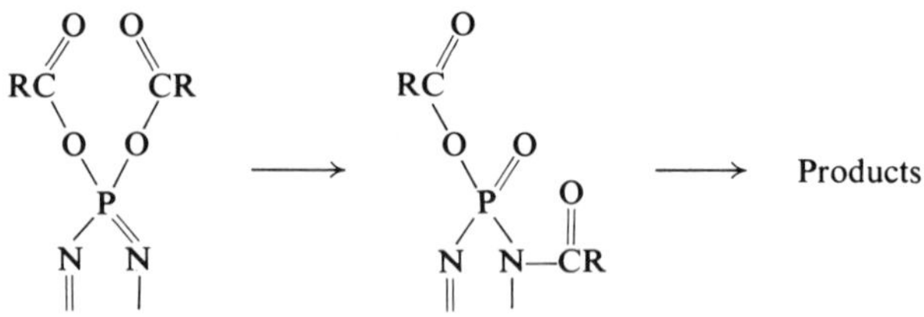

If this is so, the mechanism probably proceeds by attack by skeletal nitrogen on the carbonyl carbon. Presumably, this carbon is more activated than in alkoxyphosphazenes, and the rearrangement takes place rapidly, even at ambient temperature.

C. *Cis–Trans* Isomerizations

1. Possible Mechanisms

Cis–trans isomerizations of cyclophosphazenes take place under a variety of different conditions. With cyclic trimers the reactions can be summarized by the general process:

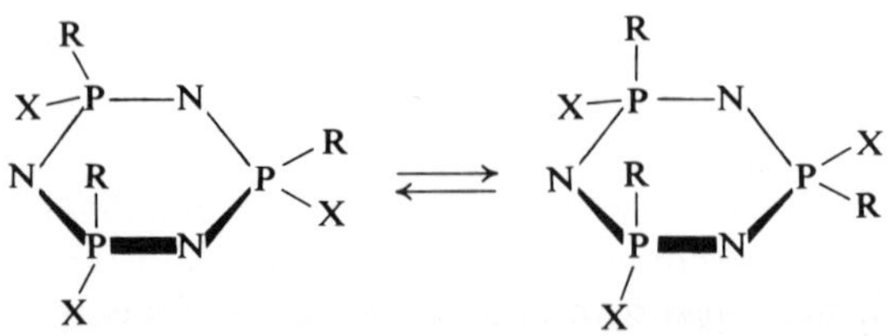

Isomerizations of this type could occur by one of three mechanisms: (1) by the breaking and reforming of a phosphorus–nitrogen skeletal bond, (2) by departure of one side group from phosphorus to leave a planar cation, followed by reattachment of the ligand, or (3) by S_N2-type attack at phosphorus to yield a pentacoordinate intermediate, followed by loss of a ligand with inversion of configuration. These three processes are depicted as follows (phosphorus and nitrogen skeletal atoms are not shown):

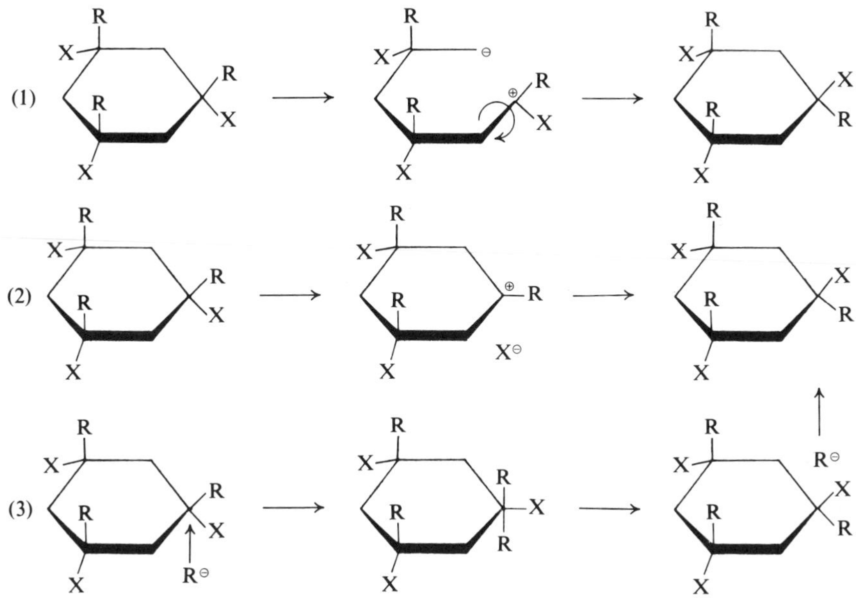

XVIII

Implicit in mechanisms (1) and (2) is the idea of *cis–trans* scrambling rather than an exclusive conversion of one isomer into another. However, mechanism (3) implies almost exclusive conversion of one form to another during the initial stages of the reaction if XVIII is a transition state, with eventual scrambling if there is no preference for attack by $R^{\ominus}$ from the side of X or R. If XVIII is a relatively stable intermediate, *cis–trans* scrambling will occur from the start.

It is doubtful if mechanism (1) would operate at moderate or low temperatures in this system, and a choice must therefore be made between mechanisms (2) and (3) for most isomerizations.

2. UNCATALYZED ISOMERIZATIONS

When non-geminal *trans*-tribromotriphenylcyclotriphosphazene, XIX, is heated in boiling acetonitrile for 1.5 hours, the *cis* isomer (XX) is formed.[8]

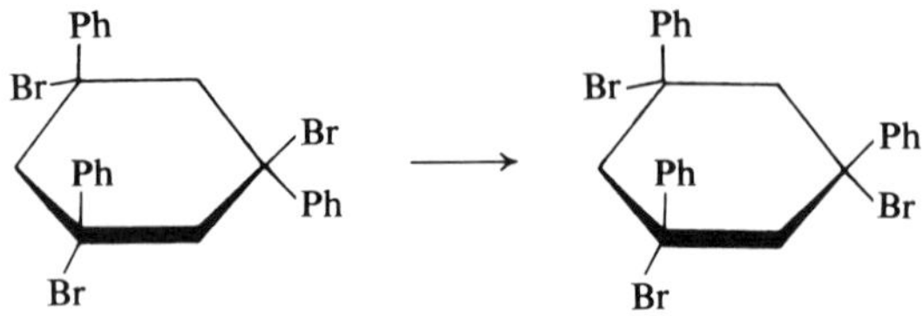

XIX XX

Furthermore, heating of the *cis* isomer (XX) in bromobenzene at 175°–180°C for 50 hours yielded the *trans* isomer (XIX).

Manhas, Chu, and Moeller[8] conclude from these results that the solvent affects the isomerization process, particularly since conversion of the *trans* species to the *cis* species could not be effected in *n*-heptane. A bromide ion ionization mechanism [mechanism (2)] is postulated for the reaction in bromobenzene.

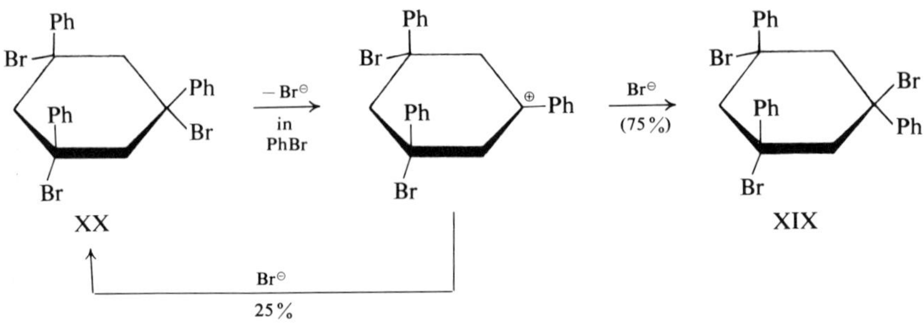

From a statistical point of view, the *trans* isomer would be favored over the *cis* by a ratio of 3:1, with three bulky phenyl groups on the same side of the ring rendering the *cis* form stereochemically less stable.[8]

A slightly different mechanism is proposed for the *trans* to *cis* conversion which occurs in acetonitrile. Ionization by bromide ion is assumed to be accompanied by a frontal interaction with acetonitrile. The displaced bromide ion then attacks by an S_N2 process to displace acetonitrile and invert the configuration to *cis*. Preferential attack away from the side of the bulky phenyl groups would favor formation of the *cis* isomer.

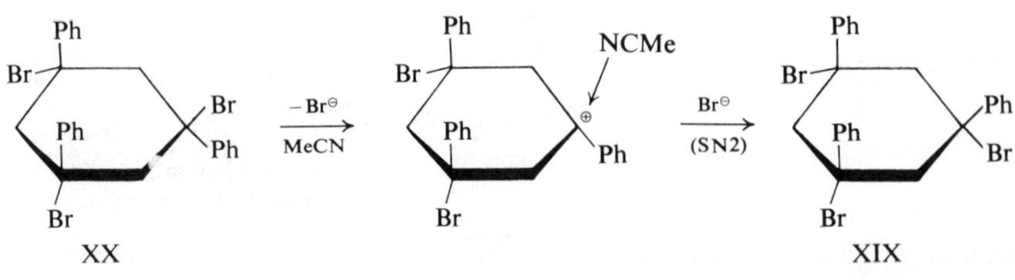

3. Isomerizations Catalyzed by Hydrogen Halides or Ammonium Halides

Non-geminal *cis*- and *trans*-tris(dimethylamino)trichlorocyclotriphosphazenes (XXI) and bis(dimethylamino)tetrachlorocyclotriphosphazenes (XXII)

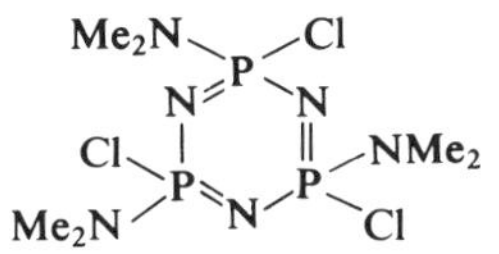

XXI

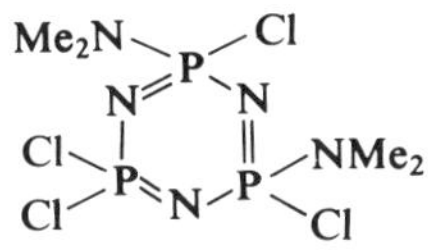

XXII

undergo a reversible *cis–trans* isomerization in the presence of hydrogen chloride.[9] The trisbromo analog of XXI behaves similarly in the presence of hydrogen bromide.

Keat and Shaw[10, 11] have studied the *cis–trans* isomerization reactions of compounds XXI and XXII, as well as the corresponding piperidinophosphazenes in the presence of alkylammonium halides. For the dimethylamino derivatives, isomerization was effected by dimethylammonium chloride in chloroform or acetonitrile, but not in benzene or ether. The *trans* isomer of XXI was converted to a 2:1 mixture of *trans* and *cis* after 8 hours at 61 °C in chloroform in the presence of dimethylammonium chloride. Inversion of the piperidino analogues of XXI and XXII proceeds at a much slower rate than the inversion of the dimethylamino derivatives, XXI and XXII.

Several observations provide a clue to the mechanism. First, the non-geminal trispiperidino derivatives undergo inversion more readily than the non-geminal bispiperidino compounds, and this suggests that the greater electron supply to the ring in the former case favors a P—Cl bond-breaking mechanism. Second, solvents of higher dielectric constant appear to favor the inversion, a fact which suggests that an ionization process is an integral part of the mechanism. Third, the dimethylamino and piperidino derivatives can be inverted by a variety of alkylammonium halides without any foreign amino groups being introduced into the resultant *cis–trans* mixture. Thus, exocyclic P—N bonds are neither broken nor formed during the reaction.

The two most plausible mechanisms involve either a bimolecular displacement of chloride by chloride ion [mechanism (3) discussed previously] or a unimolecular chloride ionization mechanism.

4. Aluminum Chloride-Catalyzed Isomerizations

When the *trans* isomer of compound XXII is heated for 5 hours at 100°C in *sym*-tetrachloroethane with aluminum chloride (in a 1:2 molar ratio) isomerization occurs to yield a mixture of the *cis* and *trans* isomers.[12] The same mixture is obtained if the *cis* isomer is the starting material. Thin-layer or column chromotography on silica gel can be used to separate the isomers. A similar *cis–trans* isomerization is observed when the non-geminal tris(dimethylamino)trichloro derivative (XXI) is heated with aluminum chloride under similar circumstances. In all these reactions, complete equilibration apparently did not occur since the starting material always predominated in the reaction mixture.

By contrast, 1-*cis*-3-dichloro-1,3,5,5-tetrakis(dimethylamino)cyclotriphosphazene (XXIII) underwent no isomerization when treated similarly. It is

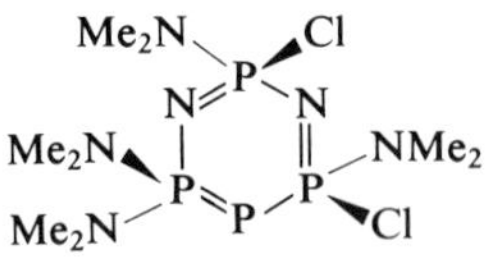

XXIII

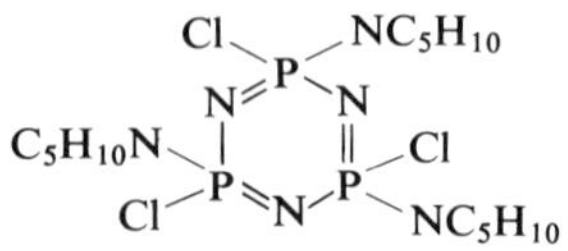

XXIV

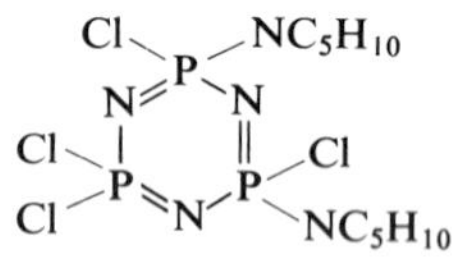

XXV

assumed, therefore, that this structure is the thermodynamically preferred configuration. The piperidinocyclotriphosphazenes, XXIV and XXV, behave in essentially the same way as the dimethylamino derivatives. However, the geminal bis and tris derivatives of these two amines do not react under these conditions.

The mechanism of these inversions probably involves electrophilic attack by aluminum chloride on a chlorine atom with the subsequent formation of a quasi-phosphazenium ion (XXVI).[12] Presumably, structure XXVI would be

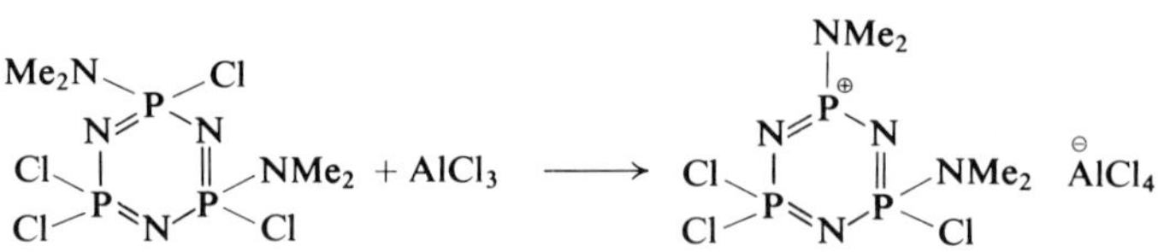

XXVI

stabilized by electron supply from the dimethylamino group and from the skeletal nitrogen atoms. If the lifetime of XXVI is long, it will have an equal probability of inversion or retention of configuration. However, if chloride ion separation is concurrent with a backside attack on phosphorus by $AlCl_4^{\ominus}$ in an S_N2 process, then inversion of configuration would result.

Finally, Grushkin *et al.*[13] have reported that the *cis* or *trans* isomers of 1,3,5-trichloro-1,3,5-triphenylcyclotriphosphazene (XXVII) readily isomerize to

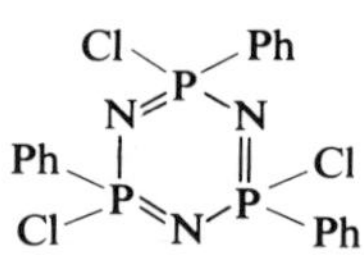

XXVII

mixtures of the two in the presence of aluminum chloride in carbon disulfide, or ferric chloride in boiling benzene. Hydrogen chloride was evolved in the latter case. Starting with the *trans* isomer, and with a 3:1 molar excess of ferric chloride, after 24 hours reaction the final mixture (after characterization by dimethylaminolysis) contained 35% *cis* and 65% *trans*. After 4 hours reaction in carbon disulfide, with a 14:1 excess of aluminum chloride, the *cis* derivative gave a 60% *cis*, 40% *trans* mixture, and the *trans* derivative gave a 25% *cis*, 75% *trans* mixture. Thus, complete equilibration did not occur.

The mechanism postulated in this case involves the formation of either a quasi-phosphazenium ion, similar to XXVI, or a metal halide–phosphazene σ complex such as XXVIII.

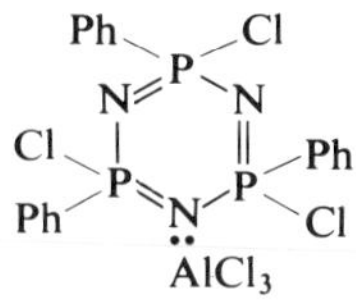

XXVIII

5. Isomerization of Azidophosphazenes

Sharts, Bilbo, and Gentry[17] have reported the *cis–trans* isomerization of the azidophenylcyclotetraphosphazene shown in XXIX. The *cis* or *trans* isomers

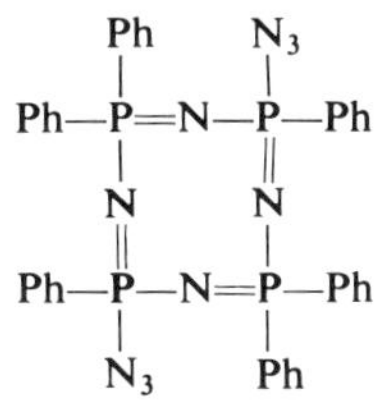

XXIX

of XXIX are converted to a 72% *trans*, 28% *cis* mixture when heated in boiling acetonitrile with lithium azide. For the change of *trans*-XXIX to *cis*-XXIX, the equilibrium constant is 2.6 and the free energy of the reaction at 355°K is 0.67 kcal/mole. The mechanism of this reaction has not been established.

References

1. B. W. Fitzsimmons and R. A. Shaw, *Proc. Chem. Soc., London* p. 258 (1961).
2. B. W. Fitzsimmons, C. Hewlett, and R. A. Shaw, *J. Chem. Soc., London* p. 4459 (1964).
3. F. Rallo, *Ric. Sci. Chim.* **8**, 1134 (1965).
4. B. W. Fitzsimmons, C. Hewlett, and R. A. Shaw, *J. Chem. Soc., London* p. 7432 (1965).
5. I. M. Filatova, E. L. Zaitseva, and A. Ya. Yakubovich, *J. Gen. Chem. USSR* **36**, 1854 (1966).

6. I. M. Filatova, E. L. Zaitseva, A. P. Simanov, and A. Ya. Yakubovich, *J. Gen. Chem. USSR* **38**, 1256 (1968).
7. I. I. Bezman and W. R. Reed, *J. Chem. Soc., London* **82**, 2167 (1960).
8. B. S. Manhas, S. K. Chu, and T. Moeller, *J. Inorg. Nucl. Chem.* **30**, 322 (1968).
9. S. N. Nabi, R. A. Shaw, and C. Stratton, *Chem. Ind. (London)* p. 166 (1969).
10. R. Keat and R. A. Shaw, *Chem. Ind. (London)* p. 1232 (1964).
11. R. Keat and R. A. Shaw, *J. Chem. Soc., London* p. 4067 (1965).
12. R. Keat, R. A. Shaw, and C. Stratton, *J. Chem. Soc., London* p. 2223 (1965).
13. B. Grushkin, M. G. Sanchez, M. V. Ernest, J. L. McClanahan, G. E. Ashby, and R. G. Rice, *Inorg. Chem.* **4**, 1538 (1965).
14. J. I. G. Cadogan, *J. Chem. Soc., London* p. 1079 (1957).
15. M. I. Kabachnik and V. A. Gilyarov, *Izv. Akad. Nauk SSSR, Otd. Khim. Nauk* p. 790 (1956).
16. A. W. Hofmann, *Chem. Ber.* **3**, 264 (1870); p. 2061 (1886).
17. C. M. Sharts, A. J. Bilbo, and D. R. Gentry, *Inorg. Chem.* **5**, 2140 (1966).

Chapter 14

SKELETAL BOND CLEAVAGE, THERMOLYSIS, AND OTHER REACTIONS

The reactions reviewed in Chapters 5 through 13 fall into well-defined reaction types. However, phosphazenes and phosphazanes undergo a number of other interactions that do not fall cleanly into one of these categories, and these are considered in this chapter. The first section deals with several reactions that lead to cleavage of the phosphorus–nitrogen skeletal system. The second covers the general behavior of phosphazenes at elevated temperatures. The third section includes a number of miscellaneous reactions, some of which offer clues to possible areas for the future development of this field.

A. Skeletal Bond Cleavage Reactions

A number of reactions are known in which an apparent nucleophilic substitution process leads to breakdown of the phosphorus–nitrogen skeleton. Included in this category are the reactions of carboxylic acids or their salts with halophosphazenes, and the reactions of catechol or *o*-aminophenol with cyclo- or polyphosphazenes. These processes will be reviewed first, to be followed by a consideration of the decomposition of alkoxycyclophosphazenes by benzoyl chloride.

1. Reactions of Chlorocyclophosphazenes with Carboxylic Acids and Their Salts

In 1960, Bezman and Reed[1] reported an unusual reaction between hexachlorocyclotriphosphazene, $(NPCl_2)_3$, and the sodium salts of benzoic, *p*-chlorobenzoic, *p*-methoxybenzoic, lauric, or acetic acids. In each case, the reactants were heated together in the absence of solvent and the products formed were the nitrile of the carboxylic acid (e.g., benzonitrile from sodium benzoate), the carboxylic acid anhydride, and a phosphorus compound believed to be a precursor of trisodium trimetaphosphate. A similar process was reported by Burg[2] when silver trifluoroacetate reacted with $(NPCl_2)_3$. Benzoic acid itself reacts similarly with $(NPCl_2)_3$ in a solvent in the presence of pyridine.[1] These reactions have been examined further by Yokoyama, Cho, and Aida.[3]

The reaction mechanism is believed to proceed through the initial transient formation of a phosphazene-benzoate-type compound (I), which rearranges

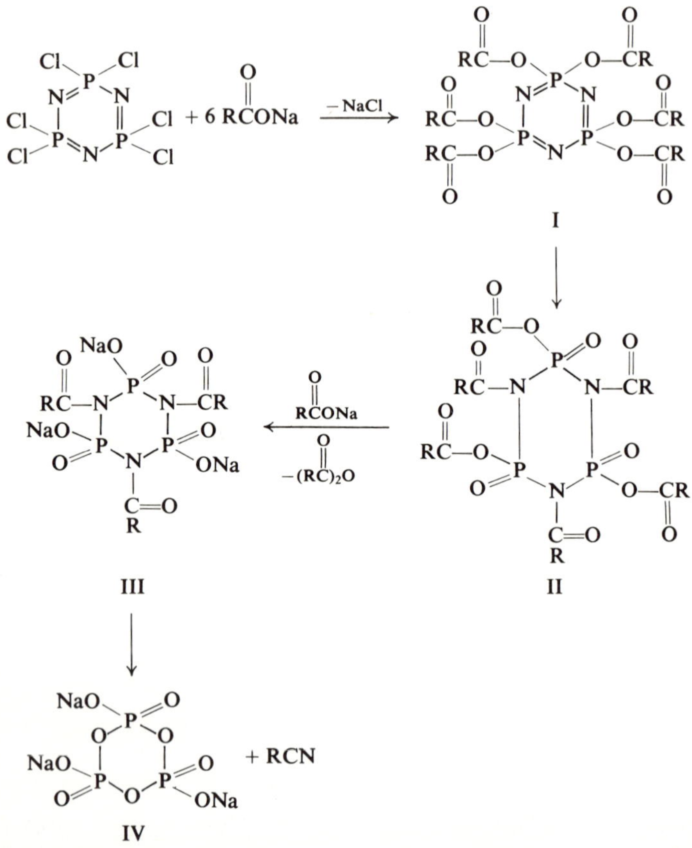

to the phosphazane (II). Reaction of this with excess sodium carboxylate splits out the acid anhydride and leaves a sodium phosphazanate (III). It is assumed that III rearranges to the trimetaphosphate (IV) with elimination of the nitrile.

The evidence for this mechanism is mainly circumstantial. Compounds I, II, or III have not been isolated from the reaction mixture, but evidence for the formation of IV has been presented.[3] The phosphazene–phosphazane rearrangement (I → II) is reminiscent of those known to take place with alkoxyphosphazenes (see Chapter 13), and the insertion of oxygen atoms into a phosphazane ring (III → IV) occurs during acid hydrolysis of cyclophosphazanes (see Chapter 5).

Hexachlorocyclotriphosphazene also functions as an activator for the formation of amides and hydrazides from carboxylic acids.[4] For example, treatment of $(NPCl_2)_3$ with a carboxylic acid or its sodium salt at room temperature in an inert solvent, followed by treatment with an excess of a primary or secondary amine, rapidly yields the amide or hydrazide.

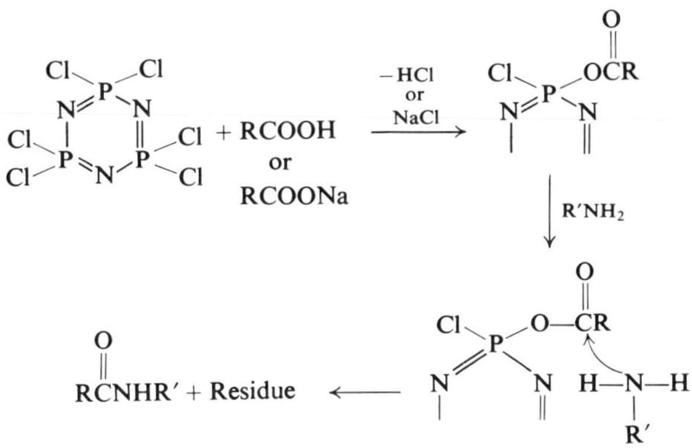

When the free carboxylic acid is used, triethylamine can be employed as a hydrohalide acceptor. A wide variety of carboxylic acids and amines participate in this reaction.[4]

2. Reactions of Phosphazenes with Catechol and *o*-Aminophenol

a. *General Features of the Reaction*

Allcock and Kugel[5–9] have investigated a series of unusual ring degradation reactions which take place when cyclo- or polyphosphazenes are treated with catechol or *o*-aminophenol. Hexachlorocyclotriphosphazene reacts with catechol in the presence of triethylamine to yield the spirocyclophosphazene V.[5,6] The reaction products always include appreciable amounts of a spirophosphorane (VI), and it can be shown that this compound is produced by a reaction between V and catechol in the presence of a tertiary amine. Ammonia is liberated during this process. The same spirophosphorane is formed when $(NPCl_2)_4$ or $(NPCl_2)_n$ high polymer is treated with catechol and triethylamine, but in these cases no phosphazenes analogous to V are isolated.[6]

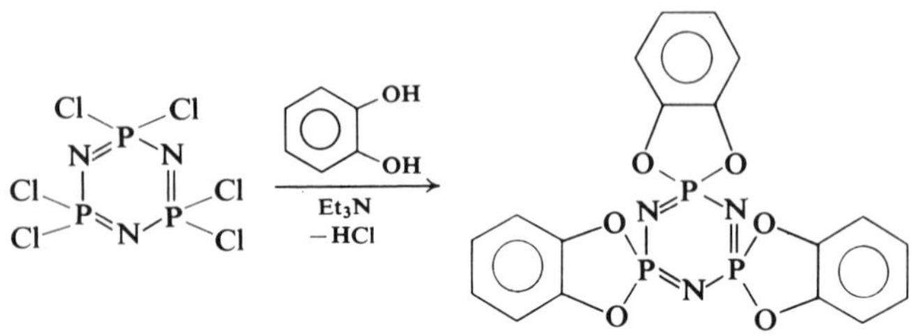

V

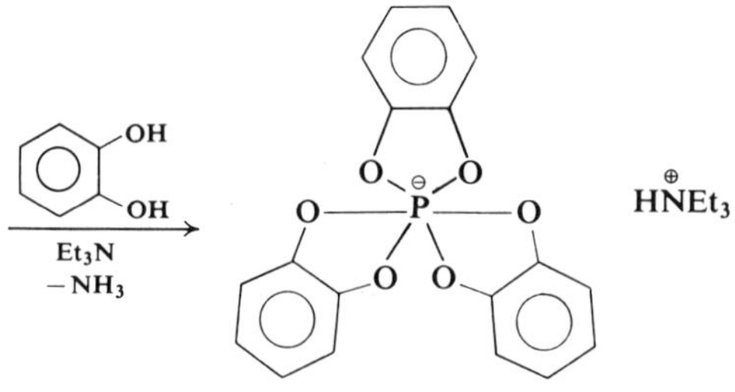

VI

A similar process occurs when $(NPCl_2)_3$, $(NPCl_2)_4$, $(NPCl_2)_n$, $(NPF_2)_3$, or $(NPBr_2)_3$ are treated with *o*-aminophenol in a hot organic solvent such as boiling xylene. Ammonium halide is evolved and a spirophosphorane (VIII)

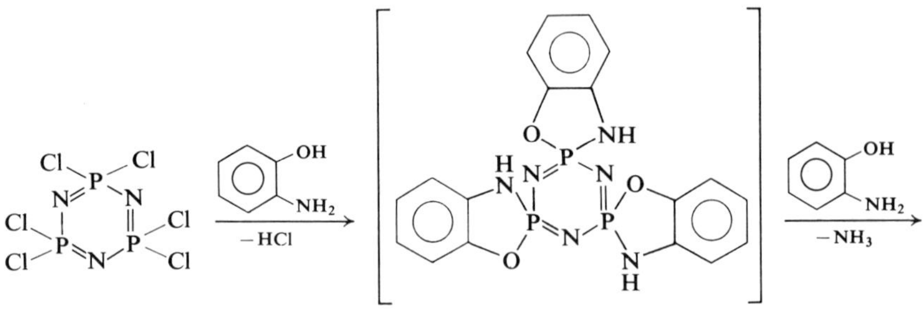

VII

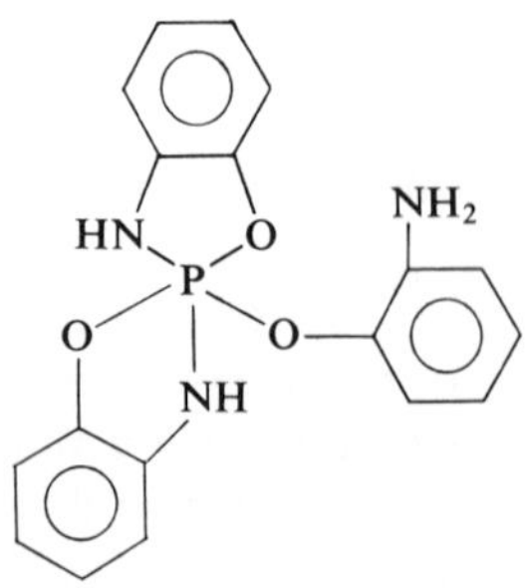

VIII

is formed. The expected spirophosphazene (VII) has not been isolated, but it is believed to be formed as a transient intermediate. Even more surprising is the fact that certain *organo*cyclophosphazenes react with *o*-aminophenol to yield the same spirophosphorane (VIII). These cyclophosphazenes are depicted as V, IX, X, and XI. The original ligands are displaced from phosphorus, and the

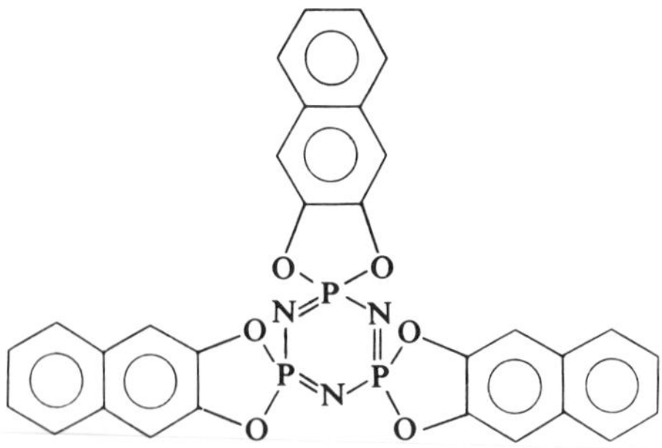

IX

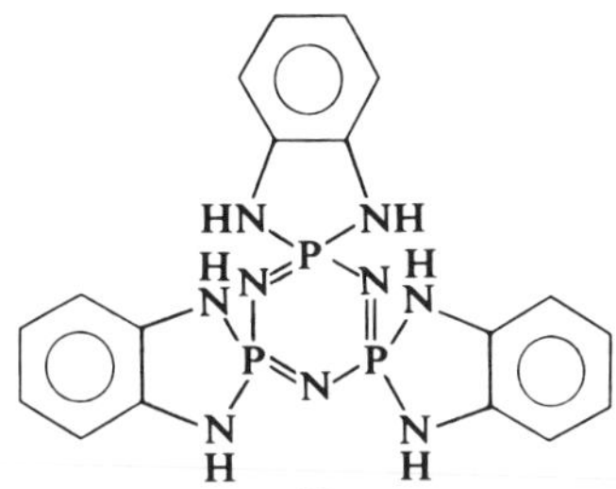

X

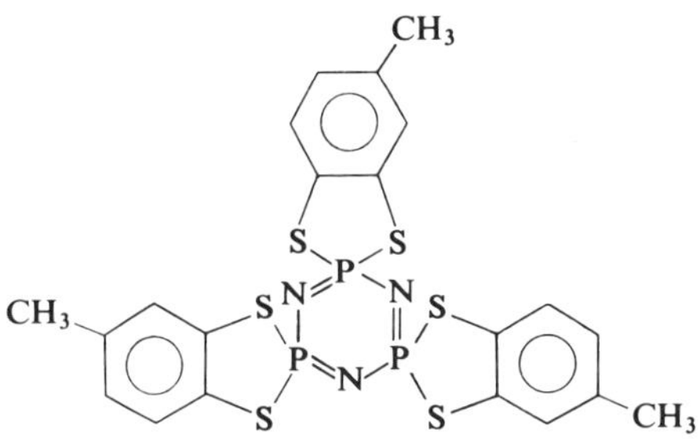

XI

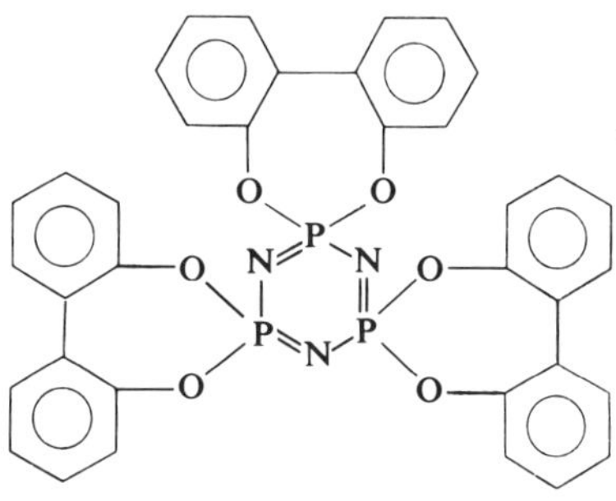

XII

o-aminophenoxy residue takes their place in the phosphorane. By contrast, other organophosphazenes do not undergo phosphazene ring degradation with *o*-aminophenol. Compound XII is inert under these conditions and $[NP(OPh)_2]_3$, $[NP(OC_6H_4NO_2\text{-}p)_2]_3$, $[NP(OCH_2CF_3)_2]_3$, and $[NP(NHPh)_2]_3$ do not yield phosphorane.

b. *Mechanism of the Reaction*

There is much evidence that the degradation of V by catechol and base, and the phosphazene ring cleavage of $(NPCl_2)_3$, V, IX, X, and XI by *o*-aminophenol follow the same general mechanism, and the following discussion covers both types of reaction. It is convenient to break up the mechanism into a number of discrete steps and these are as follows.

Formation of Spirophosphazene. It is known that spirophosphazene, V, is formed as an isolatable intermediate during the reaction of $(NPCl_2)_3$ with catechol and triethylamine.[5, 6] Similarly it is assumed that the spirophosphazene, VII, is formed as a transient intermediate when $(NPCl_2)_3$ reacts with *o*-aminophenol. The fact that this reaction does not take place at a measurable rate below ~100°C may reflect the relative slowness of this initial substitution process. The facile syntheses of V, IX, X, and XI[10] also provide evidence for this view. Thus, the subsequent steps leading to phosphorane formation from V, VII, IX, X, and XI can be considered as closely related reactions.

Nucleophilic Attack by the Reagent on the Spirophosphazene. This may be the rate-determining step in this reaction, and it can be depicted as involving the formation of a pentacoordinate phosphorus transition state (XIII). The

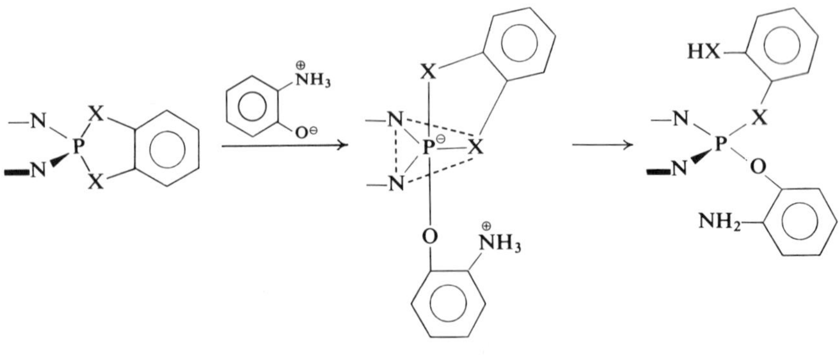

XIII XIV

catecholate mono- or dianion could attack by a similar mechanism. Some experimental evidence can be found in favor of this step. First, it is known that nucleophiles such as hydroxide, aryloxide, alkoxide ions, or amines can reversibly cleave the exocyclic rings of V by a very rapid process.[6, 9, 11, 12] The rapidity of the step reflects a release of steric strain in the five-membered exocyclic ring at phosphorus, and this presumably partially explains why substrates such as XII, or phosphazenes with two independent ligands at phosphorus, do not degrade to phosphoranes. Secondly, the rate of phosphorane formation from $(NPCl_2)_3$ and *o*-aminophenol rises with increases in solvent polarity. Although this effect could represent an influence on step 1 (initial chlorine replacement), it is also consistent with an increased charge separation during the formation of transition state XIII. The approach of the nucleophile and the retreat of the displaced ligand are assumed to occur along the axial direction (XIII), in view of the widespread evidence from phosphate hydrolysis work.[13]

Cleavage of Phosphorus–Nitrogen Bonds. A structure such as XIV, formed by the initial exocyclic ring cleavage, has two strategically placed, potential

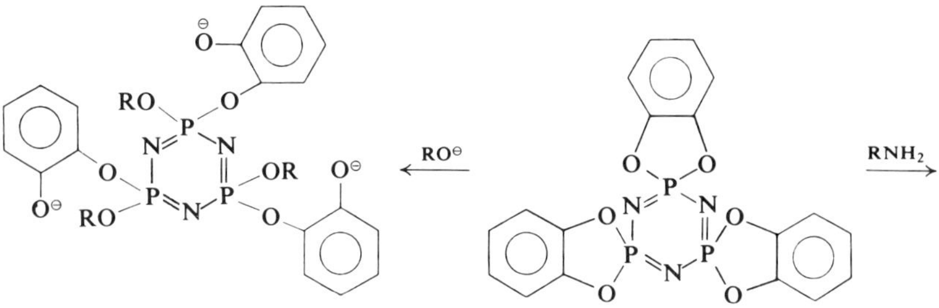

XV V

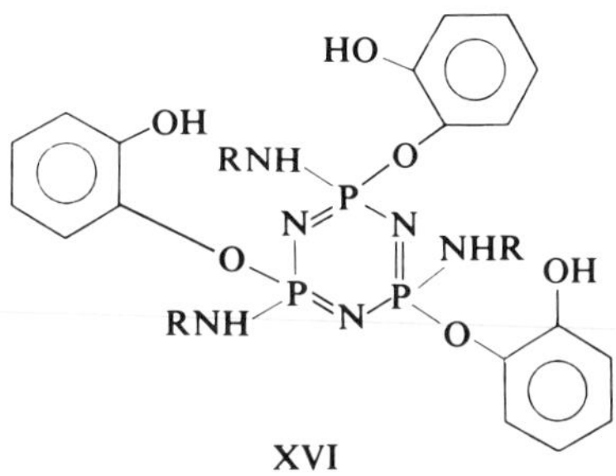

XVI

nucleophilic sites for intramolecular attack on phosphorus. It is believed that this internal nucleophilic substitution leads to cleavage of a phosphorus–nitrogen ring bond, but the details of this and subsequent steps are not clear. However, the fact that an *ortho*-difunctional nucleophile is required for degradation to phosphorane is evidence for this step. When phenol, aniline, aniline hydrochloride, or phenol–aniline mixtures are allowed to interact with V, no phosphorane is formed and the reaction does not progress beyond the reversible formation of XV or XVI.[8, 9] However, the isolation of XVII from the reaction of $(NPCl_2)_3$ with *o*-aminophenol indicates that ring cleavage occurs after two *o*-aminophenol residues have become attached to phosphorus. The cleavage of a phosphorus–nitrogen bond in XIV by internal nucleophilic attack requires either the assumption of equatorial anion attack (because of

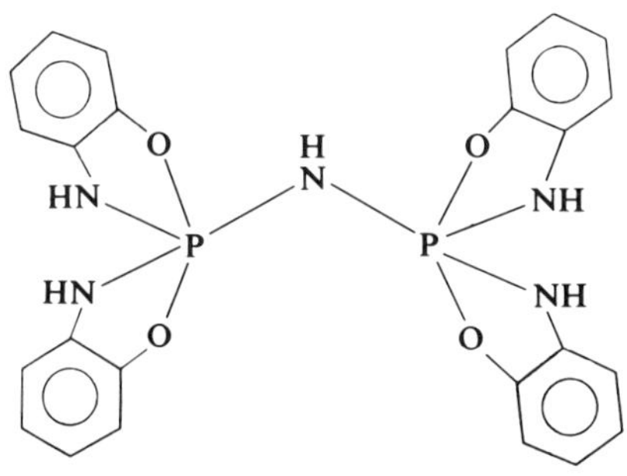

XVII

the preferred 120° N—P—N angle) or of pseudo-rotation.[13] In either case, the reaction would be facilitated by increased skeletal flexibility, and this may explain why the rate of phosphorane formation with *o*-aminophenol increases in the order: $(NPCl_2)_3 < (NPCl_2)_4 < (NPCl_2)_n$.[8]

Phosphorane Formation. The driving force for these reactions appears to be the release of exocyclic ring strain during the conversion of a spirophosphazene to a spirophosphorane. Phosphoranes which contain five-membered rings are generally stable.[5, 6, 14, 15] The stability of these compounds is probably connected with the ability of the O—P—O bond angle to approximate to 90° in a trigonal bipyramidal or octahedral structure. The O—P—O bond angle in V is 97°,[16] an appreciable narrowing from the preferred value of 103°, and the molecule can relieve much of this strain by conversion to a spirophosphorane. No such inducement exists with compounds such as XII or phosphazenes with two independent ligands at each phosphorus.

Ligand Scrambling. Perhaps the most surprising result from the mechanistic point of view is the fact that, when *o*-aminophenol is used as a reagent, the same phosphorane (VIII) is isolated irrespective of the original organic side group structure.[8] This must reflect a rapid scrambling of the ligands, with a preferential introduction of *o*-aminophenol rather than dioxyarylene, dithioarylene, or diaminoarylene. It is not clear at which point in the mechanism the scrambling occurs, and mechanisms can be formulated for displacement of the original ligands at the stage represented by XIV or at any later step, including scrambling at the phosphorane stage. However, it is known that when 2,3-dihydroxynaphthalene reacts with V in xylene in the presence of triethylamine, a mixed phosphorane, XVIII, is formed rapidly.[9] Thus, a mechanism must exist for the transfer of one ligand originally present at one phosphorus to another phosphorus atom.

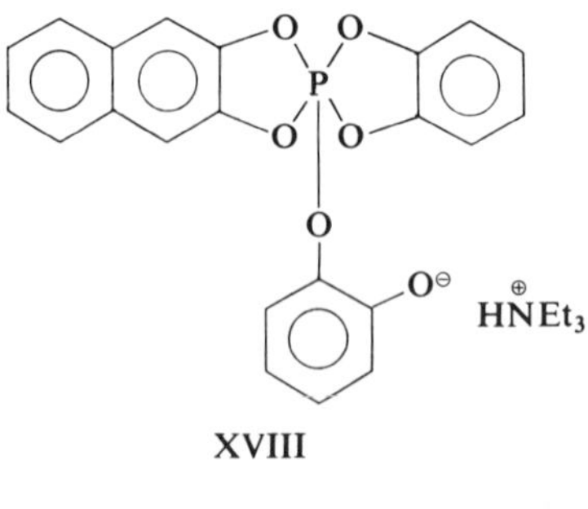

XVIII

3. Reaction of Alkoxycyclophosphazenes with Benzoyl Chloride

Fitzsimmons, Hewlett, and Shaw[17, 18] have investigated the reaction of hexakis(ethoxy)cyclotriphosphazene (XIX) or the tetramer $[NP(OEt)_2]_4$ with benzoyl chloride to yield ethyl chloride, ethyl phosphates, and triphenyl-

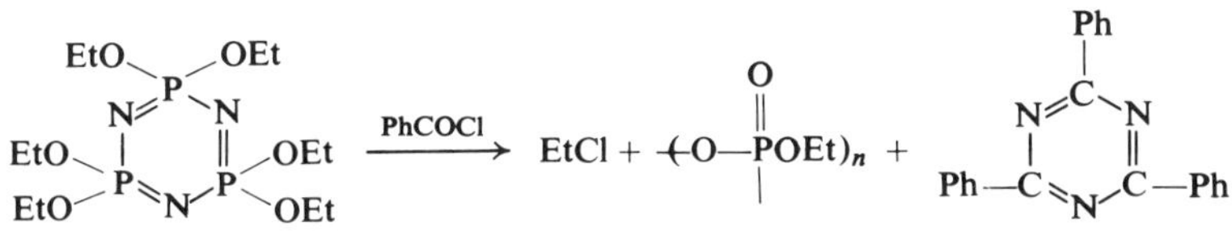

XIX

s-triazine. The reactants are heated together at 130°–140°C for 1 hour (for XIX) or at 140°–160°C for 5 hours (for the tetramer), with the mixture swept by a stream of dry nitrogen. The initial reaction step is believed to be a nucleophilic attack by a phosphazene nitrogen atom on the carbon atom of the carbonyl group (XX). One mole of ethyl chloride is eliminated for each

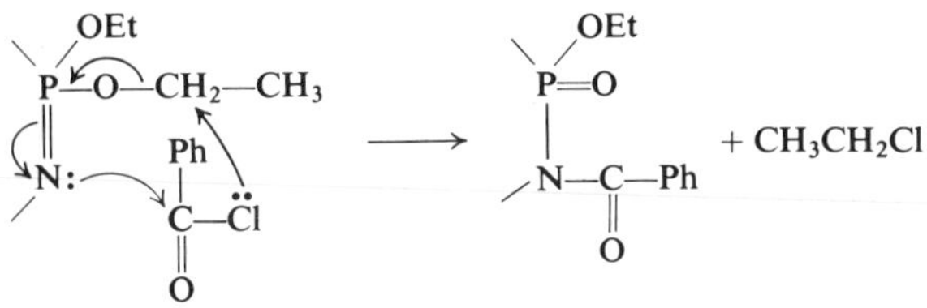

XX

phosphazene repeating unit. The related cyclophosphazane $[EtN—P(O)OEt]_3$ does not react with benzoyl chloride, and this constitutes evidence in favor of the role of skeletal nitrogen in this reaction. Subsequent steps in this reaction follow a nucleophilic attack by the carbonyl oxygen atom of the oxophosphazane on phosphorus as depicted in the scheme.

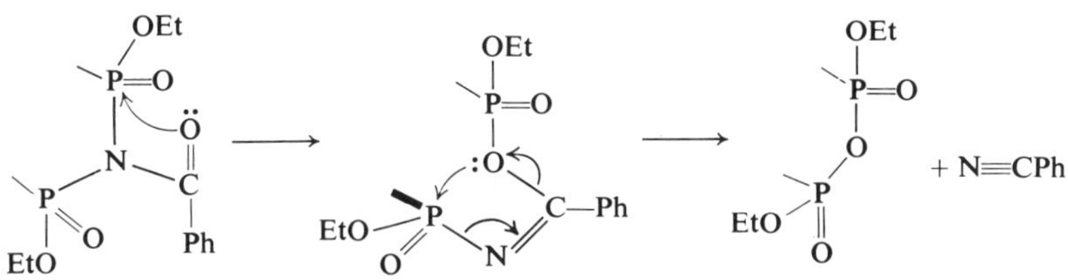

Benzonitrile is then believed to trimerize to triphenyl-*s*-triazine.

B. Thermolysis Reactions

The behavior of phosphorus–nitrogen compounds at high temperatures is a subject of some importance in high-temperature polymer technology. The polymeric aspects of the subject will be dealt with in Part III of this volume. In this section, thermal decomposition reactions will be emphasized.

1. General Thermal Stability of Phosphazenes

A number of phosphazenes are unusually stable at elevated temperatures. For example, hexakis(trifluoroethoxy)cyclotriphosphazene, $[NP(OCH_2CF_3)_2]_3$, can be heated in the atmosphere at 320°–350°C for 10 to 30 hours without decomposition occurring.[19] Even attempted combustion in a stream of oxygen over copper oxide does not result in complete oxidation, and fluoroalkoxyphosphazenes are inert to boiling nitric acid.[20] However, it has been reported[20] that hexakis(heptafluorobutoxy)cyclotriphosphazene, $[NP(OCH_2C_3F_7)_2]_3$, is converted in 50% yield to the cyclic tetramer when boiled at 340°C for 50 hours.

Phenoxycyclophosphazenes are, if anything, even more stable than the trifluoroethoxy derivatives. For example, hexaphenoxycyclotriphosphazene, $[NP(OPh)_2]_3$, is stable for long periods of time at 350°C.[21] Its stability at elevated temperatures is such that it can be used as a high-temperature standard for mass-spectrometry experiments.[22] Virtually no fragmentation is observed at 250°C. Many alkyl- and arylcyclophosphazenes are also stable at high temperatures.

On the other hand, certain classes of phosphazenes are unstable at high temperatures. The sensitivity of alkoxycyclophosphazenes to thermal rearrangement to oxophosphazenes has been discussed in Chapter 13. Isothiocyanophosphazenes, $[NP(NCS)_2]_n$, appear to decompose at high temperatures,[23] and many aminophosphazenes decompose when heated.

2. Pyrolysis of Aminophosphazenes

Aminocyclophosphazenes of formula $[NP(NH_2)_2]_3$, $[NP(NH_2)_2]_4$, or $N_4P_4Cl_4(NH_2)_4$ pyrolyze to phosphams by elimination of ammonia or hydrogen chloride.[24, 25]

$$[NP(NH_2)_2]_{3\text{ or }4} \xrightarrow{-NH_3} (NPNH)_n$$

$$N_4P_4Cl_4(NH_2)_4 \xrightarrow{-HCl} (NPNH)_n$$

Phosphams are insoluble, infusible, amorphous, gray-white solids with the general formula $(PN_2H)_n$. When pyrolyzed further they yield materials of compositions $(N_5P_3)_n$ and $(NP)_n$.[26, 27] The structures of these products are not known with certainty. It is presumed that some phosphams are made up of six- or eight-membered rings randomly linked by N—H bridges.[28]

The anilinocyclophosphazenes, $[NP(NHPh)_2]_3$ and $[NP(NHPh)_2]_4$, can also be pyrolyzed to phosphams.[29] Aniline is eliminated, presumably by processes such as

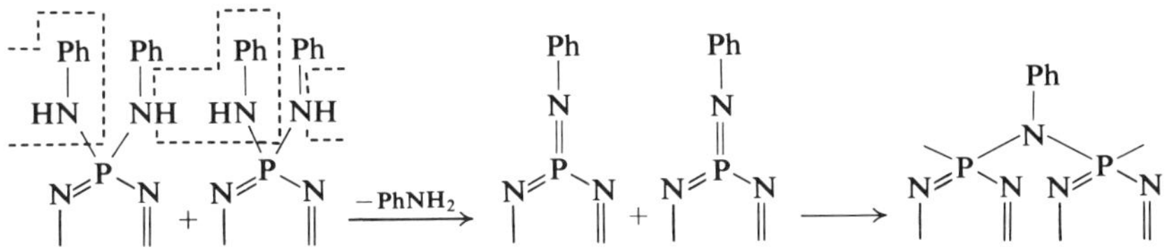

The pyrolysis products can be heated with aniline in sealed tubes to reform the original cyclophosphazene, without change in ring size.[29]

It has also been reported that the fluoroalkylaminophosphazenes, $[NP(NHCH_2CF_3)_2]_{3\text{ and }4}$, and the pyrrole derivative, $[NP(NC_4H_4)_2]_4$, decompose during 6 hours heating at 250°C.[30] The trimeric and tetrameric ethyleneiminophosphazenes $[NP(NCH_2CH_2)_2]_{3\text{ and }4}$, undergo ligand rearrangement reactions at high temperatures to yield cross-linked resinous polymers.[31] Particular instability is shown by hexakis(azido)cyclotriphosphazene, $[NP(N_3)_2]_3$, which detonates when heated or subjected to shock.[32]

3. Stability of Spirophosphazenes

Several spirophosphazenes rearrange at high temperatures to give polymeric materials. For example, tris(*o*-phenylenedioxy)cyclotriphosphazene (V) undergoes phosphazene ring opening polymerization when heated above 250°C.[33] The reaction will be discussed in Chapter 16. However, tris(2,3-dioxynaphthyl)cyclotriphosphazene (IX), and tris(3,4-dithiotolyl)cyclotriphosphazene (XI) decompose by side-group cleavage reactions in the 250°–300°C temperature range.[10] Tris(*o*-phenylenediamino)cyclotriphosphazene (X) and tris(2,2′-dioxyphenyl)cyclotriphosphazene (XII), on the other hand, are stable as solids at least up to 300°C. The stability of the latter compound suggests that steric strain in the exocyclic rings of V, IX, and XI is responsible for their instability. The implications of these results for high-polymer synthesis are discussed in Chapter 15.

4. Thermal Decomposition of Linear Phosphazenes

The phosphazene $Ph_3P{=}N{-}N{=}CPh_2$ decomposes thermally to give diphenyldiazomethane.[34] At 165°C in decalin solution, the diphenylmethyl radical can be detected by electron spin resonance techniques. Furthermore, the same phosphazene decomposes at 190°C in decalin to give benzophenone

(~5%), triphenylphosphine (~25%), tetraphenylethylene (<2%), triphenylphosphine oxide (~20%), tetraphenylethane (~20%), diphenylmethyl azine (~30%), bis(diphenylmethyl) ether (<2%), and benzhydrol (<3%). The oxygenated products result from the presence of traces of oxygen gas. A mechanism involving prior cleavage of the phosphorus–nitrogen bond has been proposed,[34] and a similar mechanism is assumed for photolytic decomposition.

C. Other Reactions

1. Halogenations

A few examples have been reported in which side groups attached to a phosphazene ring were subjected to halogenation reactions. Direct halogen addition to the phosphorus–nitrogen ring has not been reported.

Fluoroalkoxycyclophosphazenes can be chlorinated to yield α-chlorofluoroalkoxy derivatives, as illustrated by the reaction[35]

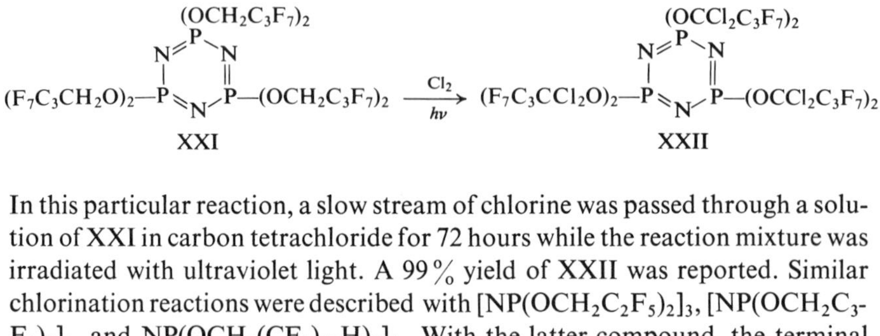

XXI XXII

In this particular reaction, a slow stream of chlorine was passed through a solution of XXI in carbon tetrachloride for 72 hours while the reaction mixture was irradiated with ultraviolet light. A 99% yield of XXII was reported. Similar chlorination reactions were described with $[NP(OCH_2C_2F_5)_2]_3$, $[NP(OCH_2C_3F_7)_2]_4$, and $NP(OCH_2(CF_2)_{10}H)_2]_4$. With the latter compound, the terminal carbon was also chlorinated.

Hexakis(allyloxy)cyclotriphosphazene, $[NP(OCH_2CH{=}CH_2)_2]_3$, has been brominated by treatment with bromine in benzene,[36] although the detailed structure of the product was not determined.

2. Cleavage of Trimethylamine by Chlorophosphazenes

Trimethylamine reacts with $(NPCl_2)_{3\text{ or }4}$ at room temperature to yield tetramethylammonium chloride and dimethylaminophosphazenes.[37] Apparently the phosphazene skeleton is unaffected by this reaction since $(NPCl_2)_3$ yields products of formula $[NPCl_{0\cdot78}(NMe_2)_{1\cdot22}]_3$. Triethylamine or pyridine do not react in this manner, and trimethylamine is almost inert to $(NPF_2)_3$.

3. The Reaction of Chlorophosphazenes with Dimethyl Sulfoxide

Hexachlorocyclotriphosphazene reacts with dimethylsulfoxide to give a moisture-sensitive chloromethyl methyl sulfide addition compound of the hydroxyoxophosphazane.[38]

$$(NPCl_2)_3 + 6\ Me_2SO \rightarrow [HN{-}P(O)OH]_3 \cdot MeSCH_2Cl + 5\ MeSCH_2Cl$$

4. Reaction of Aminophosphazenes with Phosgene

1,1-Diaminotetrachlorocyclotriphosphazene (XXIII)* reacts with phosgene

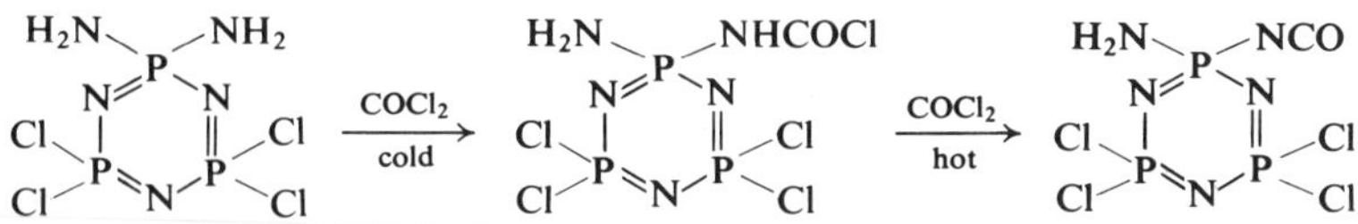

XXIII XXIV XXV

in cold chlorobenzene to yield first the carbamyl chloride (XXIV) and then, on heating, an isocyano derivative (XXV).[39] Aminopentachlorocyclotriphosphazene is also formed in this reaction. The reaction of XXV with ethanol yields a urethane adduct.

5. Reactions between Aminophosphazenes and Organic Isocyanates or Isothiocyanates

Diaminotetrakis(dimethylamino)cyclotriphosphazene (XXVI) reacts with phenyl isothiocyanate in boiling benzene to yield a monothiourea derivative (XXVII).[40]

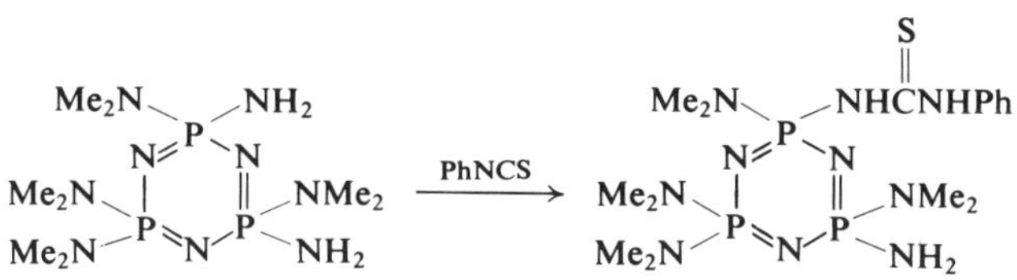

XXVI XXVII

Related products are formed when XXVI interacts with phenyl- or *n*-butyl isocyanates. Similarly, XXVI reacts with toluene-2,4-diisocyanate to yield a compound of structure XVIII.

* The original paper reported reactions with a non-geminal diamino derivative, but it seems likely from later work that the geminal isomer was actually used. This would explain the unreactivity of the second amino group.

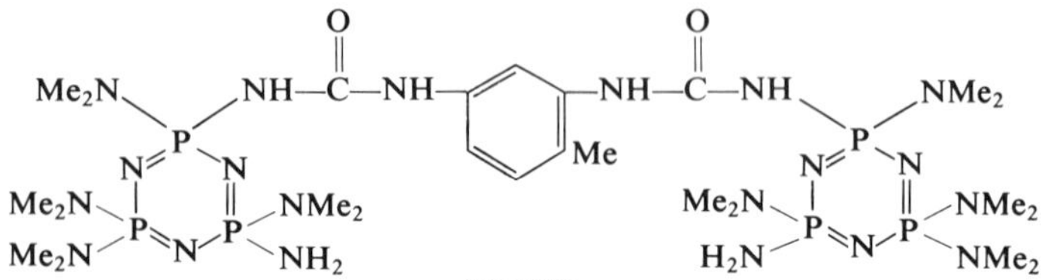

XXVIII

6. Reactions of Aminophosphazenes with Halophosphoranes

a. *Reactions with Dihalotriphenylphosphoranes*

1,1-Diamino-3,3,5,5-tetrachlorocyclotriphosphazene (XXIX) reacts with dibromotriphenylphosphorane and triethylamine in boiling carbon tetrachloride to yield an amino(triphenylphosphazenyl) derivative (XXX).[41, 42]

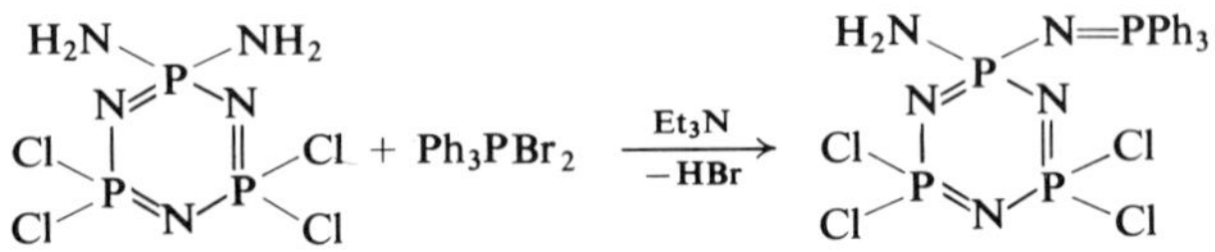

XXIX XXX

However, only one amino group undergoes reaction, even when two equivalents of phosphorane are used and, in the absence of triethylamine, only traces of XXX are formed.

With dichlorotriphenylphosphorane, substitution occurs at one amino group, but the other amino unit is replaced by a chloro group to yield XXXI.

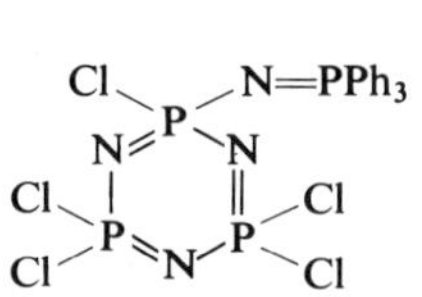

XXXI

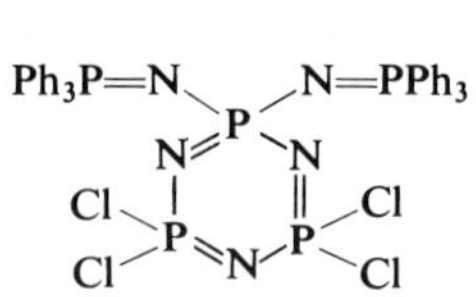

XXXII

The yield of XXXI is improved by the absence of triethylamine, but it has been established that hydrogen chloride is not responsible for the deamination process. In contrast to the reaction with dibromotriphenylphosphorane, dichlorotriphenylphosphorane reacts with XXX to yield the bis(triphenylphosphazenyl) derivative, XXXII.

Two possible mechanisms have been suggested to account for these reactions.[42] One of these requires an initial ionization of the phosphorane to give a phosphonium ion, which then acts as an electrophile toward the phosphazene amino group.

$$Ph_3PX_2 \rightleftharpoons Ph_3\overset{\oplus}{P}X + X^{\ominus}$$

$$N_3P_3Cl_4(NH_2)_2 + Ph_3\overset{\oplus}{P}X \rightarrow N_3P_3Cl_4(NH_2)\overset{\oplus}{N}H_2PPh_3X \xrightarrow{X^{\ominus}\text{ or }Et_3N}$$

$$N_3P_3Cl_4(NH_2)NHPPh_3X \xrightarrow{-X^{\ominus}}$$

$$N_3P_3Cl_4(NH_2)NHPPh_3^{\oplus} \xrightarrow{X^{\ominus}\text{ or }Et_3N} N_3P_3Cl_4(NH_2)_n{=}PPh_3$$

A second possibility is that the phosphorane does not ionize, but that the reaction proceeds via a bimolecular nucleophilic mechanism.[42] It is presumed that the lower reactivity of the second amino group results from the electron supply from the triphenylphosphazenyl group to phosphorus.

b. *Reactions with Trichlorodiphenylphosphorane*

Both amino groups in XXIX react readily with trichlorodiphenylphosphorane in boiling carbon tetrachloride to yield XXXIII, which can be treated with ammonia at room temperature to give compound XXXIV.[42] However,

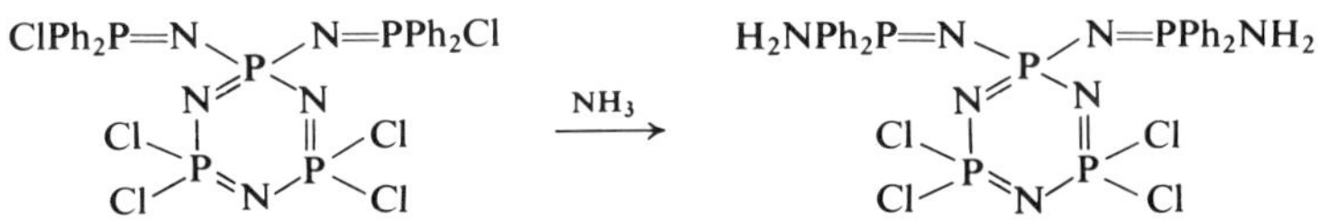

XXXIII XXXIV

the phosphazenyl side chain in XXXIV cannot be lengthened by treatment with more trichlorodiphenylphosphorane and ammonia. This result has been ascribed to the strong electron supply from the phosphazenyl unit to the amino group. However, side chain lengthening can be effected by the treatment of $N_3P_3F_5{-}N{=}PCl_3$ or $N_3P_3F_5{-}N{=}PF_3$ alternately with hexamethyldisilazane and phosphorus pentachloride[42a]

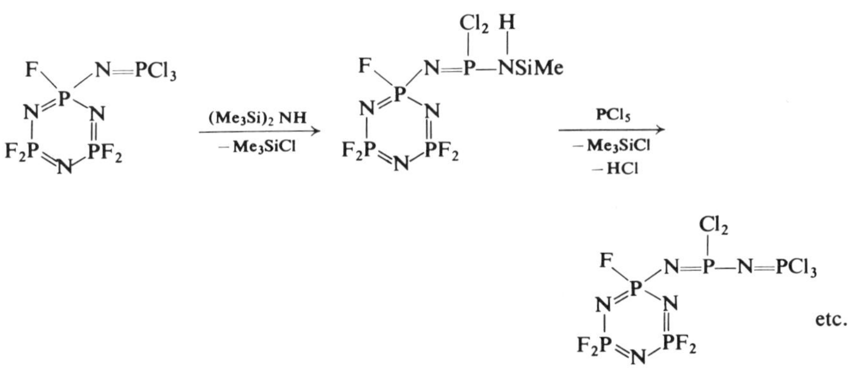

7. Reactions of Azidocyclophosphazenes

1,5-Dichlorohexaphenylcyclotetraphosphazene reacts with lithium azide in boiling acetonitrile to give *cis*- and *trans*-1,5-diazidohexaphenylcyclotetraphosphazene (XXXV). This compound reacts with triphenylphosphine at

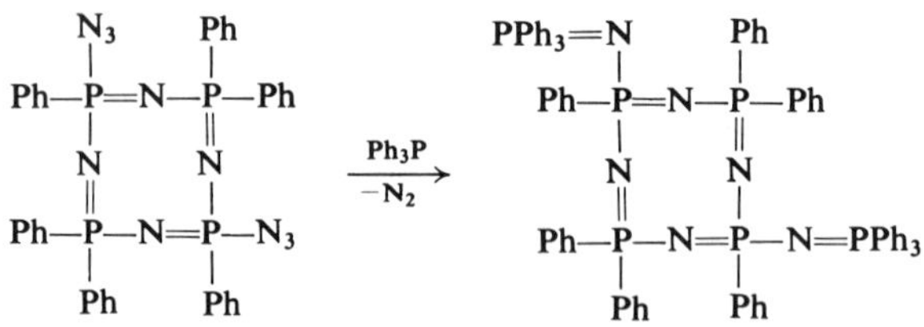

XXXV XXXVI

132°C in the melt or in chlorobenzene solution to yield a bis(triphenylphosphazenyl) derivative (XXXVI).[43] A similar reaction occurs with heptafluoropropyldiphenylphosphine. However, treatment of the *cis* isomer of XXXV with 1,4-bis(diphenylphosphino)butane, $Ph_2P(CH_2)_4PPh_2$, yielded a cage-type compound with the structure shown in XXVII.

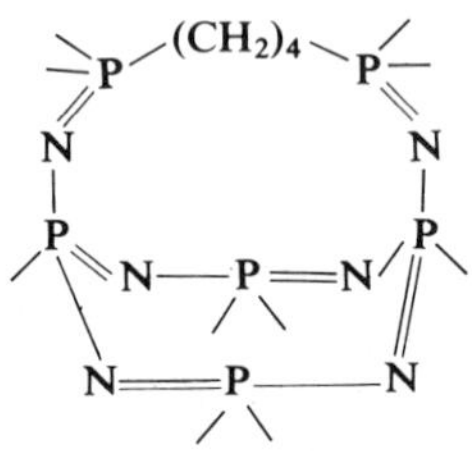

XXXVII

8. Substitution Reactions on Phenoxyphosphazenes

Several of the preceding processes have involved organic-type reactions on the side-group units of cyclophosphazenes, without interaction of the phosphazene ring. In the same vein, it is possible to perform substitution reactions on the phenyl rings of phenoxycyclophosphazenes without decomposition of the phosphazene skeleton.

Thus, hexakis(*p*-nitrophenoxy)cyclotriphosphazene, XXXVIII, can be reduced in high yield to the hexakis(*p*-aminophenoxy) derivative (XXXIX) by hydrogenation over Raney nickel in aniline.[44] This product can then be treated with phosgene in boiling *o*-dichlorobenzene to yield the hexakis(*p*-isocyanophenoxy) derivative, which can then be converted to the appropriate carbamates with methanol or 1-butanol.

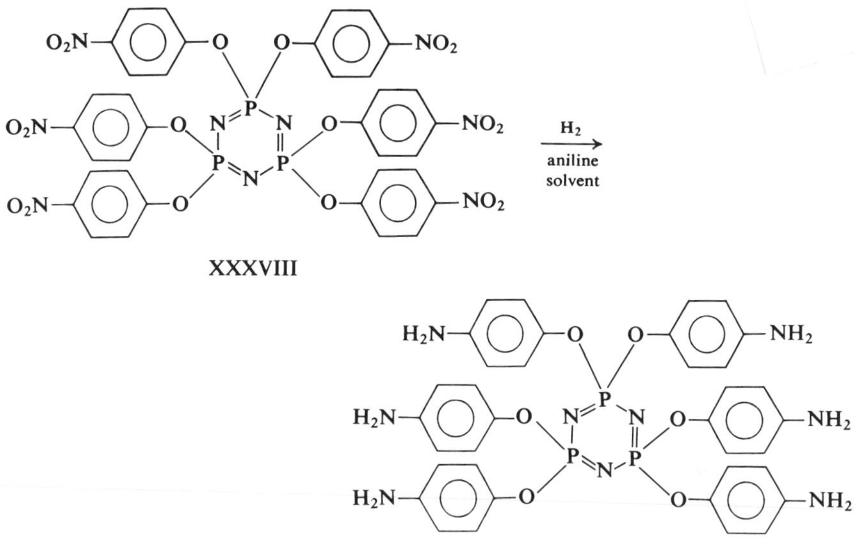

XXXVIII

XXXIX

9. Halocyclophosphazenes as Peptide Coupling Agents

Octachlorocyclotetraphosphazene, $(NPCl_2)_4$, has been employed as an amino acid coupling reagent.[45] For example, the triethylammonium salt of *N*-benzyloxycarbonyl-L-phenylalanine was treated with $(NPCl_2)_4$ in chloroform, and the methyl ester of L-alanine was added. The product was *N*-benzyloxycarbonyl-L-phenylalanyl-L-alanine (XL) in 65% yield. *N*-*t*-Butoxycarbonyl-L-tryptophyl-L-methionyl-L-aspartyl-L-phenylalanine (XLI) has been prepared by a similar procedure. Apparently the trimer $(NPCl_2)_3$ can also be used in coupling reactions of this type.

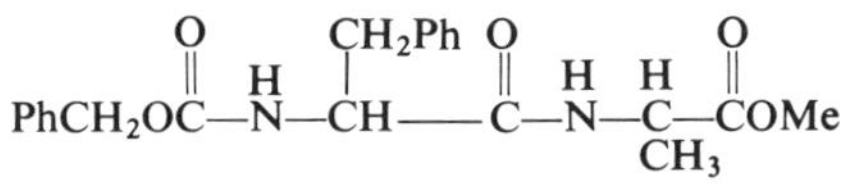

XL

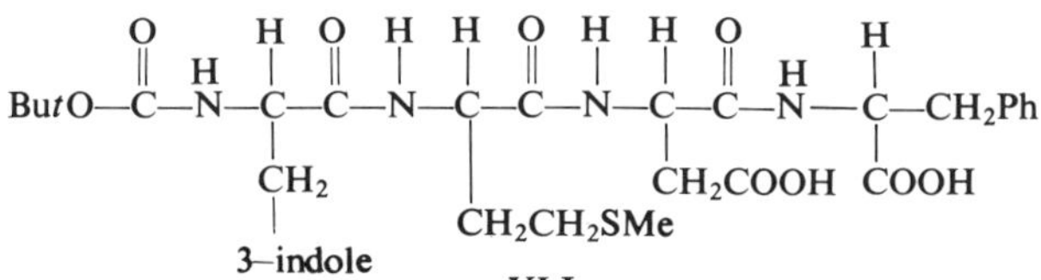

XLI

10. Chlorocyclophosphazenes as Aromatic Coupling Reagents

In the presence of chlorocyclophosphazenes, such as $(NPCl_2)_{3\text{ or }4}$, pyridine reacts with *N,N*-dialkylanilines or with indole to yield 4-substituted pyridines.[46] A cyclophosphazene pyridinum salt is postulated as an intermediate.

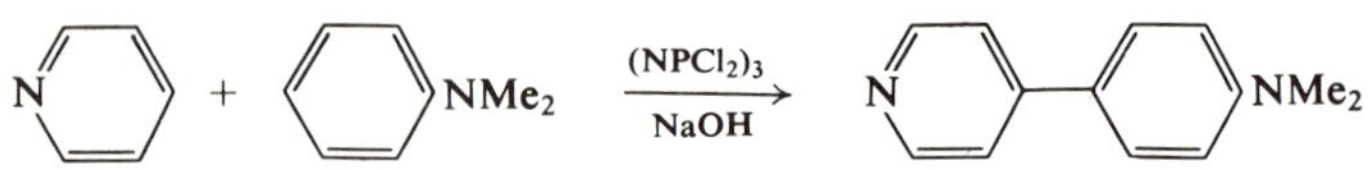

11. Reactions of Chlorophosphazenes with Dimethylformamide

Chlorocyclophosphazenes react with dimethylformamide to form "onium"-type salts.[47]

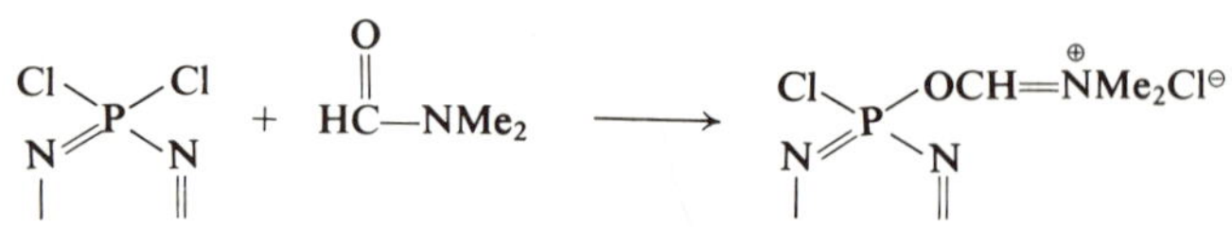

It has also been reported that $(NPCl_2)_3$ interacts with dimethylformamide in the presence of potassium iodide to yield the product shown in XLII.[48]

XLII

12. Photochemical and Electron-Induced Reactions of Chlorophosphazenes

Although films of solid $(NPCl_2)_3$ or $(NPCl_2)_4$ are unaffected after 40 hours of irradiation by the emission from a mercury arc, solutions of these phosphazenes in hexane, benzene, toluene, or decahydronaphthalene do undergo reaction.[49] The products are brown, gelatinous materials formed on the walls of the quartz reaction vessel, together with distillable, almost colorless liquids.

These latter products are organic-substituted cyclophosphazenes, possibly formed by the process[49]

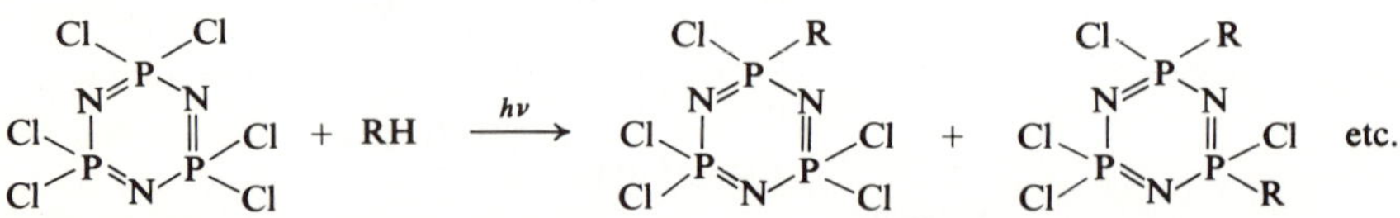

The stability of the cyclophosphazene rings during ultraviolet irradiation is noteworthy.

Octachlorocyclotetraphosphazene, $(NPCl_2)_4$, undergoes a reaction with

n-butanol under the influence of a high-energy electron beam at 30°C[50] in the presence of air. All the halogen atoms at phosphorus were replaced by butoxy groups, although the degree of polymerization of the skeleton apparently changed during the process. A mechanism for this reaction has been proposed which involves the prior formation of hydrogen radicals from the α position of butanol, followed by chlorine abstraction from the chlorophosphazene by the hydrogen radicals. Thus, the initial sequence is

$$CH_3CH_2CH_2CH_2OH \rightarrow CH_3CH_2CH_2\dot{C}HOH + H^{\bullet}$$

$$4\dot{H} + (NPCl_2)_4 \rightarrow 2\,(N\dot{P}Cl)_2 \text{ or } 4\,N\dot{P}Cl + 4\,HCl$$

Subsequent reactions of the phosphazenyl radicals with butanol could then yield butoxyphosphazene products, both oligomeric and polymeric.

It has also been noted[50] that hexachlorocyclotriphosphazene reacts with dioxane when the two are irradiated with electrons. The white, flocculent precipitate had the composition $(NPClC_4H_8O_2)_n$, but its structure was not determined.

13. Reactions of Cyclophosphazenes with Organosilicon Compounds

Hexakis(butoxy)cyclotriphosphazene reacts with triorganochlorosilanes to give butyl chloride and organosiloxycyclotriphosphazenes.[51] The triorgano-

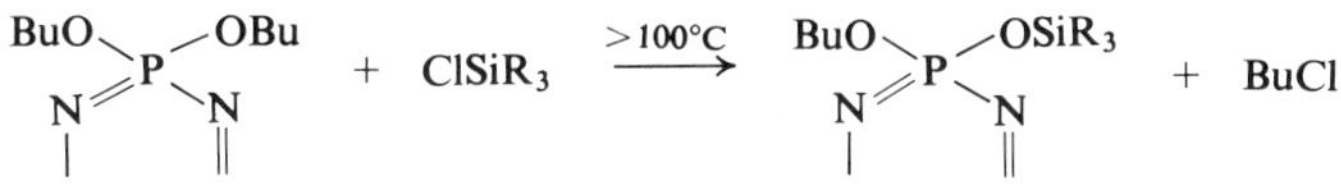

chlorosilanes used included dimethylphenylchlorosilane, methyldiphenylchlorosilane, and triphenylchlorosilane. With the first two reagents, only five or three siloxy groups, respectively, were introduced. On the basis of infrared spectra it was concluded that the initial products are phosphazenes and not phosphazanes formed by a rearrangement process (see Chapter 13). The organosiloxycyclotriphosphazenes are readily hydrolyzed in the atmosphere to form silanols or disiloxanes and hydroxycyclotriphosphazenes.

Kinetic studies have shown that the reaction of the butoxycyclophosphazene with triorganochlorosilanes is an overall first-order process, zero order with respect to the phosphazene and first order with respect to the chlorosilane.[52] A mechanism was proposed which required a prior ionization of the chlorosilane, followed by siliconium ion attack on the oxygen of a butoxy group.

$$R_3SiCl \rightleftharpoons R_3Si^+ + Cl^-$$

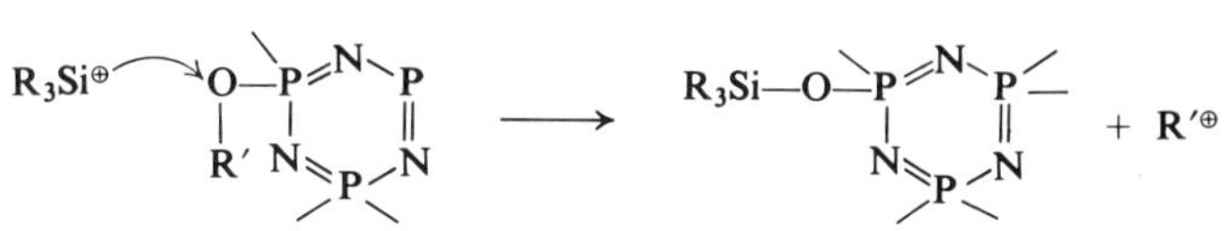

At 140°C, the order of chlorosilane reactivity is $PhMe_2SiCl > Ph_2MeSiCl > Ph_3SiCl$, whereas the reaction rates decrease with changes in alkoxy groups in the order: n-Bu $> i\text{-}C_5H_{11}O > n\text{-}C_5H_{12}O > n\text{-}C_6H_{13}O$. The reaction does not take place with fluoroalkoxycyclotriphosphazenes.[52]

Cyclolinear polymers have been prepared by the interaction of hydroxyphosphazenes and diphenyldichlorosilane.[53] These are discussed further in Chapter 16.

The reactions between (dimethylamino)trimethylsilane and fluorocyclophosphazenes have been used to prepare dimethylamino-substituted phosphazenes.[54] The general reaction sequence is illustrated by the scheme:

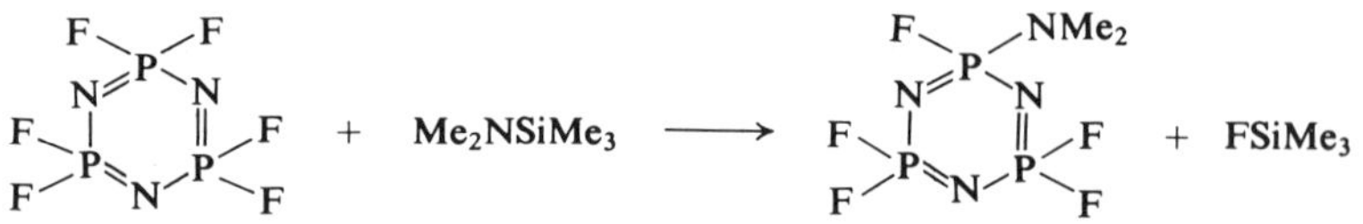

Cyclic, trimeric, tetrameric, pentameric, and hexameric fluorophosphazenes participate in this reaction, and interactions between equimolar amounts of the reagents yield monodimethylamino derivatives in all cases, bisdimethylamino compounds for $(NPF_2)_{4-6}$, and a trisdimethylamino product only for $(NPF_2)_6$. In general, the yields of dimethylaminofluorophosphazenes are higher than when dimethylamine in ether at 0°C is used as a reagent. Substitution takes place by a non-geminal pathway when both the silylamine and dimethylamine are used.

14. Reactions of Monophosphazenes and Phosphazanes with Phenyl Isocyanate

Arylmonophosphazenes react with phenyl isocyanate to yield phosphine oxides and diphenyl carbodiimide according to the reaction sequence[55]

$$R_3P{=}NPh + PhN{=}C{=}O \rightleftharpoons \begin{array}{c} PhN{=}C{-}O \\ |\quad\;\; | \\ PhN{-}PR_3 \end{array} \rightleftharpoons PhN{=}C{=}NPh + R_3P{=}O$$

The group R can be phenyl or ethyl. Monophosphazenes react 10^5–10^7 times more rapidly with phenyl isocyanate than do phosphine oxides. Furthermore, the activation energy of the monophosphazene–phenyl isocyanate reaction (9.6 kcal/mole) is independent of the substituents linked to phosphorus, in sharp contrast to reactions of phenyl isocyanate with phosphine oxides.

Monophosphazanes of formula $Ph_2P{-}NHR$ also react with isocyanates to give phosphinoureas, according to the following equation[56]

$$Ph_2P{-}NHR + PhNCO \rightarrow Ph_2P{-}NR{-}CO{-}NHPh$$

where R is methyl, ethyl, or *n*-propyl. Compounds of structure $R_2P—NHR'$ are 10^4–10^5 times more reactive in this process than are species of structure $R_2P—NR_2$. Furthermore, $(EtO)_2P—NHR$ is much more reactive than $Ph_2P—NHR$, and $R_2P—NHPr$ is more reactive than $R_2P—NHPh$. Species of structure $(EtO)_2P—NHR$ give two isomeric products, namely, $(EtO)_2P—NR—CO—NHPh$ and $(EtO)_2P—NPh—CO—NHR$.

References

1. I. I. Bezman and W. R. Reed, *J. Amer. Chem. Soc.* **82**, 2167 (1960).
2. A. B. Burg, *134th Amer. Chem. Soc. Meet.* (1958); also quoted by Bezman and Reed.[1]
3. M. Yokoyama, H. Cho, and F. Aida, *Kogyo Kagaku Zasshi* **66**, 609 (1963).
4. L. Caglioti, M. Poloni, and G. Rosini, *J. Org. Chem.* **33**, 2979 (1968).
5. H. R. Allcock, *J. Amer. Chem. Soc.* **85**, 4050 (1963).
6. H. R. Allcock, *J. Amer. Chem. Soc.* **86**, 2591 (1964).
7. H. R. Allcock and R. L. Kugel, *Chem. Commun.* p. 1606 (1968).
8. H. R. Allcock and R. L. Kugel, *J. Amer. Chem. Soc.* **91**, 5452 (1969).
9. H. R. Allcock, R. L. Kugel, and G. Moore, unpublished work (1970).
10. H. R. Allcock and R. L. Kugel, *Inorg. Chem.* **5**, 1016 (1966).
11. H. R. Allcock, R. L. Kugel, and E. J. Walsh, *Chem. Commun.* p. 1283 (1970).
12. H. R. Allcock and E. J. Walsh, *J. Amer. Chem. Soc.* **91**, 3102 (1969).
13. F. H. Westheimer, *Accounts Chem. Res.* **1**, 70 (1968).
14. F. Ramirez, *Accounts Chem. Res.* **1**, 168 (1968).
15. M. Sanchez, R. Wolf, R. Burgada, and F. Mathis, *Bull. Soc. Chim. Fr.* p. 772 (1968).
16. L. A. Siegel and J. H. Van den Hende, *J. Chem. Soc., London* p. 817 (1967).
17. B. W. Fitzsimmons, C. Hewlett, and R. A. Shaw, *Proc. Chem. Soc., London* p. 340 (1962).
18. B. W. Fitzsimmons, C. Hewlett, and R. A. Shaw, *J. Chem. Soc., London* p. 4799 (1965).
19. E. T. McBee, H. R. Allcock, R. Caputo, A. Kalmus, and C. W. Roberts, *U.S. Govt. Res. Rep.* **AD 209-669** (1958).
20. R. Rätz, H. Schroeder, H. Ulrich, E. Kober, and C. Grundmann, *J. Amer. Chem. Soc.* **84**, 551 (1962).
21. H. R. Allcock and R. J. Best, *Can. J. Chem.* **42**, 447 (1964).
22. H. R. Allcock and A. H. Struck, unpublished results (1964).
23. H. R. Allcock, G. Konopski, and R. W. Tomlinson, unpublished work (1968).
24. A. M. de Ficquelmont, *C. R. Acad. Sci.* **200**, 1045 (1935).
25. M. C. Miller and R. A. Shaw, quoted by R. A. Shaw, B. W. Fitzsimmons, and B. C. Smith, *Chem. Rev.* **63**, 247 (1963).
26. A. M. de Ficquelmont, *Ann. Chim. (Paris)* [11] **12**, 169 (1939).
27. H. Moureau and P. Rocquet, *C. R. Acad. Sci.* **198**, 1691 (1934).
28. M. C. Miller and R. A. Shaw, *J. Chem. Soc., London* p. 3233 (1963).
29. H. Bode and H. Clausen, *Z. Anorg. Allg. Chem.* **258**, 99 (1949).
30. E. T. McBee, P. Johncock, S. French, and H. P. Braendlin, *U.S. Govt. Res. Rep.* **AD 254,985** (1961).
31. R. Rätz, E. Kober, C. Grundmann, and G. Ottmann, *Inorg. Chem.* **3**, 757 (1964).
32. C. Grundmann and R. Rätz, *Z. Naturforsch. B* **10**, 116 (1955).
33. H. R. Allcock, *Inorg. Chem.* **5**, 1320 (1966).
34. D. R. Dalton and S. A. Liebman, *Tetrahedron* **25**, 3321 (1969).
35. R. Rätz, U.S. Pat. 2,876,248 (1959) (to Olin Mathieson).

36. C. Hamalainen and J. D. Guthrie, *Text. Res. J.* **26**, 141 (1956).
37. A. B. Burg and A. P. Caron, *J. Amer. Chem. Soc.* **81**, 836 (1959).
38. S. K. Ray, R. A. Shaw, and B. C. Smith, *Nature* (*London*) **196**, 372 (1962).
39. G. Tesi and R. Zimmer-Galler, *Chem. Ind.* (*London*) p. 1916 (1964).
40. G. Tesi, A. J. Matuszko, R. Zimmer-Galler, and M. S. Chang, *Chem. Ind.* (*London*) p. 623 (1964).
41. R. Keat, M. C. Miller, and R. A. Shaw, *Proc. Chem. Soc., London* p. 137 (1964).
42. R. Keat, M. C. Miller, and R. A. Shaw, *J. Chem. Soc., A* p. 1404 (1967).
42a. H. W. Roesky, W. G. Boewing, and E. Niecke, *Chem. Ber.* **104**, 653 (1971).
43. C. M. Sharts, A. J. Bilbo, and D. R. Gentry, *Inorg. Chem.* **5**, 2140 (1966).
44. G. Ottmann, H. F. Lederle, H. Hooks, and E. Kober, *Inorg. Chem.* **6**, 394 (1967).
45. K. C. Das, Y.-Y. Lin, and B. Weinstein, *Experientia* **15**, 1238 (1969).
46. B. I. Stepanov and G. I. Migachev, *J. Gen. Chem. USSR* **35**, 2245 (1965).
47. B. I. Stepanov and G. I. Migachev, *J. Gen. Chem. USSR* **38**, 195 (1968).
48. H. Hartl, H. Pritzkow, and J. Jander, *Chem. Ber.* **103**, 652 (1970).
49. B. R. Dishon and Y. Hirshberg, *J. Polym. Sci.* **4**, 75 (1949).
50. V. I. Spitsyn, N. A. Afanas'eva, A. K. Pikaev, I. D. Kolli, and P. Ya. Glazunov, *Proc. Acad. Sci. USSR, Chem. Sect.* **131**, 387 (1960).
51. S. I. Belykh, S. M. Zhivukhin, V. V. Kireev, and G. S. Kolesnikov, *Russ. J. Inorg. Chem.* **14**, 668 (1969).
52. S. I. Belykh, S. M. Zhivukhin, V. V. Kireev, and G. S. Kolesnikov, *J. Gen. Chem. USSR* **39**, 761 (1969).
53. S. I. Belykh, S. M. Zhivukhin, V. V. Kireev, and G. S. Kolesnikov, *Vysokomol. Soedin., Ser. A* **11**, 625 (1969).
54. T. Chivers and N. L. Paddock, *Chem. Commun.* p. 337 (1969).
55. G. Akones and P. Froyen, *Acta Chem. Scand.* **23**, 2697 (1969).
56. R. F. Hudson and A. Mancuso, *Chem. Commun.* p. 522 (1969).

Part III

POLYMER CHEMISTRY

Chapter 15

POLYMERIZATION AND DEPOLYMERIZATION

A. Introduction

Certain cyclophosphazenes polymerize to linear or macrocyclic polymers at elevated temperatures. The overall reaction scheme for this transformation is illustrated by the scheme

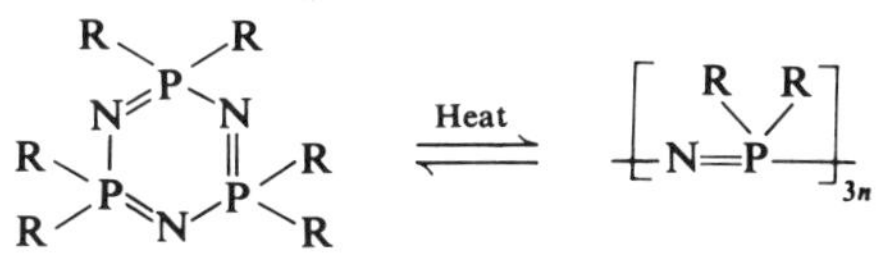

Polymerizations of this type are known when R is F, Cl, Br, or NCS, and, in a few cases, when R is an organic group. Polymerizations have been reported for cyclic trimers, tetramers, pentamers, and higher cyclic species, and the reactions have been carried out in the melt, in solution, and in the solid state. The rate of polymerization can be accelerated by catalysts, and it is also known that high-energy radiation induces polymerization of cyclophosphazenes in the crystalline state.

The high polymers formed by these and other reactions can be depolymerized to cyclic oligomers at elevated temperatures. For example, although hexachlorocyclotriphosphazene, $(NPCl_2)_3$, polymerizes at 250°C to yield a transparent, rubbery polymer, heating of the polymer to 350°C induces depolymerization to cyclic trimer, tetramer, and other cyclic oligomers. The term "depolymerization" is used here to describe the breakdown of a polymer to lower cyclic or linear homologues, whereas the term "thermal decomposition"

$$(NPCl_2)_n \xrightarrow{350°C} (NPCl_2)_3 + (NPCl_2)_4 + (NPCl_2)_5 \quad \text{etc.}$$

will be used when general fragmentation occurs. There is now convincing evidence that polymerization and depolymerization of chlorophosphazenes are different aspects of a general equilibration process. This subject will be considered in detail later. However, it is appropriate to note here the similarities between the equilibration reactions of phosphazenes and those of organosiloxanes (silicones), polyphosphates, polyoxymethylenes, and other heteroatom polymer systems.[1, 2]

The polymerization–depolymerization equilibria which occur with polyphosphazenes have an important bearing on the synthesis and thermal stability of phosphazene high polymers. Thus, every high molecular weight linear-type polyphosphazene known is synthesized by polymerization of a cyclic oligomer followed, in many cases, by subsequent substitution reactions on that polymer. Furthermore, the thermal breakdown of both halo- and organophosphazene high polymers frequently takes place by a depolymerization process rather than by general fragmentation. Thus, this subject is of interest not only from the theoretical and mechanistic points of view, but also because of its importance in high-temperature polymer technology.

In the following sections the influence of the variety of factors on phosphazene polymerization and depolymerization will be considered, including uncatalyzed equilibrations, catalysts, the influence of different ligands, ring size, temperature, pressure, reaction phase, copolymerization, and reaction mechanisms. However, in order to fully appreciate the influence of these factors it is necessary to have some understanding of the basic theories of ring–polymer equilibration. This aspect is considered in the following section.

B. Theories of Ring–Polymer Equilibria

An examination of the available data for phosphazene polymerization and depolymerization reactions raises several important questions. For example, why do some cyclophosphazenes (principally halo- and isothiocyanophosphazenes) polymerize thermally, whereas others (most organophosphazenes) do not? Why do polyphosphazenes with different substituents attached to phosphorus depolymerize at different temperatures? Which factors determine whether polymerization or depolymerization predominates at a particular temperature? Theories of ring–polymer equilibria provide partial answers to these questions. Two detailed analyses of ring–polymer equilibria in heteroatom systems have been published,[1, 2] and only a brief outline will be presented here.

1. Thermodynamic Considerations

Assume that polymerization of a cyclophosphazene and depolymerization of a high molecular weight polyphosphazene are both manifestations of a general equilibration process involving every homologue in the series. Thus, if a cyclic trimer were heated, it would undergo equilibration to cyclic tetramer, pentamer, hexamer, heptamer, and all the other homologues up to the high polymers. Similarly, equilibration of the high polymer would yield the same mixture of homologues, and each homologue would be in equilibrium with every other species. This situation would exist if each homologue in the series had a similar free energy.

If, on the other hand, one or a few homologues had a lower free energy than the others, then these more stable species would be present in greater amounts at equilibrium. It is conceivable that, in an extreme case, only one or two species would be stable and the remainder of the series may be absent at equilibrium. From a practical point of view it is necessary to calculate or guess which structural factors will raise or lower the free energy of a particular homologue relative to the others in the series. Clearly, any factor which raises the entropy or lowers the enthalpy of one homologue will favor the formation of that species at equilibrium. For example, polymerization of a trimer or tetramer to a high polymer has the effect of reducing the translational entropy but increasing the torsional and vibrational entropy in the nonrigid chain. The former factor is presumed to predominate in most cases. The enthalpy change will depend, among other things, on the nature of the substituents. The usual equation, $\Delta G = \Delta H - T\Delta S$, can be used to predict the influence of various structural features on the free energy of an equilibration process. However, in practice, polymerization is often favored at low temperatures where the $T\Delta S$ term is small, and depolymerization predominates at high temperatures where the $T\Delta S$ term is large. We shall see that this situation appears to occur with chlorophosphazene equilibrations.

In some polymer systems, the gradual shift of the equilibrium to favor low molecular weight species as the temperature is raised is broken by a sharp cutoff temperature above which no high polymer can exist. This is known as the *ceiling temperature*. A ceiling temperature results when the $T\Delta S$ term increases rapidly enough to overtake the ΔH term sharply at this temperature. However, other polymers exhibit a *floor temperature* below which polymer cannot be found. Tobolsky[3–8] and Gee[9] have developed derivations which aid in predicting the behavior of polymer equilibria at different temperatures. Thus, Gee's treatment suggests that, within specific concentration limits, a ceiling temperature will exist if both ΔH and ΔS for polymerization are both negative, and a floor temperature will be found if both ΔH and ΔS are positive. These, and other possibilities, are shown in Table 15.1. The difficulty of applying these

TABLE 15.1

PREDICTED INFLUENCE OF ΔH AND ΔS ON POLYMER CONCENTRATION AT EQUILIBRIUM[a]

ΔH	ΔS	Expected effect
<0	<0	Polymer concentration decreases with temperature and a ceiling temperature can exist.
<0	0 or >0	Polymer can exist at any temperature.
0	>0	Polymer concentration is independent of temperature.
0	0 or <0	Only cyclic oligomers can exist.
>0	>0	Polymer concentration increases with temperature, but no polymer can exist below a floor temperature.
>0	0 or <0	No polymer can exist at any temperature.

[a] From Gee.[9]

concepts to phosphazene chemistry is that relatively little experimental information is available about ΔH or ΔS values for polymerization.

2. STATISTICAL CONSIDERATIONS

Two statistical approaches to ring–polymer equilibria have been developed. In the first, the concentration of a cyclic oligomer in equilibrium with polymer is assumed to be determined by the probability of ring closure.[10] For example, there is a greater probability that linear trimers or tetramers will cyclize than that a high polymer will cyclize. In the former case, there is a higher probability that the chain ends will approach close enough for bond formation to occur. Thus, *cyclic* oligomers are likely to be found along with *linear* polymers, the relative proportions being dependent on the skeletal bond angles, the steric properties of the side groups, and on special factors such as stabilization of specific homologues by π-electron delocalization. In a generalized case, the proportions of each species formed during polymerization would depend on the ring-closure statistics of the following process (where AB is a hypothetical monomer unit):

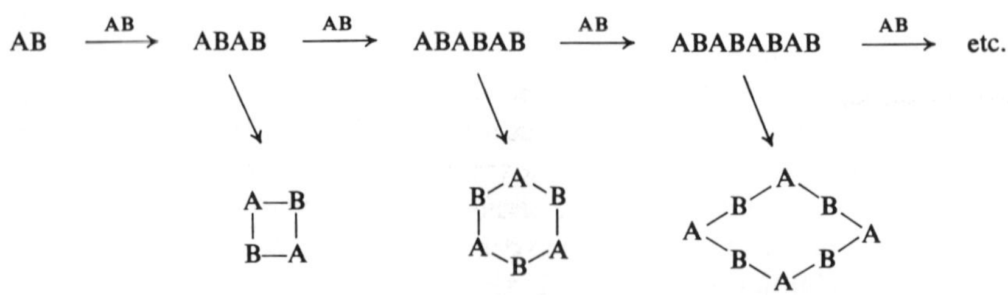

Similarly, the formation of cyclic species during depolymerization would depend on the statistics of "back-biting" reactions which split out the various homologues, or on the reverse process to that shown above.

It should be noted that, for phosphazenes, no stable monomer of structure $N{\equiv}PR_2$ has been reported, although the existence of such species as intermediates in equilibration processes cannot be discounted. Cyclic dimers of structure $(NPR_2)_2$ are assumed to play an insignificant role in phosphazene equilibration reactions because of their presumed instability. Nevertheless, cyclization statistics could play an important part in phosphazene equilibrations even if whole trimer or tetramer units are added or removed at each step. The important point to note is that, if cyclization statistics are important for these reactions, the concentrations of different homologues present at equilibrium will depend on the skeletal bond angles and perhaps also on the minimum energy conformations of the linear species.

A second statistical approach is one which has been widely used by Van Wazer and Moedritzer,[11–14] for other inorganic and organometallic polymer systems. It views equilibration as a random scrambling of chain-end units (end-capping units) and middle units to form chains or rings. In a phosphazene equilibration system, this would constitute a scrambling process between monomer units, NPR_2, and end-capping units, E, as shown:

$$n\,NPR_2 + mE$$

$$\swarrow \qquad\qquad \searrow$$

$$\underset{\text{linear}}{E{-}(NPR_2)_n{-}E} \qquad\qquad \underset{\text{cyclic}}{(NPR_2)_n} + \frac{m}{2}\,E_2$$

Qualitative information is available about the effects of end-capping species such as phosphorus pentachloride or metal halides (see Chapter 16), and there seems little doubt that scrambling statistics play a part in determining the proportions of linear and cyclic homologues, at least for the lower molecular weight fractions.

3. Molecular Structural Effects

For phosphazene equilibrations, the amounts of various homologues formed will depend on the presence or absence of appreciable ring strain and on the steric attributes of the ligands attached to phosphorus. Arguments have been made for an influence by pseudo-aromaticity in favoring the formation of specific homologues but for phosphazenes this is considered to be a minor factor. Skeletal bond energy effects probably exert only a minimal influence on phosphazene equilibrations since no striking differences are expected between the values for cyclic species or linear polymers.

The effect of ring strain on equilibration was alluded to earlier. The apparent absence of cyclic dimeric phosphazenes can be attributed to this factor. In most phosphazenes, the skeletal bond angle at phosphorus approximates to 120°, whereas the angle at nitrogen varies from ~119° to 148° depending on the substituents attached to phosphorus. However, the formation of a cyclic dimer would require an appreciable distortion of these angles from the preferred values, and the resultant enthalpy increase is apparently sufficient to eliminate the dimer as a significant contributor to the equilibrium. Minor differences are expected between the strain enthalpies of cyclic trimers, tetramers, pentamers, hexamers, heptamers, and other low molecular weight cyclic species. However,

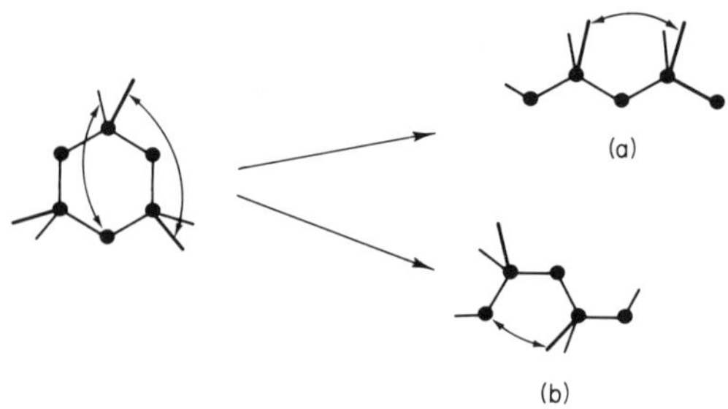

FIG. 15.1. Decreased intramolecular distances resulting from ring-opening polymerization.

the enhanced ring flexibility which accompanies increased ring size should provide a mechanism for the release of strain in species higher than the cyclic pentamer or hexamer.

Side-group steric interactions apparently play a dominant role in determining the equilibration behavior of phosphazenes[15] and other heteroatom polymer systems.[16] Consider the conversion of a cyclic trimer to a linear polymer, as illustrated in Fig. 15.1. Irrespective of the preferred conformation of the polymer, polymerization has the effect of decreasing the distance between side groups on adjacent phosphorus atoms or between side groups and nearby skeletal atoms. When the side group is small, such as fluoro, the intramolecular repulsions in the oligomer and polymer will be comparable. However, if bulky side groups are present, such as phenoxy or heptafluorobutoxy, polymerization of the trimer or tetramer will be accompanied by an enthalpy increase resulting from intramolecular repulsions within the polymer. Polymer formation will then be thermodynamically inhibited. Conversely, a polymer with bulky substituents should be susceptible to depolymerization to cyclic oligomers. Experimental evidence in favor of this effect will be discussed later.

4. Mechanistic Influences

It is quite possible for a polymerization or depolymerization to be thermodynamically feasible, yet kinetically inhibited. The activation energy for the process may be so high that equilibration does not take place except at high temperatures. If the depolymerization process is thermodynamically favored at high temperatures, and the activation energy for the equilibration steps is high, then the polymerization of a cyclic oligomer will be impossible. On the other hand, if catalysts can be found for the equilibration, it may be possible to bring about polymerization at a lower temperature. Thus, the search for catalysts for phosphazene polymerizations has become a subject of some importance. The mechanistic aspects of phosphazene equilibration will be considered in detail later (Section E).

C. General Features of Hexachlorocyclotriphosphazene Polymerization

The thermal polymerization of hexachlorocyclotriphosphazene, I, to high molecular weight poly(dichlorophosphazene), II, has been subjected to intensive study. For this reason, the general characteristics of this reaction will be considered here to provide a basis for the discussion of related polymerizations.

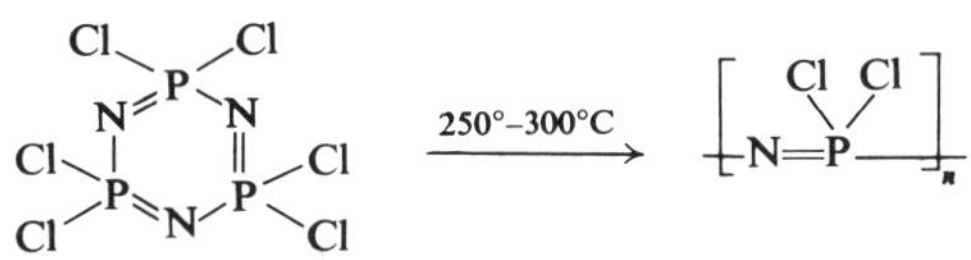

I II

1. Typical Experimental Procedure[17, 18]

The thermal polymerization of I on a laboratory scale is conducted with the use of evacuated glass tubes. A critical step in the procedure is the purification of the trimeric starting material. Hexachlorocyclotriphosphazene (I) (1000 g) (obtained commercially, or prepared as described in Chapter 4) is purified by recrystallization from warm *n*-heptane (1500 ml) at a temperature below 75°C, after decoloration with activated carbon. The first crop of crystals (~621 g) obtained after cooling to room temperature is filtered off and dried *in vacuo* or in air (m.p. 108°–113°C). Recrystallized I (200 g) is placed in a ~32.5 × 3.5 cm constricted Pyrex tube. The tube is evacuated for 30 minutes at a pressure of 0.1 mm and then isolated from the vacuum line by closing of a stopcock. The trimer is then melted and allowed to resolidify. The stopcock to the vacuum line is opened and evacuation is continued for 30 minutes. This degassing procedure may be repeated if desired. The tube is sealed at the constriction and then immersed in a constant-temperature Woods metal bath, or placed in a thermoregulated oven at 250°C for 48 hr. When the tube is cooled to 25°C, the polymer is a transparent immobile rubber in which crystallization of residual trimer and other oligomers occurs slowly. The polymer is un-cross-linked and is soluble in benzene, toluene, or tetrahydrofuran.

Several modifications to this procedure are possible. First, changes can be effected in the trimer purification step. Commercial hexachlorocyclotriphosphazene often contains appreciable quantities of the tetramer (up to ~40% tetramer in some instances) and small amounts of phosphorus pentachloride and linear phosphazenes. The latter two components may inhibit the polymerization, and they can be carried through the recrystallization step if the solvent temperature is too high. Octachlorotetraphosphazene polymerizes at a much slower rate than the trimer. Thus, starting material which contains appreciable concentrations of tetramer may require 4 to 7 days of polymerization time.[18] Care must also be taken to remove occluded solvent and air from the trimer, and this can be effected by a sublimation step. The polymerization time can be reduced appreciably by raising the temperature above 250°C. For example, at 300°C, a 50% conversion to polymer occurs in 1 hour. Temperature fluctuations during polymerization appear to favor the formation of cross-linked polymer. The properties and reactions of poly(dichlorophosphazene) are reviewed in Chapter 16.

2. The Cross-linking Reaction

The polymerization of hexachlorocyclotriphosphazene actually yields two different high polymeric products—a linear or branched polymer, and a cross-linked product. The latter is swelled by many organic solvents but it does not dissolve. The cross-linked polymer is usually isolated after long polymerization times, and often when the conversion to linear polymer has exceeded ~75–80%. Thus, it seems likely that the cross-linked polymer is derived from the un-cross-linked material rather than being a product formed directly during polymerization. This distinction between the two classes of chlorophosphazene high polymers was not recognized until 1965,[19] and much of the early work on the equilibration reaction is open to question, since conversion of the trimer or tetramer to cross-linked polymer was the only reaction studied. From a practical point of view, it should be noted that only the un-cross-linked polymer is suitable for subsequent substitution reactions[17] (see Chapter 16).

3. Kinetics and Temperature Effects

Below 230°C, the uncatalyzed bulk polymerization of hexachlorocyclotriphosphazene is extremely slow in sealed glass tubes. At 211°C the maximum initial rates are of the order of 3% conversion per day.[20] Above 250°C, the polymerization rate increases rapidly with temperature.[17, 20, 21] For example, at 250°C, a 30% conversion to polymer occurs in 21 hours, whereas at 300°C a similar conversion takes place in 0.5 hour.[17] However, the rate of cross-linking also increases at a comparable rate as the temperature is raised.[17]

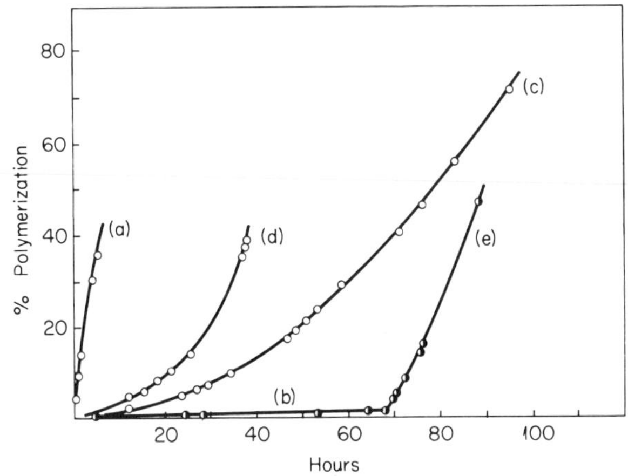

Fig. 15.2. Conversion–time curves for thermal bulk polymerization of hexachlorocyclotriphosphazene. (a) Semipure trimer at 270°C. (b) Purified trimer at 300°C. (c) Purified trimer at 320°C. (d) Purified trimer at 340°C. (e) Purified trimer at 300°C after addition of dry oxygen. From Colclough and Gee.[27]

The "uncatalyzed" polymerization of hexachlorocyclotriphosphazene has been studied kinetically by a number of investigators.[20–27] The earliest quantitative work by Patat and co-workers[22, 24] illustrated the practical difficulties involved in following this polymerization reaction, where the products include cyclic oligomers, un-cross-linked polymers and cross-linked products. A second-order process, with an activation energy of 42 kcal/mole, was reported for reactions at 300°C.[24] Similar results were described by Chakrabartty and Ghosh,[25] but the activation energy was 51.34 kcal/mole. MacCallum and Werninck[26] followed the polymerization kinetics in the temperature range of 240°–255°C and confirmed the second-order kinetics, but found an activation energy of 57 kcal/mole. Finally, Colclough and Gee[27] examined the polymerization of rigorously purified trimer at 270°C and found the rate to be considerably slower than when the vacuum-sublimed material used by other workers was employed. Even at 300°C, less than 2% of the highly purified material was polymerized after 60 hours. These results confirm the observations of other investigators[17, 20] that minute traces of impurities function as powerful polymerization catalysts. Indeed, evidence exists that at 320° and 340°C a catalytic species is formed from the purified trimer as the reaction proceeds.[27] It is interesting to note that dry oxygen gas and other materials function as catalysts (see Section C.6). Figure 15.2 illustrates the effects of temperature and degree of purification on the reaction rate.

4. Thermal Depolymerization

The thermal depolymerization of high molecular weight poly(dichlorophosphazene) has been investigated with two different types of conditions, either by continuous volatilization of the depolymerization products from the heated polymer, or by examination of the products formed after equilibration in a closed system. Nonequilibrium methods will be considered here, and equilibrations are reviewed in the following section. Four aspects of the nonequilibrium depolymerization process are of interest: (1) the qualitative facts and evidence that depolymerization does occur, (2) the molecular weight distribution of the depolymerization products, (3) the kinetics of depolymerization, and (4) the differences in depolymerization behavior of polymers prepared by different techniques. These aspects will be considered in turn.

Stokes, who was the first to report the polymerization of hexachlorocyclotriphosphazene, also noted that rapid heating of the high polymer to 350°C brought about partial depolymerization to the trimer.[28] These observations were confirmed by Schenk and Römer[29] and by de Ficquelmont.[30] Total depolymerization was effected at higher temperatures.[28] Patat and Frömbling[24] also observed the formation of cyclic oligomers when the polymer was heated in vacuum above 350°C. Approximately two-thirds of the product consisted of trimer and tetramer, and one-third of pentamer through heptamer. Patat and Derst,[31] Gimblett,[32] Yokoyama and Konya,[33] Allcock and Cook,[34] and MacCallum, Tanner, and Werninck[35, 35a] have also examined this depolymerization, although the reported results vary considerably according to the mode of preparation and purification of polymer and the depolymerization technique.

Considerable care must be exercised in the interpretation of depolymerization data. Poly(dichlorophosphazene) usually contains small amounts of trimer, tetramer, and other cyclic oligomers, either as unreacted starting materials or as equilibration products formed during polymerization. Thus, the polymer must be freed from these compounds prior to depolymerization in order to avoid misleading results. Some of the early work recorded in the literature is suspect because these precautions were not taken, and abnormally high proportions of trimer as a product were recorded. The oligomers can be removed by extraction of the cross-linked polymer with benzene or toluene, or by precipitation of the soluble (un-cross-linked) polymer from benzene solution into pentane or heptane. If this precaution is taken, it can be shown by vapor-phase chromatography techniques that the products of nonequilibrium depolymerization at 300°C are $(NPCl_2)_3$, $(NPCl_2)_4$, $(NPCl_2)_5$, $(NPCl_2)_6$, and higher cyclic species, with the tetramer predominating.[34] A small residue of nondepolymerizable polymer is often left after the oligomeric products have

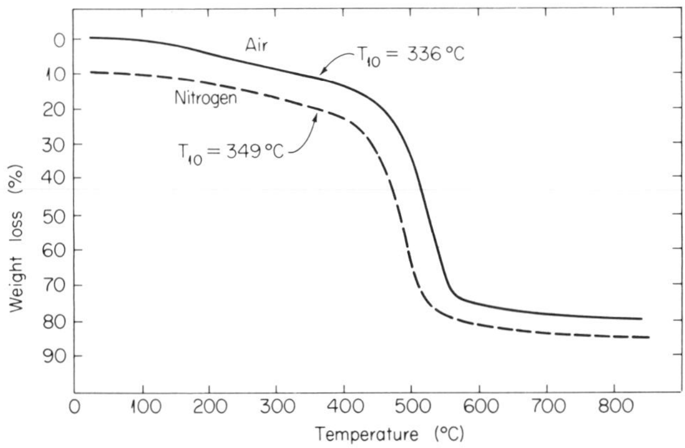

FIG. 15.3. Thermogravimetric analysis curves for $(NPCl_2)_n$.

been removed. This residue may represent highly cross-linked fractions formed during the heating process. Infrared spectra of these residues resemble those of phospham or phosphorus nitride.[33]

A number of investigators have examined the depolymerization process with the use of thermogravimetric analysis (TGA) and differential thermal analysis (DTA) techniques.[31, 33, 35, 36] Thermogravimetric analysis follows the weight loss which occurs with time at a constant rate of heating. Figure 15.3 shows typical TGA curves for $(NPCl_2)_n$ in air and nitrogen.[36] Considerable caution must be exercised in the interpretation of TGA curves. For example, the weight loss observed may merely reflect the volatilization temperatures of oligomers formed by depolymerization at lower temperatures. The characteristics of the curve may also vary with different heating rates. Thus, although depolymerization can be demonstrated by TGA experiments, more detailed data can only rarely be obtained by this technique.

The same problem complicates studies of the rates of depolymerization followed by weight loss at constant temperature. Here again, the rate of weight loss may represent the rate of sublimation or distillation of depolymerization products rather than the rate of depolymerization. Patat and Derst[31] attempted to overcome this problem by following the weight loss from thin films of $(NPCl_2)_n$ adhering to quartz sand. The first-order rate curves corrected for the presence of residual trimer in the polymer are shown in Fig. 15.4. Activation energy values of 26 ± 2 kcal/mole, 25 ± 2 kcal/mole, and 22.5 ± 2 kcal/mole[37] have been reported for this process.

In any discussion of the depolymerization of poly(dichlorophosphazene), it

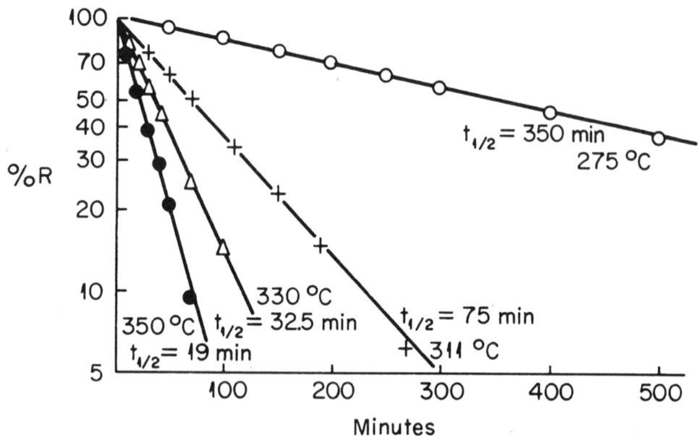

FIG. 15.4. Corrected depolymerization curves for $(NPCl_2)_n$. From Patat and Derst.[31]

is important to note that a variety of polymers with different structures exist, and that each of these species probably shows different depolymerization characteristics. For example, thermally cross-linked polymers, un-cross-linked polymers, and polymers produced by catalytic processes, would be expected to differ slightly in their depolymerization behavior. Added complications are introduced by the fact that the polymers may react with glass or quartz reaction vessels at high temperatures. Thus, careful quantitative work needs to be obtained before sound mechanistic arguments can be formulated.

5. THERMAL EQUILIBRATION

The foregoing evidence that both polymerization and depolymerization take place with chlorophosphazenes within the same temperature range suggests that both of these processes are manifestations of a general equilibration sequence. It could be argued that polymerization of the trimer and tetramer at 250°–300°C involves the establishment of an equilibrium in which substantial concentrations of high polymers are in equilibrium with small amounts of low molecular weight cyclic oligomers. Only in the later stages of polymerization may the cross-linking process force the equilibrium totally toward the high polymer. Furthermore, the almost complete conversion of the high polymer to cyclic oligomers at elevated temperatures may be a consequence of the rapid removal of the more volatile oligomers as they are formed. This continuous volatilization of oligomers would require constant reformation of oligomers from polymers in an attempt to reestablish the equilibrium, and the final effect would be a total conversion of high polymers to oligomers. These hypotheses can only be tested by experiments with reactions in closed systems of known volume, at specific temperatures and pressures.

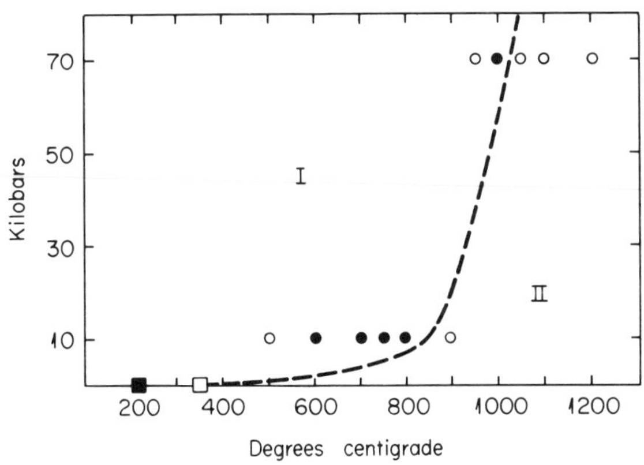

FIG. 15.5. Equilibrium of $(NPCl_2)_3$ polymerization to high polymer: ●, over 90% conversion to high polymer; ○, depolymerization in Region II, incomplete polymerization (due to slow rate) in Region I (1-atm results indicated on abscissa); ■, over 90% conversion to high polymer, data of Stokes[28] and Konecny *et al.*,[42] □, depolymerization, data of Stokes.[28] Region I: high polymer favored, region II: depolymerization occurs. From Soulen and Silverman.[40]

In 1939, Schmitz-DuMont[38, 39] conducted a series of experiments in which 1 g samples of chlorophosphazenes were heated in a 100 ml quartz tube at 600°C. The same mixture of trimer, tetramer, and higher polymers was formed when either of these homologues was used as a starting material. An increase in the concentration of trimer or tetramer, or an increase in pressure increased the amount of higher polymers in the mixture, but elevation of the temperature decreased the proportion of higher polymers. Thus, the characteristics of an equilibrium were observed, with the polymerization step being exothermic.

Further work at high pressures and high temperatures was carried out by Soulen and Silverman.[40] Pressures between 10 and 70 kbar were used with temperatures up to 1200°C. The equilibrium diagram shown in Fig. 15.5 was obtained. Increased pressure favored the formation of high polymer. This effect was consistent with the measured change in density from 52.2 ml/mole $NPCl_2$ in $(NPCl_2)_3$ to 49.5 ml/mole $NPCl_2$ in $(NPCl_2)_n$. The polymerization process was again found to be exothermic. Thus, depolymerization of the polymer occurred at 1050°C and above at 70 kbar pressure, but at 1000°C, no depolymerization took place, and over 90% of the product consisted of high polymer. At 10 kbar pressure, depolymerization occurred at 900°C, but over 90% conversion to high polymer was observed between 600° and 850°C. However, kinetic experiments indicated that the *rate* of polymerization decreased as the pressure was raised from 10 to 70 kbar at 700°C. This suggests that the transition state in the rate-determining polymerization step has a larger volume than the reactants in that step. If a first-order process is assumed and the trimer is a reactant, the data would indicate that the transition state volume is 9% greater than the molar volume of the trimer.

Evidence has also been obtained by Jacques, Mole, and Paddock[41] that in the 200°–270°C range the zinc-catalyzed polymerization of the oligomers, $(NPCl_2)_{3-7}$, is exothermic, but the exothermicity decreases as the degree of polymerization of the starting material increases. Thus, evidence for a wide range of reaction conditions indicates that ΔH_{polym} is negative. If it is assumed that the entropy change on polymerization is also negative, then a ceiling temperature might be anticipated for this equilibration. Only circumstantial evidence exists for this at present.

6. Influence of Catalysts

A number of additives or impurities accelerate the rate of polymerization of chlorocyclophosphazenes. Konecny, Douglas, and Gray[20, 42] found that carboxylic acids, ethers, ketones, alcohols, nitromethane, and metals, such as zinc, tin, or sodium enhance the rate of polymerization. The rate enhancement was such that extensive polymerization could be induced in 24 hours at 210°C, compared to only 3% conversion to polymer in the same time in the absence of the catalyst. Comparable catalytic activity was shown by all the catalysts, and it appears that reagents which facilitate removal of chloride ion from phosphorus may accelerate the polymerization rate. Sulfur also functions as a catalyst in the 215°–254°C temperature range,[43] and 2,6-di-*t*-butyl-*p*-cresol,[44, 45] paraquinone, and hydroquinone[46] have also been reported to accelerate the polymerization. Hexachlorocyclotriphosphazene has also been polymerized in the solid state *below* 114°C when irradiated with 50 kV X-rays.[47] This process will be discussed later.

The question of catalysis by oxygen, air, or the walls of the reaction vessel has been a subject of some dispute. It was initially reported that traces of oxygen catalyze the polymerization of $(NPCl_2)_3$ in both the bulk and solution states.[23] However, large amounts of air, or oxygen, appear to inhibit the reaction.[17, 23] When highly purified trimer and rigorously dried oxygen are used at 300°C, a marked catalytic effect is shown by the oxygen.[27] There seems to be little doubt that the walls of the reaction vessel affect the rate of polymerization.[25, 32]

In view of the wide range of materials which function as polymerization catalysts, it is perhaps surprising that some compounds apparently have little effect on the reaction rate. These include carbon tetrachloride, chloroform, ligroin, benzene, biphenyl, cyclohexane, ethyl bromide, phosphorus pentachloride, ammonia, water, mercuric chloride, aluminum chloride, and stannic chloride.[20] Ultraviolet light also does not induce polymerization, at least in the 120°–160°C region.[20, 48] Free-radical initiators, such as di-*tert*-butyl peroxide have no accelerating influence at 120°.[20]

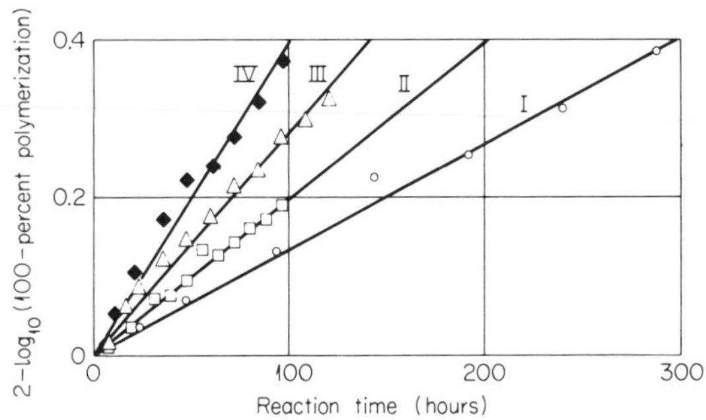

FIG. 15.6. Solution polymerization of $(NPCl_2)_3$ catalyzed by benzoic acid. Concentration of benzoic acid per gram of benzene. (I) 2.89 mg; (II) 5.70 mg; (III) 9.80 mg; (IV) 13.70 mg. From Konecny, Douglas, and Gray.[42]

The faster polymerization rates which are possible in the presence of catalysts facilitate the measurement of kinetic data. Accordingly, a number of kinetic studies have been reported with the use of benzoic acid,[26, 42, 49] sodium benzoate,[26] 2,6-di-*t*-butyl-*p*-cresol,[44] methanol,[42] ethanol,[20] ether,[20, 42] or tin.[20] Both the bulk and benzene-solution polymerization of $(NPCl_2)_3$ are first order in trimer when benzoic acid, diethyl ether, methanol, or ethanol are used as catalysts.[20, 42] With methanol and benzoic acid, but not with ether, the first-order rate constants increased linearly with catalyst concentration.[42] This is illustrated in Figs. 15.6 and 15.7.

Catalysis of melt polymerizations by benzoic acid has been studied in some detail. Gimblett[49] reported that polymerizations with a 1:0.04 weight ratio of $(NPCl_2)_3$ and benzoic acid showed an activation energy of 24.3 kcal/mole of trimer, with a frequency factor of $10^{8.44}$ min^{-1}. Oxygen had no effect on the rate at 210°C. A curious feature of the benzoic acid-catalyzed polymerization is the existence of an induction period.[26, 49] MacCallum and Werninck[26] showed

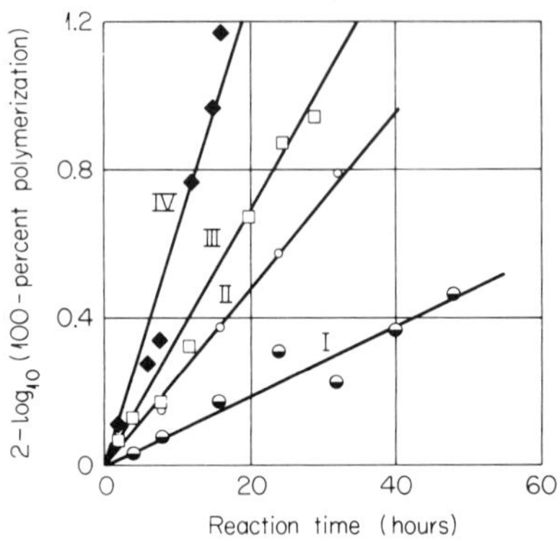

FIG. 15.7. Methanol-catalyzed bulk polymerization of the trimer. Concentration of methanol per gram of trimer: (I) 3.6 mg; (II) 11.0 mg; (III) 17.9 mg; (IV) 25.3 mg. From Konecny, Douglas, and Gray.[42]

that sodium benzoate has approximately the same catalytic activity as benzoic acid.

Evidence has been presented[49] that, for benzoic acid-catalyzed reactions, the chlorine content of the polymer decreases as polymerization proceeds. It should also be noted that the use of catalysts such as benzoic acid to accelerate the thermal polymerization of $(NPCl_2)_3$ or $(NPCl_2)_4$ favors the cross-linking of the polymer.[17] Thus, high polymer prepared by this technique is generally unsuitable for subsequent use in substitution reactions.[36]

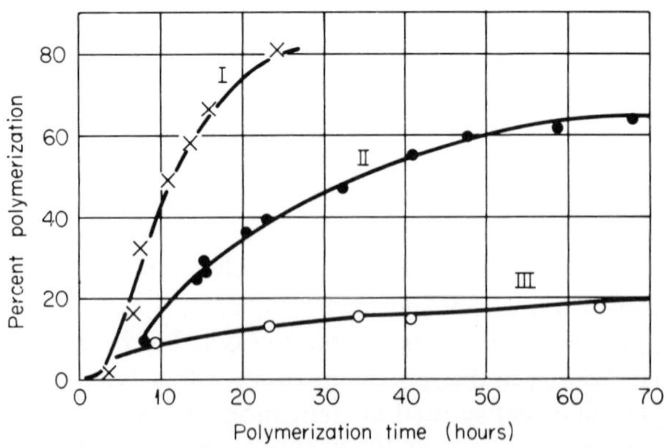

FIG. 15.8. Polymerization catalyzed by 8.4 mg of ether. (I) Trimer; (II), trimer–tetramer mixture (initial composition: 53 wt. % trimer); (III) tetramer. From Konecny and Douglas.[20]

7. Influence of Phosphazene Ring Size

The trimer, $(NPCl_2)_3$, polymerizes faster than the tetramer, $(NPCl_2)_4$, for both uncatalyzed and catalyzed polymerizations.[18, 20, 41] This is illustrated for the diethyl ether-catalyzed reaction shown in Fig. 15.8. This behavior may reflect either a greater release of ring strain during polymerization of the trimer or a greater accessibility of the planar trimer ring to attack by a propagating chain.

From a thermodynamic point of view, differences have been recorded between the heats of polymerization of the zinc-catalyzed polymerization of $(NPCl_2)_{3-7}$ in the 200°–270°C temperature range.[41] Table 15.2 shows values

TABLE 15.2

Heats of Polymerization for $(NPCl_2)_{3-7}$[a]

Phosphazene	$-\Delta H_{lc}$[b]
$(NPCl_2)_3$	1.39 ± 0.12
$(NPCl_2)_4$	0.86 ± 0.06
$(NPCl_2)_5$	0.80 ± 0.06
$(NPCl_2)_6$	0.26 ± 0.04
$(NPCl_2)_7$	0 ± 0.05

[a] From Jacques, Mole, and Paddock.[41]

[b] $-\Delta H_{lc}$ represents the heat of polymerization from the liquid to the condensed phase.

calculated from differential thermal analysis experiments. If the assumption is made that the different cyclic oligomers yield the same polymer, then the tetramer appears to have a greater heat of formation per monomer unit than the trimer.

8. Influence of Reaction Medium and Phase

Chlorophosphazenes have been polymerized in the molten state, in solution, and in the crystalline state. Most of the previous discussion has been concerned with polymerizations in the molten state. Here, the effects of solvents and solid-state effects will be considered.

The study of solution polymerizations offers a number of advantages over the same reactions carried out in the melt. In the first place, the viscosity increases which occur during polymerization are less in solution than in the bulk phase. Secondly, experience derived from organic polymer chemistry suggests that solution polymerizations are more controllable, and complications resulting from cross-linking are likely to be fewer.

Solution polymerization of $(NPCl_2)_3$ and $(NPCl_2)_4$ can be effected by heating 5–30% solutions of the chlorophosphazene in carbon tetrachloride at 270°–300°C for up to 36 hours.[22] The polymers obtained have molecular weights of up to 130,000. Conversion to polymer increases with both increased reaction temperature and increased trimer concentration. Oxygen gas was found to catalyze the reaction, and the kinetic chain length decreased with increases in oxygen concentration. Curiously, in this work, no rate differences between $(NPCl_2)_3$ and $(NPCl_2)_4$ were detected.[22]

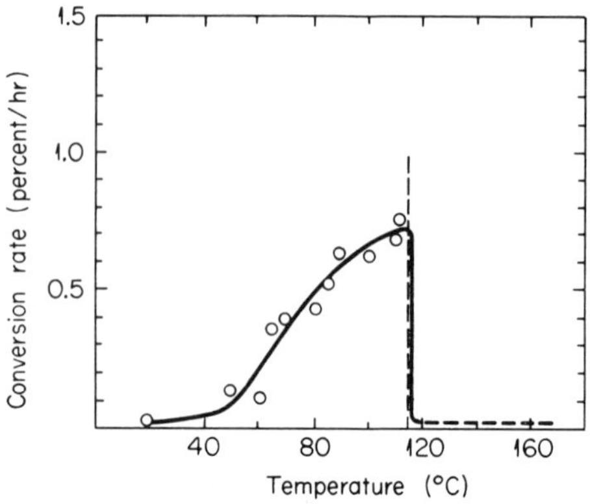

FIG. 15.9. Effect of temperature on the polymerization of hexachlorocyclotriphosphazene (dose rate, 1.2 Mr/hour; irradiation time, 10 hours). From Caglioti, Cordischi, and Mele.[47]

When hydrocarbon solvents, such as benzene, toluene, xylene, or hexane, are used for solution reactions at 300°C, the chlorophosphazenes react with the solvent to form colorless oils and gelatinous substitution products.[22] However, at lower temperatures (210°C) benzene is a satisfactory, unreactive solvent for catalyzed polymerizations.[42] Under these conditions, with benzoic acid, ether, or methanol used as catalysts, the polymerizations were found to show first-order kinetics with respect to $(NPCl_2)_3$, as illustrated in Figs. 15.6 and 15.7. In general, polymerizations in solution are one-half to one-third as rapid as those in the molten state, but polymerizations in benzene solution yield lower molecular weight products than reactions in the melt. In benzene solution, uncatalyzed polymerizations rates are difficult to reproduce, but at 210°C approximately 6% polymerization occurs in 100 hours in the absence of catalyst.

Polymerization of $(NPCl_2)_3$ in the crystalline state takes place during irradiation with 50 kV X-rays at a dose rate of 1.2 Mr/hour.[47] The conversion to alkane-insoluble polymer varies with temperature for constant irradiation time as shown in Fig. 15.9. A striking feature of this polymerization is that the conversion to polymer increases with temperature up to the melting point, but falls to zero for the molten trimer. This behavior is characteristic of many solid-state, irradiation-type polymerizations. It could result from the fact that the crystal-packing arrangement is especially favorable for bond interchange, or because unstable intermediates have a finite lifetime only in the solid state because recombination is inhibited. The polymerization appears to "saturate," or yield little additional polymer, after about 10% conversion, and the rate of polymerization is influenced by the presence of oxygen.[47]

D. Effect of Different Ligands

The foregoing facts are derived from experiments with chlorophosphazenes. Less information is available for phosphazenes which contain ligands other than chloro, and most of the available data for these species are distinctly qualitative. Nevertheless, distinct patterns and differences can be discerned and these provide information about the polymerization mechanism and about thermodynamic influences.

1. Fluorophosphazenes

The fluorophosphazenes, $(NPF_2)_3$, $(NPF_2)_4$, and $(NPF_2)_5$, have been polymerized thermally.[50-54] The trimer polymerizes when heated at 350°C

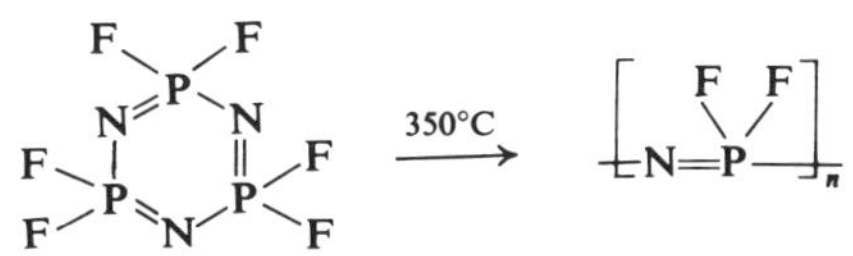

in a steel autoclave[50, 51] or in sealed glass tubes.[52, 53] Because of the high vapor pressure of $(NPCl_2)_3$ at this temperature, glass tubes must be contained within an autoclave to avoid an explosion. Pressure equalization outside the tubes is effected by pressurizing the autoclave with nitrogen.[52, 53] In contact with the steel autoclave, polymerization takes place in 15 hours at 350°C, and in sealed Pyrex tubes, 14 hours reaction is required at the same temperature. The polymer from both reactions is a colorless elastomer which is very unstable to atmospheric moisture. A noticeable difference from the polymerization of chlorophosphazenes is that appreciably higher temperatures are required before the fluorophosphazene polymerization can be induced. The tetramer, $(NPF_2)_4$, also polymerizes in sealed glass tubes within a pressurized autoclave at 350°C, but longer reaction times are required than with the trimer.[53] An incomplete polymerization of $(NPF_2)_5$ at 250°C for 65 hours has been reported.[54] The advantage of using higher cyclic fluorophosphazenes is that their vapor pressures are lower than those of the trimer or tetramer; they are, however, difficult to prepare in large quantities. Chlorofluorophosphazenes, such as $N_4P_4Cl_4F_4$ and $N_4P_4Cl_2F_6$, polymerize to rubbery polymers at 300°C.[55, 56] Polydifluorophosphazene, $(NPF_2)_n$, depolymerizes at 180°C or above, in vacuum, to yield principally the tetramer.[53] The total depolymerization of this polymer in nitrogen is illustrated by the TGA curve shown in Fig. 15.10.

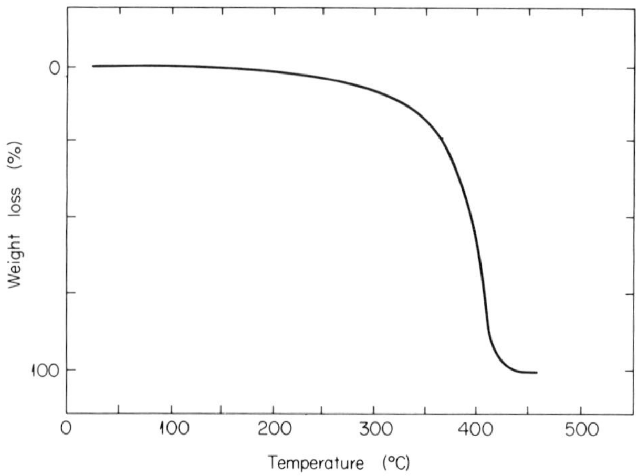

FIG. 15.10. Thermogravimetric analysis curve for $(NPF_2)_n$ in nitrogen. Heating rate, 10°C/minute.

2. BROMOPHOSPHAZENES

Hexabromocyclotriphosphazene, $(NPBr_2)_3$, polymerizes in sealed glass tubes at 220°C to give a rubbery elastomer.[53, 57, 58] The polymerization

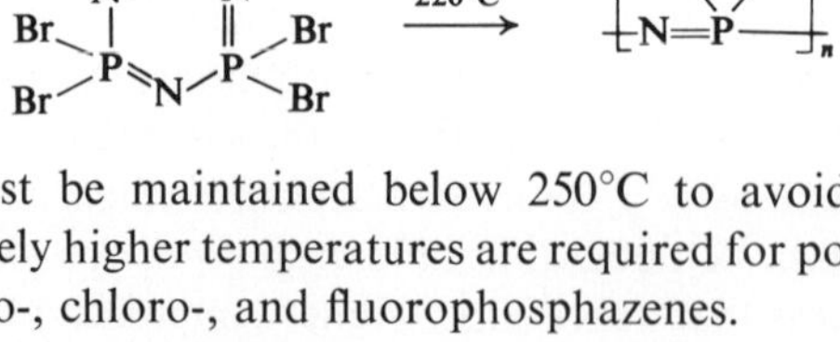

temperature must be maintained below 250°C to avoid decomposition.[53] Thus, progressively higher temperatures are required for polymerization along the series, bromo-, chloro-, and fluorophosphazenes.

Solid hexabromocyclotriphosphazene also polymerizes under the influence of γ irradiation in the 80°–150°C temperature region, and copolymers with

$(NPCl_2)_3$ can be prepared by this technique.[58] The rate of polymerization of $(NPBr_2)_3$, under these conditions decreases as $(NPCl_2)_3$ is added as a solid solution. Similarly, the polymerization rate of $(NPCl_2)_3$ is decreased by the addition of $(NPBr_2)_3$. This is in spite of the fact that both pure $(NPCl_2)_3$ and $(NPBr_2)_3$ polymerize at comparable rates.

3. Isothiocyanophosphazenes

Hexakis(isothiocyano)cyclotriphosphazene (III) and octakis(isothiocyano)-phosphazene (IV) polymerize when heated at 145°–210°C in sealed tubes to give rubbery polymers (V).[59, 60] Above 210°C, decomposition accompanies

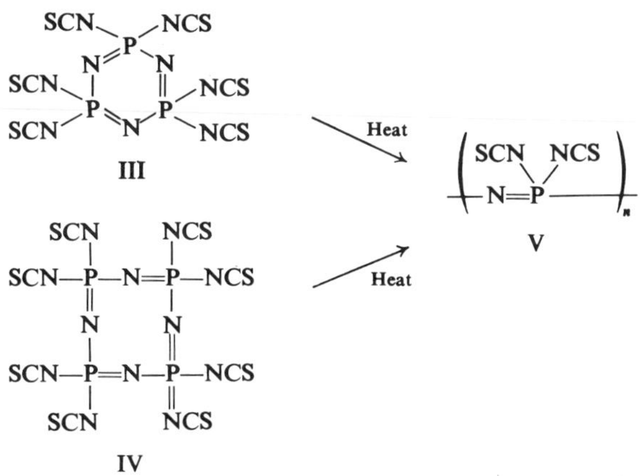

polymerization.[59] This polymerization process is complicated by the possibility that isothiocyano groups can function as cross-linkage sites. Depolymerization of the polymer can be effected in vacuum at elevated temperatures.[60]

4. Organophosphazenes

Numerous attempts have been made to thermally and catalytically polymerize organocyclophosphazenes, but most of these efforts have been unsuccessful. For example, hexaphenoxycyclotriphosphazene (VI) does not polymerize when heated to 350°C[61] or when heated with sulfur at 230°C.[43] The fluoroalkoxycyclophosphazenes, $[NP(OCH_2CF_3)_2]_3$, $[NP(OCH_2CF_3)_2]_4$, $[NP(OCH_2C_2F_5)_2]_3$, $[NP(OCH_2C_2F_5)_2]_4$, $[NP(OCH_2C_3F_7)_2]_3$, and $[NP(OCH_2C_3F_7)_2]_4$ of structures VII and VIII are stable for long periods of time at 250°–300°C.[48] These compounds are also unaffected by ultraviolet light after 12 hours irradiation at 150°C, or by treatment for 24 hours with benzoyl

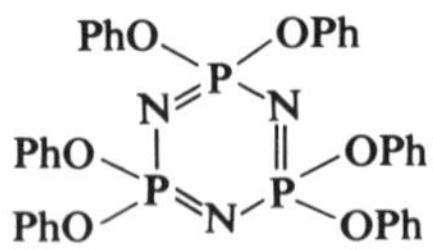

VI

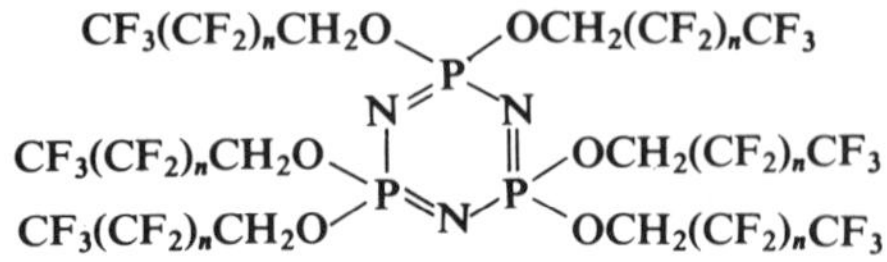

VII

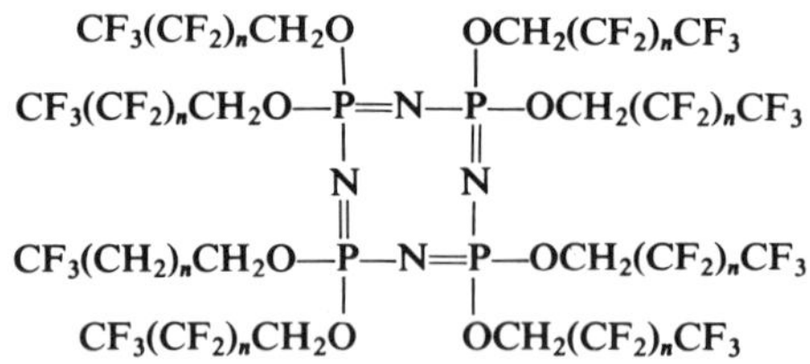

VIII

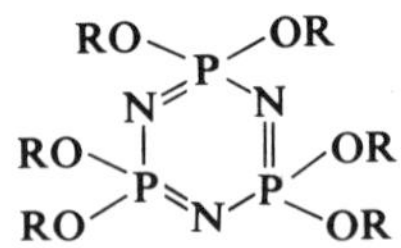

IX

peroxide in boiling benzene or with azobis(isobutyronitrile) in boiling xylene.[48] Similarly, all six compounds are unaffected when heated at 300°C for 24 hours in the presence of traces of aluminum chloride.[48] Attempts to polymerize nonfluorinated alkoxycyclophosphazenes at temperatures up to 350°C were equally unsuccessful.[62] Thus, *n*-propoxy, *n*-butoxy, and cyclohexyloxy derivatives of structure IX yielded no polymers at high temperatures or in the presence of benzoyl peroxide or aluminum chloride. Alkoxycyclophosphazenes are, of course, prone to rearrangement to phosphazanes at elevated temperatures (see Chapter 13). The methyl- and ethylcyclotriphosphazenes, $(NPMe_2)_3$ and $(NPEt_2)_3$, apparently decompose when heated, and high polymers are not obtained.[63,64]

Several cyclic phosphazenes are capable of undergoing thermal ring-expansion reactions without forming high polymers. For example, octaphenylcyclotetraphosphazene, $(NPPh_2)_4$, yields a mixture of trimer, tetramer, pentamer, hexamer, and higher homologues when heated at temperatures between 320° and 500°C.[65,66] Hexakis(heptafluorobutoxy)cyclotriphosphazene, $[NP(OCH_2C_3F_7)_2]_3$, is converted in 50% yield to the tetramer when boiled at 340°C for 50 hours.[67] In these cases, the ring-enlargement mechanism must be accessible but, nevertheless, equilibration favors the lower cyclic oligomeric homologues.

Organocyclophosphazenes which contain spirocyclic side units attached to phosphorus can also be polymerized thermally, but the products are low or medium molecular weight polymers. Tris(*o*-phenylenedioxy)cyclotriphosphazene (X) polymerizes at 300°C to yield a linear or macrocyclic polymer, apparently with structure XI.[68] However, the related compounds XII, XIII,

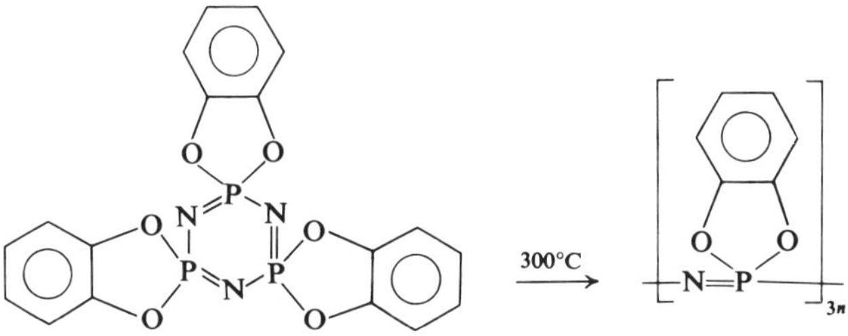

X XI

and XIV do not polymerize thermally.[68] An explanation for these differences

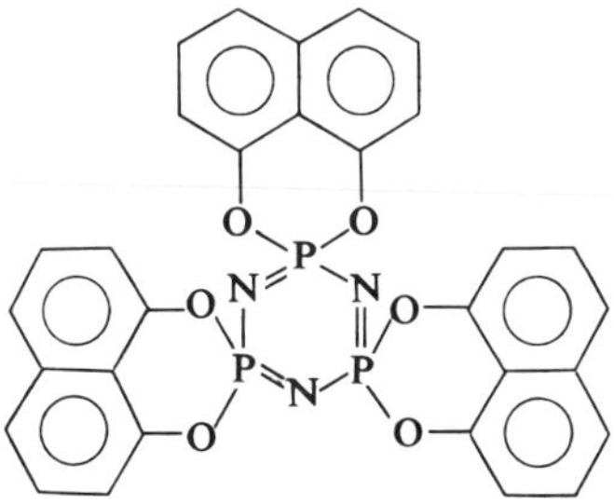

XII

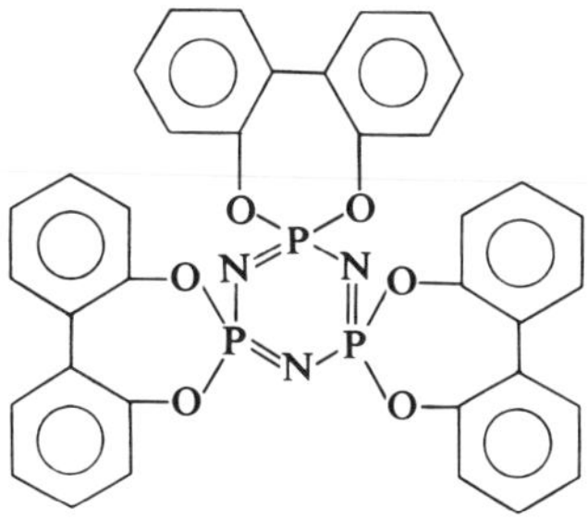

XIII

is presented later. Compounds of structure XV, yield polymers when heated to 300°C, but these may be cross-linked materials.

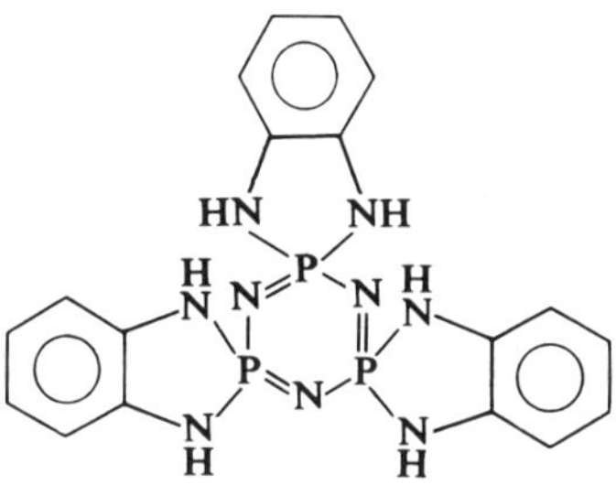

XIV

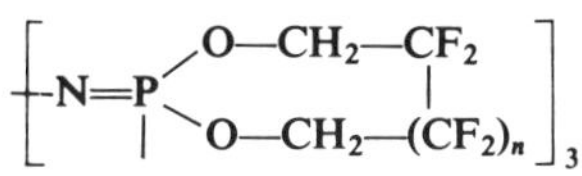

$n = 1$ or 2

XV

Although poly(organophosphazenes) cannot generally be synthesized by polymerization of the appropriate cyclic trimers and tetramers, they can be prepared by other routes (see Chapter 16). Typical polymers of this type are shown in XVI–XXI. At high temperatures, these polymers undergo depolymerization reactions, as illustrated by the following observations.

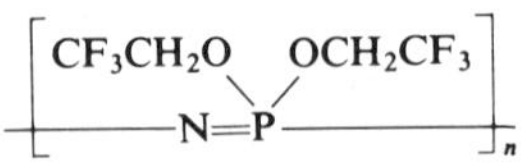

XVI

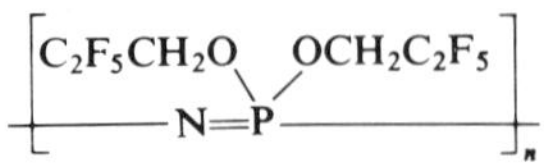

XVII

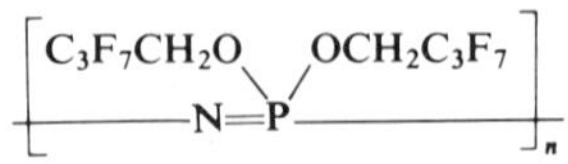

XVIII

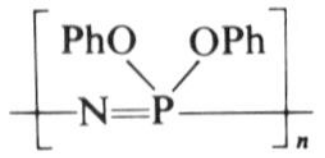

XIX

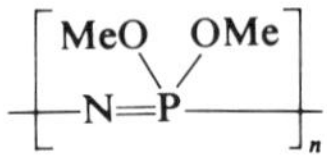

XX

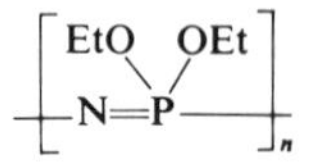

XXI

Thermogravimetric analysis curves for $[NP(OCH_2CF_3)_2]_n$ (XVI), $[NP(OCH_2C_2F_5)_2]_n$ (XVII), and $[NP(OPh)_2]_n$ (XIX) in nitrogen are shown in Fig. 15.11.[53] T_{10} weight loss values (the temperature at which 10% weight loss occurs) have been quoted for these polymers,[69] but these numbers provide very little information about the relative thermal stabilities, since they represent volatilization temperatures of low molecular weight species formed at lower temperatures. For example, the T_{10} value of $[NP(OCH_2CF_3)_2]_n$ in nitrogen is between 300°[53] and 410°C,[69] but the same polymer undergoes a sharp molecular weight decrease between 150° and 200°C.[17] Similarly, $[NP(OPh)_2]_n$, has a T_{10} value of 400°C, but a rapid molecular weight decrease takes place

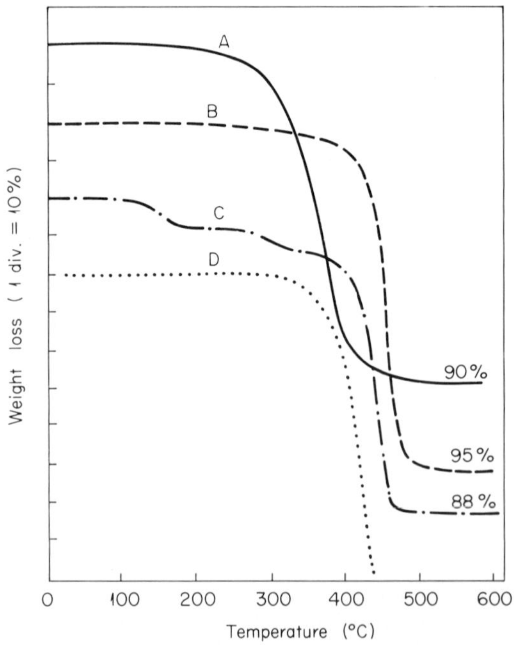

FIG. 15.11. Thermogravimetric analysis curves in nitrogen of (A), $[NP(OCH_2CF_3)_2]_n$; (B), $[NP(OCH_2C_2F_5)_2]_n$; (C), $[NP(OCH_2C_3F_7)_2]_n$; and (D), $[NP(OPh)_2]_n$. The heating rates were 10°C/minute.

below 150°C.[17] This is indicated by solution viscosity measurements on the products formed by heating the solid polymer in evacuated sealed tubes for various periods of time. The results are shown in Table 15.3. However, both the trifluoroethoxy and phenoxy derivatives retain their form and integrity as rubbery thermoplastics after 8 hours at 200°C. Similar molecular weight decreases during 8 hours heating at 255°C have been reported for $[NP(OCH_2C_6H_4CF_3\text{-}m)_2]_n$, $[NP(OC_6H_5Cl\text{-}p)_2]_n$, $[NP(OC_6H_4Cl_2\text{-}2,4)_2]_n$, $[NP(OC_6H_4C_6H_4\text{-}p)_2]_n$, and $[NP(OC_{10}H_7\text{-}\beta)_2]_n$.[69] The ethoxy derivative, $[NP(OEt)_2]_n$, appears to depolymerize slowly in ethanol solution at 25°C.[17]

TABLE 15.3

DEPOLYMERIZATION OF POLYPHOSPHAZENES, $[NP(OR)_2]_n$[a]

OR	Initial soln. viscosity (dl/g)	100°C	150°C	200°C
OCH_3	1.02	0.095	—	—
OC_2H_5	1.63	1.090	0.158	—
OCH_2CF_3	2.50	2.50	1.20	0.48
OPh	4.24	2.06	0.62	0.43

[a] From Allcock, Kugel, and Valan.[17]

Some uncertainty exists concerning the products of the pyrolysis reactions. For the polymer, $(NPMe_2)_n$, heating to 300°–500°C results in almost complete breakdown to the trimer and tetramer, $(NPMe_2)_{3\,\text{and}\,4}$.[70] Similarly, the trifluoroethoxy derivative, $[NP(OCH_2CF_3)_2]_n$, yields the trimer $[NP(OCH_2CF_3)_2]_3$ when heated in solution or in a vacuum sublimator,[34] and cyclic oligomers have also been detected among the pyrolysis products of $[NP(OPh)_2]_n$, $[NP(OC_6H_4C_6H_4\text{-}p)_2]_n$, and $(NPPh_2)_n$.[35] Thus, the predominant reaction appears to be a depolymerization process of the type

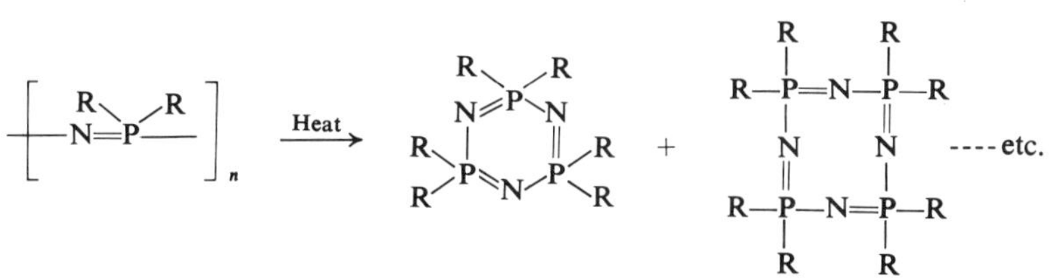

However, in some cases degradation reactions compete with depolymerization. For example, the methoxy and ethoxy derivatives, $[NP(OMe)_2]_n$, and $[NP(OEt)_2]_n$, appear to undergo fragmentation reactions at high temperatures.

These processes may be initiated by phosphazene–phosphazane-type rearrangements. Fragmentation products have also been detected from the high-temperature pyrolyses of other polymers also,[35] but it is not clear how much these products represent discontinuities in the chains (i.e., branch points or cross-linkage sites), or sites of incomplete or anomalous substitution. However, it appears likely that at high enough temperatures, particularly above 500°C, degradation processes involving the reactions of the side groups will occur.

5. Influence of Side Groups on Equilibration

The evidence presented above indicates (a) that except for unusual structures, organocyclophosphazenes do not yield high polymers when heated, whereas halogeno and isothiocyanocyclophosphazenes polymerize, and (b) that polyorganophosphazenes prepared by other routes show a marked tendency to depolymerize to cyclic oligomers at moderate or high temperatures. The limited evidence available also suggests that depolymerization of the organo-substituted polymers can occur at a lower temperature than for the chloro- and fluoro-substituted analogues.

Two explanations can be formulated for these facts. The first is that the polymerization of *organo*cyclotriphosphazenes is mechanistically inhibited. If polymerization requires the prior ionization of a side group from phosphorus (see the following section), it could be argued that organic ligands have a lower tendency to ionize and that the initiation mechanism is thereby inhibited. It will be shown later that polymerization of a cyclic trimer does not occur even when the side group (e.g., phenoxy) can ionize from phosphorus.[71] Furthermore, as mentioned earlier, phenyl- and heptafluorobutoxycyclophosphazenes equilibrate to other cyclic oligomers when heated, yet they do not yield high polymers. Finally, if the reasonable assumption is made that polymerization and depolymerization follow similar mechanisms, then it is clear that organic side groups do not inhibit the depolymerization process.

The second explanation is that the equilibration is strongly thermodynamically controlled, with the organic substituents favoring the formation of cyclic oligomers rather than high polymers. It appears that, in a very general way, there is some correlation between the *dimensions* of the side groups and the polymerization or depolymerization behavior of cyclic or polymeric phosphazenes.[15] Small groups appear to favor the polymerization process and large groups favor depolymerization. In these terms, fluoro, chloro, bromo, and isothiocyano groups have small steric dimensions, whereas phenyl, phenoxy, most amino, alkoxy, and fluoroalkoxy groups occupy a relatively large volume at the phosphorus atoms. It can, in fact, be shown[15] that the conversion of fluoro-, chloro-, bromo-, and isothiocyanocyclotriphosphazenes to high polymers results in no serious increase in intramolecular steric repulsions. In other words, the enthalpy change on polymerization should be close to zero. However, when organic substituents, such as phenyl or phenoxy, are present at phosphorus, the formation of a linear or macrocyclic polymer will give rise to serious interpenetration of the side-group torsional volumes. These intramolecular repulsions will raise the enthalpy of the polymer and strongly favor depolymerization to the trimer, tetramer, or other lower oligomers.

A rough estimate of the effects of various substituents on the destabilization of the polymer can be made on the basis of the van der Waals radii of various substituent groups and the volume swept out by these groups during their normal torsional motions.[15] In these terms it can be shown[15] that groups such as H, F, Cl, Br, I, NH_2, or SCN would have little destabilizing effect on the polymer, and CH_3 or CF_3 groups would have only a marginal influence. However, for bulkier groupings the effect on the polymer would be quite serious, with the tendency to favor depolymerization increasing in the order: Ph $\sim$ *m*-MeC_6H_4 $<$ NHMe $\approx$ NMe_2 $<$ OMe $<$ Et $<$ *o*-MeC_6H_4 $<$ OC_2H_5 $\approx$ OC_3H_7-*n* $<$ NHPh $<$ OCH_2CF_3 $<$ OPh $<$ C_4H_9. Thus, the substituent steric effect can be viewed as a tendency to lower the average molecular weight or the ceiling temperature of the species present at equilibrium until, for the bulkiest substituents, the trimer or tetramer are the only homologues present at equilibrium. Significantly, the steric configuration of *o*-phenylenedioxy substituent groups are such that the calculated intramolecular steric interactions in the polymer (XI) are comparable to those in the trimer (X), and this may explain the fact that the trimer polymerizes. By contrast, tris(2,2′-dioxybiphenylcyclotriphosphazene) (XIII) resists polymerization at high temperatures, a result which may reflect the sharp angle of twist of the biphenyl units[71] and the resultant steric repulsions in the polymer.

The implication of the foregoing discussion is that some of the poly(organophosphazenes) that have been synthesized are thermodynamically unstable. Their integrity at room temperature and above probably reflects an infinitely slow *rate* of depolymerization, but at high temperatures the depolymerization mechanism operates at an appreciable rate and breakdown of the chains to cyclic oligomers is observed.

E. The Polymerization–Depolymerization Mechanism

1. Experimental Facts

Any analysis of the polymerization or depolymerization mechanism must be based on a review of the known experimental facts. For halophosphazenes, these are as follows:

(a) The rate of uncatalyzed polymerization of $(NPCl_2)_3$ is extremely slow below 210°C, but it increases rapidly with temperature in the 220°–300°C region.[17, 20]

(b) The tetrameric compounds $(NPCl_2)_4$ and $(NPF_2)_4$ polymerize at slower rates than the corresponding trimers, $(NPCl_2)_3$ and $(NPF_2)_3$.[20, 53]

(c) Progressively higher temperatures are required to initiate polymerization along the series $(NPBr_2)_3$, $(NPCl_2)_3$, $(NPF_2)_3$.[53]

(d) The polymer formed by the initial bulk polymerization of $(NPCl_2)_{3\text{ or }4}$ at 250°–300°C undergoes a cross-linking reaction after about 75% of the trimer or tetramer has been consumed.[17]

(e) Poly(dichlorophosphazene) which has been freed from oligomers and dissolved in benzene, undergoes a slow cross-linking reaction at 25°C which suggests that latent active sites are still present at this temperature.[17]

(f) From high-pressure experiments,[40] the transition state for the rate-determining step apparently has a larger volume than the reactants in that step.

(g) Certain compounds, such as benzoic acid, sodium benzoate, methanol, or zinc accelerate the polymerization of $(NPCl_2)_{3\text{ or }4}$.[20, 42] Traces of hydrogen chloride have been detected from polymerizations catalyzed by carboxylic acids.[72] Nearly all the catalysts are capable of removing a chloride ion from phosphorus. Evidence also exists that a catalytic species is slowly formed when highly purified $(NPCl_2)_3$ is heated to high temperatures.[27]

TABLE 15.4

CONDUCTANCE MEASUREMENTS (MHO $CM^{-1} \times 10^{11}$) FOR $(NPCl_2)_3$ AND $[NP(OC_6H_5)_2]_3$[a]

$(NPCl_2)_3$		$[NP(OC_6H_5)_2]_3$	
Temp. (°C)	Conductance	Temp. (°C)	Conductance
120	1.7–0.1	120	1.5
163	1.7	165	3.5
203	33	203.5	4.3
253	6250[b]	252	17.7
202	2900[b]	163.5	4.8
		139	2.4
		350	316.0
		120	17.7

[a] From Allcock and Best.[61]

[b] The conductance rose continuously with time.

(h) Free-radical catalysts, such as benzoyl peroxide or azobis(isobutyronitrile) appear to have little effect on the rate of polymerization.[20] Ultraviolet light does not accelerate the reaction.

(i) Electron spin resonance experiments with purified molten $(NPCl_2)_3$ at 250°C and below in air or vacuum suggest the absence of free-radical species.[61]

(j) Below the temperature range where polymerization takes place (i.e., below 210°C), molten $(NPCl_2)_3$ has a low conductivity and a low dielectric constant.[61] Above this temperature the conductance rises rapidly with temperature and, when the temperature is held at 253°C, the conductance continues to rise with time. Simultaneously, as the temperature is raised from 203° to 253°C, the dielectric constant also rises rapidly, and it continues to increase with time when the temperature is maintained at 253°C. These results and comparable data for $[NP(OPh)_2]_3$ are shown in Tables 15.4 and 15.5.

TABLE 15.5

CAPACITANCE MEASUREMENTS FOR $(NPCl_2)_3$ AND $[NP(OC_6H_5)_2]_3$[a]

$(NPCl_2)_3$		$[NP(OC_6H_5)_2]_3$	
Temp. (°C)	Dielectric constant	Temp. (°C)	Dielectric constant
120	2.71 ± 0.02	120	4.32
163	2.61	163	4.11
203	2.52	203.5	3.79
253	2.56 – 3.29[b]	252	3.55
202	3.92 – 4.05[b]	163	4.11
		139	4.17
		350	3.18
		120	4.35

[a] From Allcock and Best.[61]

[b] The dielectric constant rose continuously with time.

Although the evidence is by no means complete, the above facts can be interpreted in terms of initiation, propagation, branching, cross-linking, termination, and depolymerization steps.

2. INITIATION

Three types of initiation processes can be contemplated. These are shown in the following scheme.

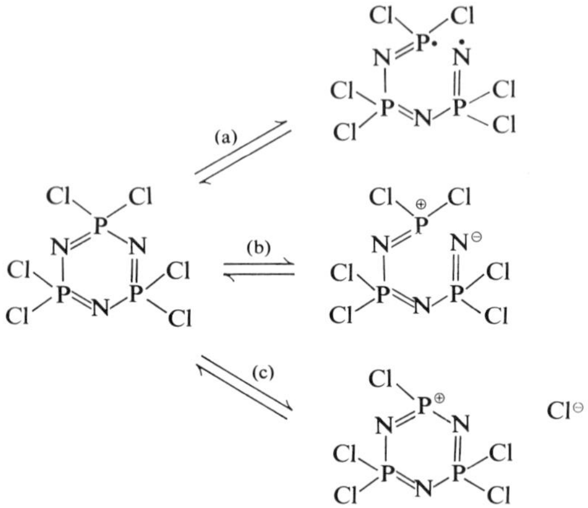

A fourth possibility—the cleavage of a chlorine radical from phosphorus—is considered to be unlikely in view of the ionic character of the P—Cl bond (the electronegativity difference is 0.9). A fifth alternative—the depolymerization of the cyclic trimer to monomer—would be preceded by steps (a) or (b).

The formation of a diradical by mechanism (a) is considered to be unlikely for three reasons. Firstly, the ionic character of the P—N bond is probably too high (electronegativity difference 0.9) to allow homolytic cleavage to occur. Secondly, free-radical sources have little effect on the reaction, and thirdly, as discussed in (i) above, radical species cannot be detected in the polymerizing mixture.

Initiation path (b) is a more reasonable possibility. The formation of zwitterionic species of this type would be consistent with the observed increase in dielectric constant and polarizibility as the reaction temperature is raised.[61] The subsequent addition of zwitterions to form polymers could be readily understood. However, such a mechanism would not explain the observed conductance increase during polymerization, nor is it consistent with the capacitance data for $[NP(OPh)_2]_3$ shown in Table 15.5. No capacitance increase is observed for $[NP(OPh)_2]_3$ over the temperature range of 120° to 350°C. It appears unlikely that $(NPCl_2)_3$ can undergo heterolytic skeletal cleavage in this temperature range if $[NP(OPh)_2]_3$ cannot.

A more plausible argument can be made for initiation mechanism (c), in which a chloride ion ionizes from phosphorus. This scheme is fully consistent with the conductance behavior of molten $(NPCl_2)_3$, and with the low conductance of $[NP(OPh)_2]_3$ at 350°C. The fact that the conductance change of molten $(NPCl_2)_3$ is not reversible with temperature decreases is a strong indication that at least one of the ionic species participates in the propagation step. The chloride ionization mechanism is also consistent with the role of catalysts in this process. Virtually the only feature which many of the catalysts have in common is their presumed ability to remove a chloride ion from phosphorus.[42, 72] Finally, the temperature required to initiate polymerization of the halophosphazenes $(NPX_2)_3$ increases in the order X = Br < Cl < F, and this follows the order of increasing strengths of the phosphorus–halogen bonds.

3. Propagation

Any postulates about the mechanism beyond the initiation step are purely speculative. However, an obvious possibility would be an electrophilic attack by the initiating cyclic cation to cleave the attacked ring and initiate a chain propagation process. This would constitute successive additions of trimer into the terminal $P^{\oplus}Cl^{\ominus}$ bond

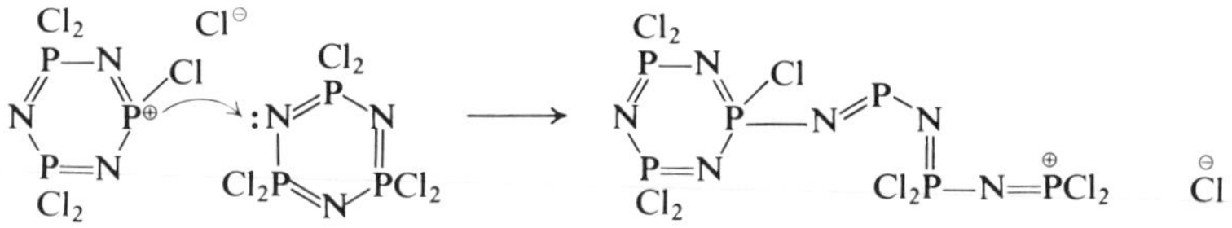

XXII XXIII

The presence of a growing, linear cation–anion complex would certainly be consistent with the observed dielectric constant increase with increasing temperature and at constant temperature.[61] The rate of dielectric constant increase at 202° and 252°C suggested an activation energy of about 19 kcal/mole,[61] but this is well below the activation energy for the overall polymerization. Since XXIII undoubtedly would have a larger molar volume than the sum of the two components in XXII,[40] the effects of pressure in slowing the polymerization can be explained. The lower rate of polymerization of cyclic tetramer compared to trimer could reflect the greater shielding of the nitrogen atoms to electrophilic attack in the puckered tetramer. Other propagation mechanisms, both cationic and ionic[26, 72] as well as free radical,[22, 24, 31, 43] have also been proposed but, in the absence of more definitive data, no definite conclusions can be drawn.

4. Branching and Cross-linking

Polydichlorophosphazene prepared from the trimer almost certainly has a branched structure, and it cross-links readily. Branching can be understood in terms of the initiation mechanism outlined above, since ionization of a chloride ion from a middle unit is feasible (XXIV). Cross-linking could occur by

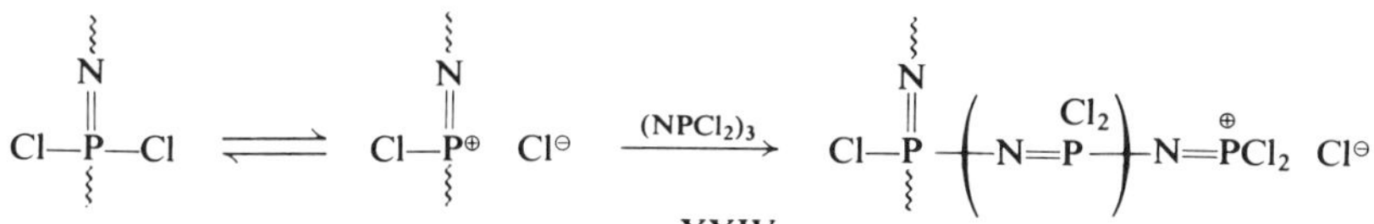

XXIV

addition of a P—Cl bond to the skeleton of another molecule, as shown in the following scheme:

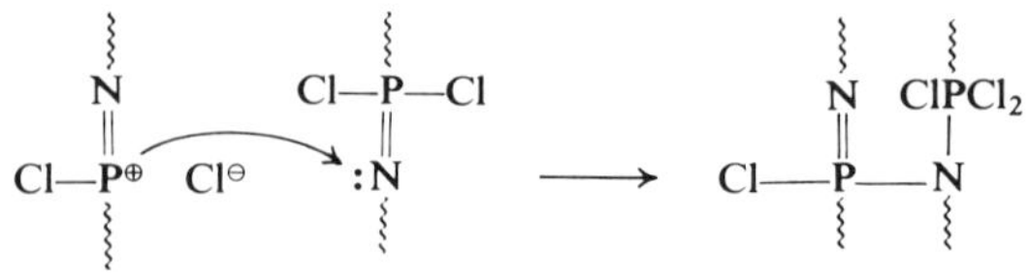

or

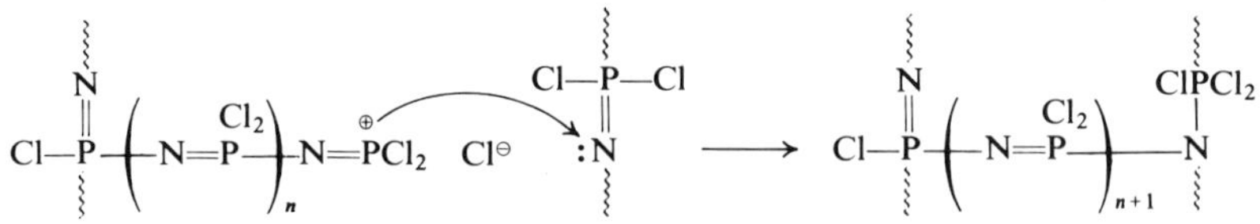

A variety of other cross-linking processes can also be envisaged, particularly when oxygen, hydrocarbons, or traces of water are present.

5. Termination

As in organic ionic polymerizations, a variety of termination processes are possible. Traces of water or ammonia, solvents, or even the hydroxyl groups at the surface of glass reaction vessels could terminate the chain growth. Alternatively, termination may result from the reduced degree of dissociation of the $P^{\oplus}Cl^{\ominus}$ ion pair as the temperature is lowered (XXV). A termination mechanism

$$\text{\textasciitilde\textasciitilde}N{=}\overset{\oplus}{P}Cl_2 \quad Cl^{\ominus} \xrightleftharpoons{\text{cool to } 25^\circ\text{C}} \text{\textasciitilde\textasciitilde}N{=}PCl_3$$

XXV

of this type would explain why polydichlorophosphazene cross-links slowly in benzene solution[17] or in the purified solid state. Presumably, the chain ends are still active at 25°C because of a small residual degree of $P^{\oplus}Cl^{\ominus}$ ionization.

6. Depolymerization

Virtually no information is available on the mechanism of depolymerization. It appears reasonable to suppose that the polymerization and depolymerization mechanisms are closely related, and a backbiting mechanism by an active chain end might initiate an unzipping-type process (XXVI). Cyclic tetramers,

XXVI

pentamers, hexamers, and a wide variety of other cyclic and linear homologues could readily be formed by this mechanism.

It must be emphasized that the experimental facts for the polymerization and depolymerization processes are still incomplete and not fully understood. Different reaction conditions, such as the presence of catalysts, oxygen, water, or solvents may alter the mechanism. Minute traces of yet unidentified impurities, such as phosphorus halides or linear phosphazenes, may function as catalysts, or inhibitors, and no account has been taken of the possible participation by monomeric units in the equilibration scheme.

References

1. H. R. Allcock, "Heteroatom Ring Systems and Polymers." Academic Press, New York, 1967.
2. H. R. Allcock, *J. Macromol. Sci., Rev. Macromol. Chem.* **4**, 3 (1970).
3. A. V. Tobolsky, *J. Polym. Sci.* **25**, 220 (1957).
4. A. V. Tobolsky, *J. Polym. Sci.* **31**, 126 (1958).
5. A. V. Tobolsky and A. Eisenberg, *J. Amer. Chem. Soc.* **81**, 780 (1959).
6. A. V. Tobolsky and A. Eisenberg, *J. Amer. Chem. Soc.* **81**, 2302 (1959).
7. A. V. Tobolsky and A. Eisenberg, *J. Amer. Chem. Soc.* **82**, 289 (1960).
8. A. V. Tobolsky, "Properties and Structure of Polymers." Wiley, New York, 1960.
9. G. Gee, *Chem. Soc., Spec. Publ.* **15**, 67 (1961).
10. H. Jacobson and W. H. Stockmayer, *J. Chem. Phys.* **18**, 1600 (1950).
11. J. R. Van Wazer, *Amer. Sci.* **50**, 450 (1962).
12. J. R. Van Wazer, *J. Macromol. Sci., Part A* **1**, 29 (1967).
13. J. R. Van Wazer and K. Moedritzer, *Angew. Chem., Int. Ed. Engl.* **5**, 341 (1966).
14. K. Moedritzer, *Organometal. Chem. Rev.* **2**, 1 (1967).
15. H. R. Allcock, *Inorg. Chcm.* **5**, 1320 (1966).
16. H. R. Allcock, *J. Polym. Sci., Part A* **1**, 355 (1967).
17. H. R. Allcock, R. L. Kugel, and K. J. Valan, *Inorg. Chem.* **5**, 1709 (1966).
18. H. R. Allcock and D. P. Mack, unpublished work (1969).
19. H. R. Allcock and R. L. Kugel, *J. Amer. Chem. Soc.* **87**, 4216 (1965).
20. J. O. Konecny and C. M. Douglas, *J. Polym. Sci.* **36**, 195 (1959).
21. M. Yokoyama, *Chem. High Polym.* **17**, 651 (1960); *Chem. Abstr.* **55**, 24354 (1961).
22. F. Patat and F. Kollinsky, *Makromol. Chem.* **6**, 292 (1951).
23. F. Patat, *Angew. Chem.* **65**, 173 (1953).
24. F. Patat and K. Frömbling, *Monatsh. Chem.* **86**, 718 (1955).
25. D. Chakrabartty and B. N. Ghosh, *J. Polym. Sci.* **62**, 5130 (1962).
26. J. R. MacCallum and A. Werninck, *J. Polym. Sci., Part A*-1 **5**, 3061 (1967).
27. R. O. Colclough and G. Gee, *J. Polym. Sci., Part C* **16**, 3639 (1968).
28. H. N. Stokes, *Amer. Chem. J.* **19**, 782 (1897).
29. R. Schenck and G. Römer, *Chem. Ber.* **57B**, 1343 (1924).
30. A. M. de Ficquelmont, *C. R. Acad. Sci.* **204**, 689, 867 (1937).
31. F. Patat and P. Derst, *Angew. Chem.* **71**, 105 (1959).
32. F. G. R. Gimblett, *Plast. Inst., Trans.* **28**, 65 (1960).
33. M. Yokoyama and S. Konya, *Kogyo Kagaku Zasshi* **69**, 1835 (1966).
34. H. R. Allcock and W. J. Cook, unpublished work (1970).
35. J. R. MacCallum and J. Tanner, *J. Macromol. Sci., Chem.* **4**, 481 (1970).

35a. J. R. MacCallum, *J. Macromol. Sci., Chem.* **A5**, 653 (1971).
36. H. R. Allcock, unpublished work (1966).
37. F. G. R. Gimblett, "Inorganic Polymer Chemistry," p. 409. Butterworth, London, 1963.
38. O. Schmitz-DuMont, *Z. Elektrochem.* **45**, 651 (1939).
39. O. Schmitz-DuMont, *Angew. Chem.* **52**, 498 (1939).
40. J. R. Soulen and M. S. Silverman, *J. Polym. Sci., Part A* **1**, 823 (1963).
41. J. K. Jacques, M. L. Mole, and N. L. Paddock, *J. Chem. Soc., London* p. 2112 (1965).
42. J. O. Konecny, C. M. Douglas, and M. Y. Gray, *J. Polym. Sci.* **42**, 383 (1960).
43. J. R. MacCallum and J. Tanner, *J. Polym. Sci., Part B* **7**, 743 (1969).
44. M. Kajiwara and H. Saito, *Kogyo Kagaku Zasshi* **66**, 621 (1963).
45. M. Kajiwara and H. Saito, *Kogyo Kagaku Zasshi* **67**, 1002 (1964).
46. H. Saito and M. Kajiwara, private communication (1970).
47. V. Caglioti, D. Cordischi, and A. Mele, *Nature (London)* **195**, 491 (1962).
48. E. T. McBee, H. R. Allcock, R. Caputo, A. Kalmus, and C. W. Roberts, *U.S. Govt. Res. Rep.* **AD 209-669** (1959).
49. F. G. R. Gimblett, *Polymer* **1**, 418 (1960).
50. F. Seel and J. Langer, *Angew. Chem.* **68**, 461 (1956).
51. F. Seel and J. Langer, *Z. Anorg. Allg. Chem.* **295**, 316 (1958).
52. H. R. Allcock, G. F. Konopski, R. L. Kugel, and E. G. Stroh, *Chem. Commun.* p. 985 (1970).
53. H. R. Allcock, R. L. Kugel, D. P. Mack, C. W. Cameron, and W. J. Cook, unpublished results (1967–1970).
54. G. Allen, C. J. Lewis, and S. M. Todd, *Polymer* **11**, 31 (1970).
55. O. Schmitz-DuMont and H. Kulkens, *Z. Anorg. Allg. Chem.* **238**, 189 (1938).
56. O. Schmitz-DuMont and A. Braschos, *Z. Anorg. Allg. Chem.* **243**, 113 (1939).
57. N. E. Bean and R. A. Shaw, *Chem. Ind. (London)* p. 1189 (1960).
58. D. Cordischi, A. D. Site, and A. Mele, *J. Macromol. Chem.* **1**, 219 (1966).
59. G. Tesi, R. J. A. Otto, F. G. Sherif, and L. F. Audrieth, *J. Chem. Soc., London* **82**, 528 (1960).
60. H. R. Allcock, G. F. Konopski, and R. W. Tomlinson, unpublished work (1969).
61. H. R. Allcock and R. J. Best, *Can. J. Chem.* **42**, 447 (1964).
62. M. Yokoyama, F. Yamada, S. Kawasaki, and K. Suzuki, *Kogyo Kagaku Zasshi* **66**, 613 (1963).
63. H. T. Searle, *Proc. Chem. Soc , London* p. 7 (1959).
64. A. J. Bilbo, *Z. Naturforsch. B* **15**, 330 (1960).
65. V. V. Korshak, I. A. Gribova, T. V. Artamonova, and A. N. Bushmarina, *Vysokomol. Soedin.* **2**, 377 (1960).
66. H. Bode and R. Thamer, *Chem. Ber.*7 **6B**, 121 (1943).
67. R. Rätz, H. Schroeder, H. Ulrich, E. Kober, and C. Grundmann, *J. Amer. Chem. Soc.* **84**, 552 (1962).
68. H. R. Allcock and R. L. Kugel, *J. Amer. Chem. Soc.* **5**, 1016 (1966).
69. G. Allen, C. J. Lewis, and S. M. Todd, *Polymer* **11**, 44 (1970).
70. H. H. Sisler, S. E. Frazier, R. G. Rice, and M. G. Sanchez, *Inorg. Chem.* **5**, 326 (1966).
71. H. R. Allcock, M. T. Stein, and J. A. Stanko, *J. Amer. Chem. Soc.* **93,** 3173 (1971).
72. F. G. R. Gimblett, *J. Polym, Sci.* **60**, 529 (1962).

Chapter 16

PHOSPHAZENE POLYMERS

A. Types of Phosphazene Polymers

Three types of polymers based on phosphazene structures have been developed. These are (a) linear-type polymers, (b) cyclolinear polymers, and (c) cross-linked cyclomatrix polymers. Each of these types will be considered first in general terms, and later sections will deal in detail with specific examples.

1. Linear-Type Polymers

Linear phosphazene polymers have the general structure shown in I. They

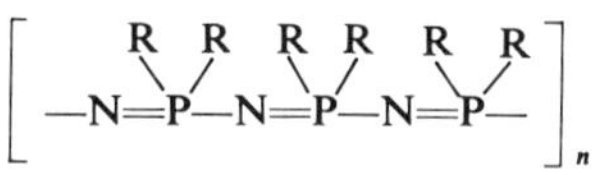

I

are produced by the thermal or catalytic polymerization of cyclic oligomers followed, in many cases, by substitution of halogen ligands by organic groups. Two modifications to the linear structure must also be considered. The first involves the presence of branches, as depicted in Fig. 16.1, and the second involves the presence of *cross-links* between chains, as illustrated in Fig. 16.2. In practice, both branches and cross-links may be present in the same structure. Most of the un-cross-linked polymers to be discussed here contain from 5,000 to 15,000 monomer units, with molecular weights up to ~4,000,000.

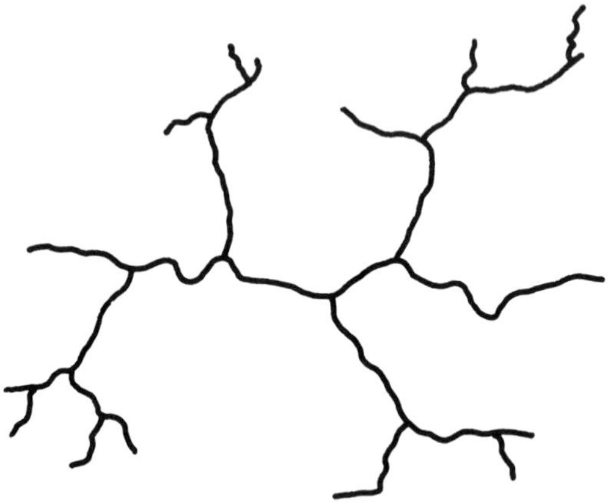

FIG. 16.1. A branched polymer.

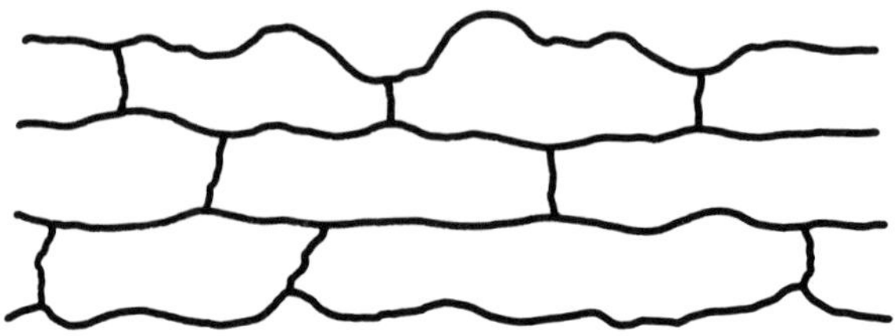

FIG. 16.2. A cross-linked linear polymer.

2. CYCLOLINEAR POLYMERS

These are prepared by the linking together of cyclic oligomeric ring systems by means of difunctional reagents. A typical cyclolinear polymer structure is shown in II. The simplest method for the preparation of polymers of this type

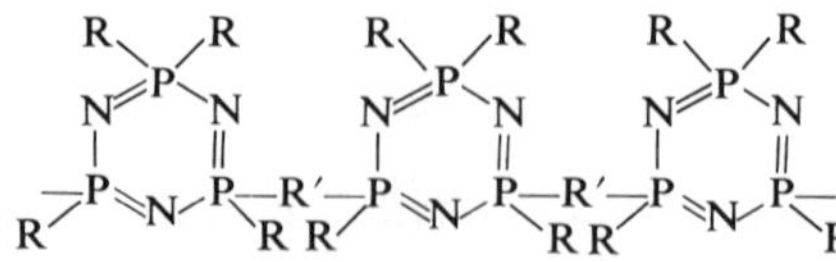

II

involves the reaction of a non-geminal dichlorotetrakis(organo)cyclotriphosphazene with the disodium salt of a diol or with a diamine. The initial step in the formation of such a polymer is illustrated by the formation of III. It is important that only two active groups (usually halogen) are present on the initial trimer to avoid the possibility of cyclomatrix polymer formation.

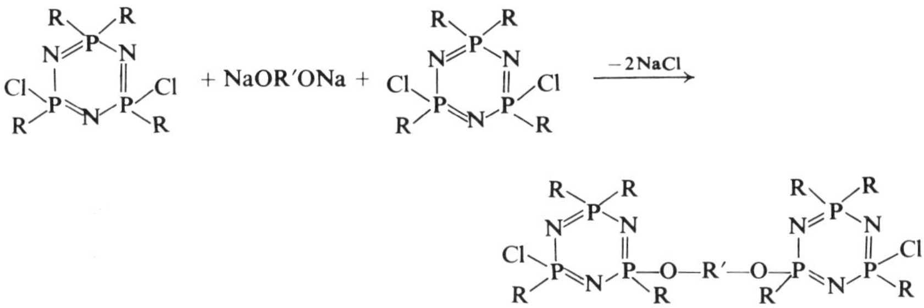

III

3. Cross-linked Cyclomatrix Polymers

The reactions of hexachlorocyclotriphosphazene or octachlorocyclotetraphosphazene with diols or diamines lead to the formation of three-dimensional, rigid, cross-linked networks which have the idealized general structure shown in IV. The degree of cross-linking is reduced if some of the halogen atoms have

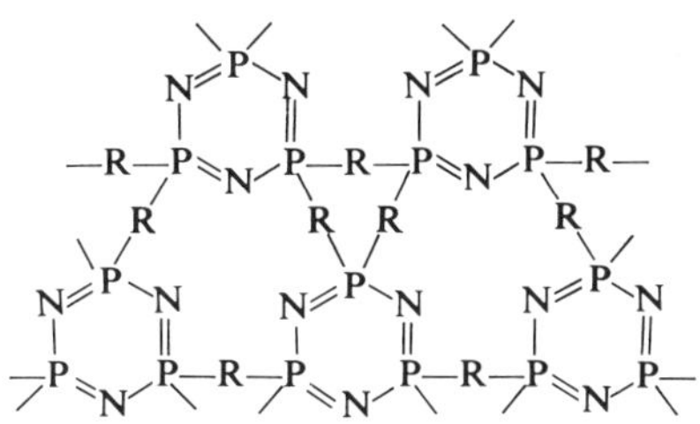

IV

Cyclomatrix polymer

previously been replaced by unreactive ligands. Related polymers can be prepared by the addition reactions of phosphazenes which contain unsaturated organic groups, or by the thermal rearrangements of some spirophosphazenes.

B. Physical Characteristics of Different Polymers

Much of the work in phosphazene chemistry has been connected with attempts to prepare new polymer systems, and it seems clear that there will be an even greater emphasis in this direction in the future. Because the techniques and terminology of polymer chemistry are often unfamiliar to the classical synthetic chemist, the following brief synopsis outlines some of the main principles of the field.

1. Physical Transitions

Low molecular weight compounds have sharp melting points, but high polymers generally do not. Consider first the effect of raising the temperature of a linear, high molecular weight polymer with negligible branching and no cross-links. At low temperatures, the polymer will be a glass. As the temperature is raised, a point will be reached at which the polymer changes from a glass to a rubbery or leathery state. This point is known as the *glass transition temperature* (T_g). The transition usually occurs over a narrow temperature range of several degrees, and it can be detected by direct thermomechanical tests, by exotherms in differential thermal analysis curves, or by the onset of line-narrowing in broad-line NMR experiments. For low molecular weight polymers, the glass transition temperature may vary with chain length but, when the chain length exceeds about a thousand linkages, the transition temperature is virtually independent of molecular weight. On a molecular level, the glass transition temperature represents the point below which torsional motions of the backbone bonds are restricted and where the chains may be entangled in a rigid network. Above this temperature, the skeletal bonds have sufficient energy to undergo appreciable torsional motions, and the polymer can be stretched or deformed. Stretching of the sample induces the chains to uncoil by torsional motions of the skeletal bonds, and a quick release of the tension will allow them to recoil to a state of higher entropy. This is the basis of rubbery or elastomeric behavior. However, if continuous tension is applied to the rubbery polymer, adjacent chains will slip past each other and the material will be permanently elongated.

Above the glass transition temperature, two alternative types of behavior are possible. Firstly, if the polymer is amorphous, temperature increases will convert it initially to a gum and then over a wide temperature range, to a liquid. Secondly, if the polymer is crystalline (because the chains can pack together in a regular manner) the rubbery phase may persist as the temperature is raised, until melting to a liquid occurs at a sharp *crystalline melting point*. Crystalline polymers often show one or more crystalline transitions below the melting temperature, and these can be detected by endotherms in differential thermal analysis curves.

The presence of chain branching has no marked effect on the glass transition, but highly branched polymers are less likely to crystallize, and crystalline melting temperatures may be difficult to detect. Cross-links have a pronounced effect on the behavior of the polymer during its rubbery phase. When the polymer is stretched, the cross-links prevent the chains from sliding past each other, and permanent elongation does not occur. Thus, light cross-linking is often necessary to provide elastomeric behavior.

Cyclolinear polymers are similar to linear species in their thermal behavior. The torsional flexibility of the chain is usually determined by the flexibility of the linking units. However, such polymers may prove difficult to crystallize because of the difficulty involved in packing the chains in an orderly crystalline array.

Cross-linked cyclomatrix polymers are rigid, intractable materials. They retain their shape and hardness up to very high temperatures, at which point thermal cleavage of the cross-links or oxidative reactions bring about decomposition. However, polymers of this type are usually much more thermally stable than linear or cyclomatrix polymers, although the uses for them may be more limited.

2. The Effect of Solvents

Linear and cyclolinear polymers, whether branched or not, usually dissolve in specific solvents. Solution may be very slow, but viscous true solutions are eventually formed. Lightly cross-linked polymers swell in aqueous or organic media, but they do not dissolve. The more highly cross-linked structures swell the least, and this can be used as a measure of the degree of cross-linking. Slightly more than one cross-link per chain is sufficient to prevent the formation of a true solution. Cross-linked cyclomatrix polymers are insoluble in all inert media, although, if the cross-linking is light, some softening or slight swelling may be observed.

True solutions of linear or branched polymers are usually viscous, with the viscosity increasing with chain length. Thus, the measured viscosity extrapolated to infinite dilution (the "intrinsic viscosity") is a measure of molecular weight. The degree of chain branching may also be estimated from solution viscosity measurements.

3. Molecular Weights of High Polymers

The molecular weights of insoluble polymers cannot be measured, since the techniques for the measurement of molecular weights of high polymers require the availability of a true solution of the polymer. These techniques include light scattering, ultracentrifugation, gel permeation chromatography, and solution viscosity measurements. Colligative methods, such as vapor pressure osmometry, boiling point elevation, freezing point lowering, or osmotic pressure changes, cannot be used with accuracy for polymers with molecular weights over 100,000.

High polymers do not consist of one discrete species with a specific molecular weight. A distribution of molecular weights is always found, as shown in Fig. 16.3. An estimate of the width and shape of the distribution curve can be obtained by comparisons of the results from number average (colligative)

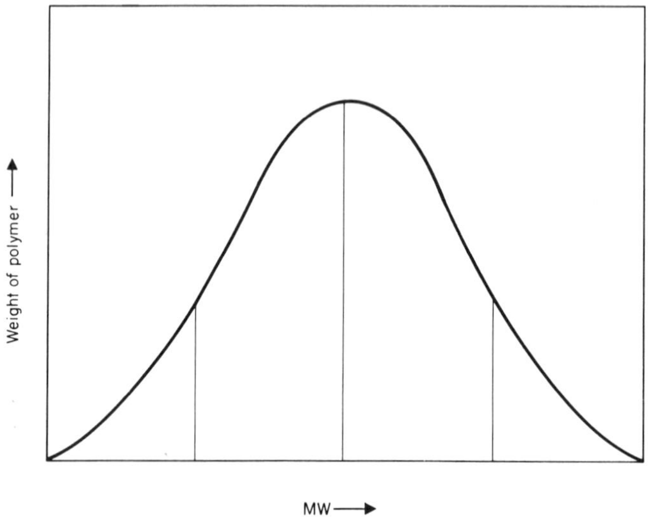

FIG. 16.3. A molecular weight distribution curve for a hypothetical polymer.

methods, which favor the lower polymers and oligomers, and weight average (e.g., light scattering) methods which are most sensitive for the higher molecular weight species. Molecular weight distribution curves can also be obtained conveniently from ultracentrifugation and gel permeation chromotography experiments.

4. X-RAY DIFFRACTION OF POLYMERS

A linear polymer will crystallize if the chains can pack together in a regular manner. In practice, crystalline segments of the polymer matrix are usually separated by amorphous, random-coil sections. Crystallization is facilitated by heating and cooling sequences above the glass transition temperature and below the crystalline melting temperature, but more especially by orientation of the polymer. Orientation occurs when a film or fiber of the polymer is stretched. Stretching or cold-drawing of the material has the effect of aligning the polymer chains along one axis, and this facilitates the formation of crystalline regions. An oriented, crystalline polymer may be subjected to X-ray diffraction investigations. The "fiber diffraction" patterns yield information about the polymer repeating distance along the fiber axis (from the layer-line separation distance), and intensity measurements often provide clues about the polymer conformation.

C. Linear Poly(dihalophosphazenes)

Polymers which fall into this category include high-polymeric poly(difluorophosphazene), $(NPF_2)_n$, polydichlorophosphazene, $(NPCl_2)_n$, and polydibromophosphazene, $(NPBr_2)_n$. The low-polymeric linear, end-capped chlorophosphazenes are considered only briefly here since they are reviewed in detail in Chapter 4. Because of its similarities to poly(dihalophosphazenes), the pseudo-halo derivative, $[NP(NCS)_2]_n$, is also discussed in this section.

1. POLY(DIFLUOROPHOSPHAZENE)

Poly(difluorophosphazene), $(NPF_2)_n$, is prepared by the thermal polymerization of the trimer or tetramer, $(NPF_2)_{3\ \text{or}\ 4}$, in a sealed high-pressure system at 350°C for 15 hours or longer.[1–3] It is a colorless or pale amber elastomer, with a temperature range of elasticity from about −70° to 250°C.[3–5] The glass transition temperature, determined by differential thermal analysis is −96°C[5] and a physical change from glass to elastomer can also be observed at the same

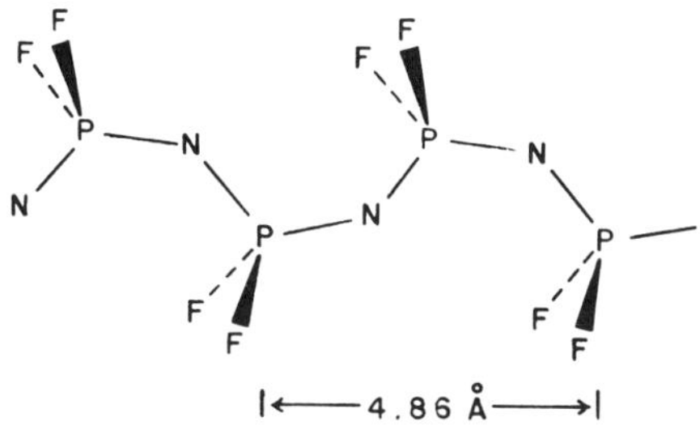

FIG. 16.4. *Cis–trans* planar conformation of poly(difluorophosphazene).

temperature. A crystalline transition is also detectable at −68°C from DTA experiments[5] and the polymer liquefies at 270°–280°C. Melting at atmospheric pressure is accompanied by loss of the volatile tetramer, $(NPF_2)_4$, presumably by depolymerization.

Poly(difluorophosphazene) is exceedingly sensitive to hydrolysis, decomposition being evident after only a few minutes exposure to a moist atmosphere. Hydrolytic attack on the phosphorus–fluorine bonds is accompanied by a loss of elasticity and eventual liquefaction.

The conformational structure of this polymer has been studied by X-ray diffraction techniques.[6] Two different crystallographic forms can be detected at low temperatures from X-ray diffraction photographs, with a transition between the two being observed near −40°C. The higher temperature form

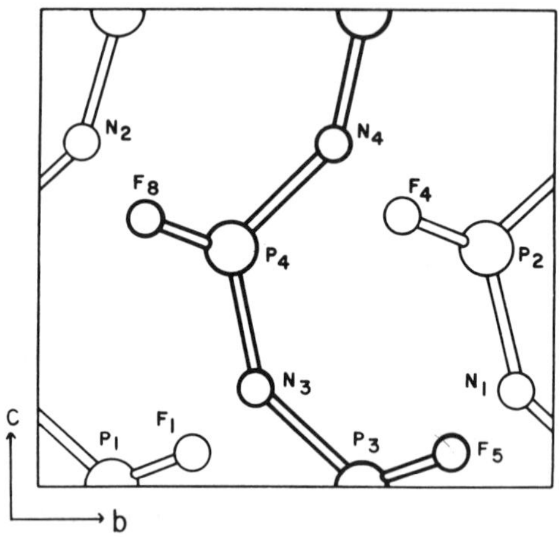

Fig. 16.5. Projection showing atomic positions in the *b–c* plane of a section of the orthorhombic unit cell of poly(difluorophosphazene). The *c*-axis repeat is 4.86 Å, and *b* is 5.38 Å. From Allcock and Stroh.[7]

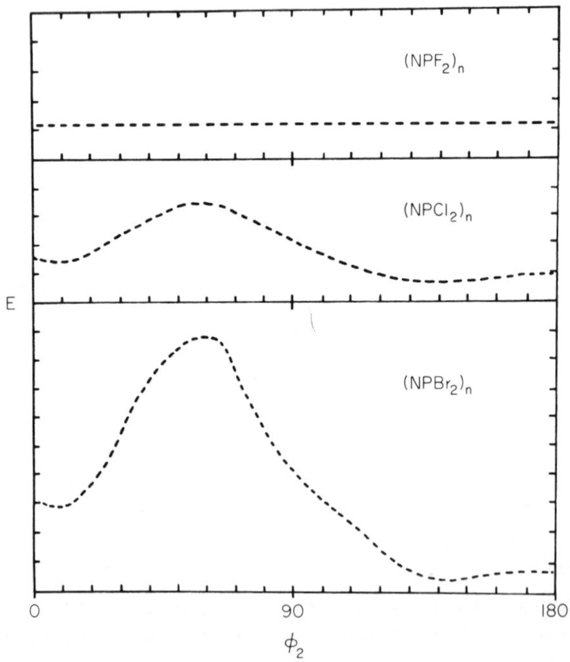

FIG. 16.6. Calculated torsional barriers for $(NPF_2)_n$, $(NPCl_2)_n$, and $(NPBr_2)_n$. E is in Kcal and ϕ_2 represents the P–N torsional angle.

(conformer A) is found between $-36°$ and 25°C. It has a chain-repeating distance of 6.49 ± 0.05 Å. At -56°C a different X-ray pattern is seen (conformer B) which indicates a fiber repeat distance of 4.86 ± 0.1 Å. The unit cell of conformer B is orthorhombic with $a = 8.69$ Å, $b = 5.38$ Å, and $c = 4.86$ Å. The space group is $Cmc2_1$. Analysis of the intensity data by Fourier techniques indicates that the polymer occupies a *cis–trans*-planar conformation (Fig. 16.4). Two monomer units occupy one chain-repeating sequence, and two chains pass through the *c*-centered cell (Fig. 16.5).[7]

One of the unusual properties of poly(difluorophosphazene) is its low-temperature elasticity. This reflects a very high degree of torsional mobility in the backbone. Thus, ^{19}F NMR experiments show that the spectrum remains a doublet at least down to -70°C, a result of the facile torsional freedom in this chain.[6] Calculations have been performed which sum the intramolecular van der Waals and Coulombic interactions for a wide range of different conformations.[8] The results suggest that the torsional barrier to 360° rotation of a skeletal bond in $(NPF_2)_n$ is less than 0.1 kcal, one of the lowest barriers known in any polymer. A comparison can be made between the calculated torsional barriers in $(NPF_2)_n$, $(NPCl_2)_n$, and $(NPBr_2)_n$, and the results are shown in Fig. 16.6. It can be seen that, if the intramolecular forces determine the barrier, then the torsional flexibility should increase for $(NPX_2)_n$ in the order Br < Cl < F, and the glass transition temperatures should rise in the same order. This is found to be the case.

2. Poly(dichlorophosphazene)

It is important to distinguish between two types of chlorophosphazene polymers—the oily materials which are formed during chlorophosphazene synthesis reactions or by the end-capping cleavage of cyclic derivatives, and the rubbery, high molecular weight polymers.

a. *Low Molecular Weight, End-Capped Polymers*

Oily, low molecular weight, linear chlorophosphazenes are formed during the reactions of phosphorus pentachloride with ammonium chloride in *s*-tetrachloroethane or chlorobenzene.[9–13] These products constitute from 10 to 25% of the reaction products, the remainder being cyclic species. As discussed in Chapter 4, these linear chlorophosphazenes have structures of the type shown in V or VI. Similar linear products are formed if cyclic or high polymeric chlorophosphazenes are treated with phosphorus pentachloride at temperatures above 300°C.[10, 14] Presumably, a complete gradation of properties, from

$$[Cl_3P{=}N{+}PCl_2{=}N{+}_n PCl_3]^{\oplus}[PCl_6]^{\ominus} \qquad [Cl_3P{=}N{+}PCl_2{=}N{+}_2 PCl_3]^{\oplus}Cl^{\ominus}$$

V VI

oils to elastomers, parallels an increase in chain length, since rubbery products have also been isolated from the chlorophosphazene synthesis reaction. If the reaction between ammonium chloride and phosphorus pentachloride is conducted in the presence of metal halides, end-capped linear products of formula $Cl{-}(PCl_2{=}N)_n\,PCl_3{-}SbCl_2$, $[Cl_3P{=}N{-}PCl_2{=}N{-}PCl_3]^+[SbCl_6]^-$, $Cl(PCl_2{=}N)PCl_3{-}AlCl_2$, $Cl(PCl_2{=}N){-}PCl_3{-}FeCl_2$, or $Cl(PCl_2{=}N){-}PCl_3{-}TlCl_2$ are formed.[15] These species are stable at high temperatures but they are sensitive to hydrolysis. Because these compounds are of low molecular weight, they do not display elastomeric or thermoplastic properties.

b. *High-Polymeric Chlorophosphazenes*

Physical Properties. The thermal polymerization of cyclic chlorophosphazenes to poly(dichlorophosphazenes) (VII) was discussed in Chapter 15. The

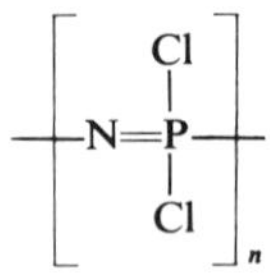

VII

polymer prepared at 250°–300°C is a soft, colorless, transparent elastomer, but prolonged polymerization times favor the formation of cross-linked material.[16, 17] Increased cross-linking yields a tougher, more rubbery polymer. The un-cross-linked polymer is soluble in benzene, toluene, or tetrahydrofuran to form highly viscous solutions. In solution, the polymer is suitable as a reagent for substitution reactions to form poly(organophosphazenes). However, the polymer may cross-link slowly in solution, but this process can be retarded by the addition of small amounts of stannic chloride.[17] The lightly cross-linked polymer is swelled considerably by many organic liquids, but it does not dissolve. In appearance and superficial physical properties, the un-cross-linked and cross-linked forms are very similar.

Poly(dichlorophosphazene) has a glass transition temperature of −63°C (DTA).[17] At low temperatures or when stretched, the polymer is crystalline. A crystalline melting temperature exists near −30°C, and the range of polymer elasticity extends from this temperature to over 300°C. The lightly cross-linked polymer shows many of the characteristics of an ideal elastomer. For example, very little viscous flow takes place when the polymer is stretched to more than twice its original length, and it returns to its original length after a 1.9 times extension for 1 minute under a stress of 1 kg/cm^2 over a temperature range of 50° to 160°C.[18] Even after several hours elongation at room temperature, little permanent elongation occurs, and prolonged stretching at high temperature is required before small amounts of viscous flow are detectable. At 25°C the elastic modulus is 2 kg/cm². The effect of heating the tensioned polymer at a constant deformation of 200% has been examined, and the results are shown in Fig. 16.7.[19] Thus, heating of the polymer from ~7° to 50°C causes an increase in stress by an amount which is more than proportional to the absolute temperature (sections II and III of Fig. 16.7). However, when the polymer is cooled, the tension does not return immediately to its former value because of crystallization and melting phenomena. Later work[18] has indicated that the elastic modulus is proportional to the absolute temperature, with values of 2.5 kg/cm² at 313°K and 3.2 kg/cm² at 413°K being reported. The density of the amorphous polymer is 1.98 at 273°K.

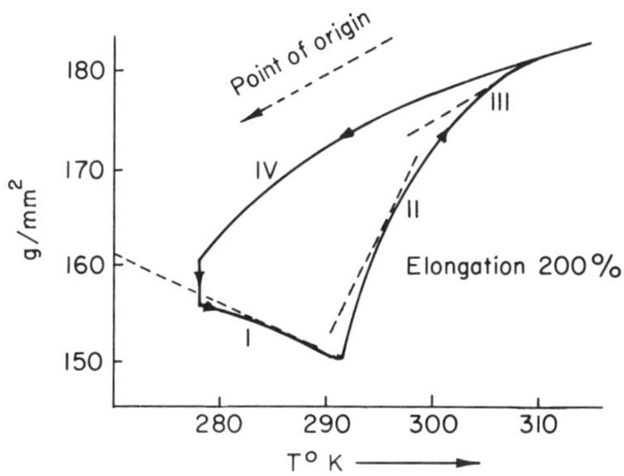

FIG. 16.7. Variation of the tension of poly(dichlorophosphazene) as a function of temperature. From Meyer, Lotmar, and Pankow.[19]

The infrared and ultraviolet spectra of poly(dichlorophosphazene) are similar to those of the cyclic trimer and tetramer. For example, peaks at 1300, 1230(s), and 745 cm^{-1} in the infrared region are assigned to skeletal vibrations.[17] There is no significant ultraviolet absorption below 220 mμ,[20] a fact which argues against a broadly delocalized π-electron structure.[17] The ^{31}P NMR spectrum consists of a peak at +17.4 ppm relative to triethyl phosphate, the positive shift being characteristic of many high-polymeric phosphazenes.[17]

The weight average molecular weight of un-cross-linked poly(dichlorophosphazene) can be as high as 1×10^6 to 2×10^6, although the molecular weight distribution is often very broad. Thus, the higher molecular weight species probably contain 15,000 or more repeating units. Most of the information about molecular weights has been derived from light-scattering data[20] and by inference from the chain length of organo-substituted derivatives.[17] The tendency of poly(dichlorophosphazene) to cross-link in solution seriously complicates attempts to obtain light-scattering data. However, Knoesel,

Parrod, and Benoit[20] have obtained light-scattering data for solutions of the polymer in toluene and chloroform. For a polymer of molecular weight 115,000 the radius of gyration was calculated to be in the region of 200 Å, and, in chloroform solution, the second virial coefficient was found to be negative. Data for the relationship between molecular weight and solution viscosity are shown in Fig. 16.8, together with related values obtained by Patat and Kollinsky[21] and

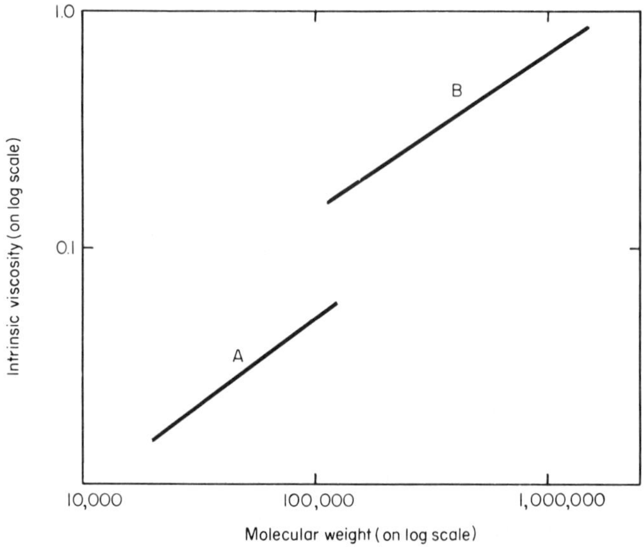

FIG. 16.8. Molecular weight–intrinsic viscosity relationship for poly(dichlorophosphazene) in toluene. Curve A shows number average molecular weight values (osmotic pressure) (from the data of Patat and Kollinsky[21]). Curve B shows weight average molecular weight values (light scattering) (from the data of Knoesel, Parrod, and Benoit.[20]).

Patat and Frömbling[22] from osmotic pressure measurements. The relationship between intrinsic viscosity, μ, and molecular weight is given by $[\mu] = KM^a$, where K is 3.89×10^{-3} in toluene or 1.32×10^{-1} in chloroform, M is the weight average molecular weight, and a is 0.71 and 0.56, respectively, in the two solvents.[20]

Molecular Conformation. The molecular structure of poly(dichlorophosphazene) has been investigated with the use of X-ray diffraction techniques.[7, 19, 23] The purified polymer, after removal of cyclic oligomers, is a colorless, transparent, rubbery material. In the unstretched state, it is amorphous to X-rays. However, when fibers of the polymer are protected from moisture and stretched, particularly at temperatures below 25°C, crystalline-type X-ray diffraction patterns can be obtained. A typical rotation-type "fiber pattern" is shown in Fig. 16.9.

The layer-line separation corresponds to a fiber repeat distance of 4.92 Å. Meyer, Lotmar, and Pankow[19] deduced an orthorhombic unit cell with the dimensions $a = 11.07$ Å, $b = 4.92$ Å, and $c = 12.72$ Å. However, a smaller unit cell ($a = 5.50$ Å, $b = 12.72$ Å, $c = 4.92$ Å) can also be justified.[7] Meyer and coworkers[19] proposed that two monomer units occupy the fiber repeating distance with four chains passing through the cell. Giglio, Pompa, and Ripamonti[23] used an optical transform method to simulate the X-ray patterns

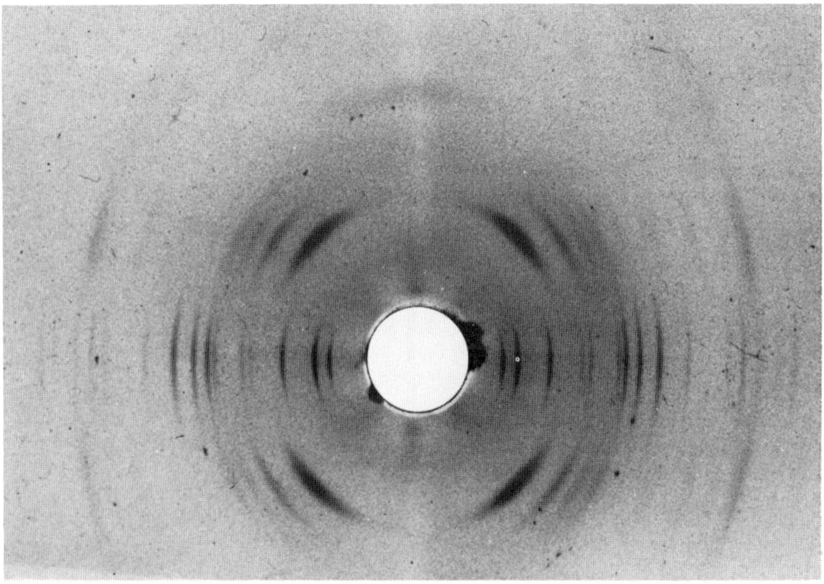

FIG. 16.9. X-Ray fiber diffraction pattern of poly(dichlorophosphazene). The layer-line separation distance corresponds to a polymer fiber axis repeating distance of 4.92 Å. From Allcock and Stroh.[7]

expected from different polymer conformations, and they concluded that a distorted *cis–trans* planar arrangement (Fig. 16.10) corresponded best to the experimental patterns.

More recent Fourier transform work[7] has shown that the observed X-ray intensities cannot be explained in terms of an exactly planar *cis–trans*-planar chain, and that a slight helical distortion appears necessary to explain the results. An analysis of infrared data for poly(dichlorophosphazene) has also provided evidence for an approximately *cis–trans*-planar type of structure.[24] It is interesting to note that identical X-ray patterns are obtained irrespective of whether the polymer is derived from the trimer, $(NPCl_2)_3$, or the tetramer, $(NPCl_2)_4$.[7] Unlike poly(difluorophosphazene), there is no evidence of a conformational change as the temperature is lowered from 25° to –80°C.[7]

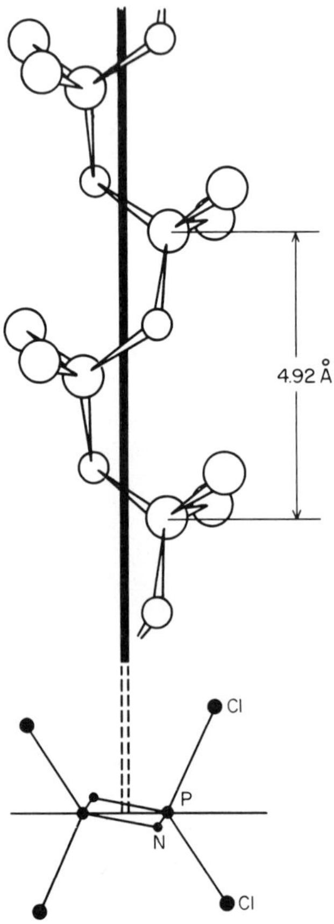

FIG. 16.10. Distorted *cis–trans* planar conformation of poly(dichlorophosphazene). From Giglio, Pompa, and Ripamonti.[23]

The underlying reasons for the backbone mobility and the conformation assumed by poly(dichlorophosphazene) have been investigated by theoretical methods based on the calculation of nonbonding intramolecular potentials.[8] Use of a 6:12 Lennard-Jones potential to sum the expected interactions within the segment, VII, for independent 360° torsion of bonds $P_1—N_2$ and $P_2—N_2$, yields the energy surface shown in Fig. 16.11. One of the deepest

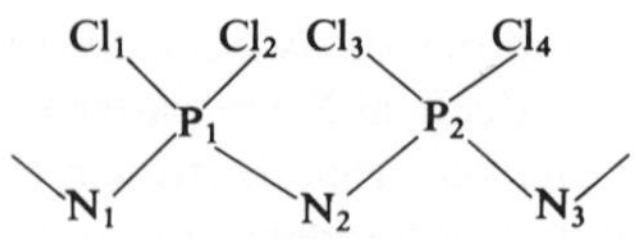

VII

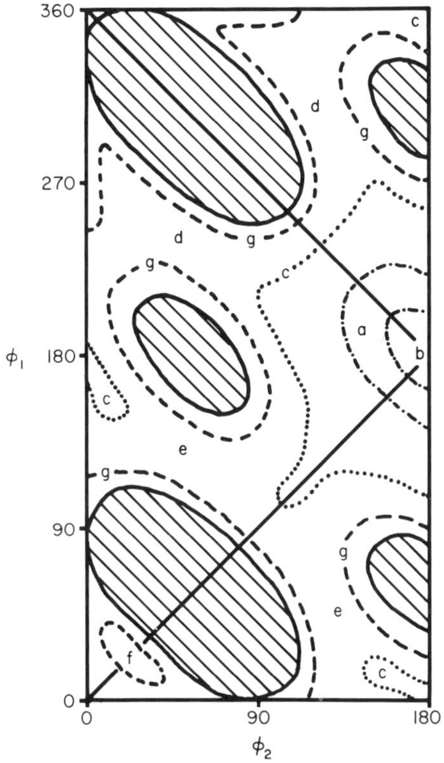

FIG. 16.11. Calculated conformational potential energy surface for poly(dichlorophosphazene), ϕ_1 and ϕ_2 refer to torsional angles of adjacent skeletal bonds from the *trans*-planar (zig-zag) conformation. The designations a, b, c, d, . . . denote points of increasing energy. The shaded areas represent especially high repulsion conformations. The experimentally determined structure corresponds to the distorted *cis-trans*-planar conformation at position c. From Allcock, Meister, and Lehnsen.[8]

minima (*c* in Fig. 16.11) on the matrix is at $\phi_1 = 156°$, $\phi_2 = 14°$, which corresponds exactly to the distorted *cis–trans*-planar conformation deduced from X-ray diffraction data.[23] The only competing minima (at *a* and *b* in Fig. 16.11) are of almost equal energy, but these represent *cis–cis* conformations which are unlikely in a long chain because of long-range repulsions. Thus, the conformation assumed by poly(dichlorophosphazene) appears to be the consequence of nonbonding intramolecular forces rather than π-bonding forces. The low glass transition temperature of this polymer (–63°C) can also be rationalized in these terms, since the calculated barrier to 360° torsion of a skeletal bond is only 2.8 kcal, and 80° torsional rocking motions between two equivalent minima (*c* to *c*) encounter a calculated barrier of only 0.01 kcal.[8] Comparisons of the barriers calculated for this polymer with those for the fluoro and bromo derivatives are shown in Fig. 16.6.

Reactions of Poly(dichlorophosphazene). Poly(dichlorophosphazene) degrades slowly in contact with water or atmospheric moisture. This so-called "aging" process is accompanied by a hardening of the elastomer and by its eventual conversion into inorganic salts. The initial hardening of the polymer has been ascribed to crystallization and to the formation of cross-links, VIII.[25, 26] The subsequent reaction is predominantly a hydrolysis process

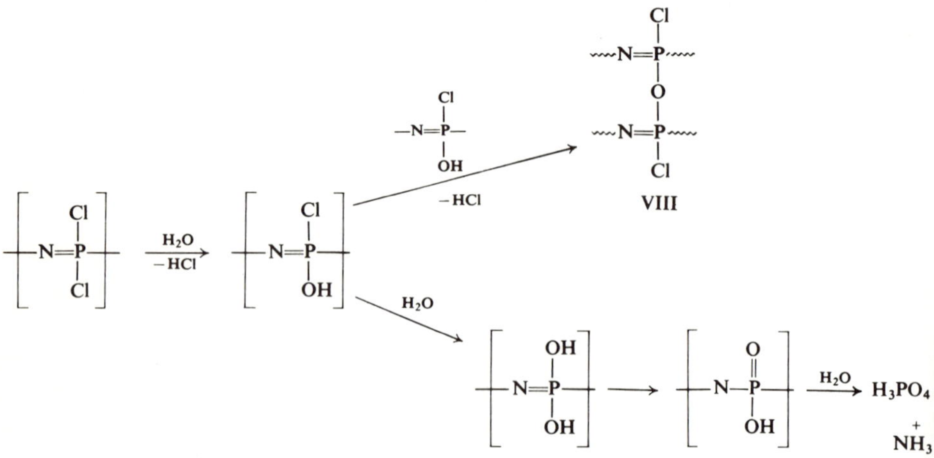

which yields hydrochloric acid, ammonium chloride, and ammonium phosphate or phosphoric acid.[25, 26] When hydrolysis is effected with the use of a 5% aqueous acetone medium, the reaction slows markedly after ~58% of the halogen has been removed.[25] It appears that chain scission does not occur until after 50% of the phosphorus–chlorine bonds have been cleaved.[25] However, in water at 100°C, hydrolysis occurs rapidly and completely.

Attempts have been made to improve the hydrolytic stability of poly(dichlorophosphazene) by radiation grafting with styrene.[27] ^{60}Co γ Radiation was employed to produce free-radical sites on the phosphazene backbone and these initiated polymerization of the styrene monomer. Up to 53.8% of polystyrene could be incorporated into the polymer in this way, but the chlorophosphazene component remained sensitive to moisture.

Poly(dichlorophosphazene) also undergoes a series of nucleophilic substitution reactions to form poly(organophosphazenes). These processes will be discussed in a following section. The thermal behavior of poly(dichlorophosphazene) is discussed in Chapter 15.

It seems unlikely that applications will be found for poly(dichlorophosphazene) except as a reaction intermediate. Various uses for this material have been proposed in the patent literature in connection with the coating of ceramics, glass, or asbestos fibers, in molding compounds,[28–30] in semiconducting copolymer compositions,[31] and in flame retardant polymer compositions,[31a] but the hydrolytic instability is a strong deterrent to the development of a large-scale direct use.

3. POLY(DIBROMOPHOSPHAZENE)

Poly(dibromophosphazene), $(NPBr_2)_n$, is an amber-colored, tough, rubbery, material, prepared by the thermal polymerization of the cyclic trimer or tetramer (see Chapter 15). The thermal range of rubbery elasticity is from the glass transition temperature near −15°C to the liquefaction point near 270°C.[5] The thermally produced polymer is amorphous to X-rays in the unstretched and stretched states over the temperature range of 30° to −30°C.[5] However, polymer prepared by the γ irradiation of crystals of the trimer showed Debye-Scherrer-type diffraction rings.[32] The most intense reflections corresponded to *d* spacings very similar to those originating from poly(dichlorophosphazene). Thus, the molecular conformations may be similar in the two cases. The hydrolytic stability of poly(dibromophosphazene) is comparable to that of poly(dichlorophosphazene).

Attempts have been made to simulate the potential-energy changes which accompany torsion of the skeletal bonds in this polymer with the use of the same techniques as those described above for the fluoro and chloro analogues.[8] The results indicate that the torsional barrier for the bromo derivative is appreciably higher than for the other two polymers. A barrier of ~8.4 kcal was calculated and this suggests a reason for the relatively high glass transition temperature observed.

4. POLY[BIS(ISOTHIOCYANO)PHOSPHAZENE]

Poly[bis(isothiocyano)phosphazene] (IX) is a yellow, rubbery elastomer

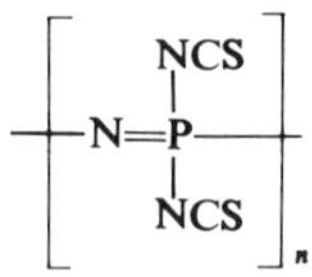

IX

which is synthesized by the polymerization of the cyclic trimer or tetramer, $[NP(NCS)_2]_{3 \text{ or } 4}$.[33, 34] It is insoluble in most organic solvents but soluble in *N,N*-dimethylacetamide, dimethylformamide, and water.[35] When freshly prepared, the polymer is soluble in tetrahydrofuran, but prolonged storage, particularly in a moist atmosphere, results in decomposition and cross-linking.[35] X-Ray diffraction experiments indicate the presence of some crystallinity in the polymer, even in the unstretched state.

Since isothiocyano groups themselves are capable of addition-type polymerization reactions, the possibility existed that isothiocyanophosphazene polymers were in reality cyclolinear species formed by the linkage of ligands without phosphazene ring cleavage. That this interpretation is incorrect has been shown by an alternative synthesis of IX by the metathetical exchange of halogen in poly(dichlorophosphazene) by isothiocyanate.[35] The isothiocyano

$$(NPCl_2)_n + KCNS \rightarrow [NP(NCS)_2]_n + KCl$$

polymer prepared by this route was of lower chain length than that produced by polymerization of the trimer, but the infrared spectra and X-ray diffraction patterns of the two polymers were very similar.[35]

The practical applications of poly[bis(isothiocyano)phosphazene] appear to be limited because of its decomposition in contact with the air. However, it is more stable than the fluoro, chloro, or bromoderivatives discussed earlier.

D. Linear Organophosphazene Polymers

A variety of different linear poly(organophosphazenes) have been described. These include poly(alkoxy- and aryloxyphosphazenes), poly(aminophosphazenes), and poly(alkyl- and arylphosphazenes). The first two types of polymers are synthesized by replacement of halogen atoms in poly(dichlorophosphazene) by alkoxy or aryloxy groups. Direct synthesis methods as well as substitution routes have been developed for polymers in the third category.

1. Poly(alkoxyphosphazenes) and Poly(aryloxyphosphazenes)

These polymers have the general structure shown in X. Attention will be

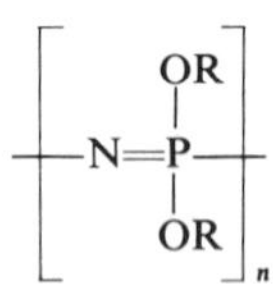

X

directed first to their synthesis, then to their chemical and physical properties, and lastly to their molecular weight distribution and molecular conformation.

a. *Synthesis*

Although the elastomeric and high-temperature properties of poly(dichlorophosphazene) have attracted wide attention, no useful applications have been found for this material, because of its hydrolytic instability. This hydrolytic sensitivity is associated with the presence of phosphorus–halogen bonds, and replacement of the halogen atoms by nonhydrolyzable organic groups should improve the stability to moisture. Early attempts to replace the halogen by treatment of poly(dichlorophosphazene) with alcohols, phenols, or their alkali metal or triethylammonium salts yielded partly substituted, hydrolyzed, or cross-linked materials.[36–40] Since phosphorus–halogen bonds often remained in these polymers, they were susceptible to hydrolytic attack.

The first synthesis of fully substituted, un-cross-linked organophosphazene polymers was reported in 1965 and 1966 by Allcock, Kugel, and Valan,[16, 17] who treated poly(dichlorophosphazene) solutions with sodium alkoxides or aryloxides.

$$(NPCl_2)_n + 2n\ NaOR \rightarrow [NP(OR)_2]_n + 2n\ NaCl$$

The important difference between this work and the earlier studies was the recognition that poly(dichlorophosphazene) could be isolated as a soluble, un-cross-linked polymer (see Section C.2). Use of this material as a substitution substrate permitted total halogen replacement to occur, whereas the use of the cross-linked poly(dichlorophosphazene) inevitably led to partial substitution, hydrolytic instability, and insolubility resulting from the presence of cross-links. The following description outlines the synthesis of poly[bis-(trifluoroethoxy)phosphazene] by a typical procedure.[17]

A solution of poly(dichlorophosphazene) (300 g, 2.58 moles) in benzene (1500 ml) is added over 3 hours to a stirred solution of sodium 2,2,2-trifluoroethoxide, prepared from sodium (120 g, 5.22 g-atoms) and trifluoroethanol (600 g, 6 moles) in tetrahydrofuran (600 ml). All solvents and reagents must be scrupulously dry. The initial reaction exotherm sustains refluxing of the solvent, and heating is eventually applied to maintain the mixture at reflux temperature for 16 hours. The reaction mixture is then cooled at 25°C, acidified with concentrated hydrochloride acid to remove excess sodium trifluoroethoxide, filtered to remove polymer and sodium chloride, and the latter is then removed by washing with water. Residual sodium chloride is removed by precipitation of the polymer from acetone into water, and substituted oligomers are removed by precipitation from acetone into benzene. The yield, which is typically about 75%, depends on the proportion of $(NPCl_2)_3$ present in the original poly(dichlorophosphazene). The properties of poly[bis(trifluoroethoxy)phosphazene] are discussed in a later section.

A variety of different polymers can be prepared by this technique with the use of different nucleophiles. Higher reaction temperatures are required to bring about total replacement of chlorine by bulky groups. For example, total substitution by phenoxy groups requires a prolonged reaction at temperatures up to 126°C.[17] Copolymers can be prepared by the interaction of poly(dichlorophosphazene) with two or more different sodium alkoxides or aryloxides. Use of a mixture of sodium trifluoroethoxide and sodium heptafluorobutoxide yields a trifluoroethoxy-heptafluorobutoxyphosphazene copolymer.[41, 42] It is possible to envisage the synthesis of a large number of alkoxy- and aryloxyphosphazene polymers by these methods, and it seems clear that these polymers will attract considerable attention in the coming years.

The following high polymers have been synthesized and characterized fully: $[NP(OMe)_2]_n$,[17] $[NP(OEt)_2]_n$,[17] $[NP(OCH_2CF_2CF_2H)_2]_n$,[5, 42] $[NP(OCH_2CF_3)_2]_n$,[17] $[NP(OCH_2C_2F_5)_2]_n$,[5, 42] $[NP(OCH_2C_3F_7)_2]_n$,[5, 42] $[NP(OCH_2CF_3)(OCH_2C_3F_7)]_n$,[41, 42] $[NP(OCH_2CF_2CF_2H)(OCH_2C_6F_{12}H)_2]_n$,[42] and $[NP(OPh)_2]_n$.[17] Other polymers have been described[43, 44] with the structures $[NP(OR)_2]_n$ where OR is $-OC_6H_4CF_3$-*m*, $-OC_6H_4Cl$-*p*, $-OC_6H_3Cl_2$-2,4, $-OC_6H_4F$-*p*, $-OC_{10}H_7$-β, $-OC_6H_4C_6H_5$-*p*, and $-OCH_2(CF_2)_6CF_3$, but elemental analyses were not reported for these compounds.[43]

b. *General Properties*

The properties of the polymers depend markedly on the nature of the substituent groups.

Poly[bis(methoxy)phosphazene], $[NP(OMe)_2]_n$, is a colorless, rubbery material, which is soluble in methanol, dioxane, tetrahydrofuran, dimethoxyethane, chloroform, pyridine, dimethylformamide, and acetonitrile, but insoluble in water, ethanol, diethylether, benzene, and acetone.[17] The glass transition temperature is −76°C. No evidence of crystallinity could be found from X-ray diffraction measurements.[5] The thermal stability of this polymer is not high since the intrinsic solution viscosity fell from 1.02 dl/g to 0.095 dl/g after 8 hours heating of the solid at 100°C. Under these conditions, the polymer changes from a rubbery material to a gum. However, the polymer is quite stable to water. The properties of this polymer are compared with those of related derivatives in Table 16.1.

Poly[bis(ethoxy)phosphazene], $[NP(OEt)_2]_n$, is a colorless elastomer which is soluble in alcohols, benzene, ethers, and ketones, but insoluble in aliphatic hydrocarbons and water.[17] Gelation of the polymer often occurs when solutions are stored for long periods of time. It is not known whether this results from cross-linking reactions or from a phosphazene–phosphazane rearrangement of the type discussed in Chapter 13. The glass transition temperature of −84°C is one of the lowest recorded for an organophosphazene. X-Ray diffraction photographs of stretched fibers show only diffuse halos which are indicative of an amorphous structure.[5] The thermal stability of this polymer is also low.[17] When the elastomer is heated at 100°C for 8 hours, the material changes to a gum, and the intrinsic viscosity falls from 1.63 dl/g to 1.09 dl/g. At 150°C the intrinsic viscosity falls to 0.16 in 8 hours. The polymer is hydrolytically stable.

Poly[bis(trifluoroethoxy)phosphazene], $[NP(OCH_2CF_3)_2]_n$,[17] is one of the

TABLE 16.1

SELECTED PROPERTIES OF SOME POLY(ALKOXY- AND ARYLOXYPHOSPHAZENES)

Compound	Molecular weight	Intrinsic viscosity (dl/g)	^{31}P NMR (ppm)	Infrared P=N peak (cm^{-1})	T_g (°C)	T_m (°C)	T_{10}	References
$[NP(OMe)_2]_n$	640,000	1.2	+3.5	1250–1330	−76	—	—	5, 17
$[NP(OEt)_2]_n$	—	2.9	+7.3	1240	−84	—	—	5, 17
$[NP(OCH_2CF_3)_2]_n$	1,700,000	2.7	+7.5	1280	−66	242	435[a]	17
$[NP(OCH_2C_2F_5)_2]_n$	—	—	—	1250	(−25)	—	260	5
$[NP(OCH_2CF_2CF_2H)_2]_n$	—	—	—	1250, 1350	—	—	—	5
$[NP(OCH_2C_3F_7)_2]_n$	—	1.65	—	1220, 1340	(14)	—	—	5
$[NP(OCH_2(CF_2)_6CF_3)_2]_n$	—	—	—	—	−40	—	425	43, 44
$[NP(OCH_2CF_3)(OCH_2C_3F_7)]_n$	>6,800,000	1.46	—	—	−77	—	—	41–42a
$[NP(OCH_2C_2F_5)(OCH_2C_3F_7)]_n$	—	1.0	—	—	—	—	—	42
$[NP(OCH_2CF_2CF_2H)(OCH_2C_6F_{12}H)]_n$	—	0.5	—	—	−60	—	—	42
$[NP(OPh)_2]_n$	3,700,000	2.7	+18.6	1240	−8	—	480[a]	5, 17
$[NP(OC_6H_4CF_3\text{-}m)_2]_n$	—	1.5	—	—	−35	330	380	43, 44
$[NP(OC_6H_4Cl\text{-}p)_2]_n$	—	1.4	—	—	−12	>350	400	43, 44
$[NP(OC_6H_4F\text{-}p)_2]_n$	—	—	—	—	−14	—	385	43, 44
$[NP(OC_6H_3Cl_2\text{-}2,4)_2]_n$	—	12.0	—	—	2	210	440	43, 44
$[NP(OC_{10}H_7\text{-}\beta)_2]_n$	—	1.9	—	—	47	>350	475	43, 44
$[NP(OC_6H_4C_6H_5\text{-}p)_2]_n$	—	1.2	—	—	43	>350	470	43, 44

[a] 10°C/mm temperature rise.

most interesting and potentially useful phosphazene high polymers yet studied. Polymers with weight average molecular weights of 1,700,000 and intrinsic viscosities of 2.5 dl/g are readily prepared. However, the molecular weight distribution is very broad with M_w/M_n values being as high as 19. The polymer can be fabricated into colorless, transparent, or opalescent films by the evaporation of solutions. The films are highly flexible and easily oriented by stretching to yield material with a high tensile strength. They resemble polyethylene films in appearance, but have a higher density (1.71). Elastic fibers can be prepared by the extrusion of solutions of the polymer into a nonsolvent.

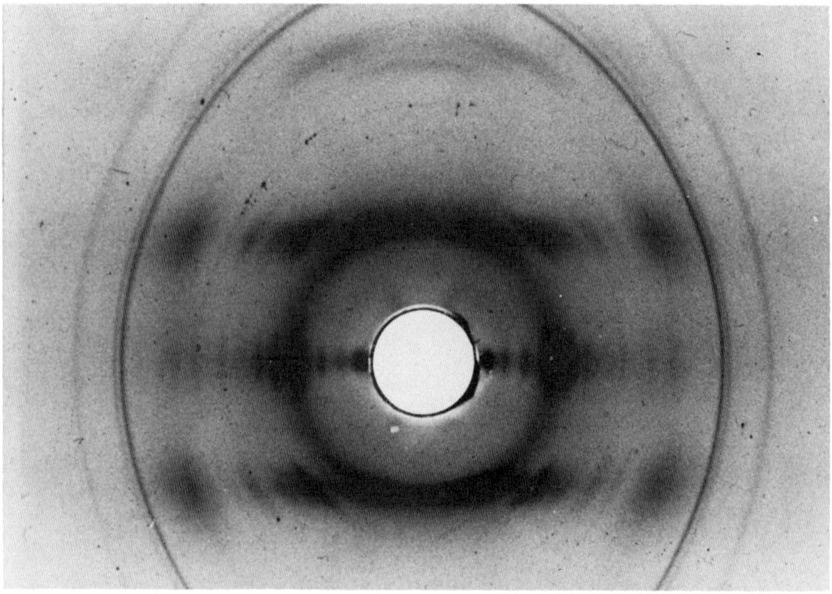

FIG. 16.12. X-Ray fiber diffraction photograph of poly[bis(trifluoroethoxy)phosphazene].[7] The layer-line separation distance corresponds to a fiber repeating distance of 4.83 Å.

The glass transition temperature is at −66°C, and the polymer is crystalline to X-rays and to optical polarization from this temperature up to the crystalline melting point at 242°–243°C. DTA, X-ray diffraction, and optical polarization experiments also reveal the presence of endothermic crystal transitions between 60° and 70° and near 220°C.

Poly[bis(trifluoroethoxy)phosphazene] is soluble in acetone, tetrahydrofuran, ethyl acetate, ethylene glycol, dimethyl ether, and methyl ethyl ketone, but insoluble in diethyl ether, dioxane, ethanol, and in aromatic or aliphatic hydrocarbons. The solid polymer is unaffected by water, glacial acetic acid, alcohols, pyridine, or concentrated sodium hydroxide solution. Films immersed in the latter reagent for 6 months remained unchanged. The polymer is also uniquely stable to prolonged exposure to high-intensity ultraviolet light. However, irradiation with 250 KVP X-rays lowers the molecular weight without the formation of cross-links. Films of the polymer show a high water repellency with a contact angle of 100° and a critical surface tension of 19.7 dyne/cm. This compares favorably with Teflon (22 dyne/cm) and with poly(dimethylsiloxane) rubber (18 dyne/cm). Unlike many organic polymers, poly[bis(trifluoroethoxy)phosphazene] does not burn in air.

The thermal stability of this polymer is appreciably higher than those of the methoxy and ethoxy derivatives. No molecular weight decrease is apparent after polymer films are heated in air, nitrogen, or vacuum at 100°C for 8 hours.[17] However, a molecular weight decrease occurs very slowly at 150°C and more rapidly at 200°C. Thermogravimetric analysis curves can provide misleading information about thermal stability, and the observed T_{10} weight loss values of 300°–410°C (see Fig. 15.11) clearly overestimate the thermal stability of this compound.

Treatment of poly[bis(trifluoroethoxy)phosphazene] in solution with the disodium salt of 2,2,3,3,4,4-hexafluoro-1,5-pentanediol induces cross-linking by the reaction[45]

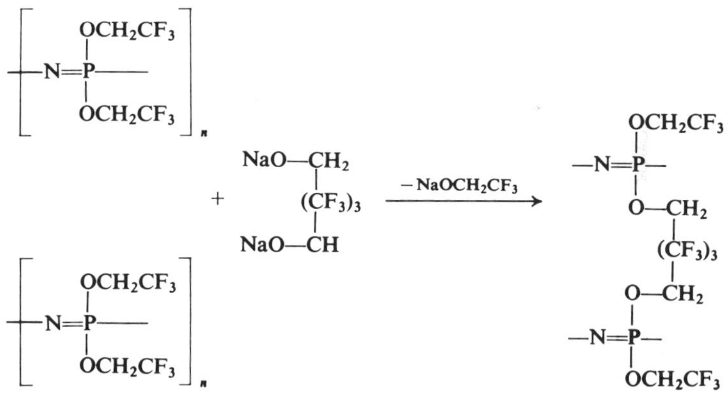

The molecular structure of poly[bis(trifluoroethoxy)phosphazene] has been examined by X-ray diffraction techniques. Oriented films of the polymer show sharp diffraction patterns (Fig. 16.12). The polymer repeating distance of 4.83 Å suggests that the conformation lies close to a *cis–trans*-planar arrangement.

Attempts have also been made to calculate the torsional potential surface for this polymer in terms of nonbonding interactions similar to those described previously for $(NPF_2)_n$, $(NPCl_2)_n$, and $(NPBr_2)_n$.[8] Using a fixed threefold isomeric model for the side groups, potential minima were obtained which corresponded to a cis–trans-planar conformation. However, it should be noted that the low glass transition value (–66°C) implies that the backbone bonds have appreciable torsional mobility in spite of the bulky nature of the trifluoroethoxy ligands.

Poly[bis(pentafluoropropoxy)phosphazene], $[NP(OCH_2C_2F_5]_n$, is formed by the interaction of poly(dichlorophosphazene) with sodium 2,2,3,3,3-pentafluoropropoxide.[5, 42] The polymer is soluble in ethyl trifluoracetate, from which it can be obtained as colorless, transparent, flexible films, similar to polyethylene in superficial appearance.[5] It is insoluble in most organic solvents and in all aqueous media. Its resistance to hydrolysis and its water repellency are comparable to those of the trifluoroethoxy analogue discussed above.[5] The TGA, T_{10}, weight loss value in nitrogen is 260°C. A glass transition may occur near –25°C (DTA), with the polymer being crystalline to X-rays at 25°C. An endothermic transition occurs at 38°C. A fiber axis repeating distance of 9.33 Å was calculated from the layer-line spacings.[5] This value is almost twice the repeating distance determined for the trifluoroethoxyphosphazene and other related polymers.

*Poly[bis(*2,2,3,3-*tetrafluoropropoxy)phosphazene]*$_n$, $[NP(OCH_2CF_2CF_2H)]_n$, is a soft, transparent, film-forming material which is soluble in tetrahydrofuran or acetone, but insoluble in aliphatic hydrocarbons.[5] It is quite stable to water and aqueous media. The polymer is crystalline to X-rays at 25°C.

Poly[bis(heptafluorobutoxy)phosphazene]$_n$, $[NP(OCH_2C_3F_7)_2]_n$, is prepared by the reaction of poly(dichlorophosphazene) with sodium 2,2,3,3,4,4,4-heptafluorobutoxide in tetrahydrofuran[5] or a benzene–tetrahydrofuran mixture.[42] It is soluble in fluorocarbon solvents such as Freon TA but not in other organic solvents.[5, 42] An exotherm in DTA curves at 14° may represent a glass transition temperature. The polymer is highly crystalline to X-rays at 25°C,[5] and it can be heated in air at temperatures below 270° without softening or obvious decomposition occurring.[42]

Poly(trifluoroethoxy-heptafluorobutoxyphosphazene) copolymer, $[NP(OCH_2CF_3)(OCH_2C_3F_7)]_n$, is a white rubbery elastomer with a glass transition temperature of –77°C.[41, 42] The copolymer is unaffected by prolonged immersion in boiling water, common organic solvents, concentrated sulfuric acid, or concentrated potassium hydroxide solution. It is soluble in mixtures of trichlorotrifluoroethane with acetone. When held in a flame, the polymer softens, but it does not burn. Softening of the material occurs near 175°C and melting takes place at 260°C. However, the rubbery properties reappear when the material is cooled. Another copolymer, $[NP(OCH_2C_2F_4H)(OCH_2C_6F_{12}H)_2]$, is reported to be soluble in acetone and ethanol, to have a glass transition temperature of –60°C, and to soften in air near 130°C. Weight average molecular weights in excess of 6×10^6 have been recorded for this polymer by light scattering techniques, with M_w/M_n values of over 40.[42a] Tentative evidence was obtained for a low degree of branching.

Poly[bis(phenoxy)phosphazene]$_n$, *[NP(OPh)$_2$]$_n$*, is prepared by the interaction of poly(dichlorophosphazene) with sodium phenoxide.[17] The appearance of the polymer depends on the method of film preparation. Films obtained by slow evaporation of solvent from viscous solutions are transparent, colorless, highly flexible materials. Rapid evaporation of solvent leaves a white, opaque, leathery, highly crystalline modification. The following data refer exclusively to the transparent form. The polymer is soluble in hot benzene, toluene, dimethylformamide, dioxane, chloroform, chlorobenzene, and tetrahydrofuran, but insoluble in acetone, dimethyl sulfoxide, hexane, ethanol, and water. Films are stable for at least 4 months to concentrated aqueous sodium hydroxide solution, and for over 6 hours in hot aqueous sulfuric acid or water.

The molecular weights of phenoxyphosphazene polymers often exceed 3.5 million, with intrinsic viscosities near 2.7 dl/g. The glass transition temperature is −8°C, and the polymer is highly crystalline at 25°C. X-Ray diffraction photographs of oriented films indicate a chain-repeating distance of 4.91 Å.[5, 17] By analogy with related polymers, this may correspond to a *cis–trans*-planar type chain arrangement. The tensile strength of this polymer is approximately one-half of that of polyethylene.[5] The electrical resistivity is lower than those of polyethylene or poly(dimethylsiloxane), but the polymer shows a remarkable electric strength (resistance to arcing breakdown). Poly[bis(phenoxy)phosphazene] is nonflammable, but it does degrade at high temperatures. Molecular weight decreases occur slowly when the polymer is heated at 100°C,[17] and more rapidly at higher temperatures.[17, 46] There is no question that this polymer is less stable thermally than the trifluoroethoxy derivative in spite of the fact that the T_{10} thermogravimetric analysis values are in the opposite order.

Other Poly[bis(aryloxy)phosphazenes] have been reported[43, 44] and a few physical characteristics of these are included in Table 16.1. The thermal instability of bis(aryloxy)phosphazene polymers follows the pattern discussed above for the phenoxy derivative. Thus, after 8 hours at 225°C, appreciable molecular weight decreases were found for *m*-CF_3-, *p*-Cl-, 2,4-Cl_2-, and *p*-Ph-phenoxy polymers.[44] Possible reasons for this phenomenon are discussed in Chapter 15. The *p*-chlorophenoxy derivative, [NP(OC_6H_4Cl-*p*)$_2$], is soluble in dimethylformamide or tetrahydrofuran, from which it can be cast to give clear, flexible films. The tensile strength of the film is slightly reduced after immersion for 8 hours in boiling water.

2. Poly(diaminophosphazenes)

a. *General Synthesis Route*

Three principal classes of poly(diaminophosphazenes) will be considered here—the bis(primary amino) derivatives (XI), bis(secondary amino) derivatives (XII), and mixed primary–secondary amino polymers (XIII). All are

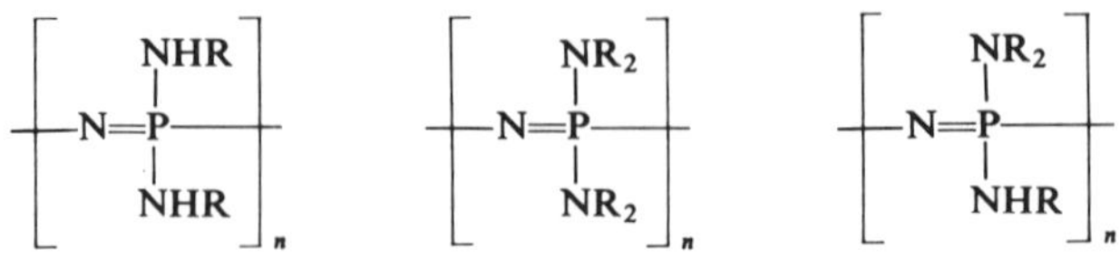

XI XII XIII

synthesized by the interaction of poly(dichlorophosphazene) with a primary or secondary amine according to the general scheme[47-50]

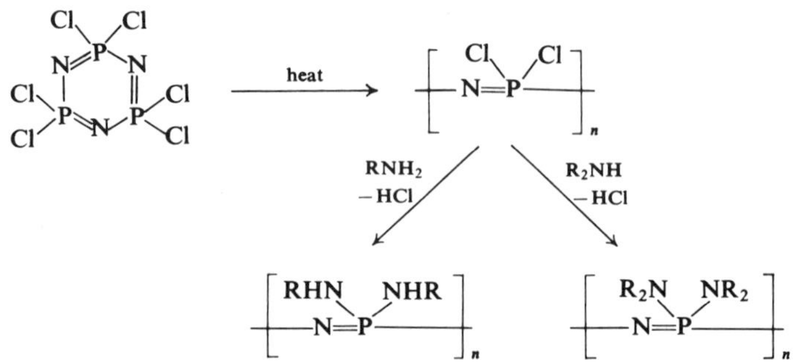

The hydrogen chloride liberated may be removed by excess of the reactant amine or by triethylamine or other tertiary amines added as hydrohalide acceptors.

The following is a typical experimental procedure for the synthesis of an aminophosphazene high polymer—in this particular case, poly[bis(phenylamino)phosphazene], $[NP(NHPh)_2]_n$.[47]

A solution of poly(dichlorophosphazene) (116 g, 1.0 unit mole) in benzene (600 ml) is added to a solution of aniline (800 g, 8.6 moles) in tetrahydrofuran (600 ml), and the mixture is then boiled at reflux for 48 hours and allowed to stand at 25°C for 1 week. The precipitated aniline hydrochloride is filtered off (174 g, 1.34 moles) and the polymer is isolated by dropwise precipitation of the filtrate into absolute ethanol (8 liters). The white, fibrous polymer (70 g, 30%) is then reprecipitated from benzene solution into 95% ethanol, from tetrahydrofuran into a large excess of water, and several times from dioxane into water. Analytical samples can be prepared by freeze-drying from dioxane solution. The intrinsic viscosity in tetrahydrofuran at 30°C is 1.44, and the molecular weight by light scattering is 1,650,000.

Specific problems exist for the synthesis of each type of aminophosphazene polymer, and these will be discussed in the appropriate sections.

b. *Primary Amino Derivatives*

Primary amino-substituted polymers of structure $[NP(NHR)_2]_n$ have been prepared in which NHR is NHMe, NHEt, $NHPr^n$, $NHBu^n$, and NHPh.[47,49]

The transient existence of the compound $[NP(NH_2)_2]_n$ has also been noted.[49] Two factors limit the number of primary aminophosphazene polymers that can be prepared. First, with primary amines, the opportunity exists for the amines to serve as cross-linking agents between chains, as illustrated in the scheme

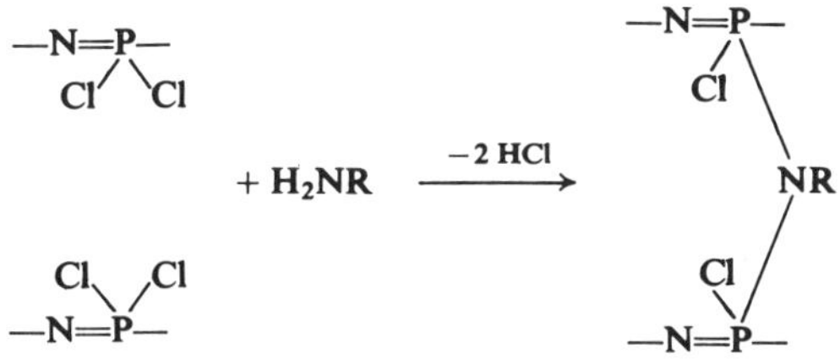

Ammonia cross-links the chains readily and methylamine behaves similarly, except under carefully controlled conditions. These cross-linked polymers are insoluble, often partly substituted materials. Ethylamine, *n*-propylamine, *n*-butylamine, and aniline show no tendency to form cross-links[47, 49] and it must be concluded that cross-linking is retarded by steric hindrance associated with the linear alkyl component. However, excessive steric hindrance in the primary amine leads to the second synthetic limitation—the prevention of total halogen replacement. Only partial substitution occurs when branched amines, such as isopropylamine, are employed.[49] This can be used as evidence that the aminolysis reaction has a substantial S_N2-type component.[47]

Some of the properties of poly[bis(primary-amino)phosphazenes] are listed in Table 16.2. In general, the primary aminophosphazene polymers are less flexible materials than the poly(alkoxyphosphazenes) discussed earlier. For example the transient $[NP(NH_2)_2]_n$ is a glass. The bis methylamino, ethylamino, *n*-propylamino, and *n*-butylamino derivatives form leathery, colorless films when cast from solution, and the phenylamino derivative forms brittle films. The physical properties are reflected by the glass transition temperatures (Table 16.2), which are appreciably higher than those of the corresponding alkoxy or aryloxy derivatives. Presumably, this is a consequence of the opportunities for intra- and intermolecular hydrogen bonding with the primary aminopolyphosphazenes.

The solubility characteristics of these polymers are unusual. The methylamino compound, $[NP(NHMe)_2]_n$, is soluble in water without decomposition.[49] The ethylamino derivative, $[NP(NHEt)_2]_n$, is soluble in dilute aqueous mineral acids, acetic acid, and trifluoroethanol, but it is insoluble in water, aqueous base, benzene, or tetrachloroethane. The propylamino analogue, $[NP(NHPr^n)_2]_n$, dissolves in trifluoroethanol, but is insoluble in methanol or water, and the butylamino compound, $[NP(NHBu^n)_2]_n$, also dissolves in trifluoroethanol but is insoluble in acetone. Poly[bis(phenylamino)phosphazene],

TABLE 16.2

PROPERTIES OF POLY(AMINOPHOSPHAZENES)[a]

Compounds	Molecular weight	Intrinsic viscosity (dl/g)	Infrared P=N peak (cm^{-1})	T_g[b]	Solubility
$[NP(NHMe)_2]_n$	80,000	—	1255, 1200	14	Water
$[NP(NHEt)_2]_n$	—	0.47	1265, 1200	30	Aqueous acid
$[NP(NHPr^n)_2]_n$	—	—	1260	—	Trifluoroethanol
$[NP(NHBu^n)_2]_n$	—	—	1250	—	Trifluoroethanol
$[NP(NHPh)_2]_n$	1,650,000[c]	1.44	1220–1180	91	Benzene
$[NP(NMe_2)_2]_n$	363,000[c]	0.89	1280, 1240	−4	Aqueous acid
$[NP(NC_5H_{10})_2]_n$	697,000[c]	1.33	1248	19	Benzene
$[NP(NC_4H_8)_2]_n$	—	—	—	—	Tetrahydrofuran
$[NP(NEt_2)Cl]_n$	—	—	1255, 1200	—	Benzene
$[NP(NEt_2)(NH_2)]_n$	—	—	—	—	Tetrahydrofuran
$[NP(NEt_2)(NHMe)]_n$	230,000[d]	1.41	1240, 1205	—	Tetrahydrofuran
$[NP(NEt_2)(NHEt)]_n$	274,000[d]	1.16	1250, 1200	—	Benzene
$[NP(NEt_2)(NHPr^n)]_n$	130,000[d]	1.20	1250, 1200	—	Benzene
$[NP(NEt_2)(NHBu^n)]_n$	155,000[d]	1.14	1265, 1200	—	Benzene
$[NP(NEt_2)(OCH_2CF_3)]_n$[e]	—	—	—	—	Benzene
$[NP(NEt_2)(NC_5H_{10})]_n$[e]	—	—	—	—	Tetrahydrofuran

[a] From Allcock, Kugel, Mack, and Cook.[47–50]
[b] By differential thermal analysis.
[c] By light scattering.
[d] By ultracentrifugation.
[e] Some MeNH present.

$[NP(NHPh)_2]_n$, is soluble in benzene, tetrahydrofuran, or dioxane, but insoluble in water or 95% ethanol. No evidence for hydrolytic instability was observed with any of these polymers. They all form stable hydrochloride salts. Ultracentrifuge experiments with $[NP(NHMe)_2]_n$ suggest a very wide molecular weight distribution with considerable chain branching present.[49]

c. *Secondary Amino Derivatives*

The following bis secondary amino polymers have been prepared and characterized[47, 49]: $[NP(NMe_2)_2]_n$, $[NP(NC_5H_{10})_2]_n$ (piperidino), and $[NP(NC_4H_8)_2]_n$ (pyrrolidino). Attempts to prepare other disubstituted secondary amino derivatives by the interaction of diethylamine, *N*-methylaniline, or diphenylamine with poly(dichlorophosphazene) were not successful. Diethylamine replaces only one-half of the available chlorine atoms,[48, 49] *N*-methylaniline replaces only 20% to 30% of the chlorine, and diphenylamine shows no evidence of any chlorine replacement. These results can be ascribed to the steric properties of these bulky amines and to a steric inhibition of S_N2-type attack by the amine on phosphorus.

The bisdimethylamino-, piperidino, and pyrrolidinophosphazene polymers are colorless, transparent, flexible, film-forming materials. The first two have glass transition temperatures of −4° and 19°C, respectively.[47, 49] The absence of hydrogen-bonding opportunities appears to enhance the chain mobility compared to the primary amino derivatives Poly[bis(dimethylamino)phosphazene] is soluble in dilute aqueous mineral acids, acetic acid, and trifluoroethanol, but insoluble in most common organic solvents or water. Although the polymer shows no evidence of hydrolysis in contact with water, the odor of dimethylamine can be detected from samples that have been stored in closed vessels for several years. The bispiperidino derivative is soluble in tetrahydrofuran, benzene, toluene, chlorinated hydrocarbons, formic and acetic acids, trifluoroethanol, and concentrated sulfuric acid.[47] It is insoluble in dilute mineral acids and is quite stable to aqueous media. Poly[bis(pyrrolidino)-phosphazene] is soluble in tetrahydrofuran, but insoluble in aliphatic hydrocarbons.[49] It too is stable to moisture. All three polymers form salts with hydrogen chloride.

d. *Mixed Diethylamino–Aminophosphazene Polymers*

Diethylamine reacts with poly(dichlorophosphazene) to replace approximately one-half of the available chlorine atoms. The remaining halogen atoms can then be substituted by a different amine to yield copolymers.[48, 49]

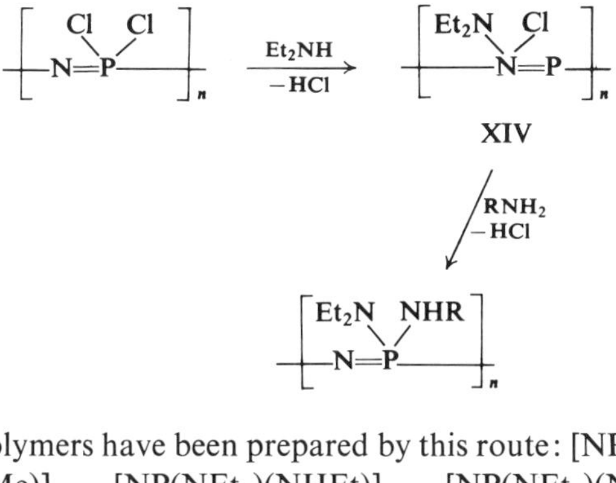

The following polymers have been prepared by this route: $[NP(NEt_2)(NH_2)]_n$, $[NP(NEt_2)(NHMe)]_n$, $[NP(NEt_2)(NHEt)]_n$, $[NP(NEt_2)(NHPr^n)]_n$, and $[NP(NEt_2)(NHBu^n)]_n$. The precurser polymer (XIV) is a clear, colorless rubbery elastomer which, in view of the presence of the P—Cl bonds, is surprisingly stable to atmospheric moisture. When freshly prepared, it is soluble in chlorinated hydrocarbons, benzene, or tetrahydrofuran, but it cross-links slowly during prolonged exposure to a moist atmosphere. The diethylamino–amino derivative is a clear, colorless, glasslike polymer which hydrolyzes slowly, with liberation of ammonia, when exposed to the atmosphere. The other four polymers are clear, colorless, flexible, film-forming materials with the flexibility increasing in the order RNH = MeNH < EtNH < Pr^nNH < Bu^nNH. They are stable to water and to prolonged exposure to the atmosphere, and they are soluble in benzene, tetrahydrofuran, carbon tetrachloride, or 2,2,2-trifluoroethanol. The polymers appear to be nongeminal derivatives, with different substituents present at each phosphorus. However, the degree of diethylamino substitution is always slightly above 50%, with 52.5% to 55% diethylamino units being typically present. This fact may reflect the presence of many branch points and terminal units where diethylamine substitution is not restricted to one site at each phosphorus. Ultracentrifuge data also suggest the presence of highly branched polymers, with a very wide molecular weight distribution. The polymers are amorphous to X-rays.

This synthetic route is not limited to the preparation of primary amino derivatives of XIV. Piperidine reacts with XIV to replace part of the remaining halogen, the residue of which can be replaced by methylamine to yield a diethylamino–piperidino–methylamino terpolymer.[49] Treatment of XIV with sodium 2,2,2-trifluoroethoxide yields a hydrolytically stable diethylamino–trifluoroethoxy copolymer $[NP(NEt_2)(OCH_2CF_3)]_n$, with properties very similar to those of the bis(trifluoroethoxy) polymer. It is soluble in benzene, from which solvent it forms clear, colorless, noncrystalline films.[49] Terpolymers which contain diethylamino, trifluoroethoxy, and methylamino groups can also be prepared.[49]

3. Poly(alkylphosphazenes) and Poly(arylphosphazenes)

a. *General Synthesis Routes*

Two synthesis routes are available for the preparation of poly(alkyl- and arylphosphazenes). One of these corresponds to the substitution route used for the preparation of alkoxy-, aryloxy-, and aminophosphazene high polymers, with the sodium alkoxide, sodium aryloxide, or amine reagent replaced by an organometallic compound. Grignard and organolithium reagents have been examined as reagents for reactions of this type. Although the overall reaction

$$\underset{}{(NPCl_2)_n} + \underset{\text{(or RLi)}}{2n\,RMgX} \rightarrow (NPR_2)_n + \underset{\text{(or LiCl)}}{2n\,MgClX}$$

has been described, there is much evidence that substitution is usually accompanied by degradative side reactions.

The second approach involves the direct synthesis of a polymer from nonphosphazene precursors. Low molecular weight materials are often obtained by this method.

b. *Poly(dimethylphosphazene)*, $(NPMe_2)_n$

Sisler and co-workers[51] have synthesized a polymeric dimethylphosphazene by the pyrolytic condensation of dimethyldiaminophosphonium chloride.

$$n[Me_2P(NH_2)_2]^{\oplus}\,Cl^{\ominus} \xrightarrow{200°C} (NPMe_2)_n + nNH_4Cl$$

The highest polymer prepared by this technique had a molecular weight of 12,500. It was soluble in benzene, and solutions could be cast into films. Melting of the polymer took place at 142°C and depolymerization occurred above 300°C. X-Ray diffraction experiments demonstrated that the polymer was crystalline. High doses of X-rays (40 Mrads) depolymerized the polymer at 150°C.

An attempt has also been made to prepare the same polymer by treatment of cross-linked poly(dichlorophosphazene) with dimethylzinc in a xylene medium.[52] Substitution occurred to give a tough, light-colored polymer. However, the exact composition of this material was not reported, although no more than 5% unreacted halogen was detected. Since the phosphazene substrate was insoluble in the medium, it is assumed that the substituted polymer was also cross-linked. An ethylphosphazene polymer was prepared by the same technique.

c. *Poly[bis(trifluoromethyl)phosphazene]*, $[NP(CF_3)_2]_n$

Polymers with the composition $[NP(CF_3)_2]_n$ have been prepared by two direct synthesis methods; first by the slow decomposition of bis(trifluoromethyl)azidophosphine,[53]

$$n(CF_3)_2PN_3 \rightarrow [NP(CF_3)_2]_n + nN_2$$

and second, by the tertiary amine-induced dehydrohalogenation of bis(trifluoromethyl)dichlorophosphinic amide,[54]

$$n(CF_3)_2P(Cl_2)NH_2 \xrightarrow{2n\,R_3N} [NP(CF_3)_2]_n + 2n\,HCl$$

The polymer prepared by these methods is insoluble in all media, and is, therefore, of unknown molecular weight. It is a white, amorphous, nonvolatile, waxy solid with a melting point of 90°–94°C, and a decomposition or depolymerization temperature near 380°C. It is stable to boiling sulfuric, nitric, or perchloric acid but is attacked by aqueous 10% sodium hydroxide solution.

d. *Poly(diphenylphosphazene)*, $(NPPh_2)_n$

Attempts have been made to replace the halogen in poly(dichlorophosphazene) by phenyl groups with the use of phenylmagnesium bromide or phenyllithium as reagents.[52, 55] When phenylmagnesium bromide was allowed to react with swelled, cross-linked poly(dichlorophosphazene), a tough thermoplastic of unspecified composition was obtained.[52] Soluble poly(dichlorophosphazene) reacted with phenyllithium to yield a solid polymer.[55] Both materials were stable to aqueous media. A direct synthesis route has been used to prepare a low-polymeric poly(phenylchlorophosphazene), $(NPClPh)_n$.[56] In this reaction, phenyldichlorophosphazene was allowed to react with sodium azide to give a ploymer with a number average molecular weight of over 5,000.

$$n\mathrm{PhPCl_2} + n\mathrm{NaN_3} \xrightarrow{-\mathrm{NaCl}} n\mathrm{PhClPN_3} \xrightarrow{-\mathrm{N_2}} (\mathrm{NPClPh})_n$$

4. Structural and Spectral Trends

The availability of a wide variety of phosphazene high polymers provides an almost unique opportunity for the study of the effects of different ligands on the properties of a polymer system. In particular, the trend in chain flexibilities and glass transition temperatures can be traced qualitatively as a function of the dimensions, polarity, and hydrogen-bonding characteristics of the ligands. The lowest glass transition temperatures recorded for poly(organophosphazenes) are for alkoxy and fluoroalkoxy derivatives (Table 16.1). Values in the –60° to –85°C range are found for $[NP(OMe)_2]_n$, $[NP(OEt)_2]_n$, $[NP(OCH_2CF_3)_2]_n$, and for copolymers which contain two different fluoroalkoxy substituents such as $[NP(OCH_2CF_3)(OCH_2C_3F_7)]_n$, $[NP(OCH_2C_2F_5)(OCH_2C_3F_7)]_n$, and $[NP(OCH_2CF_2CF_2H)(OCH_2C_6F_{12}H)]_n$. Increasing chain length of a fluoroalkoxy ligand, as in $[NP(OCH_2C_3F_7)_2]_n$, raises the glass transition temperature, but copolymers which contain two different fluoroalkoxy ligands have low glass transitions. Presumably, the resultant loss of symmetry inhibits chain packing and crystallization and allows chain torsional vibrations to persist at lower temperatures. These motions are apparently inhibited when two phenoxy groups are attached to each phosphorus ($T_g = -8$°C), a result which can be ascribed to steric interference by the ligands. Hydrogen bonding in primary amino phosphazene polymers appears to lower the chain flexibility also.

One striking characteristic of all linear phosphazene high polymers is that the ultraviolet and infrared spectra are remarkably similar to those of the corresponding cyclic trimers and tetramers.[16, 17] Electronic spectral absorptions from the polyphosphazene backbone are found only at wavelengths shorter than about 220 mμ, and no evidence is found for shifts toward the visible region as the ring or chain size increases. This provides a strong argument against a broad π-electron delocalization arrangement. From a practical point of view, this transparency to visible and near-ultraviolet radiation is the reason for the high stability of many polyphosphazenes to radiation in this frequency range. The position of the characteristic –P=N– "stretching" frequencies in the 1,200–1,300 cm^{-1} region suggests that the force constants of the skeletal bonds in the high polymers are comparable to those in cyclic trimers and tetramers.

5. Applications

Poly(organophosphazenes) show a number of unusual properties which are expected to dictate the areas of most immediate application. These properties are (1) low-temperature flexibility and elasticity; (2) high stability to visible and ultraviolet radiation; (3) resistance to water, solvents, hydrocarbon fuels, oils, etc.; (4) water and soil repellency for the fluoroalkoxy derivatives; (5) thermal stability up to 200°C, and the prospect of prolonged stability at higher temperatures; (6) water solubility for specific aminophosphazene polymers; and (7) nonflammability and flame retardancy. The most important utility of the phosphazene polymer system is the ease with which the properties can be modified by the introduction of different substituent groups. Thus, the design and synthesis of polymers for specific applications may be easier for polyphosphazenes than for almost any other polymer system.

The following are applications for which specific polyphosphazene systems are currently being developed. There is a critical need for the development of elastomers, plastics, and sealants which remain flexible at very low temperatures. Such materials are needed for use in high-altitude aircraft, space capsules, and in arctic environments. Specific poly(organophosphazenes), particularly the fluoroalkoxy and aryloxy derivatives, are highly resistant to chemical or solvent attack, and these materials show promise as seals, gaskets, and hydrocarbon-resistant elastomers. The high stability of fluoroalkoxyphosphazene polymers to visible and ultraviolet irradiation suggests applications in the areas of outdoor surface coatings and wire coatings. Fluoroalkoxy derivatives have also been proposed as textile-impregnating agents to impart water and soil repellency to fabrics.[57] Although poly(alkoxy- and aryloxyphosphazenes) show a tendency to depolymerize near 200°C, the prospect remains that stability at higher temperatures can be achieved by cross-linking of the polymers. As discussed earlier, metathetical exchange reactions with diol salts can be used to effect cross-linking.[45] Nucleophilic substitution reactions carried out on cross-linked poly(dichlorophosphazene) also yield cross-linked, substituted polymers.[40] Applications for heat-stable elastomers, coatings, and fibers exist in nearly all technological and textile fields, and the prospects for poly(organophosphazenes) appear promising.

The flame-retardant properties of phosphazenes have generated considerable interest.[58, 59] These properties result from the presence of high concentrations of phosphorus and nitrogen in the molecules. Poly(aminophosphazenes) show particular promise for this application.

E. Cyclolinear Polymers

The development of cyclolinear phosphazene polymers (II) has resulted from attempts to utilize the high thermal stability of cyclotri- or cyclotetraphosphazene rings but, at the same time, to introduce flexible or elastic properties into a polymeric material. The lower cyclophosphazene rings are rigid structures, and the rings must be selectively linked by flexible units to prevent the formation of rigid, insoluble materials. A variety of reactions have been utilized to prepare cyclolinear structures. These include condensation-type reactions, dehydrochlorinations, transesterifications, denitrogenation of azides, and radiation-induced reactions.

1. Condensation-type Polymers

When hexachlorocyclotriphosphazene is treated with ammonia, a diamino derivative, $[N_3P_3Cl_4(NH_2)_2]$, is formed. Treatment of this compound with sodium 2,2,2-trifluoroethoxide yields a tetrakis(trifluoroethoxy) diamino derivative, $N_3P_3(OCH_2CF_3)_4(NH_2)_2$.[60] Similar derivatives were formed by the use of higher fluoroalkoxides. The assumption was made that the two amino groups were located non-geminally, but more recent evidence suggests that they are geminally disposed.[61] Heating of these compounds with fluorinated polymethylene α,ω-glycols yielded elastomeric products by condensation and ammonia evolution. The process presumably involves reaction steps such as

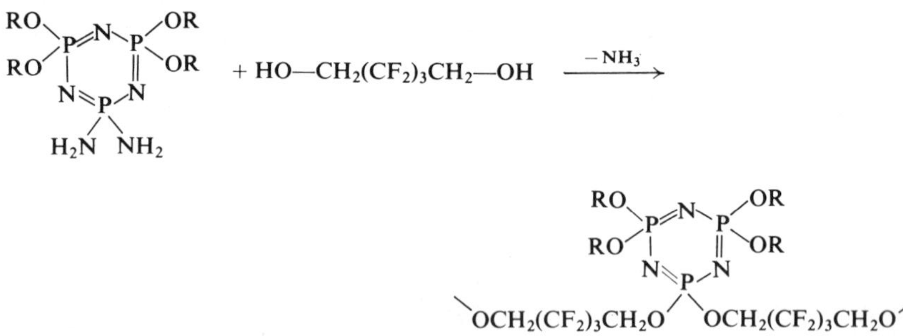

The derivative where $R = CH_2CF_2CF_2H$ was reported to be insoluble in aqueous media and common organic solvents and to be stable in air up to 250°C.

2. Dehydrochlorination-type Reactions

The hexaphenyldichlorocyclotetraphosphazene, XIV, reacts with aromatic diols at elevated temperatures to eliminate hydrogen chloride and form cyclolinear polymers (XV).[62] Reactions were performed in the melt and in solution, with 4,4′-dihydroxybiphenyl, 4,4′-dihydroxydiphenyl ether, resorcinol, and benzidine used as linking agents. With 4,4′-dihydroxybiphenyl, the product was a pale yellow glass, m.p. 185°–220°C, with a molecular weight

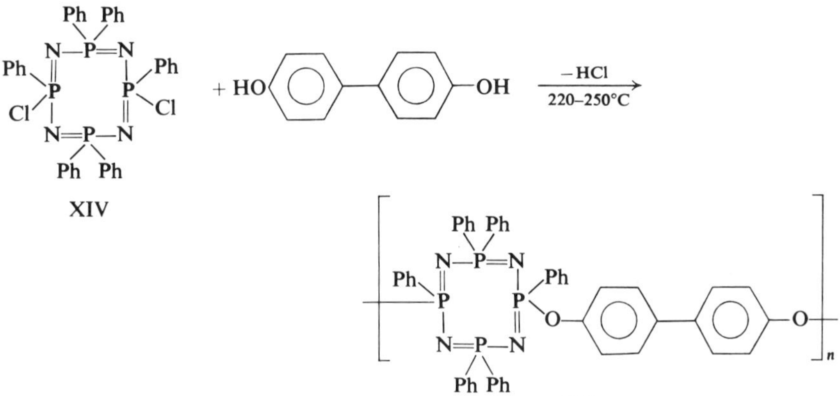

XIV

XV

near 500,000. 4,4′-Dihydroxydiphenyl ether and resorcinol yielded resinous products of lower molecular weight. Thermal decomposition occurred above 470°C for XV, and $(NPPh_2)_{3\,\text{and}\,4}$ and benzene were detected as vacuum pyrolysis products. This suggests that cleavage of the phosphazene ring may constitute the initial decomposition reaction for XV. Moisture and oxygen appear to lower the decomposition temperature. The polymer derived from 4,4′-dihydroxydiphenyl ether decomposes at a lower temperature (~360°C), probably because of cleavage of the linking groups. Tetrakis(organo)dichlorocyclotriphosphazenes react also with alkyl phosphoric acids to yield low-polymeric cyclolinear polymers[62a, b] and with the sodium salts of diols to yield soluble polymers.

Dichlorosilanes have been used as linking agents for hydroxyorganophosphazenes.[63] For example, non-geminal tris(hydroxy)tris(phenoxy)cyclotriphosphazene (XVI) reacts with diphenyldichlorosilane in the melt or in solution to liberate hydrogen chloride and form a polymer (XVII). When

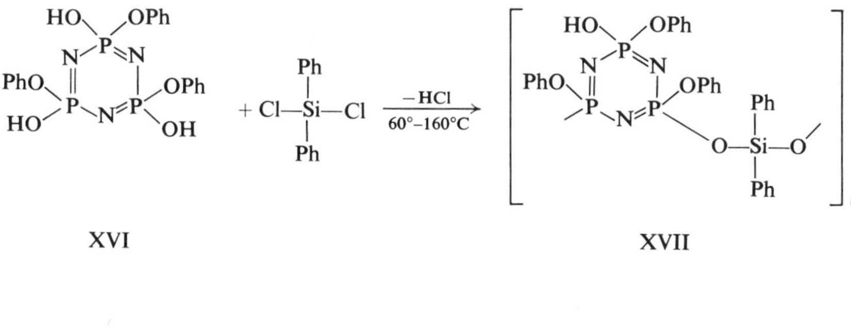

XVI XVII

a 1:1 molar ratio of phosphazene to dichlorosilane is employed, solid polymers of molecular weight 6,000 are produced. They are soluble in organic media. Even if a 2:3 molar ratio of reactants is used, the polymers remain soluble and apparently un-cross-linked. The polymers are hydrolytically stable, but thermal decomposition occurs above 200°C.

Cyclophosphazane rings have also been incorporated into cyclolinear polymer systems.[64] Horn and Becke-Goehring have shown that aromatic dihydroxy compounds interact with the cyclodiphosphazane, XVIII, with evolution of hydrogen chloride and the formation of polymer (XIX). A variety of aromatic

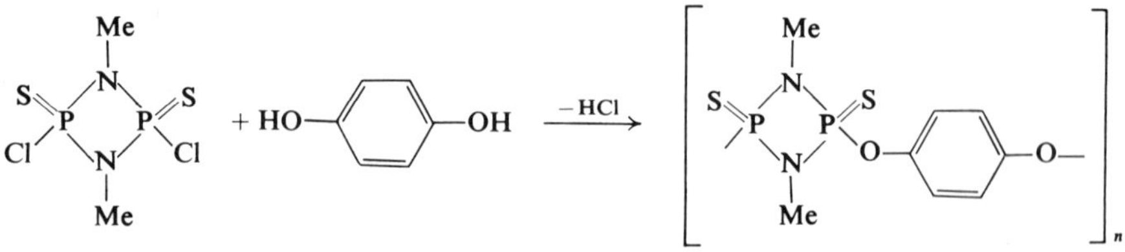

XVIII XIX

diols yield hard, brittle, nonflammable polymers. The polymer thermal stability decreases with changes in linking agent in the order: hydroquinone ≈ resorcinol > tetramethylhydroquinone ≫ tetrachlorohydroquinone ≈ tetrafluorohydroquinone. Catechol yields low molecular weight oils, possibly a consequence of the known chelating properties of this reagent.[65] The maximum thermal stability exhibited by the hydroquinone derivative was indicated by a decomposition temperature of 590°C. Infrared spectra were consistent with the cyclolinear structure.

3. Transesterification-type Reactions

Low molecular weight polymers have been prepared by Zhivukhin and coworkers by transesterification of hexakis(butoxy)cyclotriphosphazene with resorcinol or hydroquinone.[66] The products are soluble oils or solids of molecular weight 3,000–10,000, and a cyclolinear structure is assumed. These products would be formed by reactions such as

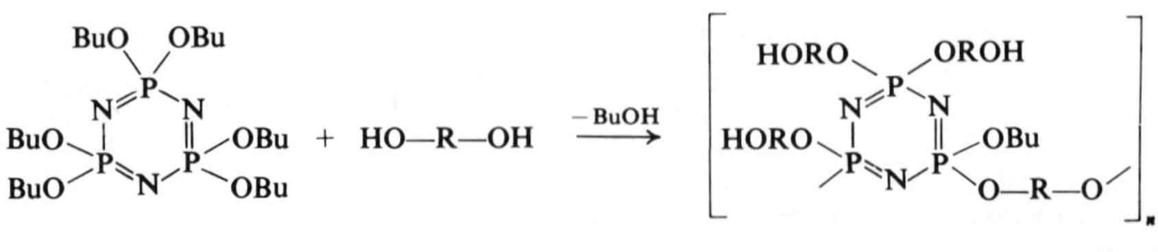

A related reaction occurs when butoxycyclophosphazenes react with diphenyldichlorosilane.[63] Butyl chloride is eliminated and soluble, glassy polymers are produced.

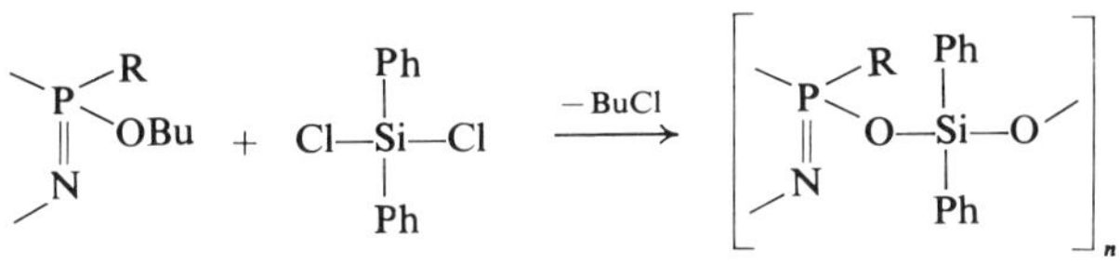

When two butoxy groups are present on each phosphorus, spirocyclic derivatives are presumed to be formed.

4. Denitrogenation Reactions

trans-1,5-Diazidohexaphenylcyclotetraphosphazene(XX)reacts with 1,4-bis-(diphenylphosphino)butane (XXI) or with 4,4′-bis(diphenylphosphino)biphenyl to yield polymers by nitrogen elimination.[67] However, the polymer

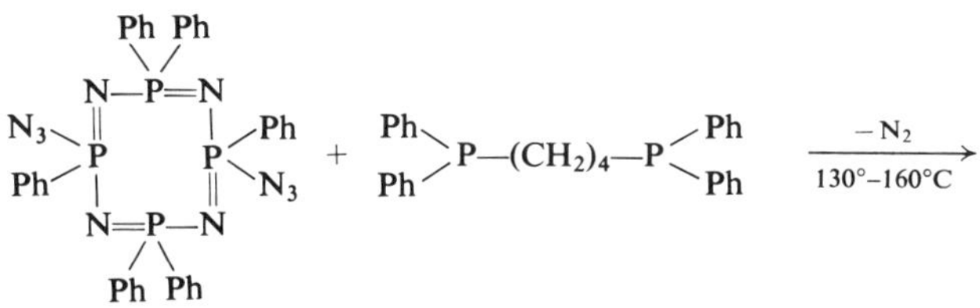

XX XXI

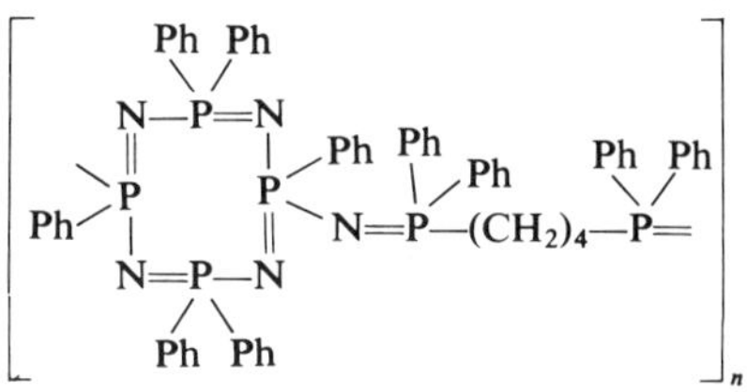

molecular weights are low (9,000–10,000), and the polymers are brittle, although thermal stability up to 400°C was observed.

Similarly, cyclophosphazane XXII reacts with bis(diphenylphosphino)-methane or bis(diphenylphosphino)ethane in chlorobenzene or chloroform with evolution of nitrogen to form soluble, cyclolinear low polymers with weight average molecular weights of up to 2,600.[67a]

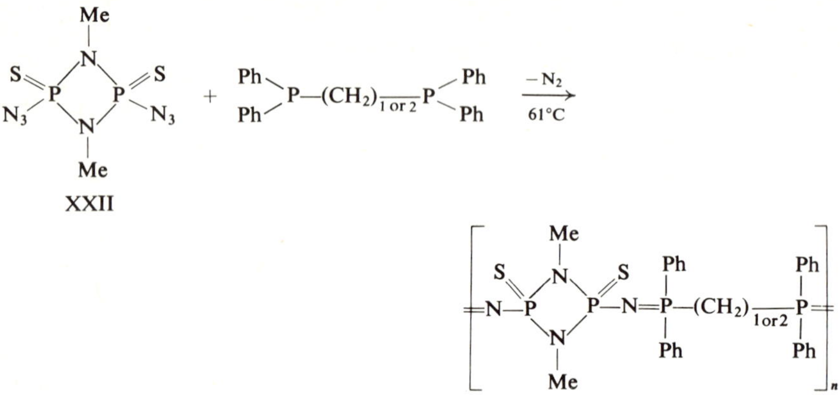

5. Radiation-Induced Reactions

A white rubbery polymer is formed when a 1 : 1 mixture of hexachlorocyclotriphosphazene and triallyl cyanurate is irradiated with 3 megarads of γ radiation.[68] The polymer was insoluble in a variety of organic solvents, and it retained its rubbery properties after being heated to 300°C. It is assumed that the material is formed by the radiation-induced addition of a P—Cl bond to an allyl group. Since the polymer is insoluble, cross-links are probably present.

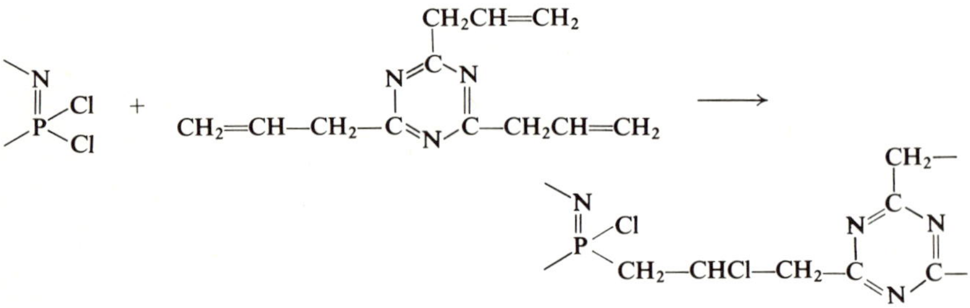

F. Cyclomatrix Polymers

Phosphazene cyclomatrix polymers are prepared by the cross-linking of cyclic trimeric or tetrameric rings to yield rigid, insoluble, high-melting resins. Such materials are used as structural plastics, surface coating, wire enamels, or textile treatment agents, and they are usually fabricated as soluble or low-melting "prepolymers" followed by a "curing" process to complete the cross-linking reaction. Three types of phosphazene cyclomatrix polymers will be considered here—those formed by the addition reactions of unsaturated side groups, those which result from ligand rearrangements, and those formed by condensation processes.

1. Polymers Formed by Addition Reactions

The sodium salt of allyl alcohol reacts with hexachlorocyclotriphosphazene to yield hexakis(allyloxy)cyclotriphosphazene (XXIII).[69, 70] This compound

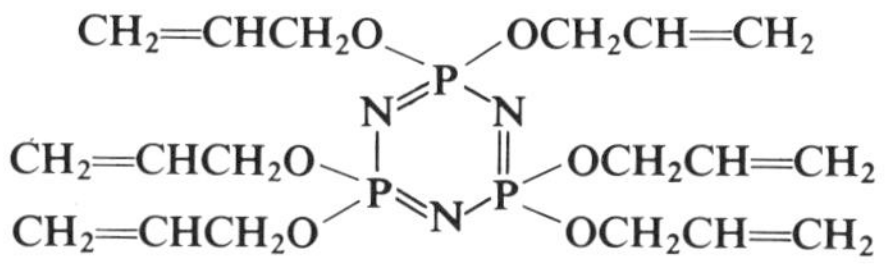

XXIII

and its brominated derivatives can be cross-linked on fabrics by heating. However, free-radical, chain-transfer reactions reduce the extent of polymerization.[71]

Allylamine reacts with $(NPCl_2)_{3 \text{ and } 4}$ to yield the allylamino derivatives (XXIV and XXV). Addition polymerization of the allyl side groups can be

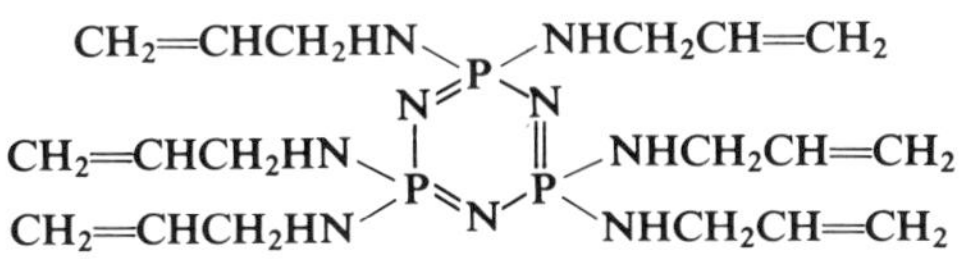

XXIV

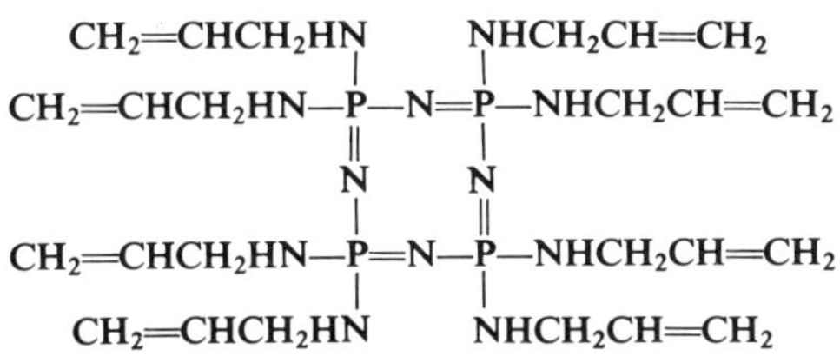

XXV

induced at 130°–220°C in the presence of free-radical initiators, such as *t*-butyl peroxide or 2,5-di-*t*-butylperoxy-*n*-hexane, to yield hard, insoluble, amber-colored, cross-linked resins.[72] The polymer T_{10} weight loss is at 400°C, and articles prepared from the resin with the use of glass cloth or other fillers are resistant to thermal degradation above 260°C in air. The trimer, XXIV, can be converted to low-melting prepolymers with the use of free-radical initiators at temperatures below 100°C. In keeping with the behavior of other allyl monomers, XXIV is of low reactivity in free-radical copolymerization reactions with methyl methacrylate and acrylonitrile.

2. Polymers Formed by Ligand Rearrangement

Insoluble, brittle, cross-linked resins are obtained when the spirocyclotriphosphazene, XXVI, is heated above 300°C. Ligand exchange presumably occurs with neighboring molecules to create a three-dimensional structure.[73]

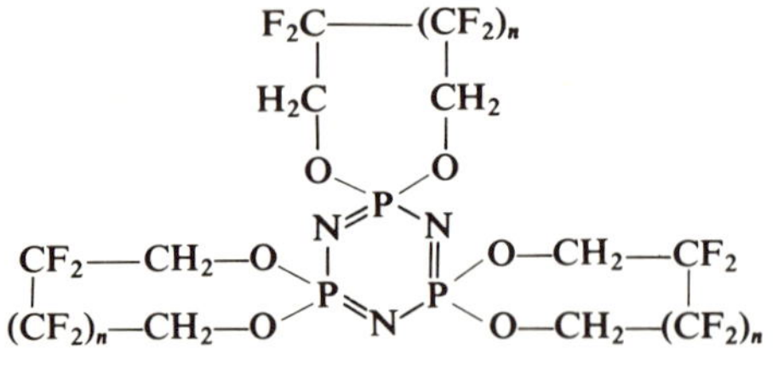

XXVI

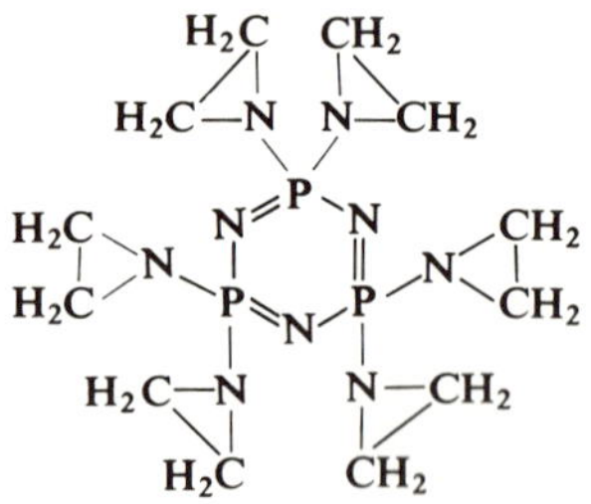

XXVII

Similarly, clear, hard, colorless, flameproof resins are obtained when ethyleneiminophosphazenes, such as XXVII, are heated.[74, 75] The ethyleneimino rings undergo cleavage and cross-linking reactions to yield species such as XXVIII. Colorless resins can also be obtained from XXVI by treatment with

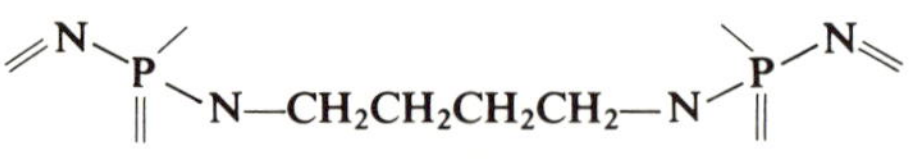

XXVIII

di- or polyfunctional alcohols or amines.[75]

Epoxyphosphazenes, of formula XXIX, have been prepared by the treatment

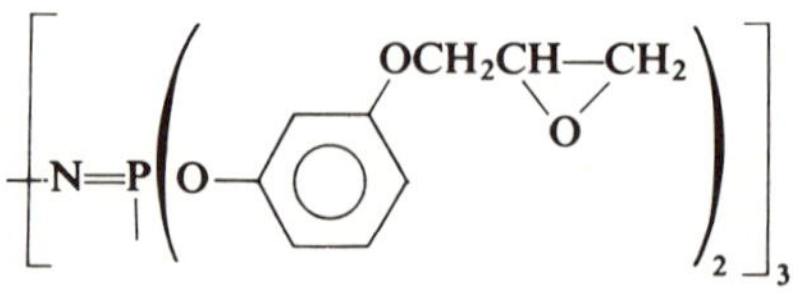

XXIX

of hexakis(*m*-hydroxyphenoxy)cyclotriphosphazene with epichlorohydrin.[76] This material (XXIX) is a brown, viscous liquid which can be stored unchanged for long periods of time. However, heating of the compound or treatment with hardeners, such as polyamines or organic acid anhydrides, brings about a conversion of the material to hard resins.

3. Resins Formed by Polycondensation

A substantial number of processes have been described for the preparation of cross-linked resins by polycondensation or related reactions. These can be categorized into reactions which involve the treatment of a chlorophosphazene with a di- or poly-hydroxy compound, or with a diamine, those which involve transesterification-type reactions, and miscellaneous reactions such as pyrolyses, methylol condensations, and elimination of silicon compounds.

a. *Reactions of Chlorophosphazenes with Di- or Polyhydroxy Compounds*

Aromatic diols, such as hydroquinone or resorcinol react with chlorocyclophosphazenes at elevated temperatures with the liberation of hydrogen chloride and the formation of cross-linked cyclomatrix resins.[77–84b] An initial stage in the cross-linking process can be depicted as

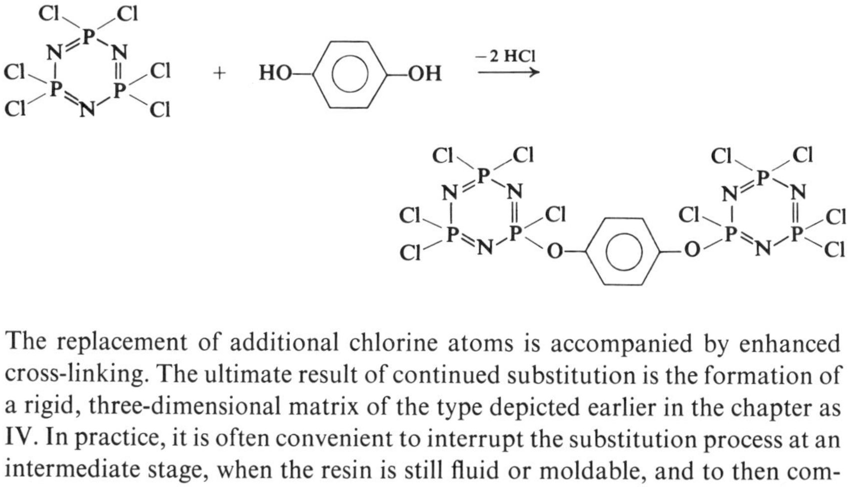

The replacement of additional chlorine atoms is accompanied by enhanced cross-linking. The ultimate result of continued substitution is the formation of a rigid, three-dimensional matrix of the type depicted earlier in the chapter as IV. In practice, it is often convenient to interrupt the substitution process at an intermediate stage, when the resin is still fluid or moldable, and to then complete the curing cycle at higher temperatures or by the use of additional cross-linking agents. Polymers of this type have been made on a commercial scale.[77–80]

A variety of hydroxyaromatic compounds participate in polymer formation, including hydroquinone, resorcinol, phloroglucinol, diphenylolpropane, pyrogallic acid, or their phosphorus esters.[77, 78] The chlorophosphazenes used can include mixtures of the lower homologues. Polymer formation may take place in the molten state with evolution of hydrogen chloride,[77] or in solution (in carbon tetrachloride) in the presence of a hydrohalide acceptor such as pyridine.[78] The latter process yields an oily "prepolymer" which can be subsequently cross-linked when applied to a substrate in butanol solution.

The details of the cross-linking process have been examined by Zhivukhin and co-workers[81-83] and by Yokoyama.[84] Although hexachlorocyclotriphosphazene does not react with hydroquinone, resorcinol, or diphenylolpropane in the melt or in organic solvents at temperatures below 170°C, hydrogen chloride is evolved in nitrobenzene solution above 200°C.[81] However, in the presence of a tertiary amine, such as quinoline or pyridine, substitution takes place rapidly even below 130°C. Soluble cyclolinear polymers are formed when equimolar amounts of chlorophosphazene and diol are used at temperatures above 200°C, or if the chlorophosphazene is in excess in the presence of tertiary amines. When the diol is in excess, cyclomatrix polymers are obtained.[81, 84] Similar results are obtained when metal salts of the diols are employed.[81] The final curing process at high temperatures can be followed by infrared techniques[82, 84] and the evidence indicates that the phosphazene rings open to form cross-linked chains above 250°C. Kajiwara and Saito[85-88] have examined the reactions of 2,6-dimethylol-*p*-cresol with hexachlorocyclotriphosphazene to yield rubbery polymers and resins.

The white prepolymers obtained by the reactions of chlorophosphazenes with aromatic di- or polyhydroxy compounds can be cross-linked by treatment with formaldehyde or hexamethylenetetramine.[77-78] The formaldehyde molecules link together residual free hydroxyl groups by the reaction

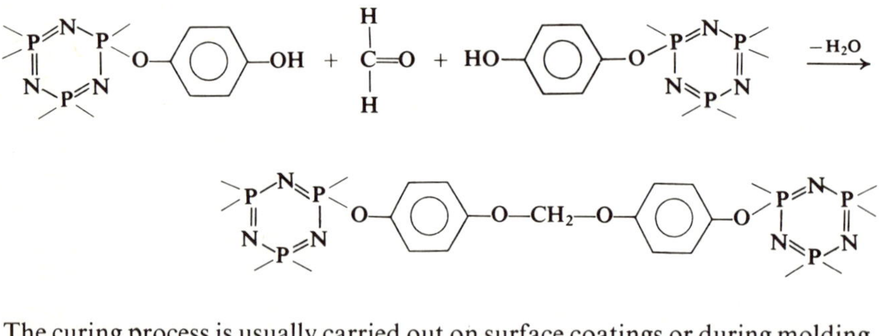

The curing process is usually carried out on surface coatings or during molding procedures. Cyclomatrix polymers prepared by these methods are described variously as rubbery solids or tough resins. They are stable for continuous use at 250°C and for intermittent use at temperatures up to 540°C. They are highly resistant to acids, alkalis, and organic liquids, and they have a high impact strength.[77-80] Uses have been proposed for these resins as structural adhesives for high-temperature or aerospace applications,[80] as heat-resistant electrical components and wire coatings, as laminating resins, and as radiation-resistant components in nuclear reactors.

Cross-linked cyclomatrix polymers have also been prepared with the use of aliphatic diols[89] and with aromatic diamines.[90]

b. *Transesterification Reactions*

When chlorocyclophosphazenes are treated with alcohols, part of the halogen is replaced by alkoxy groups and part by hydroxy units to yield products of formula $[NP(OR)_m(OH)_{2-m}]_n$, where m has a maximum value of 2.[91, 92] Cyclomatrix resins are formed when these materials are heated with polybasic inorganic acids or their esters, such as sulfuric acid, phosphoric acid, ethylphosphate, dimethyl sulfate, benzenesulfonic acids, or boric acid. Alcohol and water are presumably eliminated during the curing process. The products are useful as heat-resistant moldings, reinforced laminates, or surface coatings. Related polymers are produced by the treatment of alkoxyphosphazenes with polyhydroxy aromatic compounds, such as hydroquinone or resorcinol, followed by cross-linking with heat or hexamethylene tetramine.[81, 93] These resins also can be used as reinforced structural plastics. An advantage of transesterification-type cross-linking reactions is that, unlike the dehydrohalogenation procedures, no corrosive fumes are liberated during the curing process.

c. *Miscellaneous Condensation Reactions*

Hexachlorocyclotriphosphazene reacts with hexaethyldisiloxane or phenylethoxysilanes with the evolution of triethylchlorosilane and the formation of triethylsiloxychlorophosphazene.[94, 94a]

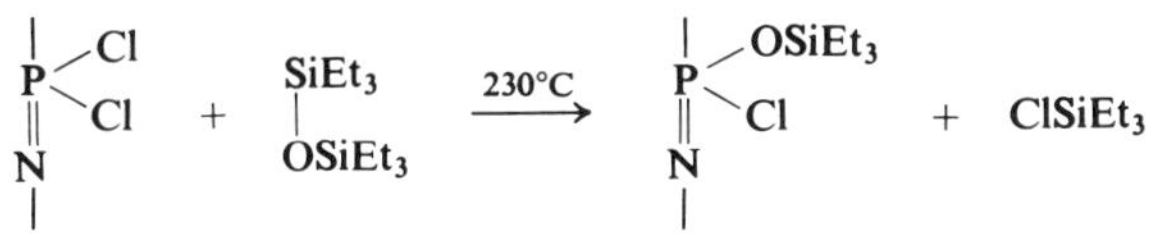

At higher temperatures (~300°C), colorless, transparent, rubbery materials or white cross-linked solids are formed by the liberation of hexaethyldisiloxane, with the resultant formation of P—O—P bridges

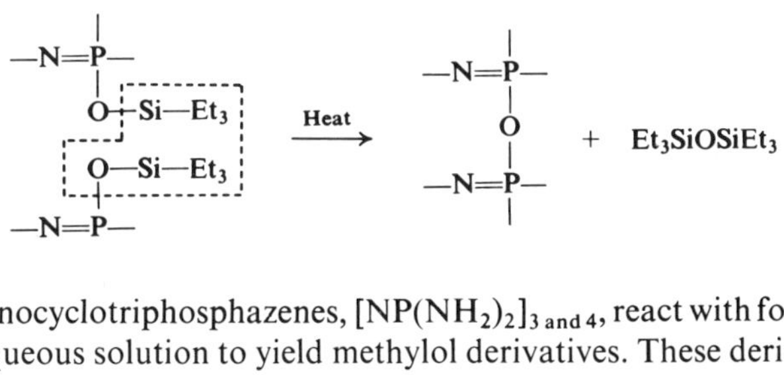

The aminocyclotriphosphazenes, $[NP(NH_2)_2]_{3\ \text{and}\ 4}$, react with formaldehyde in basic aqueous solution to yield methylol derivatives. These derivatives have the idealized structure XXX.[95] When heated at 80°C, the methylolcyclophosphazenes lose water to form heat-stable, polymeric products. By analogy with

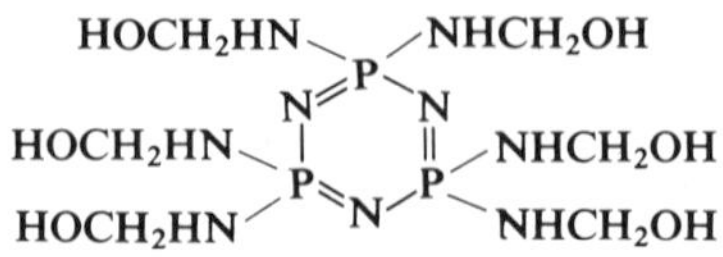

XXX

the well-known melamine-formaldehyde resins, it is presumed that polymerization involves the formation of P—$NHCH_2OCH_2NH$—P bridges between cyclophosphazene units. Similar condensations have been reported for the methylol derivatives formed from $N_3P_3(OCH_2CF_3)_3(NH_2)_3$, but the polymers decomposed at 250°C.[96] Polycondensation reactions also take place between hexakis(amino)cyclotriphosphazene and diacid chlorides, ClCOR COCl [where R is CH_2, $(CH_2)_2$, $(CH_2)_4$, $(CH_2)_8$, or C_6H_4] or phenylphosphonic dichloride.[97] Hydrogen chloride is evolved as cross-linking occurs. The polymers decompose in the 250°–350°C temperature range.

Polymers are also formed when cyclic chlorophosphazenes are treated with secondary or tertiary aromatic amines, such as acetanilide, diphenylurea, diethyldiphenylurea, or diphenylguanidine.[98] Hydrogen chloride and an organic derivative (acetyl chloride or phenyl isocyanate) are evolved, and the residual polymer is believed to contain one P═N—Aryl unit in place of each original PCl_2 group in the molecule. Butylaminophosphazenes, $[NP(NHC_4H_9)_2]$, are converted to pale yellow, hard, brittle, heat-stable, glassy resins when heated at 350°C for 1 hour.[99] The polymers are stable to acids, alkalis, water, and organic solvents. Phenylaminophosphazenes, such as $[NP(Ph)(NH_2)]_3$, also decompose at 270°–300°C with elimination of ammonia and the formation of cyclomatrix polymers.[100] Other cross-linked polymers are formed by the interaction of isothiocyanophosphazenes with diamines,[101] or aminophosphazenes with 2,4-toluene diisocyanate.[102]

References

1. F. Seel and J. Langer, *Angew. Chem.* **68**, 461 (1956).
2. F. Seel and J. Langer, *Z. Anorg. Allg. Chem.* **295**, 316 (1958).
3. H. R. Allcock, R. L. Kugel, C. W. Cameron, and W. J. Cook, unpublished work (1970).
4. C. P. Haber, *Chem. Soc., Spec. Publ.* **15**, 123 (1961).
5. H. R. Allcock, W. J. Cook, R. L. Kugel, G. F. Konopski, and E. G. Stroh, unpublished work (1968).
6. H. R. Allcock, R. L. Kugel, G. F. Konopski, and E. G. Stroh, *Chem. Commun.* p. 985 (1970).
7. H. R. Allcock and E. G. Stroh, unpublished work (1970).
8. H. R. Allcock, J. J. Meister, and J. E. Lehnsen, unpublished calculations.
9. R. Schenck and G. Römer, *Chem. Ber.* **57B**, 1343 (1924).
10. J. E. Proctor, N. L. Paddock, and H. T. Searle, Aust. Pat. 233,600 (1959).
11. N. L. Paddock, U.S. Pat. 3,026,174 (1962).

12. M. Becke-Goehring and G. Koch, *Chem. Ber.* **92**, 1188 (1959).
13. S. M. Zhivukhin, V. B. Tolstoguzov, and M. M. Levitskii, *Zh. Prikl. Khim.* **35**, 290 (1962); *Chem. Abstr.* **57**, 1838i (1962).
14. E. F. Moran, *J. Inorg. Nucl. Chem.* **30**, 1405 (1968).
15. G. M. Nichols, *Rep. Conf. High Temp. Polym. Fluid Res., 1962.* ASD-TDR-62-372 (1962).
16. H. R. Allcock and R. L. Kugel, *J. Amer. Chem. Soc.* **87**, 4216 (1965).
17. H. R. Allcock, R. L. Kugel, and K. J. Valan, *Inorg. Chem.* **5**, 1709 (1966).
18. H. Spencer, *Angew. Chem.* **65**, 299 (1953).
19. K. H. Meyer, W. Lotmar, and G. W. Pankow, *Helv. Chim. Acta* **19**, 930 (1936).
20. R. Knoesel, J. Parrod, and H. Benoit, *C. R. Acad. Sci.* **251**, 2944 (1960).
21. F. Patat and F. Kollinsky, *Makromol. Chem.* **6**, 292 (1951).
22. F. Patat and K. Frömbling, *Monatsh. Chem.* **86**, 718 (1955).
23. E. Giglio, F. Pompa, and A. Ripamonti, *J. Polym. Sci.* **59**, 293 (1962).
24. T. R. Manley and D. A. Williams, *Polymer* **10**, 307 (1969).
25. F. G. R. Gimblett, *Trans. Faraday Soc.* **56**, 528 (1960).
26. S. M. Zhivukhin, V. B. Tolstoguzov, and Y. V. Meitn, *Vysokomol. Soedin.* **3**, 414 (1961).
27. M. W. Spindler and R. L. Vale, *Makromol. Chem.* **43**, 237 (1961).
28. H. J. Kauth, U.S. Pat. 2,334,710 (1943).
29. U.S. Pat. 2,382,423 (1945).
30. Walker Chemical Co., Brit. Pat. 801,929 (1958).
31. L. Adany, Fr. Pat. 1,285,327 (1962).
31a. G. Singleton and J. A. Ross, Brit. Pat. 919,377 (1963) (to BTR Industries).
32. D. Cordischi, A. D. Site, A. Mele, and P. Porta, *J. Macromol. Chem.* **1**, 219 (1966).
33. R. J. A. Otto and L. F. Audrieth, *J. Amer. Chem. Soc.* **80**, 5894 (1958).
34. G. Tesi, R. J. A. Otto, F. G. Sherif, and L. F. Audrieth, *J. Amer. Chem. Soc.* **82**, 528 (1960).
35. H. R. Allcock, G. F. Konopski, and R. W. Tomlinson, unpublished work (1969).
36. G. Goldschmidt and B. Dishon, *J. Polym. Sci.* **3**, 481 (1948).
37. A. S. Flatau, Brit. Pat. 878,324 (1961) (to D. Napier & Son Ltd.).
38. M. V. Lenton, B. Lewis, and C. A. Pearce, *Chem. Ind. (London)* p. 1387 (1964).
39. M. E. Mirhej and J. F. Henderson, *J. Macromol. Chem.* **1**, 187 (1966).
40. R. L. Evans, U.S. Pat. 3,271,330 (1966) (to 3M Company).
41. S. H. Rose, *J. Polym. Sci., Part B* **6**, 837 (1968).
42. S. H. Rose and J. Cable, U.S. Army Materials and Mechanics Research Report, Project DAAG17-68-C-0096 P001 (1969).
42a. G. L. Hagnauer, N. Schneider, and R. E. Singler, 161st Amer. Chem. Soc. Meeting, Los Angeles (1971), *Polymer Preprints* **12**, [1], 525 (1971).
43. G. Allen, C. J. Lewis, and S. M. Todd, *Polymer* **11**, 31 (1970).
44. G. Allen, C. J. Lewis, and S. M. Todd, *Polymer* **11**, 44 (1970).
45. H. R. Allcock and G. Y. Moore, unpublished work (1971).
46. J. R. MacCallum and J. Tanner, *J. Macromol. Sci., Chem.* **4**, 481 (1970).
47. H. R. Allcock and R. L. Kugel, *Inorg. Chem.* **5**, 1716 (1966).
48. H. R. Allcock and D. P. Mack, *Chem. Commun.* p. 685 (1970).
49. H. R. Allcock, W. J. Cook, and D. P. Mack, unpublished work (1969).
50. H. R. Allcock and D. P. Mack, unpublished work (1969).
51. H. H. Sisler, S. E. Frazier, R. G. Rice, and M. G. Sanchez, *Inorg. Chem.* **5**, 326 (1966).
52. C. F. Liu and R. L. Evans, U.S. Pat. 3,169,933 (1965) (to 3M Company).
53. G. Tesi, C. P. Haber, and C. M. Douglas, *Proc. Chem. Soc., London* p. 219 (1960); U.S. Pat. 3,087,937 (1963).

54. G. Tesi and C. M. Douglas, *J. Amer. Chem. Soc.* **84**, 549 (1962).
55. J. R. MacCallum and J. Tanner, *J. Polym. Sci., Part A*-1 **6**, 3163 (1968).
56. D. L. Herring, *Chem. Ind.* (*London*) p. 717 (1960).
57. American Cyanamid Co., Neth. Pat. Appl. 6,512,493 (1966); *Chem. Abstr.* **65**, 3997b (1966).
58. W. R. Grace & Co., Brit. Pat. 961,912 (1964).
59. R. W. Stephens, U.S. Pat. 3,193,397 (1965) (to W. R. Grace & Co.).
60. M. V. Lenton and B. Lewis, *J. Chem. Soc., A* p. 665 (1966).
61. G. R. Feistel and T. Moeller, *J. Inorg. Nucl. Chem.* **20**, 2731 (1967).
62. A. J. Bilbo, C. M. Douglas, N. R. Fetter, and D. L. Herring, *J. Polym. Sci., Part A*-1 **6**, 1671 (1968).
62a. R. J. Irving, Brit. Pat. 921,034 (1963).
62b. H. Coats and P. A. Dye, Brit. Pat. 1,013,462 (1965) (to Albright and Wilson Mnfg. Ltd.).
63. S. I. Belykh, S. M. Zhivukhin, V. V. Kireev, and H. S. Kolesnikov, *Vysokomol. Soedin., Ser. A* **11**, No. 3, 625 (1969).
64. H. G. Horn and M. Becke-Goehring, *J. Polym. Sci., Part A*-1 **7**, 959 (1969).
65. H. R. Allcock, *J. Amer. Chem. Soc.* **85**, 4050 (1963); **86**, 2591 (1964).
66. S. M. Zhivukhin, V. B. Tolstoguzov, and F. I. Yakobson, *Vysokomol. Soedin.* **8**, No. 4, 727 (1966).
67. A. J. Bilbo and C. M. Sharts, *J. Polym. Sci., Part A*-1 **5**, 2891 (1967).
67a. H. G. Horn, *Makromol. Chem.* **138**, 163 (1970).
68. Quantum Inc., *U.S. Govt. Res. Rep.* **AD 231-270** (1959).
69. B. Dishon, *J. Amer. Chem. Soc.* **71**, 2251 (1949).
70. C. Hamalainen and J. D. Guthrie, *Text. Res. J.* **24**, 141 (1956).
71. S. Konya and M. Yokoyama, *Kogyo Kagaku Zasshi* **68**, 2450 (1965).
72. H. R. Allcock, P. S. Forgione, and K. J. Valan, *J. Org. Chem.* **30**, 947 (1965).
73. R. Rätz, H. Schroeder, H. Ulrich, E. Kober, and C. Grundmann, *J. Amer. Chem. Soc.* **84**, 551 (1962).
74. R. Rätz and C. Grundmann, U.S. Pat. 2,858,306 (1958).
75. E. T. McBee, L. Brinkmann, and H. P. Braendlin, *U.S. Govt. Res. Rep.* **AD 254-983** (1961).
76. A. F. Nikolaev, E. T. Wan, T. A. Zyryanova, G. A. Balaev, E. B. Lebedeva, and K. S. Afanas'eva, *Sov. Plast.* **9**, 24 (1967).
77. C. A. Redfarn, U.S. Pat. 2,866,773 (1958) (to Walker Extract and Chemical Company).
78. R. G. Rice, B. H. Geib, and L. A. Kaplan, U.S. Pat. 3,121,704 (1964) (to General Dynamics Corp.).
79. *Chem. & Eng. News* **39**, Oct. 16, p. 63 (1961).
80. El Monte Chemical Co., Product Data Sheet, 64-2 (1964).
81. S. M. Zhivukhin, V. B. Tolstoguzov, and V. V. Kireev, *Sov. Plast. No.* 7, p. 24 (1963).
82. S. M. Zhivukhin and V. V. Kireev, *Plast. Massy* No. 7, p. 24 (1964).
83. S. M. Zhivukhin, V. V. Kireev, and A. N. Zelenskii, *Zh. Prikl. Khim.* **39**, 234 (1966).
84. M. Yokoyama, H. Cho, and K. Kono, *Kogyo Kagaku Zasshi* **67**, 1378 (1964).
84a. Hitachi Corp. Brit. Pat. 1,047,052 (1966).
84b. R. G. Rice and M. V. Ernest, Fr. Pat. 1,450,474 (1966) (to W. R. Grace & Co.).
85. M. Kajiwara and H. Saito, *Kogyo Kagaku Zasshi* **67**, 1000 and 1002 (1964).
86. M. Kajiwara and H. Saito, *Kogyo Kagaku Zasshi* **68**, 1210 (1965).
87. M. Kajiwara and H. Saito, *Kogyo Kagaku Zasshi* **70**, 1101 (1967).
88. M. Kajiwara and H. Saito, *Kogyo Kagaku Zasshi* **71**, 1470 (1968).
89. A. F. Nikolaev and E. T. Wan, *J. Gen. Chem. USSR* **34**, 1846 (1964).

90. M. Yokoyama and S. Konya, *Kogyo Kagaku Zasshi* **68**, 2444 (1965).
91. Walker Chemical Company, Fr. Pat. 1,281,571 (1961).
92. M. Apley and J. R. Alexander, Brit. Pat. 933,720 (1963); U.S. Pat. 3,164,556 (1965) (to Walker Chemical Co.).
93. C. A. Redfarn, U.S. Pat. 3,194,787 (1965) (to Walker Chemical Co.).
94. S. M. Zhivukhin, V. B. Tolstoguzov, and A. I. Ivanov, *Russ. J. Inorg. Chem.* **7**, 1134 (1962).
94a. Unilever Ltd. Brit. Pat. 1,052,388 (1966).
95. R. J. Irving and J. Dewing, Brit. Pat. 903,046 (1962) (to Imperial Chemical Industries).
96. E. T. McBee, L. Brinkmann, and H. P. Braendlin, *U.S. Govt. Res. Rep.* **AD 254-982** (1961).
97. M. Yokoyama and S. Konya, *Kogyo Kagaku Zasshi* **70**, 1453 (1967).
98. C. J. Brown, U.S. Pat. 2,374,646 (1945) (to Imperial Chemical Industries).
99. D. Lipkin, U.S. Pat. 2,214,769 (1940) (to Atlantic Refining Co.).
100. W. R. Grace & Co., Brit. Pat. 981,821 (1965); Fr. Pat. 1,461,442 (1966).
101. R. Rätz, U.S. Pat. 3,074,989 (1963).
102. S. M. Zhivukhin and I. M. Bublevskii, USSR Pat. 170,684 (1965).

Appendix I

BOND ANGLES AND INTERATOMIC DISTANCES FOR PHOSPHORUS–NITROGEN COMPOUNDS

X-Ray diffraction methods were used for the structural determinations, except for the two electron-diffraction studies listed. Where several structural determinations for the same compound are reported, the most recent or most accurate values are listed first. Note that the skeletal atom numbering system used in the body of the book is used in this Appendix also, even though this system differs from conventional crystallographic practice.

Compounds	Structure	Bond angles (degrees)		Bond lengths (Å)		Comments	Ref.
Hexafluorocyclotriphosphazene	$[N{=}PF_2]_3$	N—P—N P—N—P F—P—F	119.5, 119.3 119.6, 121.1 <99.9	P—N P—F	1.57 (average) 1.52	Planar ring. Orthorhombic cell; s.g. *Pnma*, $Z=4$	1
		N—P—N P—N—P F—P—F	121.4 118.6 98.8	P—N P—F	1.59 1.54	Electron-diffraction study	2
Octafluorocyclotetra-phosphazene	$[N{=}PF_2]_4$	N—P—N P—N—P F—P—F	122.7 ± 1.0 147.2 ± 1.4 99.9 ± 1.0	P—N P—F	1.51 ± 0.02 1.51 ± 0.02	Planar ring (D_{4h}) Monoclinic cell; s.g. $P2_1/a$, $Z=2$	3 4
Poly(difluorophosphazene)	$[N{=}PF_2]_n$	N—P—N P—N—P F—P—F	119 136 95	P—N P—F	1.52 1.47	*Cis–trans* planar chain. Orthorhombic cell; s.g. $Cmc2_1$	5, 6
Fluoropentachlorocyclo-triphosphazene	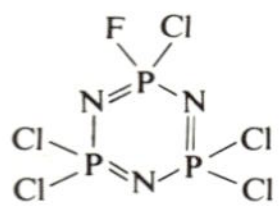	N—P—N P—N—P Hal—P—Hal	118.5 ± 0.5 121.5 ± 0.4 100.1 ± 0.2 101.1 ± 0.2	P—N	1.563 ± 0.007	Planar ring. Orthorhombic cell; s.g. *Pnma* or $Pn2_1a$. P—Hal lengths uncertain, $Z=4$	7
Hexachlorocyclotri-phosphazene (Phosphonitrilic chloride trimer)	$[N{=}PCl_2]_3$	N—P—N P—N—P Cl—P—Cl	118.4 ± 0.3 121.4 ± 0.4 101.4 ± 0.2	P—N P—Cl	1.581 1.993	Almost planar ring. Orthorhombic cell; s.g. *Pnma*, $Z=4$	8, 9
		Cl—P—Cl	107–110	P—N P—Cl	1.65 ± 0.03 1.97 ± 0.03	Electron-diffraction study "Benzene-type" ring Slight chair configuration Nonequivalent P—N bonds	10 11 12
Octachlorocyclotetra-phosphazene	$[N{=}PCl_2]_4$	N—P—N P—N—P Cl—P—Cl	120.5 ± 0.7 133.6 ± 0.8 137.6 ± 0.8 103.1 ± 0.2	P—N P—Cl	1.559 ± 0.012 1.992 ± 0.004	Puckered (chair) ring: stable T form. Tetragonal cell; s.g. $P4_2/n$, $Z=4$	13
		N—P—N P—N—P Cl—P—Cl	121.2 ± 0.5 131.3 ± 0.6 102.8 ± 0.2	P—N P—Cl	1.57 ± 0.01 1.99 (average)	Puckered (tub) ring: metastable K form Tetragonal cell; s.g. $P4_2/n$, $Z=2$	14

		N—P—N P—N—P Cl—P—Cl	117 123 105.5	P—N P—Cl	1.67[5] 1.99[5]	Puckered ring: metastable K form. Tetragonal cell; s.g. $P4_2/n$	15
Decachlorocyclopenta-phosphazene	$[N{=}PCl_2]_5$	N—P—N P—N—P Cl—P—Cl	118.4 148.6 102.2	P—N P—Cl	1.521 1.961	Nearly planar ring: orthorhombic cell; s.g. $P(2_12_12_1)$, $Z=4$	16
Poly(dichlorophosphazene)	$[N{=}PCl_2]_n$	N—P—N P—N—P	119 127	P—N	1.60	Closer to *cis–trans* planar than to uniform helix model. Twofold repeating conformation. Structure solved by optical transform technique	17
		N—P—N P—N—P	124 124			Helical chain with 4.92±0.05 Å repeating distance	18
Hexabromocyclotri-phosphazene	$[N{=}PBr_2]_3$	N—P—N P—N—P Br—P—Br	115.8, 118.3 126.8, 118.6 103.4, 101.7	P—N P—Br	1.55, 1.60 2.18 (average)	Slight chair conformation. Orthorhombic cell; s.g. $Pcmn$, $Z=4$	19, 20
Hexakis(isothiocyanato)-cyclotriphosphazene	$[N{=}P(NCS)_2]_3$	N—P—N (ring) P—N—P N(exo)—P—N (exo)	119±3 121±4 100±1	P—N (ring) P—N (exo) N—C C—S	1.58 ±0.05 1.63 ±0.01 1.12±0.04 1.59±0.04	Slightly puckered ring. Triclinic cell; s.g. $P\bar{1}$, $Z=2$	21
Octamethylcyclotetra-phosphazene	$[N{=}PMe_2]_4$	N—P—N P—N—P C—P—C	119.8 131.9 104	P—N P—C	1.596, ±0.003 1.805, ±0.004	Puckered ring. Tetragonal cell; s.g. $I4_1/a$, $Z=4$	22
1,1-Dimethylhexafluoro-cyclotetraphosphazene	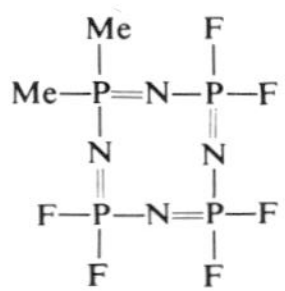	N—P—N[a] P—N—P[a] C—P—C F—P—F	116.9, 126.1, 124.6 146.7, 143.3 105.7 93.2, 94.1	P—N[a]	1.584, 1.470, 1.532, 1.487	Puckered (saddle) ring. Monoclinic cell; s.g. $I2/a$, $Z=4$. Alternation of ring bond lengths away from methyl substituents. The 1.47 Å P—N bond is one of the shortest known.	23, 23a

Compounds	Structure	Bond angles (degrees)		Bond lengths (Å)		Comments	Ref.
1,1,5,5-Tetramethyltetra-fluorocyclotetraphos-phazene	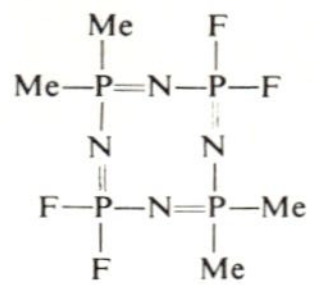	N—P—N P—N—P C—P—C F—P—F	125.9[b], 117.5 134.6 106.9 96.4, 97.0	P—N	1.59[b], 1.54	Slightly puckered (saddle) ring: monoclinic; s.g. $C2/c$, $Z=4$	23, 23b
Bis(octamethyl-N-hydrocyclo-tetraphosphazenium) tetrachlorocobaltate(II)	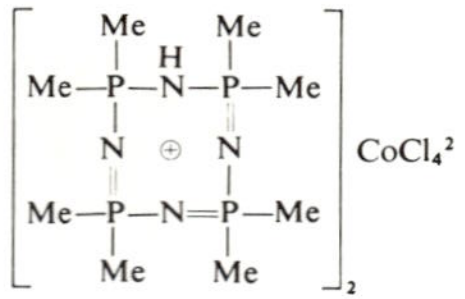	N—P—N P—N—P C—P—C	109.9–118.7 125.3[c]–135.5 104.0–109.6	P—N P—C	1.537–1.694[d] 1.795	Both rings are puckered (one tub, one saddle). The association between the phosphazene rings and the $CoCl_4^{2-}$ units is ionic. Monoclinic cell; s.g. $P2_1/n$, $Z=4$	24, 25
Octamethyl-N-hydrocyclotetra-phosphazenium trichloro copper (II)	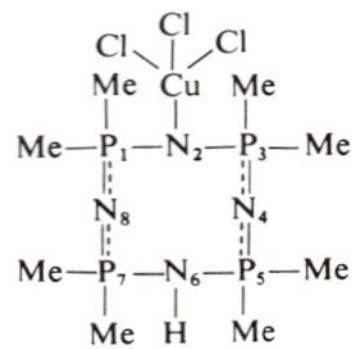	$N_2—P_1—N_8$ $N_2—P_3—N_4$ $N_8—P_7—N_6$ $N_4—P_5—N_6$ $P_1—N_2—P_3$ $P_1—N_8—P_7$ $P_3—N_4—P_5$ $P_5—N_6—P_7$	117.3 112.7 111.7 114.5 123.2 134.0 131.5 127.3	$P_1—N_2$ $P_3—N_2$ $P_1—N_8$ $P_3—N_4$ $P_7—N_8$ $P_5—N_4$ $P_7—N_6$ $P_5—N_6$ $N_1—Cu$ P—C	1.631 1.638 1.602 1.591 1.563 1.544 1.670 1.665 2.042 1.782 (average)	Puckered (Tub) ring. Bonding between Cu and N is covalent. Orthorhombic cell; s.g. $Pbca$, $Z=8$	26
Octakis(dimethylamino)-cyclotetraphosphazene		N—P—N (ring) P—N—P N—P—N (exo)	120 133 104	P—N (ring) P—N (exo)	1.58 1.67, 1.69	Puckered ring. Short P—N exo bond denotes some π-character. Tetragonal cell; s.g. $I\bar{4}$, $Z=2$	27
Dodeca(dimethylamino)-cyclohexaphosphazene	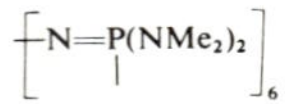	N—P—N (ring) P—N—P N—P—N (exo)	120±0.5 147.5±0.7 102.9±0.5	P—N (ring) P—N (exo)	1.563 ±0.01 1.669 ±0.01	Puckered ring (pseudo-boat), with conformation resulting from steric effects. Cubic cell: s.g. $Pa3$, $Z=4$	28

Tetrakis(isopropylamino)-dichlorocyclotriphosphazene hydrochloride	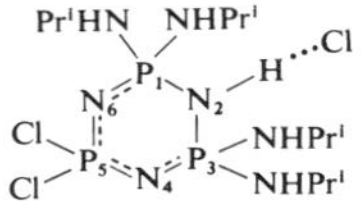	N_2—P_1—N_6 N_4—P_5—N_6 N_2—P_3—N_4 P_1—N_2—P_3 P_1—N_6—P_5 P_3—N_4—P_5 Cl—P—Cl N—P—N (exo)	107.5±0.2 120.1±0.2 107.5±0.2 131.7±0.3 125.4±0.3 125.4±0.3 100.3±0.1 103.2 (average) ±0.2	P_1—N_2 P_3—N_2 P_1—N_6 P_3—N_4 P_3—N_4 P_5—N_4 P—N (exo) P—Cl P—Cl	1.66 ±0.005 1.583 ±0.005 1.558 ±0.005 1.60–1.62 2.00 2.00	Puckered (boat) ring. Note the long P—N bonds to the protonated nitrogen. Triclinic cell; s.g. $P\bar{1}$, $Z=2$	29
Octamethoxycyclotetra-phosphazene	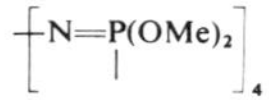	N—P—N P—N—P O—P—O	121.0±0.6 132.2±0.7 102.8–106.8	P—N P—O	1.55–1.60 1.54–1.61	Puckered (saddle) ring. Orthorhombic cell; s.g. *Pbca*. $Z=8$	30
Hexadeca(methoxy)-cyclooctaphosphazene	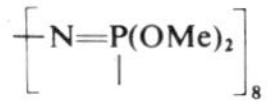	N—P—N P—N—P O—P—O	116.7 (average)[e] 136.7 (average)[e] 99.2–105.6	P—N P—O	1.561 (average) 1.576 (average)	Molecular conformation consists of two planar, parallel segments, joined by a step. Triclinic cell; s.g. $P1$ or $P\bar{1}$, $Z=1$	31
Hexaphenoxycyclotri-phosphazene	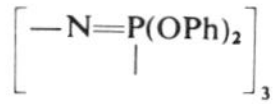	N—P—N P—N—P O—P—O	117.3±3 121.9±3 94.1±2 100.1±2	P—N P—O	1.575±0.002 1.582±.0002	Slightly nonplanar phosphazene ring. Monoclinic cell; s.g. $P2_1/n$; $Z=4$	31a
Tris (*o*-phenylenedioxy)-cyclotriphosphazene (as channel inclusion adduct host)	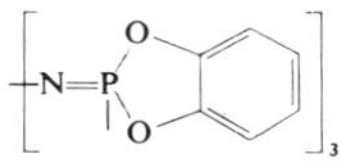	N—P—N P—N—P O—P—O P—O—C	117.5 122.5 97 109	P—N P—O	1.575±0.002 1.594	Slight (6°) chair distortion of the phosphazene ring from planar. Phenyl groups slightly twisted from plane which bisects phosphazene ring. Hexagonal cell; s.g. $P6_3$, $Z=2$	32
Tris(2,3-dioxynaphthyl)-cyclotriphosphazene (as channel inclusion adduct host)	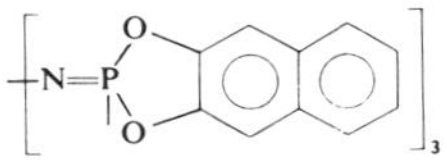	—	—	—	—	Hexagonal cell.	33
Tris (2,2′-dioxybiphenyl)-cyclotriphosphazene	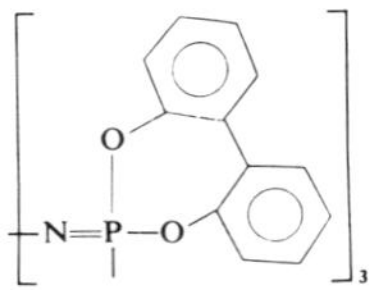	N—P—N P—N—P O—P—O	118.6±0.3 120.8±0.3 121.2±0.3 102.4±0.3 103.0±0.2	P—N P—O	1.572 (average) 1.584±0.005	Phosphazene ring has slight boat distortion from planar. Two of the side groups are twisted, propellor-like in one direction, and the other is twisted in the opposite direction. Monoclinic cell; s.g. $C2/c$, $Z=4$	34

Compounds	Structure	Bond angles (degrees)		Bond lengths (Å)		Comments	Ref.
1,1-Diphenyl-3,3,5,5-tetrafluorocyclotri-phosphazene	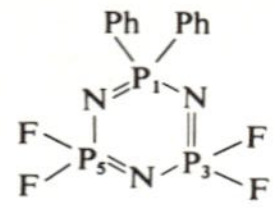			P_3—N_2 P_3—N_4	1.54 ±0.01 1.56 ±0.01	The phosphorus atom which bears the phenyl groups lies above the plane of the other ring atoms. Orthorhombic cell; s.g. *Pnma* or *Pn*2_1*a*, *Z*=8	35
1,1-Diphenyl-3,3,5,5-tetrachlorocyclotri-phosphazene	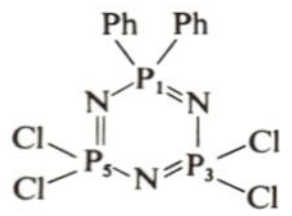	N—P_1—N N—P—N P_1—N—P_3 P_3—N—P_5 Cl—P—Cl Ph—P—Ph	115.2 ±0.2 119.75±0.3 122.0 ±0.3 119.2 100.3 ±0.1 104.4 ±0.3	P—N^f	1.615, 1.578 1.555, (all ±0.005)	Phosphazene ring has slight chair form. P—N bonds unequal. Monoclinic cell; s.g. *P*2_1/*n*, *Z*=4	36
1,1-Dichloro-3,3,5,5-tetraphenylcyclotri-phosphazene	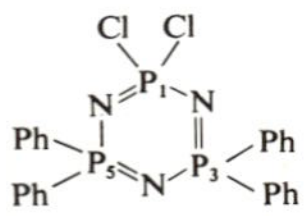	N_2—P_1—N_6 N_2—P_3—N_4 P_1—N_2—P_3 P_3—N_4—P_5 Cl—P_1—Cl C—P_3—C } C—P_5—C }	120.7±0.5 115.5±0.4 121.0±0.5 124.9±0.5 98.5±0.2 104.4±0.5	P_1—N_2 P_3—N_4 P_3—N_2	1.556 1.578 1.609	Puckered (boat) ring. Monoclinic cell; s.g. *C*2/*c*, *Z*=16	37
Hexaphenylcyclotri-phosphazene	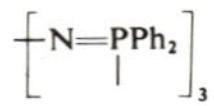	N—P—N P—N—P C—P—C	117.8±0.4 122.1±0.4 103.8±0.3	P—N P—C	1.597±0.006 1.804±0.007	Phosphazene ring has an irregular slight chair form. Triclinic cell; s.g. $P\bar{1}$; *Z*=2	38
cis-1,3,5,7-Tetrachloro-1,3,5,7-tetraphenylcyclo-tetraphosphazene	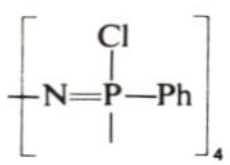	N—P—N P—N—P Cl—P—Ph	121 (average) 137 (average) 103 (average)	N—P P—Cl P—C	1.57 2.04 1.78	Phosphazene ring has crown shape. Triclinic cell; s.g. *P*1; *Z*=2	39
β-trans-1,3,5,7-Tetrachloro-1,3,5,7-tetraphenylcyclo-tetraphosphazene	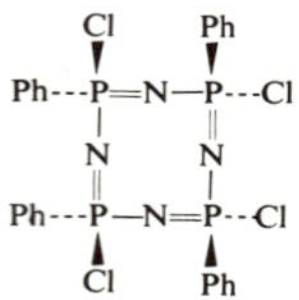	N—P—N P—N—P Cl—P—C	119.5, 121.3 132.4, 138.6 102.2, 105.1	P—N P—Cl P—C	1.55–1.60 1.80, 2.04 1.77, 1.80	Phosphazene ring puckered in chair shape. Triclinic cell; s.g. $P\bar{1}$, *Z*=1	40

β-trans-1,3,5,7-tetrakis-(methylamino)-1,3,5,7-tetraphenylcyclotetra-phosphazene	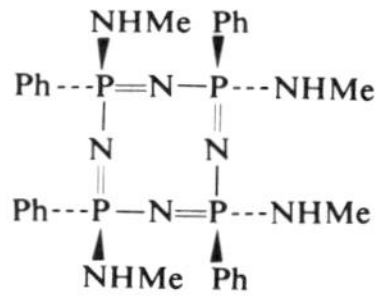	N—P—N P—N—P N(exo)—P—C	118.2, 121.0 124.6, 131.2 105.6, 106.2	P—N P—N (exo) P—C	1.60 1.67–1.69 1.81	Phosphazene ring puckered in chair shape. Triclinic cell; s.g. $P\bar{1}$, $Z=1$	40
1,3,5-Trimethoxy-1,3,5-trioxo-2,4,6-trimethylcyclo-triphosphazane	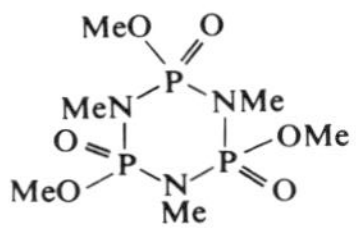	N—P—N P—N—P O═P—O	106.2, 105.6, 103.8, ±0.6 121.7 (average) 117.7, 117.1, 113.4, ±0.7	P—N P═O P—O	1.66 (average) 1.45 (average) 1.56 (average)	Phosphazane ring has twisted boat shape. Monoclinic cell; s.g. $P2_1$, $Z=2$	41
Trisodium 1,1,3,3,5,5-hexakis-(oxo)-2,4,6-trihydrocyclotriphos-phazanate (tetrahydrate) (Sodium trimetaphosphi-mate)	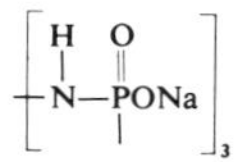	N—P—N P—N—P O—P—O O—P—N	104.5 123 118 106–109	P—N P—O	1.68 1.50	Chair configuration. Monoclinic cell; s.g. $P2_1/n$, $Z=4$	42, 43
1,3,5,7-Tetrahydroxy-1,3,5,7-tetrakis-(oxo)-2,4,6,8-tetrahydro-cyclotetraphosphazane (dihydrate) (Tetrameta-phosphimic acid)	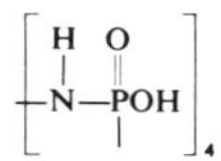	N—P—N P—N—P O—P—O O—P—N	107.3 125.6±0.5 116.1 105–111	P—N P—O	1.66±0.01 1.52, 1.48±0.01	Longer P—O bonds involved in hydrogen bonding. Orthorhombic cell; s.g. $P2_12_12$, $Z=2$	44, 45
Tetrapotassium 1,1,3,3,5,5,7,7-octakis-(oxo)-2,4,6,8-tetrahydrocyclo-tetraphosphazanate (tetrahydrate). (Potassium tetrametaphosphimate)	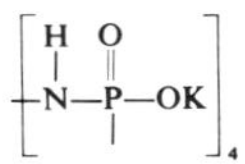	N—P—N P—N—P O—P—O	108.1 131.9 117.8	P—N P—O	1.673 1.500	Chair conformation. Monoclinic cell; s.g. $P2_1/c$, $Z=2$	46
Tetracesium 1,1,3,3,5,5,7,7-octakis-(oxo)-2,4,6,8-tetrahydrocyclotetraphos-phazanate (hexahydrate). (Cesium tetrametaphos-phimate)	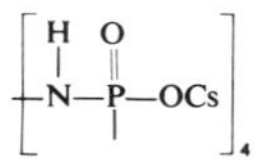	N—P—N P—N—P O—P—O	106.9 128.9 117.7	P—N P—O	1.676 1.498	Saddle conformation. Tetragonal cell; s.g. $P4_2/nmc$, $Z=2$	46

Compounds	Structure	Bond angles (degrees)		Bond lengths (Å)		Comments	Ref.
1,1,1,3,3,3-Hexachloro-2,4-dimethylcyclodiphosphazane	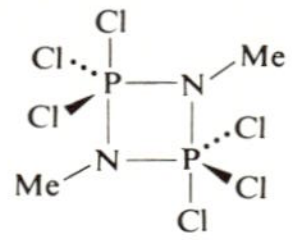	N—P—N P—N—P Cl—P—Cl	80.5 ± 0.4 99.5 ± 0.4 90.3, 110.0	P—N (axial) P—N (equat) P—Cl (axial) P—Cl (equat) N—C	1.769 ± 0.007 1.635 ± 0.007 2.133 ± 0.003 2.026 ± 0.004 1.475 ± 0.010	Planar ring with a trigonal bipyramidal arrangement at each phosphorus, and with bonds to nitrogen occupying one axial and one equatorial position. Monoclinic cell; s.g. $P2_1/n$, $Z=4$	47
		N—P—N P—N—P Cl—P—Cl	81.7 ± 0.5 98.3 ± 0.5 90.5, 109.8	P—N (axial) P—N (equat) P—Cl (axial) P—Cl (equat) N—C	1.776 ± 0.01 1.629 ± 0.01 2.152 ± 0.005 2.020 ± 0.005 1.476 ± 0.016		48
1,1,3,3-Tetrafluoro-1,3-diphenyl-2,4-dimethylcyclodiphosphazane	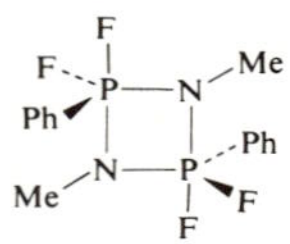	N—P—N P—N—P F—P—F Ph—P—F	80.6 99.4 87.0 91.8, 109	P—N (axial) P—N (equat) P—F (axial) P—F (equat) P—C (equat) N—C	1.78 ± 0.02 1.64 ± 0.02 1.62 ± 0.02 1.57 ± 0.02 1.79 ± 0.02 1.44	Phosphazane ring is planar, with trigonal bipyramidal arrangement at phosphorus (phenyl group and one nitrogen equatorial). Orthorhombic cell; s.g. $Pbca$, $Z=4$	49
1,3-dichloro-1,3-dithio-2,4-dimethylcyclodiphosphazane	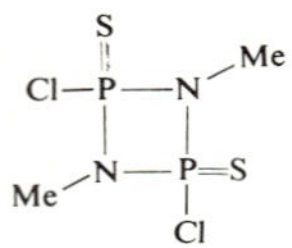	N—P—N P—N—P S—P—Cl	84.0 96.0 115.0	P—N P—S } P—Cl } N—C	1.67 1.93 1.46	Tetragonal cell, s.g. $P4_2/mnm$, $Z=2$	50

Compound	Structure	Angle	Degrees	Bond	Å	Remarks	Ref.
	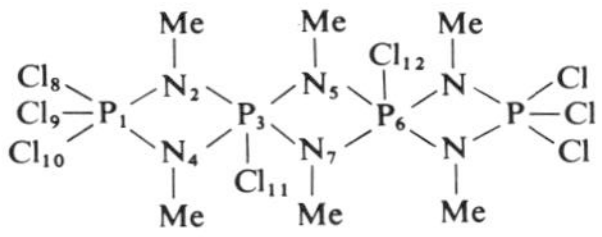	N_2—P_1—N_4 N_2—P_3—N_4 N_5—P_3—N_7 Cl_8—P_1—Cl_9 Cl_8—P_1—Cl_{10} Cl_9—P_1—Cl_{10}	80.7 79.0 80.0 91.0 89.3 109.4	P—N P—Cl N—C	1.64–1.77 2.05, 2.16, 2.10 1.45–1.52	Phosphazane skeleton occupies a distorted spiro-type conformation, with trigonal bipyramidal-type arrangements at phosphorus. Monoclinic cell; s.g. $P2_1/c$, $Z=2$	51
1,1,3,5-Tetrachloro-3,5-dioxocyclotri-1-phospha-3,5-thiazatriene	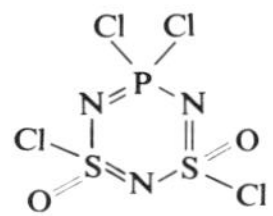	N—P—N N—S—N S—N—S P—N—S Cl—P—Cl Cl—S—O	115.3±0.7 115.0±0.7 120.3±0.8 123.5, 120.6±0.8 104.4±0.3 105.7±0.6 106.3±0.5	P—N S—N P—Cl S—Cl S—O	1.585 ±0.013 1.578, 1.540±0.013 1.957 ±0.006 2.018 ±0.006 1.421 ±0.013	Ring has deformed chair shape. Monoclinic cell; s.g. $P2_1/n$, $Z=4$	52
	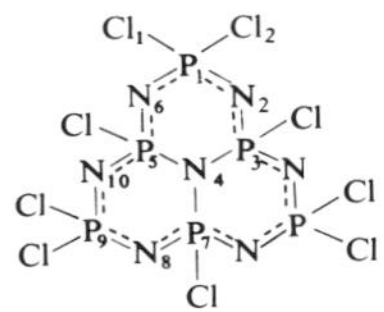	N_6—P_1—N_2 N_4—P_5—N_{10} N_6—P_5—N_{10} P_1—N_2—P_3 Cl_1—P—Cl_2 Cl—P_5—N_4	116.9 110.7 112.5 125.5 101.6 104.0	P_1—N_2 P_3—N_2 P_3—N_4 P_1—Cl_1	1.571 1.558 1.723 1.979	Molecule as a whole deviates from planarity. Central P—N bond is unusually long. Monoclinic cell; s.g. $C2/c$,$=Z=8$	53

[a] In sequence from Me-substituted phosphorus.

[b] At P—Me phosphorus atoms.

[c] The lower value represents the angle of the P—NH—P unit.

[d] The longest P—N bonds are connected to the nitrogen atom that bears the proton, with an alternation of short and longer bond lengths occurring from this point.

[e] One N—P—N angle is significantly larger (121.2°), and one P—N—P angle is larger (141°).

[f] In sequence from P_1.

References

1. M. W. Dougill, *J. Chem. Soc., London* p. 3211 (1963).
2. M. I. Davis and J. W. Paul, *Bull. Amer. Phys. Soc.* (II) **13**, 832 (1968).
3. H. McD. McGeachin and F. R. Tromans, *Chem. Ind.* (*London*) p. 1131 (1960).
4. H. McD. McGeachin and F. R. Tromans, *J. Chem. Soc., London* p. 4777 (1961).
5. H. R. Allcock, G. F. Konopski, R. L. Kugel, and E. G. Stroh, *Chem. Commun.* p. 985 (1970).
6. H. R. Allcock, R. L. Kugel, and E. G. Stroh, unpublished work (1970).
7. R. Olthof, *Acta Crytsallogr., Sect. B* **25**, 2040 (1969).
8. G. J. Bullen, *J. Chem. Soc., A* p. 1450 (1971).
9. A. Wilson and D. F. Carroll, *J. Chem. Soc., London* p. 2548 (1960).
10. L. O. Brockway and W. M. Bright, *J. Amer. Chem. Soc.* **65**, 1551 (1943).
11. F. Pompa and A. Ripamonti, *Ric. Sci.* **29**, 1516 (1959).
12. E. Giglio, *Ric. Sci.* **30**, 721 (1960).
13. A. J. Wagner and A. Vos, *Acta Crystallogr., Sect. B* **24**, 707 (1968).
14. R. Hazekamp, T. Migchelson, and A. Vos, *Acta Crystallogr.* **15**, 539 (1962).
15. J. A. A. Ketelaar and T. A. de Vries, *Rec. Trav. Chim. Pays-Bas* **58**, 1081 (1939).
16. A. W. Schlueter and R. A. Jacobson, *J. Chem. Soc., A* p. 2317 (1968).
17. E. Giglio, F. Pompa, and A. Ripamonti, *J. Polym. Sci.* **59**, 293 (1962).
18. K. H. Meyer, W. Lotmar, and G. W. Pankow, *Helv. Chim. Acta* **19**, 930 (1936).
19. H. Zoer, D. A. Koster, and A. J. Wagner, *Acta Crystallogr., Sect A* **25**, S107 (1969).
20. E. Giglio and R. Puliti, *Acta Crystallogr.* **22**, 304 (1967).
21. J. B. Faught, T. Moeller, and I. C. Paul, *Inorg. Chem.* **9**, 1656 (1970).
22. M. W. Dougill, *J. Chem. Soc., London* p. 5471 (1961).
23. W. C. Marsh, T. N. Ranganathan, J. Trotter, and N. L. Paddock, *Chem. Commun.* p. 815 (1970).

23a. W. C. Marsh and J. Trotter, *J. Chem. Soc., A* p. 573 (1971).

23b. W. C. Marsh and J. Trotter, *J. Chem. Soc., A* p. 569 (1971).

24. J. Trotter, S. H. Whitlow, and N. L. Paddock, *Chem. Commun.* p. 695 (1969).
25. J. Trotter and S. H. Whitlow, *J. Chem. Soc., A* p. 460 (1970).
26. J. Trotter and S. H. Whitlow, *J. Chem. Soc., A* p. 455 (1970).
27. G. J. Bullen, *J. Chem. Soc., London* p. 3193 (1962).
28. A. J. Wagner and A. Vos, *Acta Crystallogr., Sect. B* **24**, 1423 (1968); *Sect. B* **27**, 51 (1971).
29. N. V. Mani and A. J. Wagner, *Chem. Commun.* p. 658 (1968).
30. G. B. Ansell and G. J. Bullen, *J. Chem. Soc., A* p. 2498 (1971).
31. N. L. Paddock, J. Trotter, and S. H. Whitlow, *J. Chem. Soc., A* p. 2227 (1968).

31a. W. C. Marsh and J. Trotter, *J. Chem. Soc., A* p. 169 (1971).

32. L. A. Siegel and J. H. van den Hende, *J. Chem. Soc., A* p. 817 (1967).
33. H. R. Allcock, M. T. Stein, and J. A. Stanko, unpublished work (1971).
34. H. R. Allcock, M. T. Stein, and J. A. Stanko, *Chem. Commun.* p. 944 (1970); *J. Amer. Chem. Soc.* **93**, 3173 (1971).
35. C. W. Allen, I. C. Paul, and T. Moeller, *J. Amer. Chem. Soc.* **89**, 6361 (1967).
36. N. V. Mani, F. R. Ahmed, and W. H. Barnes, *Acta Crystallogr.* **19**, 693 (1965).
37. N. V. Mani, F. R. Ahmed, and W. H. Barnes, *Acta Crystallogr.* **21**, 375 (1966).
38. F. R. Ahmed, P. Singh, and W. H. Barnes, *Acta Crystallogr., Sect. B* **25**, 316 (1969).
39. G. J. Bullen and P. A. Tucker, *Chem. Commun.* p. 1185 (1970).
40. G. J. Bullen and P. R. Mallinson, *Chem. Commun.* p. 691 (1969).

41. G. B. Ansell and G. J. Bullen, *J. Chem. Soc., A* p. 3026 (1968).
42. R. Hazekamp and A. Vos, *Acta Crystallogr.* **16**, A38 (1963).
43. R. Olthof, T. Migchelsen, and A. Vos, *Acta Crystallogr.* **19**, 596 (1965).
44. T. Migshelsen and A. Vos, *Acta Crystallogr.* **16**, A38 (1963).
45. T. Migchelsen, R. Olthof, and A. Vos, *Acta Crystallogr.* **19**, 603 (1965).
46. D. Mootz and B. Berking, *Angew. Chem.* **9**, 78 (1970).
47. L. G. Hoard and R. A. Jacobson, *J. Chem. Soc., A* p. 1203 (1966).
48. H. Hess and D. Forst, *Z. Anorg. Allg. Chem.* **342**, 240 (1966).
49. J. W. Cox and E. R. Corey, *Chem. Commun.* p. 123 (1967).
50. J. Weiss and G. Hartmann, *Z. Naturforsch. B* **21**, 891 (1967).
51. J. Weiss and G. Hartmann, *Z. Anorg. Allg. Chem.* **351**, 152 (1966).
52. J. C. van de Grampel and A. Vos, *Acta Crystallogr., Sect. B* **25**, 651 (1969).
53. W. Harrison, R. T. Oakley, N. L. Paddock, and J. Trotter, *Chem.Commun.* p. 357 (1971).

Appendix II

CHARACTERIZATION DATA FOR PHOSPHAZENES AND PHOSPHAZANES

This table lists melting points, boiling points, and other characteristic information for a wide variety of different phosphorus–nitrogen compounds. The tabulation for cyclo- and polyphosphazenes and phosphazanes is comprehensive, with different classes of compounds categorized according to their substituent groups.

Representative monophosphazenes and linear phosphazenes are also listed at the end of the table. The compounds included in this list were selected to illustrate the diversity of structures and the trends in physical properties, as well as to suggest sources for data on related compounds. In many cases, specific monomeric or linear compounds not listed in the table can be traced through the references listed here and in the appropriate sections of the text.

In the table, ^{31}P shifts are quoted relative to aqueous 85% phosphoric acid, refractive index values are indicated by n_D and the measurement temperature, and IR (P=N) refers to the characteristic skeletal infrared frequency. Where two different melting point or spectral values are reported, the alternative data are given in parentheses. The lettered superscript notations are as follows:

[a] Isomer not specified.
[b] Value varies with molecular weight.
[c] Mixture of isomers.
[d] The position of the substituent on the aryl ring is not specified.
[e] Two different crystalline forms exist.
[f] Melts, resolidifies, and then melts again at a higher temperature.
[g] The boiling points and refractive index values of a wide variety of derivatives with this general formula can be located from this reference.

The table has been compiled in such a way that a substituent group does not appear on the table *before* the section allotted to that group, but it may appear in a later section as a subsidiary substituent. For example, no amino derivatives appear in the table before the "Amino Phosphazene" section, but amino compounds also appear in later sections, such as in "Alkoxy Phosphazenes," when both alkoxy and amino substituents are present. The substituent group categories appear in the following order.

Compound	M.p. (°C)	Property Data	Ref.
Halogeno Phosphazenes			
$(NPF_2)_3$	28	B.p., 50°C/760 mm; density, 2.237 g/ml (20°C); ^{31}P, −13.9 ppm; ΔH_{vap}, 7.65 kcal/mole; IR (P=N), 1297–1305 cm^{-1}	1–5
$(NPF_2)_4$	30.4	B.p., 89.7°C/760 mm; density, 2.239 g/ml (20°C); n_D^{20}, 1.379–1.463; IR (P=N), 1435–1438 cm^{-1}; ΔH_{vap}, 8.91 kcal/mole	1–4, 6, 7
$(NPF_2)_5$	−50	B.p., 120.1°C/760 mm; density, 1.8259 g/ml (20°C); n_D^{20}, 1.3482; IR (P=N), 1425–1439 cm^{-1}; ΔH_{vap}, 9.80 kcal/mole	2, 4
$(NPF_2)_6$	−45.5	B.p., 147.2°C/760 mm; density, 1.8410 g/ml (20°C); n_D^{20}, 1.3533; IR (P=N), 1408 cm^{-1}	2
$(NPF_2)_7$	−61	B.p., 170.7°C/760 mm; density, 1.8496 g/ml (20°C); n_D^{20}, 1.3570; IR (P=N), 1400 cm^{-1}; ΔH_{vap}, 11.6 kcal/mole	2
$(NPF_2)_8$	−16.9	B.p., 192.8°C/760 mm; density, 1.8567 g/ml (20°C); n_D^{20}, 1.3602; IR (P=N), 1386 cm^{-1}; ΔH_{vap}, 12.0 kcal/mole	2
$(NPF_2)_9$	<−78	B.p., 214.4°C/760 mm; density, 1.8589 g/ml (20°C); n_D^{20}, 1.3622; IR (P=N), 1375 cm^{-1}; ΔH_{vap}, 12.7 kcal/mole	2
$(NPF_2)_{10}$	−51	B.p., 230.8°C/760 mm; density, 1.8638 g/ml (20°C); n_D^{20}, 1.3633; IR (P=N), 1363 cm^{-1}; ΔH_{vap}, 13.5 kcal/mole	2
$(NPF_2)_{11}$	<−78	B.p., 246.7°C/760 mm; density, 1.8644 g/ml (20°C); n_D^{20}, 1.3645; IR (P=N), 1357 cm^{-1}; ΔH_{vap}, 14.6 kcal/mole	2
$(NPF_2)_{12}$	—	B.p., 143–148°C/126.5 mm; IR (P=N), 1351 cm^{-1}	2
$(NPF_2)_{13}$	—	B.p., 154–159°C/126.5 mm; IR (P=N), 1346 cm^{-1}	2
$(NPF_2)_{14}$	—	B.p., 166–170°C/126.5 mm; IR (P=N), 1340 cm^{-1}	2
$(NPF_2)_n$	−68	Tg, −96°C	8
$N_3P_3F_5Cl$	−30	B.p., 81°C/760 mm; density, 1.7594 g/ml (20°C); IR (P=N), 1280 cm^{-1}; ΔH_{vap}, 8.7 kcal/mole	9–13

Compound	M.p. (°C)	Property Data	Ref.
$N_3P_3F_4Cl_2$ (*gem*)	−46	B.p., 114.7 ± 0.2°C/760 mm; density, 1.7650 g/ml (20°C); n_D^{20} 1.4055; IR (P=N), 1277–1253 cm^{-1}; ΔH_{vap}, 9.6 kcal/mole	1, 9, 10, 14
$N_3P_3F_3Cl_3$ (*gem*)	−35	B.p., 150°C/760 mm; density, 1.7588 g/ml (20°C); IR (P=N), 1255–1258 cm^{-1}; ΔH_{vap}, 10.2 kcal/mole	9, 10, 14, 15
$(NPFCl)_3$	—	B.p., 140°C/760 mm	9, 15
$N_3P_3F_2Cl_4$ (*gem*)	18	B.p., 181.6° ± 0.2°C/760 mm; density, 1.7851 g/ml (20°C); IR (P=N), 1244, 1246 cm^{-1}; ΔH_{vap}, 11.0 kcal/mole	9, 10, 14, 16, 17
$N_3P_3F_2Cl_4$ (non-*gem*)	—	B.p., 170°C/760 mm	9, 14, 15
$N_3P_3FCl_5$	50	B.p., 215°C/760 mm; density, 1.981 g/ml (20°C); IR (P=N), 1230 cm^{-1}; ^{31}P, −14.4, −23.0 ppm; ΔH_{vap}, 12.5 kcal/mole	9, 10, 13, 18
$N_4P_4F_7Cl$	−5	B.p., 117°C/760 mm; density, 1.8020 g/ml (20°C); IR (P=N) 1408 cm^{-1}; ΔH_{vap}, 9.5 kcal/mole	9, 13
$N_4P_4F_6Cl_2$ (*gem*)	−21	B.p., 147°C/760 mm; density, 1.8061 g/ml (20°C); IR (P=N), 1400 cm^{-1}; ΔH_{vap}, 10.0 kcal/mole	9, 13
$N_4P_4F_6Cl_2$ (non-*gem*)	−12.4 to −12.1	B.p., 105.8°C/760 mm; density, 1.8742 g/ml (20°C)	3, 19, 20
$N_4P_4F_5Cl_3$ (*gem*)	−28	B.p., 178°C/760 mm; density, 1.8197 g/ml (20°C); IR (P=N), 1390 cm^{-1}; ΔH_{vap}, 11.0 kcal/mole	9, 13
$N_4P_4F_4Cl_4$ (*gem*)	−23	B.p., 205°C/760 mm; density, 1.8343 g/ml (20°C); IR (P=N), 1377 cm^{-1}; ΔH_{vap}, 12.2 kcal/mole	9, 13
$N_4P_4F_4Cl_4$ (non-*gem*)	−25.2 to −24.9	B.p., 130.5°C/760 mm; density, 1.9266 g/ml (22°C)	3, 21
$N_4P_4F_3Cl_5$ (*gem*)	−11 (+10)	B.p., 232°C/760 mm; density, 1.8420 g/ml (20°C); IR (P=N), 1362 cm^{-1}; ΔH_{vap}, 13.0 kcal/mole	9, 13
$N_4P_4F_2Cl_6$ (*gem*)	23	B.p., 267°C/760 mm; density, 2.060 g/ml (20°C); IR (P=N), 1346 cm^{-1}; ΔH_{vap}, 13.6 kcal/mole	9, 13
$N_4P_4FCl_7$	63	B.p., 301°C/760 mm; density, 2.049 g/ml (20°C); IR (P=N), 1330 cm^{-1}; ΔH_{vap}, 14.0 kcal/mole	9, 13
$N_3P_3F_5Br$	—	B.p., 97°C/760 mm; IR (P=N), 1263 cm^{-1}	11, 12
$N_3P_3F_4Br_2$[a]	—	IR (P=N), 1250 cm^{-1}	22

$N_3P_3F_3Br_3$[a]	—	IR (P=N), 1235 cm^{-1}	22
$N_3P_3F_2Br_4$[a]	—	IR (P=N), 1220 cm^{-1}	22
$N_3P_3FBr_5$	—	IR (P=N), 1198 cm^{-1}	22
$(NPCl_2)_3$	114	B.p., 256°C/760 mm; IR (P=N), 1218 cm^{-1}; ^{31}P, −19.7 ± 0.3 ppm; ΔH_{vap}, 13.2 kcal/mole	4, 6, 23–27
$(NPCl_2)_4$	123.5	B.p., 328.5°C/760 mm; IR (P=N), 1315 cm^{-1}; ^{31}P, +7.4 ± 0.3 ppm; ΔH_{vap}, 15.5 kcal/mole	4, 6, 23, 25, 28, 29
$(NPCl_2)_5$	41.3	B.p., 223°–224°C/13 mm; IR (P=N), 1298, 1354 cm^{-1}; ^{31}P, +17 ppm	4, 23, 30, 31
$(NPCl_2)_6$	92.3	B.p., 281°–282°C/760 mm; IR (P=N), 1325 cm^{-1}; ^{31}P, +16 ppm	23, 30, 32
$(NPCl_2)_7$	8–12 (−18)	B.p., 289°–294°C/13 mm; IR (P=N), 1310 cm^{-1}; ^{31}P, +18 ppm	6, 23, 30
$(NPCl_2)_8$	57–58	IR (P=N), 1305 cm^{-1}; ^{31}P, +18 ppm; density, 2.15	23, 30, 32
$(NPCl_2)_n$		g/ml; n_D^{20}, 1.5878; IR (P=N), 1260–1350 cm^{-1}; ^{31}P, +17.4 ppm	26, 33, 34
$N_3P_3Cl_5Br$	122.5	IR (P=N), 1205–1212 cm^{-1}; ^{31}P, δ(PCl_2), −17.7 ppm; δ(PBrCl), +7.8 ppm	35–38
$N_3P_3Cl_4Br_2$[a]	134.5–135	IR (P=N), 1200–1203 cm^{-1}; ^{31}P, δ(PCl_2), −16.1 ppm; δ(PBrCl), +8.7 ppm; δ(PBr_2), +38.6 ppm	36–39
$N_3P_3Cl_3Br_3$[a]	143	IR (P=N), 1200 cm^{-1}; ^{31}P, δ(PCl_2), −14.0 ppm; δ(PClBr), +10.0 ppm; δ(PBr_2), +39.8 ppm	37–39
$N_3P_3Cl_2Br_4$[a]	165–167	IR (P=N), 1192 cm^{-1}; ^{31}P, δ(PCl_2), −13.9 ppm; δ(PClBr), +12.1 ppm; δ(PBr_2), +41.3 ppm	37, 38, 40
$N_3P_3ClBr_5$	175	IR (P=N), 1184 cm^{-1}; ^{31}P, δ(PBrCl), +14.0 ppm; δ(PBr_2), +42.5 ppm	37–39
$(NPBr_2)_3$	191–192	IR (P=N), 1126, 1175–1184 cm^{-1}; ^{31}P, +45.4 ppm	4, 41–47
$(NPBr_2)_4$	202	IR (P=N), 1272–1280 cm^{-1}; ^{31}P, +71.8 ppm	4, 36, 47, 48
$(NPBr_2)_5$	104.5	IR (P=N), 1265, 1330 cm^{-1}	47, 49
$(NPBr_2)_6$	153–154	IR (P=N), 1305 cm^{-1}	50
$(NPBr_2)_n$	—	IR (P=N), 1302 cm^{-1}	48

Compound	M.p. (°C)	Property Data	Ref.
Pseudohalogeno Phosphazenes			
$[NP(NCS)_2]_3$	42	IR (P═N), 1200–1216 cm^{-1}; ^{31}P, -30 ± 1 ppm	37, 51–53
$[NP(NCS)_2]_4$	90	—	53
$N_3P_3Cl_4(NCS)_2$[a]	53–54	—	54
$N_4P_4Cl_6(NCS)_2$[a]	—	B.p., 116°C/0.01 mm; IR (P═N), 1312 cm^{-1}	54
$[NP(N_3)_2]_3$	<−20	IR (P═N), 1200 cm^{-1}	37, 55
Alkyl Phosphazenes			
$(NPMe_2)_3$	195–196	IR (P═N), 1180 cm^{-1}	4, 56
$(NPMe_2)_4$	163–164	IR (P═N), 1180–1220 cm^{-1}	4, 56
$(NPMe_2)_n$	136–146[b]	IR (P═N), 1160, 1200, 1270, 1290 cm^{-1}	57
$N_3P_3Cl_3Me_3$ (non-*gem*)	156–157	IR (P═N), 1198 cm^{-1}	58
$N_3P_3Cl_4Me_2$[a]	174–176	IR (P═N), 1170, 1220 cm^{-1}	59
$(NPEt_2)_3$	117.5–119.0	IR (P═N), 1157 cm^{-1}	4, 60
$(NPEt_2)_4$	—	IR (P═N), 1231 cm^{-1}	4
$N_3P_3F_4Bu^n_2$[a]	—	B.p., 87°–94°C/1 mm	61
$N_3P_3F_5Bu^n$	—	B.p., 73°–76°C/8–10 mm	61
$N_4P_4F_6Bu^n_2$	—	B.p., 111°–113°C/2–2.5 mm	61
$N_4P_4F_6Bu^n$	—	B.p., 71°–72°C/6 mm	61
$[NP(CF_3)_2]_3$	64	IR (P═N), 1300 cm^{-1}	62
$[NP(CF_3)_2]_4$	109	IR (P═N), 1412 cm^{-1}	62, 63
$[NP(CF_3)_2]_n$	90–94	—	64, 65
$[NP(C_3F_7)_2]_3$	85	IR (P═N), 1328 cm^{-1}	62, 63
Aryl Phosphazenes			
$(NPPh_2)_3$	235	IR (P═N), 1300–1250, 1190 cm^{-1}; ^{31}P, -14.3 ± 0.2 ppm	4, 51, 66, 67

$(NPPh_2)_4$	319.5–321.0	IR (P=N), 1213 cm^{-1}	4, 67–71
$N_3P_3F_2Ph_4$ (*gem*)	137–138	B.p., 270°C/0.12 mm[c]; IR (P=N), 1218 cm^{-1}; ^{31}P, δ(PF_2), −6.2 ppm; δ(PPh_2), −27.3 ppm	5, 72
$N_3P_3F_3Ph_3$ (*gem*)	79.5–80.5	B.p., 235°C/0.12 mm[c]; IR(P=N), 1230 cm^{-1}; ^{31}P, δ(PFPh), −31.2 ppm; δ(PPh_2), −27.3 ppm; δ(PF_2), −8.6 ppm	5, 72
$N_3P_3F_4Ph_2$ (*gem*)	68.5–69.5	B.p., 165°–166°C/0.12 mm[c]; IR (P=N), 1250 cm^{-1}; ^{31}P, δ(PF_2), −12.1 ppm; δ(PPh_2), −30.4 ppm	5, 72
$N_3P_3F_4Ph_2$ (non-*gem-cis*)	74–75	B.p., 165°–166°C/0.12 mm[c]; IR (P=N), 1240–1255 cm^{-1}; ^{31}P, δ(PFPh), −38.4 ppm; δ(PF_2), −12.4 ppm	72, 73
$N_3P_3F_4Ph_2$ (non-*gem-trans*)	78–79	B.p., 165°–166°C/0.12 mm[c]; IR (P=N), 1245–1255 cm^{-1}; ^{31}P, δ(PFPh), −38.3; δ(PF_2), −10.8 ppm	72, 73
$N_3P_3F_5Ph$	—	B.p., 183°–184°C/760 mm; IR (P=N), 1270 cm^{-1}	73
$N_4P_4F_2Ph_6$[a]	176–177	—	74
$(NPFPh)_n$	—	IR (P=N), 1176, 1227, 1300–1325 cm^{-1}	67
$N_3P_3ClPh_5$	150–152	^{31}P, δ(PPh_2) −17.6 ppm, δ(PClPh) −28.7 ppm	75
$N_3P_3Cl_2Ph_4$ (*gem*)	145	IR (P=N), 1208, 1162 cm^{-1}; ^{31}P, −17.2 ppm	51, 68, 76–78
$N_3P_3Cl_2Ph_4$ (non-*gem-cis*)	195–196	^{31}P, δ(PPh_2), −19.0 ppm; δ(PClPh), −30.2 ppm	75
$N_3P_3Cl_2Ph_4$ (non-*gem-trans*)	169–170	^{31}P, δ(PPh_2), −19.0; δ(PClPh), −28.5 ppm	75
$(NPClPh)_3$ (non-*gem-cis*)	191–192	IR (P=N), 1180 cm^{-1}; ^{31}P, δ(PClPh), −29.4 ppm	75, 79
$(NPClPh)_3$ (non-*gem-trans*)	158–159	IR (P=N), 1180 cm^{-1}; ^{31}P, δ(PClPh), −30.3	79
$N_3P_3Cl_4Ph_2$ (*gem*)	97–97.5	IR (P=N), 1190, 1210 cm^{-1}; ^{31}P, −20.6, −18.3 ppm	24, 75, 80–82
$N_4P_4Cl_2Ph_6$ (non-*gem*: 1,3-Cl, *cis* and *trans* isomers)	192–193 154–155	IR (P=N), 1266 cm^{-1}	68, 83
$N_4P_4Cl_2Ph_6$ (non-*gem*-1,5)	294	IR (P=N), 1280 cm^{-1}	83
$N_4P_4Cl_4Ph_4$ (*gem*-1,3)	205	IR (P=N), 1250, 1300 cm^{-1}	24, 68, 69
$N_4P_4Cl_4Ph_4$ (*gem*-1,5)	212.5	IR (P=N), 1250, 1300 cm^{-1}	24, 69, 84
$(NPClPh)_4$ (non-*gem-cis*)	225.5–226	IR (P=N), 1270, 1295, 1310 cm^{-1}	85, 86
$(NPClPh)_4$ (non-*gem-α-trans*)	155–156	IR (P=N), 1270–1275, 1291, 1306 cm^{-1}	85–87
$(NPClPh)_4$ (non-*gem-β-trans*)	260–263	IR (P=N), 1260–1282, 1293, 1305–1313 cm^{-1}	85–87
$(NPClPh)_n$	61–66	IR (P=N), 1170, 1205, 1290 cm^{-1}	88

Compound	M.p. (°C)	Property Data	Ref.
$(NPBrPh)_3$ (*cis*)	194–195	IR (P═N), 1126, 1180 cm^{-1}; ^{31}P, −16.4 ppm	79, 89, 90
$(NPBrPh)_3$ (*trans*)	152–153[e]	IR (P═N), 1117, 1183 cm^{-1}; ^{31}P, −20.1 and −18.0 ppm	79, 89
	162–163[e]		
$N_4P_4Br_2Ph_6$	254	—	74
$N_4P_4Br_4Ph_4$[a]	227–229	IR (P═N), 1295 cm^{-1}	89
$N_3P_3ClBr_2Ph_3$	155–159[c]	IR (P═N), 1120, 1180 cm^{-1}	91
	190–195	IR (P═N), 1130, 1180 cm^{-1}	
$N_3P_3(NCS)_4Ph_2$	176	—	54
$N_3P_3Cl_2(NCS)_2Ph_2$[a]	199	—	54
$N_4P_4Ph_6(N_3)_2$ (non-*gem*-1,5-*cis*)	147.5–148	IR (P═N), 1250 cm^{-1}	92
$N_4P_4Ph_6(N_3)_2$ (non-*gem*-1,5-*trans*)	207–208	IR (P═N), 1250 cm^{-1}	92, 93
$N_3P_3Ph_4MeH$ (*gem*)	152	IR (P—H), 2310, 2360 cm^{-1}; ^{31}P, $\delta(PPh_2)$, −12.9 ppm, $\delta(PMeH)$, −69 ppm	94
$N_3P_3Ph_4EtH$ (*gem*)	127–129	IR (P—H), 2300, 2350 cm^{-1}; ^{31}P, $\delta(PPh_2)$, −15.5 ppm, $\delta(PEtH)$, −12.5 ppm	94
$N_4P_4Ph_6(OH)_2$ (non-*gem*-1,5)	272	IR (POH), 2620 cm^{-1}	83
$N_3P_3ClPh_3(OH)_2$ (non-*gem*)	294	—	95
$[NP(C_6F_5)_2]_3$	307–309	IR (P═N), 1200 cm^{-1}	95a
$[NP(C_6H_4Cl\text{-}p)_2]_3$	233–235	—	76
$N_3P_3Cl_2(C_6H_4Cl\text{-}p)_4$ (*gem*)	189–191	—	76
$N_3P_3Cl_4(C_6H_4Cl\text{-}p)_2$ (*gem*)	138–139 (118–120)	—	76
$N_3P_3Cl_5(C_6H_4Cl\text{-}p)$	88–90	—	76
$N_3P_3Cl_4(C_6H_4Br)_2$[a,d]	133–135	—	76
$N_3P_3F_4(C_6H_4Me\text{-}o)_2$ (non-*gem*-*cis*)	64–65	IR (P═N), 1245–1280 cm^{-1}	73
$N_3P_3F_5(C_6H_4Me\text{-}m)$	—	B.p., 79°C/0.12 mm	72
$N_3P_3F_5(C_6H_4Me\text{-}p)$	—	B.p., 83°C/0.12 mm	72
$N_3P_3Cl_2(C_6H_4Me)_4$[d]	168–179[c]	—	76
$N_3P_3Cl_4(C_6H_4Me)_2$ (*gem*)[d]	124–126	—	76

$N_3P_3Ph_2(C_6H_4Me)_4$ (*gem*)	228–240[c]	—	76
Amino Phosphazenes			
$[NP(NH_2)_2]_3$	—	IR (P=N), 1170–1175 cm^{-1}; ^{31}P, -18.8 ± 0.2 ppm	4, 37, 51, 96, 97
$[NP(NH_2)_2]_4$	—	IR (P=N), 1240 cm^{-1}	4, 96, 97
$N_3P_3(NH_2)_2Cl_4$ (*gem*)	163–165	^{31}P, $\delta(PCl_2)$, −18.3 ppm; $\delta[P(NH_2)_2]$, −9.03 ppm	98–100
$N_3P_3(NH_2)Cl_5$	138–139	IR (P=N), 1185, 1232 cm^{-1}; ^{31}P, $\delta(PCl_2)$, −20.4 ppm; $\delta(PClNH_2)$, −19.0 ppm	98
$N_3P_3(NH_2)_4Ph_2$ hygroscopic compd.	106	—	101, 102
nonhygroscopic compd.	275		
$[NP(NH_2)Ph]_3$ (*cis*)	274	IR (P=N), 1126, 1180 cm^{-1}	89
$[NP(NH_2)Ph]_3$ (*trans*)	202–203	IR (P=N), 1120, 1180 cm^{-1}	89
$[NP(NH_2)Ph]_4$	229	—	85
$N_4P_4(NH_2)_2Ph_6$	205–207	—	74
$N_3P_3(NH_2)_2Cl_2Ph_2$ (non-*gem*: 1,3-NH_2 1,3-Cl 5,5-Ph)	162–163	—	82, 101, 103–105
$N_3P_3(NH_2)_2ClPh_2$ (*gem*)	165	—	105
$N_3P_3(NH_2)_2(N_3)_4$[a]	81–82	IR (P=N), 1200 cm^{-1}	106
$[NP(NHMe)_2]_3$	259	IR (P=N), 1175–1180 cm^{-1}	37, 80, 107–109
$[NP(NHMe)_2]_4$	206	IR (P=N), 1215 cm^{-1}	110, 111
$N_3P_3(NHMe)_2Cl_4$ (*gem*)	131	IR (P=N), 1170–1173, 1188, 1225 cm^{-1}; ^{31}P, −20.1 and −12.3 ppm	51, 80, 99, 112
$N_3P_3(NHMe)_2Cl_4$ (non-*gem-cis*)	149		99
$N_3P_3(NHMe)_2Cl_4$ (non-*gem-trans*)	103	IR (P=N), 1200, 1210–1216 cm^{-1}; ^{31}P, −22.2 ppm	51, 80, 99, 112
$N_3P_3(NHMe)Cl_5$	69	IR (P=N), 1201, 1225 cm^{-1}; ^{31}P, -22.9 ± 0.02 ppm	51, 112–114
$N_4P_4(NHMe)_2Cl_6$ (non-*gem-trans*)	143	IR (P=N), 1310 cm^{-1}	115–119
$N_4P_4(NHMe)Cl_7$	67–67.5	—	116
$N_3P_3(NHMe)_4Ph_2$ (*gem*)	178	IR (P=N), 1190 cm^{-1}	80, 120
$N_3P_3(NHMe)_4Ph_2$ (non-*gem*)	125	IR (P=N), 1170, 1190 cm^{-1}	80

Compound	M.p. (°C)	Property Data	Ref.
$[NP(NHMe)Ph]_3$ (*trans*)	112–113	IR (P=N), 1122, 1185 cm^{-1}	89
$[NP(NHMe)Ph]_3$ (*cis*)	167	IR (P=N), 1125, 1185 cm^{-1}	85, 89
$N_3P_3(NHMe)_2Ph_4$ (*gem*)	190–192	^{31}P, −18 and −19.5 ppm	51, 120
$[NP(NHMe)Ph]_4$	113, 130–154[f]	IR (P=N), 1240 cm^{-1}	85
(*cis*)	131	IR (P=N), 1220–1260 cm^{-1}	
$[NP(NHMe)Ph]_4$ (α-*trans*)	119–122	—	86
$[NP(NHMe)Ph]_4$ (β-*trans*)	151–152	—	86
$N_4P_4(NHMe)_2Ph_6$[a]	192–194	—	74
$N_3P_3(NHMe)_2Cl_2Ph_2$ (*gem*)	103	—	105, 120
$N_3P_3(NHMe)_2Cl_2Ph_2$ (1,3-NHMe 1,3-Cl 5,5-Ph)	170–171	—	105, 120
$N_3P_3(NHMe)_2Cl_2Ph_2$ (1,3-NHMe 1,3-Ph 5,5 Cl)	145	IR (P=N), 1170, 1205 cm^{-1}	80
$N_3P_3(NHMe)_2Cl_3Ph$ (1,3-NHMe 3,5,5-Cl 1-Ph)	138	IR (P=N), 1180, 1205 cm^{-1}	80
$N_3P_3(NHMe)ClPh_4$ (*gem*)	165–166	—	120
$N_3P_3(NHMe)Cl_3Ph_2$ (1-NHMe 1,3,3-Cl 5,5-Ph)	151	—	120
$N_3P_3(NHMe)_4(NH_2)_2$ (*gem*)	204	—	99, 102, 107
$N_3P_3(NHMe)_4(NH_2)_2$ (non-*gem*)	161.5	—	99, 102, 107
$N_3P_3(NHMe)_2(NH_2)_2Cl_2$ (1,1-NH_2 3,5-NHMe)	140	^{31}P, −23 ± 0.3 ppm, −16.5 ± 0.3 ppm	51, 107
$N_3P_3(NHMe)_2(NH_2)_2Ph_2$ (1,3-NHMe 1,3-NH_2 5,5-Ph)	140	—	101
$N_3P_3(NHMe)F_5$	—	B.p., 151°C/760 mm; IR (P=N), 1262 cm^{-1}	12
$[NP(NHEt)_2]_3$	119	—	108, 109, 121
$[NP(NHEt)_2]_4$	122	IR (P=N), 1118, 1262, 1392 cm^{-1}	110, 111
$[NP(NHEt)_2]_n$ ($n \approx 15{,}000$)	—	T_g, 30°C; IR (P=N), 1200–1265 cm^{-1}; ^{31}P, +26.6 ppm	122
$N_3P_3(NHEt)_4Cl_2$ (*gem*)	126	—	123
$N_3P_3(NHEt)_4Cl_2$ (non-*gem*)	109	—	123
$N_3P_3(NHEt)_3Cl_3$ (non-*gem*)	131	—	123
$N_3P_3(NHEt)Cl_5$	—	B.p., 105°C/0.05 mm; n_D^{25}, 1.5414; I.R (P=N), 1213 cm^{-1}	114, 124

$N_4P_4(NHEt)_2Cl_6$[a]	114–114.5	IR (P=N), 1310 cm^{-1}	29, 115, 116
$N_4P_4(NHEt)Cl_7$	—	B.p., 133°–134°C/0.3 mm; n_D^{30}, 1.5629	116
$N_3P_3(NHEt)_4Ph_2$ (*gem*)	97	—	120
$N_3P_3(NHEt)_2Ph_3$ (non-*gem-cis*)	126–127	IR (P=N), 1122–1175 cm^{-1}	89
$N_3P_3(NHEt)_3Ph_3$ (non-*gem-trans*)	91–92	IR (P=N), 1117–1180 cm^{-1}	89
$N_3P_3(NHEt)_2Ph_4$ (*gem*)	165	—	120
$N_4P_4(NHEt)_4Ph_4$ (non-*gem*, two isomers)	101 122	—	85
$N_3P_3(NHEt)_2Ph_2Cl_2$ (*gem*)	93	—	105, 120
$N_3P_3(NHEt)_2Ph_2Cl_2$ (1,3-NHEt 1,3-Ph 5,5-Cl) *cis* and *trans*[a]	125 130	—	125
$N_3P_3(NHEt)Ph_2Cl_3$ (1-NHEt 1,3,3-Cl 5,5-Ph)	96–97	—	120
$[NP(NHPr^n)_2]_3$	59	IR (P=N), 1183–1195 cm^{-1}	37, 108, 109, 126
$[NP(NHPr^n)_2]_4$	98	IR (P=N), 1108, 1266, 1394	110, 111
$N_3P_3(NHPr^n)_4Cl_2$ (non-*gem*)	93	—	108
$N_3P_3(NHPr^n)Cl_5$	—	B.p., 137°C/1.75 mm; n_D^{25}, 1.5362; IR (P=N), 1208–1226 cm^{-1}	114, 124
$N_4P_4(NHPr^n)_2Cl_6$	110–110.5	—	116
$N_4P_4(NHPr^n)Cl_7$	—	B.p., 157°C/1.0 mm; n_D^{30}, 1.5595	116
$N_3P_3(NHPr^n)_3Ph_3$ (non-*gem-cis*)	128–129	IR (P=N), 1128–1190 cm^{-1}	89, 91
$N_3P_3(NHPr^n)_3Ph_3$ (non-*gem-trans*)	92–93	IR (P=N), 1120–1190 cm^{-1}	89, 91
$N_4P_4(NHPr^n)_4Ph_4$ (non-*gem*)	98	—	85
$[NP(NHPr^i)_2]_3$	81	—	108, 127
$[NP(NHPr^i)_2]_4$	170	—	111
$N_3P_3(NHPr^i)_4Cl_2$ (*gem*)	126	—	123, 127, 128
$N_3P_3(NHPr^i)_2Cl_4$[a]	114–115	—	127, 128
$N_3P_3(NHPr^i)Cl_5$	54–55	—	127
$N_4P_4(NHPr^i)_4Cl_4$[a]	187	—	111
$N_4P_4(NHPr^i)_2Cl_6$	120.5–121	—	116
$N_4P_4(NHPr^i)Cl_7$	—	B.p., 128°–129°C/0.1 mm; n_D^{30}, 1.5527	116

Compound	M.p. (°C)	Property Data	Ref.
$N_4P_4(NHPr^i)_4Ph_4$ (non-*gem*), two isomers	133, 140	—	85
$[NP(NHCH_2CH{=}CH_2)_2]_3$	90–91	IR (P=N), 1177 cm^{-1}	129, 130
$[NP(NHCH_2CH{=}CH_2)_2]_4$	77–78	IR (P=N), 1220 cm^{-1}	130
$N_3P_3(NHCH_2CHCH_2)Cl_5$	—	B.p., 115°–130°C/0.5 mm; n_D^{25}, 1.5480; IR (P=N), 1215 cm^{-1}	130
$[NP(NHBu^n)_2]_3$	48 (30)	IR (P=N), 1192–1195 cm^{-1}	37, 109, 126
$[NP(NHBu^n)_2]_4$	86	IR (P=N), 1122, 1260, 1395 cm^{-1}	110, 111
$N_3P_3(NHBu^n)_5Cl$	127–128	IR (P=N), 1115, 1235–1335 cm^{-1}	131
$N_3P_3(NHBu^n)Cl_5$	—	B.p., 110°C/0.05 mm; n_D^{25}, 1.5312; IR (P=N), 1205–1222 cm^{-1}	114, 124
$[NP(NHBu^n)Ph]_3$ (*cis*)	97–98	IR (P=N), 1125–1186 cm^{-1}	89, 91
$[NP(NHBu^n)Ph]_3$ (*trans*)	78–79	IR (P=N), 1123–1185 cm^{-1}	89, 91
$[NP(NHBu^n)Ph]_4$	252	—	85
$[NP(NHBu^s)_2]_4$	111	—	111
$N_3P_3(NHBu^s)_4Cl_2$ (non-*gem*)	71	—	108
$[NP(NHBu^i)_2]_3$	59	—	108
$[NP(NHBu^i)_2]_4$	94	—	111
$[NP(NHBu^t)_2]_3$	280–282	—	132
$[NP(NHBu^t)_2]_4$	300 (decomp.)	—	111
$N_3P_3(NHBu^t)_4Cl_2$ (*gem*)	156	—	123, 132
$N_3P_3(NHBu^t)_4Cl_2$ (non-*gem*)	155.5	—	108
$N_3P_3(NHBu^t)_2Cl_4$ (*gem*)	120–122	—	132
$N_3P_3(NHBu^t)Cl_5$	−10 to −11	B.p., 91°C/0.01 mm	132
$N_4P_4(NHBu^t)_2Cl_6$ [a]	124	IR (P=N), 1312 cm^{-1}	111, 115
$N_3P_3(NHBu^t)_4Ph_2$ [a]	132–133	—	120
$N_3P_3(NHBu^t)_2Ph_4$ (*gem*)	162–163	—	120
$N_3P_3(NHBu^t)_2Ph_2Cl_2$ (*gem*)	108	—	105, 120
$N_3P_3(NHBu^t)_4(NHMe)_2$ (1,3,5,5-NHBu-1,3-NHMe)	199		108

$N_3P_3(NHBu^t)(NHEt)Cl_4$ (*gem*)	104–105	—	133
$N_3P_3(NHBu^t)(NHEt)Cl_4$ (non-*gem*)	68–70	—	133
$[NP(NH(CH_2)_4CH_3)_2]_3$	38	IR (P═N), 1183–1192 cm^{-1}	37, 109
$[NP(NH(CH_2)_4CH_3)_2]_4$	79	IR (P═N), 1120, 1265, 1395 cm^{-1}	110
$N_3P_3[NH(CH_2)_4CH_3]Cl_5$	—	B.p., 115°C/0.05 mm; n_D^{25}, 1.5265; IR (P═N), 1208–1220 cm^{-1}	114, 124
$[NP(NH(CH_2)_5CH_3)_2]_3$	49	IR (P═N), 1185–1192 cm^{-1}	37, 109
$[NP(NH(CH_2)_5CH_3)_2]_4$	70	IR (P═N), 1120, 1265, 1395 cm^{-1}	110
$[NP(NHCH_2CF_3)_2]_3$	70	IR (P═N), 1153–1280 cm^{-1}	131
$[NP(NHCH_2CF_3)_2]_4$	61–62	IR (P═N), 1164, 1275 cm^{-1}	131
$[NP(NHCH_2CH_2Cl)_2]_3$	82–82.5	—	134
$[NP(NHCH_2Ph)_2]_4$	121.5	IR (P═N), 1300 cm^{-1}	110
$N_3P_3(NHCH_2Ph)_4Ph_2$ (*gem*)	92.5	—	120
$N_3P_3(NHCH_2Ph)_2Ph_4$	133	—	120
$[NP(NHPh)_2]_3$	272–273	—	46, 135–138
$[NP(NHPh)_2]_4$	244, 261	IR (P═N), 1214, 1260 cm^{-1}	29, 85, 115, 137–139
$[NP(NHPh)_2]_n$ ($n \approx 15,000$)	—	T_g, 91°C; IR (P═N), 1180–1220 cm^{-1}, 925–960 cm^{-1}; ^{31}P, +14.0 ppm	122
$N_3P_3(NHPh)_4Cl_2$[c]	196	—	103, 123
$N_3P_3(NHPh)_2Cl_4$ (*gem*)	207.5–208.5	^{31}P, $\delta[P(NHPh)_2]$, −6.37, $\delta(PCl_2)$, −16.38 ppm	140
$N_4P_4(NHPh)_2Cl_6$[a]	166	IR (P═N), 1287, 1225 cm^{-1}	29, 115, 117
$N_3P_3(NHPh)_4Ph_2$[a]	198–199	—	103
$N_3P_3(NHPh)_2Ph_4$ (*gem*)	242	—	78
$N_3P_3(NHPh)_4(NH_2)_2$[a]	256–257	—	103
$N_3P_3(NHPh)_2Ph_2Cl_2$ (*gem*)	198	—	78, 125
$N_3P_3(NHPh)_2Ph_2Cl_2$[a]	193	—	103
$N_3P_3(NHPh)Ph_2Cl_3$ (*gem*)	118	—	78
$N_3P_3(NHPh)_2(NH_2)_2Ph_2$ (*gem*)	115	—	78
$N_3P_3(NHPh)_2(NH_2)_2Ph_2$ (1,3-NHPh-1,3-NH_2 5,5 Ph)	218	—	103
$[NP(NHC_6H_4Me\text{-}o)_2]_3$	241–242	—	141, 142

Compound	M.p. (°C)	Property Data	Ref.
$[NP(NHC_6H_4Me\text{-}p)_2]_3$	243	—	108, 136
$[NP(NHC_6H_4Me\text{-}o)_2]_4$	229 (decomp.)	IR (P=N), 1292 cm^{-1}	110
$[NP(NHC_6H_4Me\text{-}m)_2]_4$	199	IR (P=N), 1270, 1328 cm^{-1}	110
$[NP(NHC_6H_4Me\text{-}p)_2]_4$	207	IR (P=N), 1265, 1325 cm^{-1}	110
$N_3P_3(NHC_6H_4Me\text{-}p)_4Cl_2$[c]	174	—	103, 108, 123
$N_3P_3(NHC_6H_4Me\text{-}p)_2Cl_4$[a]	138	—	143
$N_4P_4(NHC_6H_4Me\text{-}o)_2Cl_6$[a]	172	IR (P=N), 1232, 1258 cm^{-1}	29, 115
$N_4P_4(NHC_6H_4Me\text{-}m)_2Cl_6$[a]	204	IR (P=N), 1220, 1275 cm^{-1}	115
$N_4P_4(NHC_6H_4Me\text{-}p)_2Cl_6$[a]	196	IR (P=N), 1220, 1275 cm^{-1}	115
$[NP(NHC_6H_4CF_3)_2]_3$	192.0–193.5	IR (P=N), 1228 cm^{-1}	28
$[NP(NHC_6H_4NH\text{-}o)]_3$ (spirocyclic)	>300	IR (P=N), 1175 cm^{-1}; ^{31}P, –23.0 ppm	144
$N_3P_3(NHC_6H_4NH\text{-}o)_2Cl_2$ (spirocyclic)	>300	—	144
$N_3P_3(NHC_6H_4NH\text{-}o)Cl_4$ (spirocyclic)	325	IR (P=N), 1190 cm^{-1}; ^{31}P, –20.0 ppm	144
$N_3P_3(NHC_6H_4NH\text{-}o)(NH_2)Cl_2$[a]	160–161	—	103
$N_3P_3[NHC_6H_3(CH_3)NH]Cl_4$ (1-Me 2,3-NH-aryl)	211	—	103
$N_3P_3[NH(CH_2)_2NH]Cl_4$	199	—	82
$N_3P_3[NH(CH_2)_2NH]Cl_2Ph_2$ (1,3-$NH(CH_2)_2NH$ 1,3-Cl 5,5-Ph)	142.5	—	82
$N_3P_3[NH(CH_2)_3NH]_2Cl_4$	166	—	82
$N_3P_3[NH(CH_2)_3NH]Cl_2Ph_2$ [1,3-$NH(CH_2)_3NH$ 1,3-Cl 5,5-Ph]	152	—	82
$N_3P_3[NH(CH_2)_3NH](NH_2)_2Ph_2$ [1,3-$NH(CH_2)_3NH$ 1,3-NH_2 5,5-Ph]	185	—	82
$N_3P_3[NH(CH_2)_4NH]Cl_4$ [1,3-$NH(CH_2)_4NH$ 1,3,5,5-Cl]	187	—	82
$[NP(NHC_6H_{11})_2]_3$	166	—	108, 121
$N_3P_3(NHC_6H_{11})_4Ph_2$[a]	138	—	120
$N_3P_3(NHC_6H_{11})_2Ph_4$ (*gem*)	122–123	—	120

$N_4P_4(NHC_6H_{11})_4Ph_4$ (non-*gem*)	132	—	85
$N_3P_3(NHC_6H_{11})_2Cl_2Ph_2$ (*gem*)	144–145	—	105, 120
$N_3P_3[NH(CH_2)_2NH_2]_2Cl_4$[a]	188	—	103, 104
$(NP[NH(CH_2)_2SH]_2)_3$	290 (decomp.)	IR (P=N), 1220 cm^{-1}	134
$(NP[NH(CH_2)_2SPh]_2)_3$	—	n_D^{30}, 1.6462; IR (P=N), 1190 cm^{-1}	134
$(NP[NH(CH_2)_2OSO_2H]_2)_3$	—	IR (P=N), 1220 cm^{-1}	134
$[NP(NHCH_2COOEt)_2]_3$	72–72.5	—	145, 146
$N_3P_3(NHCH_2COOEt)_4Cl_2$[a]	74–75	—	145, 147
$N_3P_3(NHCH_2COOEt)_2Cl_4$ (*gem*)	83–84	—	145–147
$N_3P_3(NHCH_2COOEt)Cl_5$	74–74.5	—	145
$[NP(NHCH_2CH_2COOCH_3)_2]_3$	—	Oil	146
$N_3P_3[NH(CH_2)_2COOCH_3]_4Cl_2$	42	—	146
$N_3P_3[NH(CH_2)_2COOCH_3]_2Cl_4$[a]	59–60	—	146
$N_3P_3[NH(CH_2)_2COOCH_3]Cl_5$	58.0–59.5	—	146
$[NP(NH(CH_2)_2COOPh)_2]_3$	—	IR (P=N), 1185 cm^{-1}	134
$[NP(NHSCNH(CH_2)_3CH_3)_2]_3$	155	—	53
$[NP(NHSCNH(CH_2)_3CH_3)_2]_4$	105	—	53
$[NP(NHSCNHPh)_2]_3$	151	—	53
$[NP(NHSCNHPh)_2]_4$	139	—	53
$[NP(NHCSNH_2)_2]_3$	190 (decomp.)	—	53
$[NP(NHSCNH_2)_2]_4$	120 (decomp.)	—	53
$[NP(NHSCOEt)_2]_3$	174	—	53
$[NP(NHSCOEt)_2]_4$	189	—	53
$N_3P_3(NHCSOEt)_4Ph_2$ (*gem*)	—	IR (P=N), 1200 cm^{-1}	54
$N_3P_3(NHPOCl_2)_2Cl_4$ (*gem*)	>360	^{31}P, −21.4, +13.1, −0.7 ppm	99
$[NP(NMe_2)_2]_3$	104	IR (P=N), 1195 cm^{-1}; ^{31}P, −25 ppm	4, 27, 37, 107, 108, 121, 148, 149
$[NP(NMe_2)_2]_4$	237 (decomp.) (218)	IR (P=N), 1265 cm^{-1}	4, 111, 121, 150
$[NP(NMe_2)_2]_5$	230–236	—	151
$[NP(NMe_2)_2]_6$	255–260	—	151

Compound	M.p. (°C)	Property Data	Ref.
$[NP(NMe_2)_2]_7$	230–233	—	151
$[NP(NMe_2)_2]_n$ ($n \approx 15{,}000$)	—	IR (P=N), 1240–1280 cm^{-1}; ^{31}P, −5.3, +0.7 ppm	122
$N_3P_3(NMe_2)_2F_4$ (non-*gem-cis*)	—	IR (P=N), 1253 cm^{-1}	152
$N_3P_3(NMe_2)_2F_4$ (non-*gem-trans*)	—	IR (P=N), 1253 cm^{-1}	152
$N_3P_3(NMe_2)F_5$	—	B.p., 140°C/760 mm; IR (P=N), 1268 cm^{-1}	12
$N_4P_4(NMe_2)_4F_4$ (non-*gem*)[c]	138	—	153
$N_4P_4(NMe_2)_3F_5$ (non-*gem*)[c]	—	Liquid	153
$N_3P_3(NMe_2)_4Cl_2$ (non-*gem-cis*)	104	IR (P=N), 1193–1198, 1220–1225 cm^{-1}; ^{31}P, −23.9, −30.5 ppm	27, 81, 108, 123, 148, 154, 155
$N_3P_3(NMe_2)_3Cl_3$ (*gem*)	71	IR (P=N), 1185–1235 cm^{-1}; ^{31}P, −21.6 ppm	27, 123, 148, 154–158
$N_3P_3(NMe_2)_3Cl_3$ (non-*gem-cis*)	152	IR (P=N), 1180–1210; ^{31}P, −27.6 ppm	27, 81, 123 148, 154, 156, 157, 159
$N_3P_3(NMe_2)_3Cl_3$ (non-*gem-trans*)	107 (105)	Density, 1.3975 g/ml; n_D^{25}, 1.546; IR (P=N), 1180–1220 cm^{-1}; ^{31}P, −26.1 ppm (−21 ppm)	27, 81, 102, 107, 108, 123, 148, 154–159
$N_3P_3(NMe_2)_2Cl_4$ (*gem*)	62	IR (P=N), 1180–1228 cm^{-1}	27, 148, 154, 156–158, 160
$N_3P_3(NMe_2)_2Cl_4$ (non-*gem-cis*)	86	B.p., 112°–114.5°C/760 mm; density, 1.5055 g/ml; n_D^{25}, 1.5440; IR (P=N), 1201–1218 cm^{-1}; ^{31}P, (PCl_2), −21.6 ppm, [$P(NMe_2)Cl$], −24.9 ppm	27, 148, 152, 154–158, 160
$N_3P_3(NMe_2)_2Cl_4$ (non-*gem-trans*)	103	Density, 1.4798 g/ml; n_D^{25}, 1.5544; IR (P=N), 1180–1215 cm^{-1}; ^{31}P, (PCl_2), −20.5 ppm, [$P(NMe_2)Cl$], −24.5 ppm	27, 81, 148, 152, 154–158, 160
$N_3P_3(NMe_2)Cl_5$	16 (11–13)	B.p., 70°–72°C/0.01 mm, 152–159/15 mm; density,	18, 27, 108,

		1.6139 g/ml; n_D^{25}, 1.5377; IR (P=N), 1198–1217 cm^{-1}; ^{31}P, −22.7, −20.6 ppm	113, 114, 148, 155, 158
$N_4P_4(NMe_2)_6Cl_2$ (1,1,3,5,5,7-NMe_2 *trans*)	168	—	111, 150
$N_4P_4(NMe_2)_6Cl_2$ (1,1,3,3,5,7-NMe_2 *trans*)	83	—	111, 150
$N_4P_4(NMe_2)_5Cl_3$ (1,1,3,5,5-NMe_2)	146	B.p., 174°C/0.6 mm	111, 150
$N_4P_4(NMe_2)_5Cl_3$ (1,1,3,3,5-NMe_2)	−2	—	150
$N_4P_4(NMe_2)_5Cl_3$ (1,1,3,5,7-NMe_2)	99–99.5	—	111, 150
$N_4P_4(NMe_2)_4Cl_4$ (*gem*)	110 (103)	—	111, 150
$N_4P_4(NMe_2)_4Cl_4$ (1,1,3,5-NMe_2)	148	—	111, 150
$N_4P_4(NMe_2)_4Cl_4$ (non-*gem*)	200	—	111, 150
$N_4P_4(NMe_2)_3Cl_5$ (*gem*)	103	—	150, 161
$N_4P_4(NMe_2)_3Cl_5$ (non-*gem*)	108	—	150, 161
$N_4P_4(NMe_2)_2Cl_6$ (*gem*)	67	—	161
$N_4P_4(NMe_2)_2Cl_6$ (non-*gem*-α-*cis*)	87	—	161
$N_4P_4(NMe_2)_2Cl_6$ (non-*gem*-α-*trans*)	92	—	161
$N_4P_4(NMe_2)_2Cl_6$ (non-*gem*-β-*cis*)	85	—	161
$N_4P_4(NMe_2)_2Cl_6$ (non-*gem*-β-*trans*)	171 (subl.)	IR (P=N), 1310 cm^{-1}	29, 111, 115, 150, 161
$N_4P_4(NMe_2)Cl_7$	61 (52)	—	111, 161
$N_3P_3(NMe_2)_3Br_3$ (*gem*)	86–87	IR (P=N), 1167–1221 cm^{-1}	155, 162
$[NP(NMe_2)Br]_3$ (*cis*)	170–172	—	163
$[NP(NMe_2)Br]_3$ (*trans*)	122–123	IR (P=N), 1196–1205 cm^{-1}; ^{31}P, −15.4 ppm	45, 155, 162–164
$N_3P_3(NMe_2)_2Br_4$ (*gem*)	—	IR (P=N), 1153–1210 cm^{-1}	155
$N_3P_3(NMe_2)_2Br_4$ (non-*gem-cis*)	138.5–139.5	IR (P=N), 1180–1201 cm^{-1}; ^{31}P, +36.8, −9.2 ppm	45, 155, 162, 164
$N_3P_3(NMe_2)_2Br_4$ (non-*gem-trans*)	135–136	IR (P=N), 1179–1200 cm^{-1}; ^{31}P, +36.6, −10.0 ppm	45, 155, 162, 164
$N_3P_3(NMe_2)Br_5$	112.5–113.5	IR (P=N), 1178–1197 cm^{-1}; ^{31}P, +39.3, −4.5 ppm	45, 155, 164
$N_3P_3(NMe_2)_4(NCS)_2$ (non-*gem*)	—	B.p., 175°C/0.1 mm (decomp.)	165
$N_3P_3(NMe_2)_3Me_3$	—	B.p., 146°–147°C/3.0 mm; IR (P=N), 1190 cm^{-1}	58
$N_3P_3(NMe_2)_4Ph_2$ (*gem*)	120–122	IR (P=N), 1190, 1170 cm^{-1}	78, 81, 101, 102, 120, 166

Compound	M.p. (°C)	Property Data	Ref.
$N_3P_3(NMe_2)_4Ph_2$ (non-*gem-cis*)	83–86	IR (P═N), 1200, 1170 cm^{-1}	81, 166
$[NP(NMe_2)Ph]_3$ (*cis*)	113.5–115	—	75
$[NP(NMe_2)Ph]_3$ (*trans*)	93.5–94.5	—	75
$N_3P_3(NMe_2)_2Ph_4$ (*gem*)	145	—	78, 120
$N_3P_3(NMe_2)_2Ph_4$ (non-*gem-cis*)	145.5	—	75
$N_3P_3(NMe_2)_2Ph_4$ (non-*gem-trans*)	123–124	—	75
$N_3P_3(NMe_2)Ph_5$	142–144	—	75
$N_4P_4(NMe_2)_4Ph_4$	139–141 (136–138)	—	85, 86, 167, 168
$N_4P_4(NMe_2)_4Ph_4$[a]	154–155.5	—	85, 86, 167, 168
$N_4P_4(NMe_2)_4Ph_4$[a]	178–180	—	85, 86, 167, 168
$N_4P_4(NMe_2)_2Ph_6$[a]	189–190	—	74
$N_3P_3(NMe_2)_4(NH_2)_2$ (non-*gem-cis*)	117	—	169
$N_3P_3(NMe_2)_4(NH_2)_2$ (non-*gem-trans*)	157 (145)	IR (P═N), 1185 cm^{-1}	81, 107, 165, 169
$N_3P_3(NMe_2)_4(NHMe)_2$[a]	125	—	107
$[NP(NMe_2)(NHMe)]_3$	119	—	107
$N_3P_3(NMe_2)_2(NHMe)_4$ (non-*gem*)	124	—	107
$N_3P_3(NMe_2)_2(NHMe)_2(NH_2)_2$	76	—	107
$N_3P_3(NMe_2)_5(NHPr^i)$	74–75	—	127
$N_3P_3(NMe_2)_4(NHPr^i)_2$ (non-*gem-cis*)	45	—	127
$N_3P_3(NMe_2)_4(NHPr^i)_2$ (non-*gem-trans*)	79–80	—	127
$N_3P_3(NMe_2)_2(NHPr^i)_4$ (*gem*)	42	—	127
$N_3P_3(NMe_2)_2(NHPr^i)_4$ (non-*gem-cis*)	55	—	127
$N_3P_3(NMe_2)_2(NHPr^i)_4$ (non-*gem-trans*)	64	—	127
$N_3P_3(NMe_2)_4(NHBu^t)_2$ (*gem*)	72	—	132
$N_3P_3(NMe_2)_4(NHBu^t)_2$ (non-*gem*)	—	B.p., 110°C/0.001 mm	132

$N_3P_3(NMe_2)_2(NHBu^t)_4$ (*gem*)	67	—	132
$N_3P_3(NMe_2)_2(NHBu^t)_4$ (non-*gem*)	106	—	132
$N_3P_3(NMe_2)_2F_2Cl_2$ (1,3-NMe_2 1,3-F 5,5-Cl: *cis*)	54–55	IR (P=N), 1215–1249 cm^{-1}	152
$N_3P_3(NMe_2)_2F_2Cl_2$ (1,3-NMe_2 1,3-F 5,5-Cl: *trans*)	—	IR (P=N), 1215–1249 cm^{-1}	152
$N_3P_3(NMe_2)F_2Cl_2$ (1,3-NMe_2 5,5-F 1,3-Cl: *cis*)	60–61	—	152
$N_3P_3(NMe_2)_2F_2Cl_2$ (1,3-NMe_2 5,5-F 1,3-Cl: *trans*)	84–85	—	152
$N_3P_3(NMe_2)_3Cl_2Ph$ (*gem*)	60	—	125
$N_3P_3(NMe_2)_2ClPh_3$[a]	124.5–125.5	—	75
$N_3P_3(NMe_2)_2Cl_2Ph_2$ (1,3-NMe_2 5,5-Cl 1,3-Ph: *cis*)	99	IR (P=N), 1180–1230 cm^{-1}	81, 125, 166
$N_3P_3(NMe_2)_2Cl_2Ph_2$ (1,3-NMe_2 5,5-Ph 1,3-Cl)	144	IR (P=N), 1180–1210 cm^{-1}	78, 81, 105, 120
$N_3P_3(NMe_2)_2Cl_3Ph$	–9	—	125
$N_3P_3(NMe_2)ClPh_4$[a]	164–165	—	120
$N_3P_3(NMe_2)Cl_2Ph_3$[a]	110–115	—	75
$N_3P_3(NMe_2)Cl_3Ph_2$[a]	110	—	78, 120
$N_4P_4(NMe_2)Cl_2Ph_5$ (1-NMe_2 1-Ph)	171	IR (P=N), 1240 cm^{-1}	170
$N_3P_3(NMe_2)_2(C_6H_3Me_2)_2Cl_2$ (1,3-NMe_2 1,3-$C_6H_3Me_2$ 5,5-Cl: *cis*)	122	IR (P=N), 1170–1210 cm^{-1}	81
$N_3P_3(NMe_2)(NH_2)_2Cl_3$ (1-NMe_2 3,5-NH_2 1,3,5-Cl)	87	—	102, 107
$N_3P_3(NMe_2)_2(NHMe)Cl_3$ (1,3-NMe_2 5-NHMe 1,3,5-Cl)	82	—	102, 107
$N_3P_3(NMe_2)_2(NHPr^i)_2Cl_2$[a]	120	—	125, 127
$N_3P_3(NMe_2)_3(NHPh)Cl_2$ (*gem*)	149	—	125
$N_3P_3(NMe_2)_2(NH_2)_2Ph_2$ (*gem*)	137	—	78, 102, 171
$N_3P_3(NMe_2)_2(NH_2)_2Ph_2$ (1,3-NMe_2 1,3-NH_2 5,5-Ph)	171	—	78

Compound	M.p. (°C)	Property Data	Ref.
$N_3P_3(NMe_2)_3(NHPh)Ph_2$ (*gem*)	122–123	—	78
$N_3P_3(NMe_2)_2(NHPh)_2Ph_2$ (*gem*)	153	—	78
$N_3P_3(NMe_2)(NHPh)Ph_2Cl_2$ (1-NMe 1-NHPh 3,3-Cl 5,5-Ph)	143	—	78, 125
$N_3P_3(NMe_2)_4(NHCOCH_3)_2$ (non-*gem*)	191–192	—	165
$N_3P_3(NMe_2)_4(NHCOC_6H_4\text{-}p\text{-}NO_2)_2$ (non-*gem*)	203.5–204.5	—	165
$N_3P_3(NMe_2)_4(NHCONHPh)(NH_2)$[a]	178–179	—	165
$N_3P_3(NMe_2)_4(NHCSNHPh)_2$ (non-*gem*)	144–145	—	165
$N_3P_3(NMe_2)_4(NHCSNHPh)(NH_2)$	137.5	—	165
$[NP(NEt_2)_2]_3$	205	—	108
$[NP(NEt_2)_2]_4$	200	—	111
$N_3P_3(NEt_2)_4Cl_2$ (non-*gem*)	—	B.p., 125°C/0.02 mm; n_D^{22} 1.5103	108
$N_3P_3(NEt_2)_3Cl_3$[a]	162	B.p., 170°C/5 mm; n_D^{23}, 1.5100; IR (P=N), 1195, 1255 cm^{-1}	108, 121
$N_3P_3(NEt_2)_2Cl_4$[a]	134	IR (P=N), 1226–1255 cm^{-1}	121
$N_3P_3(NEt_2)Cl_5$	—	B.p., 81°–83°C/0.01 mm; n_D^{25}, 1.5287; IR (P=N), 1200–1220 cm^{-1}	108, 114
$N_4P_4(NEt_2)_4Cl_4$[a]	172	—	111
$N_3P_3(NEt_2)_4Ph_2$[a]	76–77	—	120
$N_3P_3(NEt_2)_2Ph_4$[a]	122–123	—	120
$[NP(NEt_2)Ph]_4$	>320	—	85
$N_4P_4(NEt_2)_2Ph_6$[a]	260–265 (subl.)	—	74
$N_3P_3(NEt_2)Ph_4Cl$[a]	147.5	—	120
$N_3P_3(NEt_2)_2Ph_2Cl_2$ (1,3-NEt_2, 1,3-Cl 5,5-Ph)	66	—	105, 120
$N_3P_3(NEt_2)_2PhCl_3$ (non-*gem*)	−9	—	125
$N_3P_3(NPr^n_2)Cl_5$	—	B.p., 112°C/0.01 mm; n_D^{25}, 1.5218; IR (P=N), 1210–1220 cm^{-1}	114
$[NP(NBu^n_2)Cl]_3$	—	B.p., 170°C/0.01 mm	108
$N_3P_3(NBu^n_2)Cl_5$	—	B.p., 130°–131°C/0.01 mm; n_D^{25}, 1.5148; IR (P=N), 1205-1218 cm^{-1}	114

$N_3P_3(NBu^i_2)_2Cl_4$ (non-*gem*)	73	—	108
$[NP(NC_4H_4)_2]_3$ (pyrrole deriv.)	198–200	—	131
$[NP(NC_4H_4)_2]_4$	195–196	IR (P═N), 1335, 1290, 1235, 1200 cm^{-1}	131
$N_4P_4(NMePh)_4Cl_4$[a]	146	IR (P═N), 1275–1330 cm^{-1}	110
$N_4P_4[NMe(Ph)]_2Cl_6$[a]	146	IR (P═N), 1308 cm^{-1}	29, 115
$N_4P_4[NMe(Ph)]_2Ph_6$	183–184	—	74
$[NP(NC_2H_4)_2]_3$ (ethyleneimino)	154	IR (P═N), 1200 cm^{-1}	134, 172–174
$[NP(NC_2H_4)_2]_4$	262	—	134
$N_3P_3(NC_2H_4)_5Cl$	122 (109)	—	143, 172
$N_3P_3(NC_2H_4)_4Cl_2$ (*gem*)	130–131	^{31}P, -24 ± 1 ppm (PCl_2), -35 ± 1 ppm $[P(NR_2)_2]$	143, 147
$N_3P_3(NC_2H_4)_4Cl_2$[a]	114–115	—	172
$N_3P_3(NC_2H_4)_4Cl_2$[a]	99–101	—	172
$N_3P_3(NC_2H_4)_3Cl_3$ (isomers: 1,1,3 and unknown positions)	68–70, 172	—	172, 175
$N_3P_3(NC_2H_4)_2Cl_4$ (*gem*)	105	^{31}P, $\delta(PCl_2)$, -22 ppm, $\delta(P(NR_2)_2)$, -34 ppm	143, 147, 172, 175
$N_3P_3(NC_2H_4)Cl_5$	69–70	—	143, 172
$N_3P_3(NC_2H_4)_5(NH_2)$	132–133	—	134
$N_3P_3(NC_2H_4)_5(NHMe)$	128–129.5	^{31}P, $+29.7$, $+37.3$ ppm	175
$N_3P_3(NC_2H_4)_4(NHC_6H_4Me\text{-}p)_2$[a]	154	—	143
$N_3P_3(NC_2H_4)_2(NHC_6H_4Me\text{-}p)_4$[a]	222	—	143
$N_3P_3(NC_2H_4)_5NHCH_2COOEt$	93.5–95.0	—	145
$N_3P_3(NC_2H_4)_4(NHCH_2COOEt)_2$ (*gem*)	99–100	—	145, 147
$N_3P_3(NC_2H_4)_3(NHCH_2COOEt)_3$[a]	93–94.5	—	145
$N_3P_3(NC_2H_4)_2(NHCH_2COOEt)_4$ (*gem*)	77–78	—	145, 147
$N_3P_3(NC_2H_4)_5NMe_2$	116.5–117.5	—	175
$N_3P_3(NC_2H_4)_4(NMe_2)_2$ (*gem*)	129–130	—	143
$N_3P_3(NC_2H_4)_4(NMe_2)_2$ (non-*gem*)	99–101	—	143
$N_3P_3(NC_2H_4)_3(NMe_2)_3$ (*gem*)	178.5–180	—	175
$[NP(NC_2H_4)(NMe_2)]_3$	83–84	—	175
$N_3P_3(NC_2H_4)_2(NMe_2)_4$ (*gem*)	93	—	175
$N_3P_3(NC_2H_4)(NMe_2)_5$	80–80.5	—	175

Compound	M.p. (°C)	Property Data	Ref.
$N_3P_3(NC_2H_4)_2(NMe_2)_2Cl_2$[a]	121	—	143
$N_3P_3(NC_2H_4)(NMe_2)_3Cl_2$[a]	90–91	—	143
$N_3P_3(NC_2H_4)_5CH_2SMe$	106.5–107.5	—	175
$[NP(NHCH_2CMe_2)_2]_3$	—	n_D^{25}, 1.5110	134
$[NP(NC_4H_8)_2]_3$ (pyrrolidino)	211–213	—	176, 177
$N_3P_3(NC_4H_8)_5Cl$	182–185	—	176, 177
$N_3P_3(NC_4H_8)_4Cl_2$[a]	137.5–138 (132–134)	—	176, 177
$N_3P_3(NC_4H_8)_4Cl_2$[a]	121–122	—	176, 177
$N_3P_3(NC_4H_8)_3Cl_3$ (non-*gem-cis-cis*)	74–75	—	176–178
$N_3P_3(NC_4H_8)_3Cl_3$ (1,3-*cis*-3-*trans*)	174–175	—	176, 178
$N_3P_3(NC_4H_8)_3Cl_3$ (1,3-*cis*-5-*trans*)	82.5–83.5	—	176–178
$N_3P_3(NC_4H_8)_2Cl_4$ (non-*gem-cis*)	55.5–56.5	—	176–178
$N_3P_3(NC_4H_8)_2Cl_4$ (non-*gem-trans*)	94.5–95.5	—	176–178
$N_3P_3(NC_4H_8)Cl_5$	65.5–66.5	—	176–177
$N_3P_3(NC_4H_8)_4(NC_2H_4)_2$	97–98	—	179
$N_3P_3(NC_4H_8)_4(NC_2H_4)_2$ (non-*gem-trans*)	118–119	—	179
$N_3P_3(NC_4H_8)_3(NC_2H_4)_3$ (1,3-*trans*-5 and another isomer)	103–105, 85–86	—	179
$N_3P_3(NC_4H_8)_2(NC_2H_4)_4$ (*gem*)	89–90	—	179
$N_3P_3(NC_4H_8)_2(NC_2H_4)_4$ (non-*gem-cis*)	96–98	—	179
$N_3P_3(NC_4H_8)_2(NC_2H_4)_4$ (non-*gem-trans*)	69–70	—	179
$N_3P_3(NC_4H_8)(NC_2H_4)_5$	104.5–105.5	—	179
$N_3P_3(NC_4H_8)_3(NC_2H_4)Cl_2$[a]	65–65.6	—	179
$N_3P_3(NC_4H_8)_2(NC_2H_4)Cl_2$ (two isomers)	133–135, 117–119	—	179
$N_3P_3(NC_4H_6Me_2\text{-}2,2)_2Cl_4$[a] (dimethylpyrrolidino derivative)	101–103	—	179
$N_3P_3(NC_4H_6Me_2\text{-}2,2)Cl_5$	67–68.5	—	179

Compound	M.p. (°C)	Data	References
$N_3P_3(NC_4H_6Me_2\text{-}2,2)(NC_2H_4)_5$	53–54	—	179
$N_3P_3(NC_4H_6Me_2\text{-}2,2)_2(NC_2H_4)_2$[a]	120.5–122	—	179
$N_3P_3(NC_4H_7Ph\text{-}2)(NC_2H_4)_5$ (phenylpyrrolidino derivative)	88–89	—	179
$[NP(NC_5H_{10})_2]_3$ (piperidino)	231–236 (decomp.) (226)	—	103, 104, 141, 142
$[NP(NC_5H_{10})_2]_4$	231	IR (P=N), 1326, 1281 cm^{-1}	110, 138
$[NP(NC_5H_{10})_2]_n$ ($n \approx 15{,}000$)	—	IR (P=N), 1248 cm^{-1}; ^{31}P, +7.7 ppm	122
$N_3P_3(NC_5H_{10})_5Cl$	121.0–123.5	—	146
$N_3P_3(NC_5H_{10})_4Cl_2$ (non-*gem*)	111–112	—	123, 146, 154, 157
$N_3P_3(NC_5H_{10})_3Cl_3$ (*gem*)	–17	—	123, 154, 156, 157, 169
$[NP(NC_5H_{10})Cl]_3$ (*cis*)	190	—	123, 154, 156, 157
$[NP(NC_5H_{10})Cl]_3$ (*trans*)	114.5 (104.5)	—	123, 146, 154, 156, 157
$N_3P_3(NC_5H_{10})_2Cl_4$ (non-*gem-cis*)	129	—	154, 156, 157
$N_3P_3(NC_5H_{10})_2Cl_4$ (non-*gem-trans*)	105	—	146, 154, 156, 157, 180
$N_3P_3(NC_5H_{10})Cl_5$	68–69	—	146, 169, 180
$N_4P_4(NC_5H_{10})_4Cl_4$[a]	204	IR (P=N), 1305 cm^{-1}	110
$N_4P_4(NC_5H_{10})_2Cl_6$[a]	194	IR (P=N), 1310 cm^{-1}	115
$N_3P_3(NC_5H_{10})_4Ph_2$ (*gem*)	195–197	—	120
$N_3P_3(NC_5H_{10})_3Ph_3$	187	—	85
$N_3P_3(NC_5H_{10})_2Ph_4$ (*gem*)	180–181	—	120
$[NP(NC_5H_{10})Ph]_4$ (three isomers)	224, 240, 255	—	85
$N_4P_4(NC_5H_{10})_2Ph_6$[a]	290–292	—	74
$N_3P_3(NC_5H_{10})_3Ph_2Cl$[a]	166	—	125
$N_3P_3(NC_5H_{10})_2Ph_2Cl_2$ (*gem*)	131	—	105, 120
$N_3P_3(NC_5H_{10})_2Ph_2Cl_2$ (1,3-NC_5H_{10} 1,3-Ph 5,5-Cl)	136	—	125

Compound	M.p. (°C)	Property Data	Ref.
$N_3P_3(NC_5H_{10})_2Ph_2Cl_2$[a]	145–146	—	105, 120
$N_3P_3(NC_5H_{10})_2PhCl_3$[a]	132	—	125
$N_3P_3(NC_5H_{10})Ph_4Cl$[a]	175	—	120
$N_3P_3(NC_5H_{10})Ph_2Cl_3$[a]	125–126	—	120, 181
$N_3P_3(NC_5H_{10})_4(NMe_2)_2$ (non-*gem-cis*)	136	—	169
$N_3P_3(NC_5H_{10})_4(NMe_2)_2$ (non-*gem-trans*)	141	—	169
$N_3P_3(NC_5H_{10})_3(NMe_2)_3$ (*gem*)	87–88	—	169
$[NP(NC_5H_{10})(NMe_2)]_3$ (*cis*)	156	—	169
$[NP(NC_5H_{10})(NMe_2)]_3$ (*trans*)	—	—	169
$N_3P_3(NC_5H_{10})_2(NMe_2)_4$ (non-*gem-cis*)	96	—	169
$N_3P_3(NC_5H_{10})_2(NMe_2)_4$ (non-*gem-trans*)	107	—	169
$N_3P_3(NC_5H_{10})(NMe_2)_5$	56–57	—	169
$N_3P_3(N(C_6H_{11})_2)_2Cl_4$ (non-*gem*) bicyclohexylamino deriv.	200	—	108, 121
$[NP(NMeC_6H_{11})Cl]_4$	126.5–128	—	182
$[NP(NMeC_6H_{11})(NHMe)]_4$	176–177	—	182
$N_4P_4(NMeC_6H_{11})_4(NHMe)_3(NMeCONHPh)$ (non-*gem*: 1,3,5,7-$NMeC_6H_{11}$)	134–135	—	182
$N_4P_4(NMeC_6H_{11})_4(NHMe)_2(NMeCONHPh)_2$[a]	216–217	—	182
$N_3P_3(NC_4H_8O)_4Cl_2$ (non-*gem-cis*) (morpholino)	159–160	—	183–185
$N_3P_3(NC_4H_8O)_4Cl_2$ (non-*gem-trans*)	147–148	—	183–185
$N_3P_3(NC_4H_8O)_3Cl_3$ (1,1,3-isomer)	101.5–102.5	—	183, 185
$[NP(NC_4H_8O)Cl]_3$ (1,3-*cis*-5-*cis*)	100–101	—	184
$[NP(NC_4H_8O)Cl]_3$ (1,3-*cis*-5-*trans*)	114–115	—	183, 184
$N_3P_3(NC_4H_8O)_2Cl_4$ (non-*gem-cis*)	107–108	—	183–185
$N_3P_3(NC_4H_8O)_2Cl_4$ (non-*gem-trans*)	202–203	—	184, 185
$N_3P_3(NC_4H_8O)Cl_5$	92.5–93.5	—	183
$N_4P_4(NC_4H_8O)_2Cl_6$	205	IR (P=N), 1315 cm^{-1}	115
$N_3P_3(NC_4H_8O)_4(NC_2H_4)_2$ (*gem*)	161.5–163.5	—	185

$N_3P_3(NC_4H_8O)_4(NC_2H_4)_2$ (non-*gem*) (two isomers)	150–151.5, 163–165	—	185
$N_3P_3(NC_4H_8O)_3(NC_2H_4)_2Cl$ (1,1,3-NC_4H_8O 5,5-NC_2H_4 3-Cl)	144–146	—	185
$N_3P_3(NC_4H_8NMe)Ph_2Cl_3$[a] (4′-methylpiperazino-1-yl)	102.5–103.5	—	120
$N_4P_4(NMeCONHPh)(NHMe)_3Ph_4$ (non-*gem-cis*)	125–129	—	86
$N_4P_4(NMeCONHPh)(NHMe)_3Ph_4$ (non-*gem-β-trans*)	165–167	—	86
$[NP(NHSCNHNHPh)_2]_3$	165 (decomp.)	—	53
$[NP(NHNH_2)_2]_3$	>360	IR (P=N), 1218 cm^{-1}	37, 186
$N_3P_3(NHNH_2)(NC_2H_4)_5$	157–158	^{31}P, +37.6, +30.6 ppm	175
$[NP(NHNHPh)_2]_3$	200	—	137, 141, 142
$N_4P_4(NHNHPh)_4Ph_4$	176	—	85
$N_3P_3(NHNCHC_6H_4OH)_6$	225 (decomp.)	IR (P=N), 1218 cm^{-1}	186
$[NP(NHSCNHNHPh)_2]_4$	153 (decomp.)	—	53
Alkoxy Phosphazenes			
$[NP(OMe)_2]_3$	48	B.p., 114°–128°C/0.1 mm; IR (P=N), 1225–1240 cm^{-1}; ^{31}P, −21.7 ppm	4, 26, 28, 30, 36, 187–190
$[NP(OMe)_2]_4$	41	B.p., 125°C/0.1 mm; IR (P=N), 1320–1325, 1337 cm^{-1}	4, 30, 117, 188, 189
$[NP(OMe)_2]_5$	—	B.p., >200°C/1 mm; IR (P=N), 1340 cm^{-1}	30, 151
$[NP(OMe)_2]_6$	48	IR (P=N), 1335 cm^{-1}	30, 151
$[NP(OMe)_2]_7$	—	B.p., >200°C/1 mm; IR (P=N), 1315 cm^{-1}	30, 151
$[NP(OMe)_2]_8$	81	IR (P=N), 1285 cm^{-1}	30, 151
$[NP(OMe)_2]_n$	—	T_g, −76°C; IR (P=N), 1250, 755 cm^{-1}; ^{31}P, +3.5 ppm; intrinsic viscosity, 1.22 dl/g[b]	26
$N_3P_3(OMe)Cl_5$	—	^{31}P, (PCl_2), −22.5 ppm, (PClOMe), −16.7 ppm	18
$N_3P_3(OMe)_4Ph_2$ (*gem*)	99–99.5	—	191
$N_3P_3(OMe)_2Ph_4$ (*gem*)	109–110	—	191

Compound	M.p. (°C)	Property Data	Ref.
$[NP(OMe)Ph]_4$	142	—	85
$N_3P_3(OMe)(NMe_2)_4(ONCMe_2)$	—	B.p., 122°–123°C/0.08 mm; density, 1.1418 g/ml (20°C); n_D^{20}, 1.4970; IR (P=N), 1210 cm^{-1}	192
$N_3P_3(OMe)_2(NC_2H_4)_4$ (*gem*)	119–120	^{31}P, 38.0, 20.1 ppm	175
$N_3P_3(OMe)(NC_2H_4)_5$	125–126	^{31}P, 38.2, 29.6 ppm	175
$[NP(OEt)_2]_3$	—	B.p., 114°–116°C/0.1 mm; n_D^{24}, 1.4522; IR (P=N), 1225–1240 cm^{-1}; ^{31}P, –15.3 ppm	4, 26, 28, 36, 188, 189, 190, 193–195
$[NP(OEt)_2]_4$	45–47	B.p., 128°C/0.001 mm; IR(P=N), 1320–1325 cm^{-1}	4, 188, 189, 193, 194
$[NP(OEt)_2]_n$	—	T_g, –84°C; IR (P=N), 1240 cm^{-1}; ^{31}P, 7.3 ppm	26, 196
$N_3P_3(OEt)Cl_5$	—	^{31}P, (PCl_2), –21.3 ppm, (PClOEt), –13.6 ppm	18
$N_3P_3(OEt)_4Ph_2$ (*gem*)	63–64	—	191
$N_3P_3(OEt)_2Ph_4$ (*gem*)	131–133	—	191
$N_4P_4(OEt)_4Ph_4$ (non-*gem*) (two isomers)	104, 122	—	85
$N_4P_4(OEt)_2Ph_6$ (non-*gem*)	183–185	—	83
$N_3P_3(OEt)(NMe_2)_4(ONCMe_2)$	—	B.p., 126°–128°C/0.15 mm; density, 1.1309 g/ml (20°C); n_D^{20}, 1.4953	192
$[NP(OPr^n)_2]_3$	—	B.p., 146°–148°C/0.1 mm; n_D^{25}, 1.4494; IR (P=N), 1225–1240 cm^{-1}	28, 188–190
$[NP(OPr^n)_2]_4$	38–38.5	B.p., 176°–178°C/0.01 mm; IR (P=N), 1320–1325 cm^{-1}	188, 189
$N_3P_3(OPr^n)(NMe_2)_4(ONCMe_2)$[a]	—	B.p., 128°–130°C/0.1 mm; density, 1.1056 g/ml (20°C); n_D^{20}, 1.4929	192
$[NP(OPr^i)_2]_3$	28	B.p., 104°–106°C/0.005 mm; IR (P=N), 1225–1240 cm^{-1}	189, 190
$[NP(OPr^i)_2]_4$	74–75	B.p., 130°C/0.01 mm; IR (P=N), 1320–1325 cm^{-1}	189
$N_3P_3(OPr^i)Cl_5$	—	^{31}P, (PCl_2), –21.7 ppm, ($PClOPr^i$), –12.6 ppm	18
$N_3P_3(OPr^i)_4Ph_2$ (*gem*)	75	—	191
$N_3P_3(OPr^i)(NMe_2)_4(ONCMe_2)$[a]	—	B.p., 118°–119°C/0.08 mm; density, 1.1138 g/ml (20°C); n_D^{20}, 1.4918	192

$N_3P_3(OCH_2CHCH_2)(NMe_2)_4(ONCMe_2)$[a]	—	B.p., 123°–124.5°C/0.1 mm; density, 1.1267 g/ml (20°C); n_D^{20}, 1.4979	192
$[NP(OBu^n)_2]_3$	—	B.p., 170°–171°C/0.03 mm, 162°–164°C/0.01 mm; n_D^{25}, 1.4330; IR (P=N), 1225–1240 cm^{-1}	187–190
$[NP(OBu^n)_2]_4$	—	B.p., 196°–198°C/0.005 mm; n_D^{25}, 1.4565; IR (P=N), 1320–1325 cm^{-1}	188, 189, 197
$N_3P_3(OBu^n)Cl_5$	—	B.p. 58°–59°C/2 × 10^{-4} mm; IR (P=N), 1230 cm^{-1}	198
$N_3P_3(OBu^n)(NMe_2)_4(ONCMe_2)$ (1-OBu 1,3,5,5-NMe_2 3-$ONCMe_2$)	—	B.p., 124°–125°C/0.08 mm; density, 1.1085 g/ml (20°C); n_D^{20}, 1.4908	192
$N_3P_3(OBu^i)(NMe_2)_4(ONCMe_2)$ (1-OBu 3-$ONCMe_2$)	—	B.p., 132°–133°C/0.1 mm; density, 1.1030 g/ml (20°C); n_D^{20}, 1.4911	192
$N_3P_3(OBu^s)(NMe_2)_4(ONCMe_2)$ (1-OBu 3-$ONCMe_2$)	—	B.p., 124°–125°C/0.08 mm; density, 1.1131 g/ml (20°C); n_D^{20}, 1.4929	192
$N_3P_3(OC_5H_{11})(NMe_2)_4(ONCMe_2)$ (1-OBu 3-$ONCMe_2$)	—	B.p., 133°–136°C/0.1 mm; density, 1.0988 g/ml (20°C); n_D^{20}, 1.4895	192
N_3P_3(O-iso-C_6H_{11})$(NMe_2)_4(ONCMe_2)$ (1-OR 3-$ONCMe_2$)	—	B.p., 130°–133°C/0.15 mm; density, 1.0869 g/ml (20°C); n_D^{20}, 1.4873	192
$N_3P_3(OC_6H_{13})(NMe_2)_4(ONCMe_2)$ (1-OR 3-$ONCMe_2$)	—	B.p., 133°–135°C/0.05 mm; density, 1.0756 g/ml (20°C); n_D^{20}, 1.4843	192
$N_3P_3(OC_7H_{15})(NMe_2)_4(ONCMe_2)$ (1-OR 3-$ONCMe_2$)	—	B.p., 140°–143°C/0.1 mm; density, 1.0829 g/ml (20°C); n_D^{20}, 1.4889	192
$[NP(OCH_2CH_2O)]_3$ (spirocyclic)	>300 (200 decomp.)	IR (P=N), 1250 cm^{-1}	199, 200
$[NP(OCH_2CH_2O)]_4$ (spirocyclic)	210 (decomp.)	—	200
$[NP(OCH_2CHMeO)]_3$ (two isomers)	130 (decomp.) 250 (decomp.)	—	200
$[NP(OCH_2CHMeO)]_4$	220 (decomp.)	—	200
$[NP(OCHMeCHMeO)]_3$	180 (decomp.)	—	200
$[NP(OCMe_2CMe_2O)]_3$	201	—	200
$[NP(OCH_2CH_2CH_2O)]_3$	280	—	200
$[NP(O(CH_2)_4O)]_4$	210	—	200
$[NP(OC_6H_{10}O)]_3$ (cyclohexyl)	228	—	200

Compound	M.p. (°C)	Property Data	Ref.
$\{NP[OCH_2C(CH_2ONO_2)_2CH_2O]\}_3$ (spirocyclic)	139–140	IR (P=N), 1250, 1275 cm^{-1}	201
$\{NP[OCH_2C(CH_2ONO_2)_3]_2\}_3$	100	IR (P=N), 1278 cm^{-1}	201
$[NP(OCH_2Ph)_2]_3$	51.5	—	189
$[NP(OCH_2Ph)_2]_4$	38–39	—	189
$[NP(OCH_2CF_3)_2]_3$	49–50	B.p., 248°C/743 mm, 115°–116°C/3 mm; IR (P=N), 1240–1242 cm^{-1}; ^{31}P, −17.7 ppm	26, 28, 202, 203
$[NP(OCH_2CF_3)_2]_4$	65	B.p., 139°–140°C/3mm; IR (P=N), 1272 cm^{-1}	28, 202–204
$[NP(OCH_2CF_3)_2]_5$	—	B.p., 206°C/30 mm	151
$[NP(OCH_2CF_3)_2]_6$	189	—	151
$[NP(OCH_2CF_3)_2]_8$	—	B.p., 254°C/30 mm	151
$[NP(OCH_2CF_3)_2]_n$	242–243	T_g, −66°C; density, 1.71 g/ml; IR (P=N), 1270–1280 cm^{-1}; ^{31}P, +7.5 ppm	26, 196
$N_3P_3(OCH_2CF_3)_5Cl$	—	B.p., 98°–100°C/0.05 mm; n_D^{25}, 1.3730	28
$N_3P_3(OCH_2CF_3)_3Cl_3$[a]	—	B.p., 161°–162°C/46 mm; n_D^{20}, 1.3994	205
$N_3P_3(OCH_2CF_3)Cl_5$	—	^{31}P, (PCl_2), −22.7 ppm, (PClR), −16.5 ppm	18
$[NP(OCH_2CF_3)Ph]_3$ (*trans*)	78–80	—	71
$N_3P_3(OCH_2CF_3)_4Ph_2$ (1,1,3,3-OR 5,5-Ph)	51	—	71
$N_3P_3(OCH_2CF_3)_4(NH_2)_2$ (*gem*)	61–62	IR (P=N), 1230 cm^{-1}	205, 206
$N_3P_3(OCH_2CF_3)_3(NH_2)_3$ (three isomers, structures not reported)	144–146, 167–169, 186–187	—	205
$N_3P_3(OCH_2CF_3)_4(NHPh)_2$ (*gem*)	100–101	B.p., 187°–193°C/0.1–0.3 mm; ^{31}P, −6.37, −16.38 ppm	140
$N_3P_3(OCH_2CF_3)_4(NHC_6H_4Cl\text{-}m)_2$ (*gem*)	102–104	B.p., 172°–185°C/0.2–0.9 mm; ^{31}P, −6.94, −16.2 ppm	140
$N_3P_3(OCH_2CF_3)_4(NHC_6H_4Me\text{-}m)_2$[a]	102.5–104	B.p., 186°–191°C/0.08–0.1 mm	140
$N_3P_3(OCH_2CF_3)_4(NMePh)_2$[a]	—	B.p., 170°–203°C/0.08–0.2 mm; density, 1.41 g/ml; n_D^{25}, 1.4572	207
$N_4P_4(OCH_2CF_3)_6(NMePh)_2$[a]	—	B.p., 153°–154°C/0.03–0.05 mm; density, 1.47 g/ml; n_D^{25}, 1.4331	207
$\{NP[OCH_2(CF_2)_2H]_2\}_3$	41	B.p., 168°–172°C/0.55 mm	208

$\{NP[OCH_2(CF_2)_2H]_2\}_4$	162–163	—	208
$N_3P_3[OCH_2(CF_2)_2H]_4(NH_2)_2$ (*gem*)	62–63	IR (P=N), 1225 cm^{-1}	206
$N_3P_3[OCH_2(CF_2)_2H]_4(NHPh)_2$ (*gem*)	98–100	B.p., 209°–214°C/0.2–0.7 mm; density, 1.45 g/ml; n_D^{25}, 1.4504	140, 207
$[NP(OCH_2C_2F_5)_2]_3$	33–34	B.p., 136.5°C/6 mm; IR (P=N), 1200, 1250 cm^{-1}; n_D^{28}, 1.3365	28, 202–204
$[NP(OCH_2C_2F_5)_2]_4$	128	B.p., 142°–144°C/3 mm; n_D^{27}, 1.3530; IR (P=N), 1200 cm^{-1}	28, 202–204
$N_3P_3(OCH_2C_2F_5)_4(NH_2)_2$[a]	66–67	IR (P=N), 1215 cm^{-1}	206
$[NP(OCH_2C_3F_7)_2]_3$	20–23	B.p., 154°C/3 mm; n_D^{25}, 1.3298; IR (P=N), 1227 cm^{-1}	28, 202–204
$[NP(OCH_2C_3F_7)_2]_4$	108	B.p., 169.5°–171.5°C/3 mm; IR (P=N), 1220 cm^{-1}	28, 202–204
$[NP(OCH_2C_3F_7)(OCH_2CF_3)]_n$	—	T_g, −77°C; intrinsic viscosity, 1.46 dl/g	209
$N_3P_3(OCH_2C_3F_7)_4(NH_2)_2$[a]	70–72	IR (P=N), 1230 cm^{-1}	206
$[NP(OCH_2(CF_2)_4H)_2]_3$	—	B.p., 208°–212°C/0.45 mm; n_D^{25}, 1.3507; density, 1.85 g/ml	208
$[NP(OCH_2(CF_2)_4H)_2]_4$	—	B.p., 233°–236°C/0.51 mm; density, 1.78 g/ml; n_D^{25}, 1.3536	208
$N_3P_3[OCH_2(CF_2)_4H]_4(NH_2)_2$[a]	—	B.p., 201°–203°C/0.25 mm; IR (P=N), 1240 cm^{-1}	206
$N_3P_3[OCH_2(CF_2)_4H]_4(NHPh)_2$[a]	—	B.p., 217°–236°C/0.4–3.0 mm; density, 1.57 g/ml; n_D^{25}, 1.4199	207
$N_3P_3[OCH_2(CF_2)_4H]_5(NMePh)$	—	B.p., 203°–225°C/0.2 mm; density, 1.66 g/ml; n_D^{25}, 1.3856	207
$N_3P_3[OCH_2(CF_2)_4H]_5(NMeC_6H_4Cl\text{-}m)$	—	B.p., 147°–182°C/0.3 mm; density, 1.51 g/ml; n_D^{25}, 1.4235	207
$[NP(OCH_2(CF_2)_6H)_2]_3$	33–36	B.p., 170°–172°C/0.8 mm, 258°–260°C/2 mm	202, 203, 210
$[NP(OCH_2(CF_2)_6H)_2]_4$	—	B.p., 265°C/1 mm, 398°–401°C/750 mm; density, 1.86 g/ml; n_D^{25}, 1.3468	202, 203, 208
$[NP(OCH_2(CF_2)_8H)_2]_3$	90–94	—	208
$[NP(OCH_2(CF_2)_8H)_2]_4$	89.5–90.5	—	208
$[NP(OCH_2(CF_2)_{10}H)_2]_3$	103–105	B.p., 320°–324°C/2 mm	202, 203
$[NP(OCH_2(CF_2)_{10}H)_2]_4$	102–105	B.p., 312°–320°C/0.1 mm	202, 203
$[NP(OCH_2(CF_2)_2CH_2O)]_3$ (spirocyclic)	300	—	203
$[NP(OCH_2(CF_2)_3CH_2O)]_3$ (spirocyclic)	242–244	—	203
$[NP(OCCl_2CF_2CF_3)_2]_3$	147–148	B.p., 176°C/5 mm	203, 211
$[NP(OCCl_2C_3F_7)_2]_3$	94	B.p., 194°–197°C/4 mm	203, 211
$[NP(OCCl_2C_3F_7)_2]_4$	129–132	B.p., 234°C/3 mm	203, 211

Compound	M.p. (°C)	Property Data	Ref.
$[NP(OCCl_2C_4F_9)_2]_4$	129–132	B.p., 234°C/3 mm	203
$[NP(OCCl_2(CF_2)_{10}Cl)_2]_4$	145–149	—	203, 211
Aryloxy Phosphazenes			
$[NP(OPh)_2]_3$	115–115.5	B.p., 280°–290°C/0.1–0.14 mm; IR (P=N), 1160–1200, 1250–1280 cm^{-1}; ^{31}P, –9 ppm	26, 46, 188, 189, 194, 212–215
$[NP(OPh)_2]_4$	85–86	IR (P=N), 1320–1350, 1191 cm^{-1}	188, 189, 194, 214, 215
$[NP(OPh)_2]_5$	129–130	—	151
$[NP(OPh)_2]_6$	320–324	—	151
$[NP(OPh)_2]_7$	340–344	—	151
$[NP(OPh)_2]_8$	338–342	—	151
$[NP(OPh)_2]_n$	—	T_g, –8°C; IR (P=N), 1240, 775 cm^{-1}; ^{31}P, +18.6	26, 215
$N_3P_3(OPh)_5Cl$	69–71	—	189, 213, 216
$N_3P_3(OPh)_4Cl_2$ (non-*gem-cis*)	75	—	213, 216, 217
$N_3P_3(OPh)_4Cl_2$[a]	—	B.p., 165°–166°C/8.5 × 10^{-5} mm	217, 218
$N_3P_3(OPh)_3Cl_3$[a]	68	—	216, 217
$N_3P_3(OPh)_3Cl_3$	—	B.p., 148°–149.5°C/5 × 10^{-5} mm, 160°C/0.05 mm; n_D^{25}, 1.578	216–218
$N_3P_3(OPh)_2Cl_4$ (non-*gem*)	—	B.p., 130°C/0.02 mm; n_D^{25}, 1.568	216–218
$N_3P_3(OPh)_2Cl_4$[a]	—	B.p., 169°–170°C/10^{-4} mm; density, 1.4973 g/ml; n_D^{20}, 1.5716	198, 217
$N_3P_3(OPh)Cl_5$	48	B.p., 83°–83.5°C/5 × 10^{-5} mm	216, 218
$N_3P_3(OPh)_4Ph_2$[a]	97–99	B.p., 305°C/0.21 mm	215
$N_3P_3(OPh)Ph_4H$ (*gem*)	153	^{31}P, (PPh_2), –14.5 ppm, (OPhHP), –5.8 ppm	94
$N_4P_4(OPh)_4Ph_4$ (non-*gem*)	133.5	—	85
$N_4P_4(OPh)_2Ph_6$ (non-*gem*)	129–130	—	83

Compound	M.p. (°C)	Other	Ref.
$N_3P_3(OPh)_2Ph_2Cl_2$ (1,3-OPh 5,5-Ph 1,3-Cl: *cis*)	140.5–143	—	213
$N_3P_3(OPh)_2Ph_2Cl_2$ (1,3-OPh 1,3-Ph 5,5-Cl)[a]	143 (104)	—	219
$N_3P_3(OPh)_4(NH_2)_2$ (*gem*)	104–106	—	213
$N_3P_3(OPh)_4(NH_2)_2$ (non-*gem-cis*)	118–120	—	213
$N_3P_3(OPh)_4(NH_2)_2$ (non-*gem-trans*)	110–111.5	—	213
$N_3P_3(OPh)_3(NH_2)_3$ (*gem*)	135.5–137	—	213
$[NP(OPh)(NH_2)]_3$ (*cis*)	189–190	—	213
$[NP(OPh)(NH_2)]_3$ (*trans*)	210–212	—	213
$N_3P_3(OPh)_2Ph_2(NH_2)_2$ (*gem*)	150–152	—	213
$N_3P_3(OPh)_2(NH_2)_2Ph_2$ (1,3-OPh 1,3-NH_2 5,5-Ph: *cis*)	156–157.5	—	213
$N_3P_3(OPh)_2(NH_2)_2Ph_2$ (1,3-OPh 1,3-NH_2 5,5-Ph: *trans*)	161–163	—	213
$N_3P_3(OPh)_2Ph_2(NHMe)_2$ (1,3-OPh 1,3-Ph 5,5-NHMe)[a]	162, 140	—	219
$N_3P_3(OPh)_2(NHMe)Ph_2Cl$ (*cis*: 1,3-OPh 1,3-Ph 5-Cl 5-NHMe)	110	—	219
$N_3P_3(OPh)_5(NMe_2)$	−11	—	217
$N_3P_3(OPh)_4(NMe_2)_2$ (non-*gem-cis*)	−30	—	217
$N_3P_3(OPh)_4(NMe_2)_2$ (non-*gem-trans*)	65–66	—	217
$N_3P_3(OPh)_3(NMe_2)_3$ (*gem*)	65–66	—	217
$[NP(OPh)(NMe_2)]_3$ (*cis*)	82	—	217
$[NP(OPh)(NMe_2)]_3$ (*trans*)	90	—	217, 220
$N_3P_3(OPh)_2(NMe_2)_4$ (non-*gem-cis*)	—	B.p., 200°C/0.01 mm	217
$N_3P_3(OPh)_2(NMe_2)_4$ (non-*gem-trans*)	69	—	217
$N_3P_3(OPh)(NMe_2)_5$	52	—	217
$N_3P_3(OPh)_2(NMe_2)_3Cl$ (two isomers)	58, 94	—	217
$N_3P_3(OPh)(NMe_2)_2Cl_3$ (1-OPh 3,5-NMe_2 1,3,5-Cl: *trans*)	55	—	160
$N_3P_3(OPh)_2Ph_2(NMe_2)_2$ (1,3-OPh 1,3-Ph 5,5-NMe_2)[a]	96, 108	—	219
$N_3P_3(OPh_2)(NMe_2)Ph_2Cl$ (1,3-OPh 1,3-Ph 5-NMe_2 5-Cl)	76	—	219

Compound	M.p. (°C)	Property Data	Ref.
$[NP(OC_{10}H_7)_2]_3$ (naphthyloxy)	168–169	IR (P=N), 1250–1280 cm^{-1}	189
$[NP(OC_{10}H_7)_2]_4$	126–127	IR (P=N), 1330–1350 cm^{-1}	189
$[NP(O_2C_6H_4\text{-}o)]_3$ (spirocyclic)	244–245	IR (P=N), 1220 cm^{-1}; ^{31}P, −11.8 ppm	144, 200, 212, 221
$[NP(O_2C_{10}H_6\text{-}2,3)]_3$ (spirocyclic naphthyldioxy)	335–336	Density, 1.341 g/ml; IR (P=N), 1210 cm^{-1}	46, 144
$[NP(O_2C_{10}H_6\text{-}1,8)]_3$ (spirocyclic naphthyldioxy)	358–359	—	221a
$[NP(O_2C_{10}H_6\text{-}1,8)]_4$ (spirocyclic naphthyldioxy)	>360	—	221a
$[NP(O_2C_{12}H_8\text{-}2,2')]_3$ (spirocyclic biphenylenedioxy)	>300	IR (P=N), 1170 cm^{-1}; ^{31}P, −9.0 or −10.0	144
$[NP(O_2C_{12}H_8\text{-}2,2')]_4$ (spirocyclic biphenylenedioxy)	>340	—	221a
$[NP(OC_6H_4Cl\text{-}p)_2]_3$	155–156, (151)	IR (P=N), 1181 cm^{-1}	214, 222
$[NP(OC_6H_4Cl\text{-}p)_2]_4$	83–85	IR (P=N), 1199 cm^{-1}	214
$N_3P_3(OC_6H_4Cl\text{-}o)(OPh)_5$	106.5–107.5	B.p., 305°C/0.21 mm	222
$N_3P_3(OC_6Cl_5)_3Cl_3$[a]	211–215	—	205
$[NP(OC_6H_4Br\text{-}p)_2]_3$	175	—	216
$N_3P_3(OC_6H_4Br\text{-}p)_5Cl$	106.5–108	—	216
$N_3P_3(OC_6H_4Br\text{-}p)_4Cl_2$	87	—	216, 217
$N_3P_3(OC_6H_4Br\text{-}p)_3Cl_3$	129–130	—	216, 217
$N_3P_3(OC_6H_4Br\text{-}p)_2Cl_4$	—	B.p., 210°C/0.01 mm	216, 217
$N_3P_3(OC_6H_4Br\text{-}p)Cl_5$	60	—	216
$N_3P_3(OC_6H_4Br\text{-}p)_4(NMe_2)_2$[a]	148–149	—	217
$N_3P_3(OC_6H_4Br\text{-}p)_4(NMe_2)_2$[a]	101–102	—	217
$N_3P_3(OC_6H_4Br\text{-}p)_3(NMe_2)_3$ (two isomers)	174–175, 180–181	—	217
$N_3P_3(OC_6H_4Br\text{-}p)_2(NMe_2)_4$[a]	100–101	—	217
$N_3P_3(OC_6H_4Br\text{-}p)_2(NMe_2)_4$[a]	88–89	—	217
$[NP(OC_6H_4NO_2\text{-}p)_2]_3$	263–264	—	46, 214, 223, 224

$[NP(OC_6H_4NO_2\text{-}p)_2]_4$	299–301	—	214, 223, 224
$N_3P_3(OC_6H_4NO_2\text{-}p)_5Cl$	183–184	—	223, 224
$N_3P_3(OC_6H_4NO_2\text{-}p)_3Cl_3$[a]	225–226	^{31}P, −17.4 ppm	223, 224
$N_3P_3(OC_6H_4NO_2\text{-}p)Cl_5$	95–97	B.p., 165°C/0.02 mm	223, 224
$[NP(OC_6H_4NO_2\text{-}m)_2]_3$	141.5	IR (P=N), 1185 cm^{-1}	224a
$[NP(OC_6H_4NO_2\text{-}o)_2]_3$	155.5	IR (P=N), 1185 cm^{-1}	224a
$[NP(OC_6H_4Me\text{-}p)_2]_3$	122–124	IR (P=N), 1181 cm^{-1}	214
$[NP(OC_6H_4Me\text{-}p)_2]_4$	90	IR (P=N), 1194, 1205, 1215	214
$[NP(OC_6H_4OMe\text{-}p)_2]_3$	103–104	IR (P=N), 1250–1280 cm^{-1}	189, 214, 224
$[NP(OC_6H_4OMe\text{-}p)_2]_4$	92.5–94.5 73–74	IR (P=N), 1190, 1330–1350 cm^{-1}	189–214
$N_3P_3(OC_6H_4Ph\text{-}p)(OPh)_5$	96–98	B.p., 360°C/0.60 mm	215
$N_3P_3(OC_6H_4OPh\text{-}m)(OPh)_5$	—	B.p., 316°–345°C/0.18 mm	215, 222
$N_3P_3[m\text{-}OC_6H_4C(O)Me](OCH_2CF_3)_5$	—	B.p., 162°–167°C/0.02–0.07 mm; density, 1.52 g/ml; n_D^{25}, 1.4115	225
$[NP(OC_6H_4CF_3\text{-}m)_2]_3$	—	B.p., 250°C/0.2 mm; density, 1.50 g/ml; n_D^{25}, 1.4741; IR (P=N), 1205 cm^{-1}	222, 226, 227
$[NP(OC_6H_4CF_3\text{-}m)_2]_4$	54–55	B.p., 280°C/0.2 mm; density 1.51 g/ml; n_D^{25}, 1.4950; IR (P=N), 1370 cm^{-1}	226
$[NP(OC_6H_4CF_3\text{-}m)_2]_5$	—	B.p., 306–346°C/0.2 mm; n_D^{25}, 1.499	226
$[NP(OC_6H_4CF_3\text{-}m)_2]_6$	—	B.p., 347°–375°C/0.2 mm; n_D^{25}, 1.501	226
$N_3P_3(OC_6H_4CF_3\text{-}m)(OPh)_5$	—	B.p., 252°–270°C/0.2 mm	222
$N_4P_4(OPh)_a[OCH_2(CF_2)_xH]_6$[g]	—	—	208
$N_3P_3(OC_6H_2\text{-}2\text{-}R_1\text{-}4\text{-}R_2\text{-}5\text{-}R_3)_a[OCH_2(CF_2)_xH]_b$[g]	—	—	208
$N_3P_3(OC_6H_2\text{-}2\text{-}R_1\text{-}4\text{-}R_2\text{-}5\text{-}R_3)_a[OCH_2(CF_2)_xF]_b$[g]	—	—	225, 228
$N_4P_4(OC_6H_2\text{-}2\text{-}R_1\text{-}4\text{-}R_2\text{-}5\text{-}R_3)_a[OCH_2(CF_2)_xF]_b$[g]	—	—	225, 228
$N_3P_3(OC_6\text{-}2\text{-}H\text{-}3\text{-}R_1\text{-}4\text{-}R_2\text{-}5\text{-}R_3\text{-}6\text{-}R_4)_a[OCH_2(CF_2)_xF]_b$[g]	—	—	225, 228

Compound	M.p. (°C)	Property Data	Ref.
Thioalkyl Phosphazenes			
$N_3P_3(SMe)(C_6H_4OMe\text{-}p)Ph_4$ (*gem*)	129–130	^{31}P, $\delta(PPh_2)$, −14.1 ppm; $\delta[P(SMe)C_6H_4OMe]$, −27.0 ppm	229
$[NP(SEt)_2]_3$	35.5	B.p., 196°C/0.06 mm; IR (P=N), 1150 cm^{-1}; ^{31}P, −45.7 ± 0.3 ppm	25, 230–232
$N_3P_3(SEt)_4Cl_2$ (*gem*)	—	B.p., 174°C/0.07 mm	231
$N_3P_3(SEt)_3Cl_3$ (*gem*)	—	B.p., 148°–150°C/0.1 mm	231, 232
$N_3P_3(SEt)_2Cl_4$ (*gem*)	—	B.p., 128°C/0.2 mm; IR (P=N), 1190 cm^{-1}; ^{31}P, $\delta(PCl_2)$, −17.7 ± 0.3 ppm, $\delta[P(SR)_2]$, −51.7 ± 0.3 ppm	25, 230–232
$N_4P_4(SEt)_4Cl_4$ (*gem*)	105–106.5	B.p., 148°–150°C/0.2 mm; ^{31}P, $\delta(PCl_2)$, +9.7 ± 0.1 ppm, $\delta[P(SR)_2]$, −28.6 ± 0.1 ppm	25, 230, 232
$[NP(SPr^n)_2]_3$	—	B.p., 209°C/0.02 mm; n_D^{20}, 1.5813	231, 232
$N_3P_3(SPr^n)_2Cl_4$ (*gem*)	—	B.p., 124°C/0.1 mm	231, 232
$N_4P_4(SPr^n)_4Cl_4$[a]	74.5–75	B.p., 144°C/0.1 mm	232
$[NP(SPr^i)_2]_3$	—	B.p., 175°C/0.03 mm	231, 232
$N_3P_3(SPr^i)_4Cl_2$[a]	47–49	B.p., 160°C/0.09 mm	231, 232
$N_3P_3(SPr^i)_2Cl_4$ (*gem*)	53–54	B.p., 116°–118°C/0.03 mm	231, 232
$[NP(SBu^n)_2]_3$	—	B.p., 238°C/0.04 mm; n_D^{20}, 1.5645	231, 232
$N_3P_3(SBu^n)_2Cl_4$ (*gem*)	—	B.p., 150°C/0.08 mm	231, 232
$N_4P_4(SBu^n)_4Cl_4$[a]	—	B.p., 148°–150°C/0.2 mm	232
$N_3P_3(SBu^i)_2Cl_4$ (*gem*)	—	B.p., 138°C/0.07 mm	231, 232
$N_4P_4(SBu^i)_3Cl_5$[a]	—	B.p., 175°–180°C/0.04 mm	232
$[NP(SCH_2Ph)_2]_3$	77.5–78.5	—	231, 232
$[NP(SC_6H_{11})_2]_3$	99–100	—	231, 232
$N_3P_3(SC_6H_{11})_4Cl_2$	80–81	—	231, 232
$\{NP[SC(NH)NH_2]_2\}_3$	156	—	233
$\{NP[SC(NH)R]_2\}_3$	175	—	233
R = 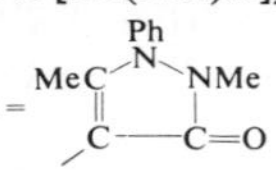			
$\{NP[SC(NH)C(NH)S]\}_3$ (spirocyclic)	>300	—	233
$\{NP[SC(NH)(NH)_3C(NH)S]\}_3$ (spirocyclic)	275 (decomp.)	—	233

Thioaryl Phosphazenes			
$[NP(SPh)_2]_3$	153.5	—	230–232
$N_3P_3(SPh)_2Cl_4$ (*gem*)	107	B.p., 180°C/0.01 mm; ^{31}P, (PCl_2), -18.6 ± 0.1 ppm, $[P(SPh)_2]$, -46.8 ± 0.1 ppm	25, 230–232
$N_4P_4(SPh)_4Cl_4$ (*gem*)	156	—	230, 232
$N_3P_3(SPh)(SMe)Ph_4$ (*gem*)	100–103	—	229
$[NP(S_2C_6H_3Me\text{-}3,4,1)]_3$ (spirocyclic)	256–263	IR (P=N), 1170 cm^{-1}	46
Phosphazenyl Cyclophosphazenes			
$N_3P_3(N{=}PCl_3)_2Cl_4$ (*gem*)	57–59	IR (P=N), 1192, 1210, 1290 cm^{-1}; ^{31}P, $\delta(PCl_2)$, -17.7 ppm, $\delta[P(NPCl_3)_2]$, $+20.6$ ppm, $\delta(NPCl_3)$, $+13.5$ or $+12.9$ ppm	51, 98, 99
$N_3P_3(N{=}PCl_3)Cl_5$	39.5–41	IR (P=N), 1200, 1253, 1335 cm^{-1}; ^{31}P, $\delta(PCl_2)$, -20.5 ppm, $\delta(PClNPCl_3)$, $+2.20$ ppm, $\delta(NPCl_3)$, $+3.29$ ppm	98
$N_3P_3(NPCl_2O)_2Cl_4$ (*gem*)	—	^{31}P, (PCl_2), -21.4 ppm, $\delta[P(NPCl_2O)_2]$, -13.1 ppm, $\delta(PCl_2O)$, -0.7 ppm	51
$N_3P_3Ph(N{=}PPh_3)Cl_4$ (1-Ph 1-N=PPh_3)	174–176	IR (P=N), 1285, 1260 cm^{-1}; ^{31}P, $\delta(PCl_2)$, -15.8 ppm $\delta[P(NPPh_3)_2]$, -14.6 ppm, $\delta(NPPh_3)$, -2.4 ppm	234
$N_3P_3(N{=}PPh_3)_2Cl_4$ (structure uncertain)	198–199	IR (P=N), 1205 cm^{-1}; ^{31}P, $\delta(PCl_2)$, -13.4 ppm, $\delta[P(NPPh_3)_2]$, $+11.6$ ppm, $\delta(NPPh_3)$, -5.5 ppm	234
$N_3P_3(N{=}PPh_3)Cl_5$	211–212	IR (P=N), 1270, 1215, 1175–1180, 1108 cm^{-1}; ^{31}P, $\delta(PCl_2)$, -20 ppm; $\delta(NPPh_3)$, -15 ppm	24, 234
$N_3P_3(N{=}PPh_3)PhCl_4$ (*gem*)	181	IR (P=N), 1180, 1215 cm^{-1}	24
$N_3P_3(N{=}PPh_3)(NH_2)Cl_4$ (*gem*)	173	IR (P=N), 1109, 1155, 1200, 1300 cm^{-1}; ^{31}P, $\delta(PCl_2)$, -15.6 ppm; $\delta[P(NH_2)NPPh_3]$, $+1.2$ ppm; $\delta(PPh_3)$, -10.6 ppm	51, 99, 234
$N_3P_3(N{=}PPh_3)(NPCl_3)Cl_4$ (*gem*)	—	^{31}P, $\delta(PCl_2)$, -18.5 ± 0.5 ppm; $\delta[P(NPCl_3)NPPh_3]$, $+20.3$ ppm; $\delta(PCl_3)$, $+11.8 \pm 0.2$ ppm; $\delta(PPh_3)$, -19.9 ppm	51, 99
$N_3P_3[N{=}P(C_6H_4Me\text{-}p)_3]_2Cl_4$[a]	214–214.5	IR (P=N), 1200, 1116 cm^{-1}	234
$N_3P_3[N{=}P(N{=}PPh_3)_2Ph]Cl_5$	152	^{31}P, $\delta(PCl_2)$, $+93.2 \pm 0.5$ ppm; $\delta(PPh_3)$, $+105.4 \pm 0.7$ ppm; $\delta[P(NPR_3)_2Ph]$, $+118.1 \pm 0.5$ ppm;	235

Compound	M.p. (°C)	Property Data	Ref.
		δ{PCl[NP(NPR$_3$)$_2$Ph]}, +122.6 ± 1.0 ppm	
N$_4$P$_4$(NMePPh$_2$)(NHMe)$_3$Ph$_4$ (non-*gem*: 1-NRPR$_2$ 1,3,5,7-Ph 3,5,7-NHMe: *β-trans*)	220–222	—	86
Cyclolinear Species			
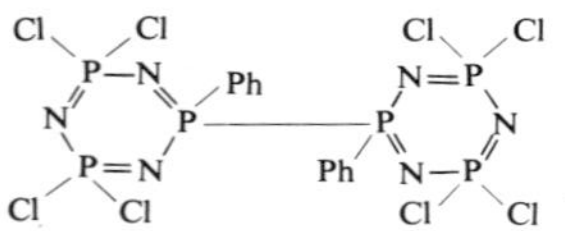	250	IR (P═N), 1215 cm^{-1}	24
(N$_3$P$_3$Cl$_5$)-*m*-OPhO-(N$_3$P$_3$Cl$_5$)	71–72	IR (P═N), 1225 cm^{-1}	236
(N$_3$P$_3$Cl$_5$)-*p*-OPhO-(N$_3$P$_3$Cl$_5$)	114–116	IR (P═N), 1225 cm^{-1}	236
(N$_3$P$_3$Cl$_5$)-*p*-OC$_6$H$_4$CMe$_2$C$_6$H$_4$O-(N$_3$P$_3$Cl$_5$)	102–103	IR (P═N), 1225 cm^{-1}	236
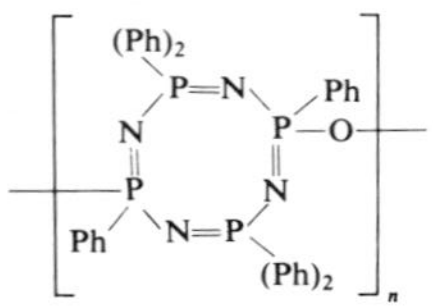		Softens at 220°–300° or 100°–120°C according to the method of preparation	83
Coordination Complexes			
(NPF$_2$)$_3$·2SbF$_5$	183–188	IR (P═N), 1320 cm^{-1}	11
(NPF$_2$)$_4$·2SbF$_5$	143–145	IR (P═N), 1350, 1550 cm^{-1}	11
(NPF$_2$)$_5$·2SbF$_5$	80–81	IR (P═N), very broad, 1300–1700 cm^{-1}	11
(NPF$_2$)$_6$·2SbF$_5$	92–94	IR (P═N), 1330–1600 cm^{-1}	11
[N$_3$P$_3$(NMe$_2$)$_4$(NMe$_3$)$_2$]$^{2+}$ 2BF$_4^-$	207–209 (decomp.)	—	237

$(NPMe_2)_3 \cdot SnCl_4$	320	IR (P=N), 1390 cm^{-1}	238
$(NPMe_2)_3 \cdot TiCl_4$	315 (decomp.)	IR (P=N), 1380 cm^{-1}	238
$(NPCl_2)_3 \cdot AlBr_3$	174	IR (P=N), 1310, 1280, 1215 cm^{-1} (ring stretch)	239
$(NPBr_2)_3 \cdot AlBr_3$	180–182	IR (P=N), 1180, 1240, 1290 cm^{-1}	239
$(NPMe_2)_4 \cdot Mo(CO)_4$	—	IR (P=N), 1146, 1165, 1225 cm^{-1}	240
$Be(OPPh_2O)(OPPh_2(NPPh_2)_3O)_2$	—	IR (P=N), 1375, 1290, 1233, 1208, 1140 cm^{-1}	241
Hydrohalide Salts			
$[NP(NMe_2)_2]_3 \cdot HCl$	198–200	—	163
$[NP(NMe_2)_2]_3 \cdot HBr$	216–217	—	163
$[NP(NMe_2)_2]_3 \cdot HI$	220–222	—	163
$[NP(NMe_2)_2]_3 \cdot HI_3$	144–145	—	163
$N_3P_3(NMe_2)_4(NH_2)_2 \cdot 2HCl$	176–177 (with decomp.)	—	165
$N_3P_3(NMe_2)_2(NHPr^i)_4 \cdot HCl$ (*gem*)	168–169	—	127, 128
$N_3P_3(NMe_2)_2(NHPr^i)_4 \cdot HCl$ (non-*gem-cis*)	112	—	127
$N_3P_3(NMe_2)_2(NHPr^i)_4 \cdot HCl$ (non-*gem-trans*)	123–124	—	127, 128
$N_3P_3(NMe_2)_2(NHBu^t)_4 \cdot HCl$ (*gem*)	259	—	132
$N_3P_3(NMe_2)_2(NHBu^t)_4 \cdot HCl$ (non-*gem*)	240–243	—	132
$N_3P_3(NHPr^n)_6 \cdot 2HCl$	178	—	242
$[NP(NHPr^i)_2]_3 \cdot HCl$	203	—	127
$[NP(NHBu^n)_2]_3 \cdot 2HCl$	164	—	242
$[NP(NHC_5H_{11})_2]_3 \cdot 3HCl$	120	—	242
$N_3P_3(NHPr^i)_4Cl_2 \cdot HCl$ (*gem*)	220–225	—	128
$N_3P_3(NHPr^i)_4Cl_2 \cdot HCl$ (non-*gem*)	195	—	127, 128
$N_3P_3(NHPr^i)_2Cl_4 \cdot 2HCl$ (*gem*)	—	B.p., 150°C/0.05 mm	128
$N_3P_3(NC_4H_8NMe)Ph_2Cl_3 \cdot HCl$	211–213	—	120
$N_3P_3(NHMe)_4Ph_2 \cdot HCl$	195	—	120
$N_3P_3(NHEt)_4Ph_2 \cdot HCl$	217	—	120
$N_3P_3(NHBu^t)_4Ph_2 \cdot HCl$	290 (decomp.)	—	120
$N_3P_3(NHC_6H_{12})_4Ph_2 \cdot HCl$	243	—	120

Compound	M.p. (°C)	Property Data	Ref.
Cyclophosphazanes			
$(MeN\text{-}PF_3)_2$	−8.3 to −10.3	B.p., 91.6°C/757 mm; density, 1.532 g/ml; n_D^{25}, 1.3323; IR (P═N), 847, 614, 535 cm^{-1}; ^{31}P, +71.5 ppm	243–247
$(MeN\text{-}PCl_3)_2$	173–176	IR (P═N), 847 cm^{-1}; ^{31}P, +78.2 ppm	246, 248–252
$(ClCH_2N\text{-}PCl_3)_2$	140–141	IR (P═N), 821 cm^{-1}; ^{31}P, +80 ppm	253
$(EtN\text{-}PCl_3)_2$	—	^{31}P, +79.0 ppm	251
$(Pr^nN\text{-}PCl_3)_2$	—	^{31}P, +79.0 ppm	251
$(MeN\text{-}PF_2CH_2Cl)$	—	^{31}P, +56.3	243
$(MeN\text{-}PF_2Et)_2$	—	^{31}P, +45.6 ppm	243
$(MeN\text{-}PF_2Ph)_2$	162	—	243
$(PhCH_2N\text{-}PNHCH_2Ph)_2$	94	IR (P═N), 816, 855, 893, 934 cm^{-1}	254
$(MeN\text{-}P(O)Cl)_2$ (two isomers: *cis* and *trans*)	101–103	IR (P═N), 852 cm^{-1}; ^{31}P, +3.0 (+5.3), 0.0 ppm	249, 251
$(EtN\text{-}P(O)Cl)_2$ (two isomers: *cis* and *trans*)	65–68	IR (P═N), 886 cm^{-1}; ^{31}P, +5.7 ppm (1st isomer) +3.1 (2nd isomer)	251
$(Pr^nN\text{-}P(O)Cl)_2$ (*cis* and *trans* isomers)	—	B.p., 110°C/0.1 mm; IR(P═N), 844, 926 cm^{-1}; ^{31}P, (1st isomer), +2.4 ppm, (2nd isomer), +1.1 ppm	251
$[PhN\text{-}P(O)Cl]_2$	144.5–145	IR (P═N), 893 cm^{-1}	251
$[MeN\text{-}P(O)NMe_2]_2$	172–173	IR (P═N), 871 cm^{-1}; ^{31}P, −7.0 ppm	251
$[EtN\text{-}P(O)NMe_2]_2$	73–75	IR (P═N), 910 cm^{-1}	251
$[Pr^iN\text{-}P(O)NMe_2]_2$	54–55	IR (P═N), 851 cm^{-1}	251
$[Pr^iN\text{-}P(O)NEt_2]_2$	92	—	255
$[Pr^iN\text{-}P(O)NHPr^i]_2$	270	—	255
$[Bu^tN\text{-}P(O)NHBu^t]_2$	279	—	255
$[MeN\text{-}P(O)SMe]_2$	176–178	IR (P═N), 855, 847 cm^{-1}	251
$[MeN\text{-}P(S)Cl]_2$	120–122	^{31}P, −51.5 ppm	249
$[MeN\text{-}P(S)NHMe]_2$	224	^{31}P, −60 ppm	249
$[PrN\text{-}P(S)NHPr^n]$	157	—	255
$[MeN\text{-}P(S)NEt_2]_2$	169	^{31}P, −62 ppm	249

$[MeN\text{-}P(S)NHPh]_2$	146	^{31}P, −54 ppm	249
$(Et\text{-}PCl)_3$	—	B.p., 128°–130°C/0.2 mm; density, 1.371 g/ml (20°C); n_D^{20}, 1.5732	256
$(EtN\text{-}P(O)OEt)_3$	74–75	—	257
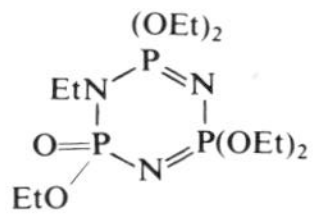	—	B.p., 90°–100°C/0.01 mm	257
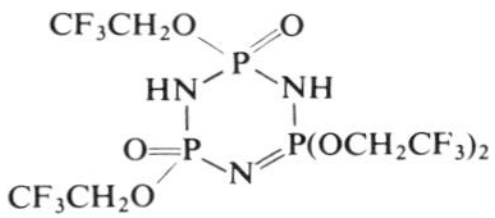	173–174	—	258
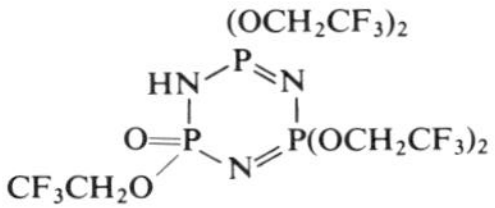	56–57	—	258
$[Pr^iN\text{-}P(O)OEt]_3$	150–152	B.p., 130°C/0.01 mm	257
$[PhCH_2N\text{-}P(O)OH]_3$	185 (decomp.)	—	191
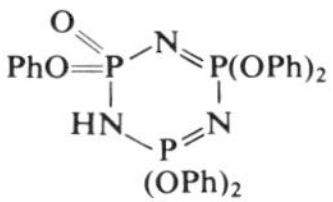	163–165	IR (P═N), 1240, 1280 cm^{-1}	191
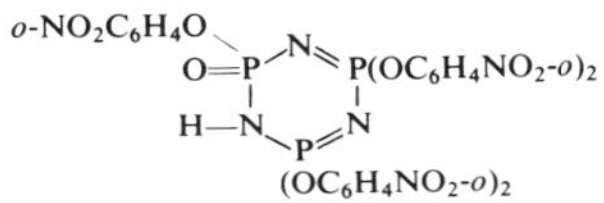	130	IR (P—N—H), 1300 cm^{-1}	258

Compound	M.p. (°C)	Property Data	Ref.
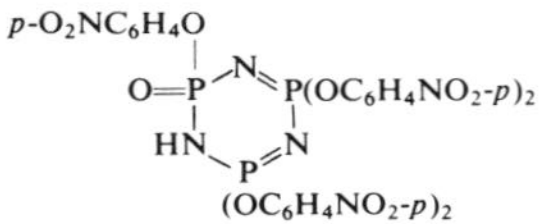	225	IR (P=N), 1210 cm^{-1}, (P—N—H), 2650 cm^{-1}	258
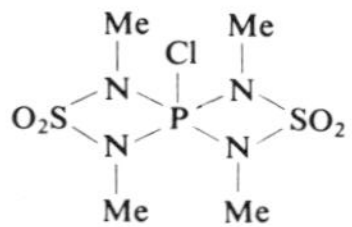	170–171.5 (decomp.)	IR (P=N), 866, 853 cm^{-1}; ^{31}P, +86.0 ± 0.5 ppm	248
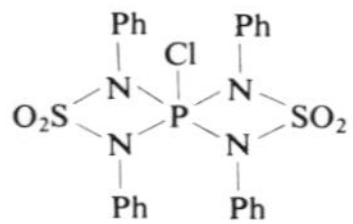	188 (decomp.)	—	248
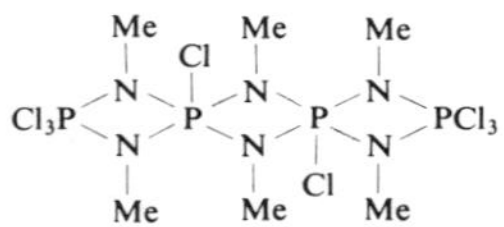	395 (decomp.)	IR (P=N), 877, 843 cm^{-1}; ^{31}P, +74.5 ppm	249
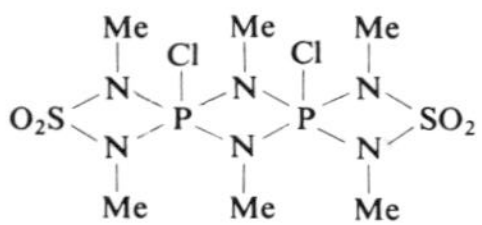	220.5–221	IR (P=N), 875, 837 cm^{-1}; ^{31}P, +81.2 ± 1 ppm	248
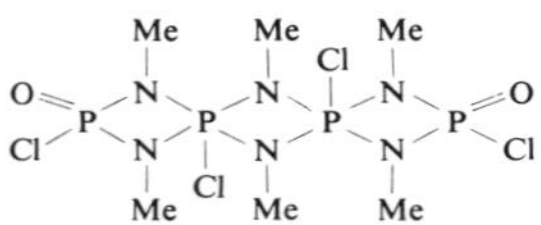	185–220 (decomp.)	IR (P=N), 872, 832 cm^{-1}; ^{31}P, +73, −4 ppm	249

Compound	M.p. (°C)	Data	Ref.
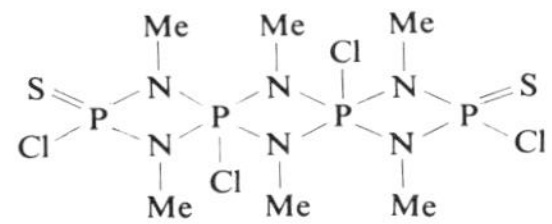	230 (decomp.)	IR (P=N), 890, 845 cm^{-1}; ^{31}P, −55.3, +68 ppm	249
Monophosphazenes			
$FSO_2N{=}PF_3$	−26.0 ± 0.5	IR (P=N), 1357 cm^{-1}	259
$Cl_3CN{=}PCl_3$	—	B.p., 56°C/10^{-2} mm; IR (P=N), 1360 (1450) cm^{-1}	260, 261
$Me_3CN{=}PCl_3$	—	IR (P=N), 1370 cm^{-1}	260
$(CF_3)_3CN{=}PCl_3$	—	B.p., 134°–135°C/760 mm; density 1.7600 g/ml (20°C); n_D^{20}, 1.3723; IR (P=N), 1500, (1640) cm^{-1}	260
$PhN{=}PCl_3$	180–182	May be dimeric	262
$CCl_3C(O)N{=}PCl_3$	—	IR (P=N), 1320, 1370 cm^{-1}	260
p-$NO_2PhC(O)N{=}PCl_3$	—	IR (P=N), 1295 cm^{-1}; ^{31}P, −18.7 ppm	263
$CF_3C(O)N{=}PCl_3$	—	IR (P=N), 1390 cm^{-1}	260
p-$MePhSO_2N{=}PCl_3$	103–104	IR (P=N), 1199 cm^{-1}	264
$FSO_2N{=}PBr_3$	57–58	B.p., 106°–107°C/0.05 mm; IR (P=N), 1190 cm^{-1}	265
$PhN{=}PPhCl_2$	119–120	—	266
$PhC(O)N{=}PMeCl_2$	63–65	IR (P=N), 1307 cm^{-1}	263
$PhN{=}PPh_2Cl$	96–97	IR (P=N), 1357 cm^{-1}	264
$PhN{=}PBu_3$	—	B.p., 89°–94°C/0.01 mm; IR (P=N), 1339 cm^{-1}	264
$HN{=}PPh_3$	128	—	267
$MeN{=}PPh_3$	92–94	IR (P=N), 1230 cm^{-1}	264
$PhN{=}PPh_3$	131–132	IR (P=N), 1344 cm^{-1}	264
$PhC(O)N{=}PPh_3$	196–197	IR (P=N), 1332 cm^{-1}	264
$PhC(O)NHN{=}PPh_3$	202–203	IR (P=N), 1333 cm^{-1}	268
$Et_2NSO_2N{=}PPh_3$	127	—	269
p-$MePhSO_2N{=}PPh_3$	193–194	IR (P=N), 1147 cm^{-1}	264
$PhN{=}PPh_2(NHPh)$	178–179	IR (P=N), 1304, 1316 cm^{-1}	264
$MePhN{=}PPhEt_2$	65–67	—	270
$PhN{=}P(OMe)_3$	—	B.p., 150°–154°C/10 mm	264

Compound	M.p. (°C)	Property Data	Ref.
$PhN{=}P(OPh)_3$	—	B.p., 222°–224°C/0.05 mm; density, 1.2085 g/ml (20°C); n_D^{20}, 1.6005	271
$PhC(O)N{=}P(NHPh)_3$	112–115	—	272
$Me_3SiN{=}PMe_3$	3–4	B.p., 169°C/760 mm; IR (P═N), 1286 cm^{-1}	273
$(Me_3Al)(Me_3Si)N{=}PMe_3$	79–80	B.p., 120°C/1 mm; IR (P═N), 1071 cm^{-1}	273
$(HN{=}PPh_3)_2 \cdot Mo(CO)_4$	127–140 (decomp.)	Red crystals	267
Linear Diphosphazenes			
$Cl_3P{=}N{-}P(O)Cl_2$	35.7	n_D^{25}, 1.5313; ^{31}P, δ(PCl_3), −1 ppm; δ($POCl_2$), +14 ppm	274–277
$(Cl_3P{=}N{-}PCl_3)^+(PCl_6)^-$	310–315 (decomp.)	^{31}P, δ(PCl_3), −21.4 ppm; δ(PCl_6), +305 ppm	277
$(Me_2ClP{=}N{-}PMe_2Cl)^+Cl^-$	157	^{31}P, −60.7 ppm	278
$Ph_3P{=}N{-}PPh_2$	124–125	IR (P═N), 1168 cm^{-1}; ^{31}P, δ(PPh_2), −39.4 ppm, δ($Ph_3P{=}N$), −15.9 ppm	279
$(S{=}PF_2N{=}PPh_3)$	147	IR (P═N), 1305.5 cm^{-1}	280
$(S{=}PF_2N{=}PF_2NH_2)$	—	B.p., 26°C/0.01 mm; IR (P═N), 1370 cm^{-1}	281
$[Me_2(NH_2)P{=}NPMe_2(NH_2)]^+Cl^-$	108–109	IR (P═N), 1241 cm^{-1}	278
$[(NH_2)_3P{=}NP(NH_2)_3]^+Cl^-$	174–176	IR (P═N), 1265, (1560) cm^{-1}; ^{31}P, −15.6 ppm	282
$[Ph_2(NH_2)P{=}NPPh_2(NH_2)]^+Cl^-$	245–246	—	68
$(MeO)_3P{=}NP(O)Et_2$	—	B.p., 85°–88°C/0.04 mm; density, 1.1711 g/ml (20°C); n_D^{20}, 1.4525; IR (P═N), 1320 cm^{-1}	283
$Me_2(MeS)P{=}NP(S)Me_2$	72–74	IR (P═N), 1208 cm^{-1}; ^{31}P, δ(PMe_2S), −45.3 ppm; δ(PMe_2SMe), −34.6 ppm	284
$Et_3P{=}NP(O)(OEt)_2$	—	^{31}P, δ[$P(O)(OEt)_2$], −1.3 ppm; δ(PEt_3), −32 ppm	285
$Ph_3P{=}NP(O)Ph_2$	152–153	IR (P═N), 1161 cm^{-1}	264
$Et_2Pr^nP{=}NP(O)Pr^n{}_2$	—	B.p., 119°–120°C/0.03 mm; density, 1.0417 (20°C); n_D^{20}, 1.4510; IR (P═N), 1300 cm^{-1}	283
$PhN{=}P(NEt_2)_2{-}N{=}P(OEt)_2NHPh$	94–95	—	286

Higher Linear Phosphazenes			
$Cl_3P{=}NPCl_2N{=}POCl_2$	32–33	^{31}P, $\delta(PCl_3)$, −7.6 ppm, $\delta(NPCl_2)$, +19.5 ppm, $\delta(NPOCl_2)$, +12.9 ppm	276, 277, 287
$Cl_3P{=}(NPCl_2)_2\text{-}N{=}POCl_2$	—	^{31}P, $\delta(\alpha$ or $PCl_3)$, −10.5 ppm, $\delta(\beta)$, +14.5 ppm, $\delta(\gamma)$, +19.5, $\delta(\delta$ or $POCl_2)$, +11.0	276
$[Cl(PCl_2{=}N)_2PCl_3]^+(PCl_6)^-$	—	^{31}P, $\delta(PCl_3)$, −12.7, $\delta(PCl_2)$, +13.8 ppm, $\delta(PCl_6)$, +305 ppm	277, 287, 288
$[Cl(PCl_2{=}N)_3PCl_3]^+(PCl_6)^-$	95–96	^{31}P, $\delta(PCl_3)$, −11.3 ppm, $\delta(PCl_2)$, +13.5 ppm, $\delta(PCl_6)$, +300 ppm	288
$[Cl(PCl_2{=}N)_4PCl_3]^+(PCl_6)^-$	98–100	^{31}P, $\delta(PCl_3)$, −11.5 ppm, $\delta(PCl_2)$, +13.4 ppm, $\delta(PCl_6)$, +297 ppm	23, 250, 288
$[Cl(PCl_2{=}N)_nPCl_3]^+Cl^-$ ($n \approx 10$)	—	^{31}P, +18, +15.3, −10 ppm	23, 288
$Ph_2(O)P{-}N{=}PPh_2\text{-}NHP(O)Ph_2$	145–146	IR (P═N), 1316, 1227, 1176 cm^{-1}	289
$Ph(Ph_2P{=}N)_4\text{-}P(O)Ph_2$	157–160	IR (P═N), 1272, 1258, 1227 cm^{-1}	289
$HO(Ph_2P{=}N)_2P(O)Ph_2$	148–149	—	290
$Ph_2(O)P{-}NH{-}(PPh_2{=}N)_2P(O)Ph_2$	171	—	68
$Cl[Ph(NH_2)P{=}N]_5Ph(NH_2){=}NH$	80–120	IR (P═N), 1205, 1280, 1440 cm^{-1}	59

References

1. R. Rätz and C. Grundmann, *J. Inorg. Nucl. Chem.* **16**, 60 (1960).
2. A. C. Chapman, N. L. Paddock, D. H. Paine, H. T. Searle, and D. R. Smith, *J. Chem. Soc., London* p. 3608 (1960).
3. F. Seel and J. Langer, *Z. Anorg. Allg. Chem.* **295**, 316 (1958).
4. R. A. Chittenden and L. C. Thomas, *Spectrochim. Acta* **22**, 1449 (1966).
5. C. W. Allen, F. Y. Tsang, and T. Moeller, *Inorg. Chem.* **7**, 2183 (1968).
6. L. F. Audrieth, R. Steinman, and A. D. F. Toy, *Chem. Rev.* **32**, 109 (1943).
7. O. Schmitz-DuMont and M. Walther, *Z. Anorg. Allg. Chem.* **298**, 193 (1959).
8. H. R. Allcock, R. L. Kugel, and E. G. Stroh, unpublished work (1970).
9. J. Emsley and N. L. Paddock, *J. Chem. Soc., A* p. 2590 (1968).
10. J. Emsley, *J. Chem. Soc., A* p. 109 (1970).
11. T. Chivers and N. L. Paddock, *J. Chem. Soc., A* p. 1687 (1969).
12. O. Glemser, E. Niecke, and H. W. Roesky, *Chem. Commun.* p. 282 (1969).
13. G. Allen, M. Barnard, J. Emsley, N. L. Paddock, and R. F. M. White, *Chem. Ind. (London)* p. 952 (1963).
14. A. C. Chapman, D. H. Paine, H. T. Searle, D. R. Smith, and R. F. M. White, *J. Chem. Soc., London* p. 1768 (1961).
15. B. Green and D. B. Sowerby, *Chem. Commun.* p. 628 (1969).
16. M. L. Heffernan and R. F. M. White, *Proc. Chem. Soc., London* p. 9 (1961).
17. M. L. Heffernan and R. F. M. White, *J. Chem. Soc., London* p. 1382 (1961).
18. F. Heatley and S. M. Todd, *J. Chem. Soc., A* p. 1152 (1966).
19. O. Schmitz-DuMont and H. Külkens, *Z. Anorg. Allg. Chem.* **238**, 189 (1938).
20. O. Schmitz-DuMont, *Angew. Chem.* **50**, 415 (1937).
21. O. Schmitz-DuMont and A. Braschos, *Z. Anorg. Allg. Chem.* **243**, 113 (1939).
22. E. Steger and D. Klemm, *J. Inorg. Nucl. Chem.* **29**, 1812 (1967).
23. L. G. Lund, N. L. Paddock, J. E. Proctor, and H. T. Searle, *J. Chem. Soc., London* p. 2542 (1960).
24. M. Biddlestone and R. A. Shaw, *Chem. Commun.* p. 407 (1968).
25. N. Boden, J. W. Emsley, J. Feeney, and L. H. Sutcliffe, *Chem. Ind. (London)* p. 1909 (1962).
26. H. R. Allcock, R. L. Kugel, and K. J. Valan, *Inorg. Chem.* **5**, 1709 (1966).
27. R. Keat and R. A. Shaw, *J. Chem. Soc., London* p. 2215 (1965).
28. E. T. McBee, H. R. Allcock, R. Caputo, A. Kalmus, and C. W. Roberts, *U.S. Govt. Res. Rep.* AD 209,666 and AD 209,669 (1959).
29. K. John, *WADC Tech. Rep.* **58–51**, Part III, 53 (1960).
30. F. Rallo, *Chimica* **8**, 1134 (1965).
31. A. C. Chapman and N. L. Paddock, *J. Chem. Soc., London* p. 635 (1962).
32. M. Becke-Goehring and W. Lehr, *Z. Anorg. Allg. Chem.* **327**, 128 (1964).
33. S. M. Zhivukhin, V. B. Tolstoguzov, V. V. Kireev, and K. G. Kuznetsova, *Zh. Neorg. Khim.* **10**, 333 (1965); *Chem. Abstr.* **62**, 11407a (1965).
34. T. Manley and D. A. Williams, *Polymer* **10**, 307 (1969).
35. R. G. Rice, L. W. Daasch, J. R. Holden, and E. J. Kohn, *J. Inorg. Nucl. Chem.* **5**, 190 (1958).

36. B. Lakatos, A. Hesz, Zs. Vetessy, and G. Horvath, *Acta Chim. Acad. Sci. Hung.* **60**, 309 (1969).
37. P. Pulay, B. Lakatos, G. Toth, A. Hesz, and Zs. Vetessy, *Acta Chim. Acad. Sci. Hung.* **60**, 333 (1969).
38. G. Engelhardt, E. Steger, and R. Stahlberg, *Z. Naturforsch. B* **21**, 1231 (1966).
39. G. E. Coxon and D. B. Sowerby, *Inorg. Chim. Acta* **1**, 381 (1967).
40. G. E. Coxon, T. F. Palmer, and D. B. Sowerby, *Inorg. Nucl. Chem. Lett.* **2**, 215 (1966).
41. N. E. Bean and R. A. Shaw, *Chem. Ind. (London)* p. 1189 (1960).
42. A. Besson, *C. R. Acad. Sci.* **114**, 1479 (1892).
43. H. Bode, *Z. Anorg. Allg. Chem.* **252**, 113 (1943).
44. K. John and T. Moeller, *J. Amer. Chem. Soc.* **82**, 2647 (1960).
45. G. Engelhardt, E. Steger, and R. Stahlberg, *Z. Naturforsch. B* **21**, 586 (1966).
46. H. R. Allcock and R. L. Kugel, *J. Amer. Chem. Soc.* **91**, 5452 (1969).
47. G. E. Coxon, D. B. Sowerby, and G. C. Tranter, *J. Chem. Soc., London* p. 5697 (1965).
48. K. John and T. Moeller, *J. Inorg. Nucl. Chem.* **22**, 199 (1961).
49. G. E. Coxon and D. B. Sowerby, *Spectrochim. Acta, Part A* **24**, 2145 (1968).
50. G. E. Coxon and D. B. Sowerby, *J. Chem. Soc., A* p. 1566 (1967).
51. H. P. Latscha, *Z. Anorg. Allg. Chem.* **362**, 7 (1968).
52. L. F. Audrieth and R. J. A. Otto, *J. Amer. Chem. Soc.* **80**, 5894 (1958).
53. R. J. A. Otto, G. Tesi, F. G. Sherif, and L. F. Audrieth, *J. Amer. Chem. Soc.*, **82**, 528 (1960).
54. G. Tesi, A. F. Vandi, and K. John, *WADC Tech. Rep.* **58–51**, Part III, 46 (1960).
55. C. Grundmann and R. Rätz, *Z. Naturforsch. B* **10**, 116 (1955).
56. H. T. Searle, *Proc. Chem. Soc., London* p. 7 (1959).
57. H. H. Sisler, S. E. Frazier, R. G. Rice, and M. G. Sanchez, *Inorg. Chem.* **5**, 326 (1966).
58. G. Tesi and P. J. Slota, *Proc. Chem. Soc., London* p. 404 (1960).
59. S. E. Frazier and H. H. Sisler, *Inorg. Chem.* **5**, 925 (1966).
60. A. J. Bilbo, *Z. Naturforsch. B* **15**, 330 (1960).
61. T. Moeller, A. Failli, and F. Y. Tsang, *Inorg. Nucl. Chem. Lett.* **1**, 49 (1965).
62. G. Tesi and C. M. Douglas, *J. Amer. Chem. Soc.* **84**, 549 (1962).
63. G. Tesi and C. M. Douglas, U.S. Pat. 3,065,266 (1962).
64. G. Tesi, C. P. Haber, and C. M. Douglas, *Proc. Chem. Soc., London* p. 219 (1960).
65. G. Tesi, C. M. Douglas, and C. P. Haber, U.S. Pat. 3,087,937 (1963).
66. R. Ramain, Y. Runavot, and P. Schneebeli, *J. Chim. Phys.* **56**, 659 (1959).
67. R. Schmutzler, *Z. Naturforsch. B* **19**, 1101 (1964).
68. D. L. Herring and C. M. Douglas, *Inorg. Chem.* **3**, 428 (1964).
69. H. Bode and R. Thamer, *Chem. Ber.* **76B**, 121 (1943).
70. I. I. Bezman, U,S. Pat. 3,098,871 (1963).
71. H. R. Allcock and W. J. Birdsall, *Inorg. Chem.* **10**, 2495 (1971).
72. T. Moeller and F. Y. Tsang, *Chem. Commun.* p. 361 (1962).
73. C. W. Allen and T. Moeller, *Inorg. Chem.* **7**, 2177 (1968).
74. A. J. Bilbo, C. M. Grieve, D. L. Herring, and D. E. Salzbrunn, *Inorg. Chem.* **7**, 2670 (1968).
75. B. Grushkin, M. G. Sanchez, M. V. Ernest, J. L. McClanahan, G. E. Ashby, and R. G. Rice, *Inorg. Chem.* **4**, 1538 (1965).
76. K. G. Acock, R. A. Shaw, and F. B. G. Wells, *J. Chem. Soc., London* p. 121 (1964).
77. C. D. Schmulbach and C. Derderian, *J. Inorg. Nucl. Chem.* **25**, 1395 (1963).
78. V. B. Desai, R. A. Shaw, and B. C. Smith, *J. Chem. Soc., A* p. 1977 (1969).
79. B. Grushkin, M. G. Sanchez, and R. G. Rice, *Inorg. Chem.* **3**, 623 (1964).
80. C. T. Ford, F. E. Dickson, and I. I. Bezman, *Inorg. Chem.* **4**, 890 (1965).

81. C. T. Ford, F. E. Dickson, and I. I. Bezman, *Inorg. Chem.* **3**, 177 (1964).
82. M. Becke-Goehring and B. Boppel, *Z. Anorg. Allg. Chem.* **322**, 239 (1963).
83. D. L. Herring and C. M. Douglas, *Inorg. Chem.* **4**, 1012 (1965).
84. R. A. Shaw and M. Biddlestone, *Chem. Commun.* p. 205 (1965).
85. R. A. Shaw and C. Stratton, *J. Chem. Soc., London* p. 5004 (1962).
86. B. Grushkin, A. J. Berlin, J. L. McClanahan, and R. G. Rice, *Inorg. Chem.* **5**, 172 (1966).
87. R. A. Shaw and C. Stratton, *Chem. Ind. (London)* p. 52 (1959).
88. R. H. Kratzer and K. L. Paciorek, *Inorg. Chem.* **4**, 1767 (1965).
89. P. Nannelli and T. Moeller, *Inorg. Chem.* **2**, 896 (1963).
90. B. S. Manhas, S. K. Chu, and T. Moeller, *J. Inorg. Nucl. Chem.* **30**, 322 (1968).
91. T. Moeller and P. Nannelli, *Inorg. Chem.* **1**, 721 (1962).
92. C. M. Sharts, A. J. Bilbo, and D. R. Gentry, *Inorg. Chem.* **5**, 2140 (1966).
93. A. J. Bilbo and C. M. Sharts, *J. Polym. Sci., Part A*-1 **5**, 2891 (1967).
94. A. Schmidpeter and J. Ebeling, *Angew. Chem., Int. Ed. Engl.* **7**, 209 (1968).
95. H. Bode and H. Bach, *Chem. Ber.* **75B**, 215 (1942).
95a. D. D. Magnelli, G. Tesi, J. U. Lowe, and W. E. McQuiston, *Inorg. Chem.* **5**, 457 (1966).
96. L. F. Audrieth and D. B. Sowerby, *Chem. Ind. (London)* p. 748 (1959).
97. D. B. Sowerby and L. F. Audrieth, *Chem. Ber.* **94**, 2670 (1961).
98. G. R. Feistel and T. Moeller, *J. Inorg. Nucl. Chem.* **29**, 2731 (1967).
99. W. Lehr, *Z. Anorg. Allg. Chem.* **350**, 18 (1967).
100. A. M. de Ficquelmont, *Ann. Chim. (Paris)* [11] **12**, 169 (1939).
101. M. Becke-Goehring and K. John, *Z. Anorg. Allg. Chem.*, **304**, 126 (1960).
102. M. Becke-Goehring and K. John, *Angew. Chem.* **70**, 657 (1958).
103. H. Bode, K. Bütow, and G. Lienau, *Chem. Ber.* **81**, 547 (1948).
104. H. Bode, *Angew. Chem.* **60**, 67 (1948).
105. D. Feakins, S. N. Nabi, R. A. Shaw, and P. Watson, *J. Chem. Soc., A* p. 10 (1968).
106. M. S. Chang and A. J. Matuszko, *J. Amer. Chem. Soc.* **82**, 5756 (1960).
107. M. Becke-Goehring, K. John, and E. Fluck, *Z. Anorg. Allg. Chem.* **302**, 103 (1959).
108. S. K. Ray and R. A. Shaw, *J. Chem. Soc., London* p. 872 (1961).
109. S. G. Kokalis, K. John, T. Moeller, and L. F. Audrieth, *J. Inorg. Nucl. Chem.* **19**, 191 (1961).
110. K. John, T. Moeller, and L. F. Audrieth, *J. Amer. Chem. Soc.* **83**, 2608 (1961).
111. S. K. Ray, R. A. Shaw, and B. C. Smith, *J. Chem. Soc., London* p. 3236 (1963).
112. W. Lehr, *Z. Anorg. Allg. Chem.* **352**, 27 (1967).
113. J. M. E. Goldschmidt and J. Weiss, *J. Inorg. Nucl. Chem.* **26**, 2023 (1964).
114. T. Moeller and S. Lanoux, *Inorg. Chem.* **2**, 1061 (1963).
115. K. John, T. Moeller, and L. F. Audrieth, *J. Amer. Chem. Soc.*, **82**, 5616 (1960).
116. G. Mattogno and A. Monaci, *Ric. Sci. Parte* 1 **8**, 1139 (1965).
117. R. A. Shaw, *Chem. Ind. (London)* p. 54 (1959).
118. S. K. Ray, R. A. Shaw, and B. C. Smith, *J. Chem. Soc., London* p. 3236 (1963).
119. K. John, *WADC Tech. Rep.* **58–51**, Part III, 53 (1960).
120. K. Hills and R. A. Shaw, *J. Chem. Soc., London* p. 130 (1964).
121. S. K. Ray and R. A. Shaw, *Chem. Ind. (London)* pp. 53 and 54 (1959).
122. H. R. Allcock and R. L. Kugel, *Inorg. Chem.* **5**, 1716 (1966).
123. D. Feakins, W. A. Last, S. N. Nabi, and R. A. Shaw, *J. Chem. Soc., A* p. 1831 (1966).
124. T. Moeller and S. Lanoux, *J. Inorg. Nucl. Chem.* **25**, 229 (1963).
125. D. Feakins, R. A. Shaw, P. Watson, and S. N. Nabi, *J. Chem. Soc., A* p. 2468 (1969).
126. S. G. Kokalis and T. Moeller, *WADC Tech. Rep.* **58–51**, Part III, 59 (1960).
127. S. K. Das, R. Keat, R. A. Shaw, and B. C. Smith, *J. Chem. Soc., A* p. 1677 (1966).

128. S. K. Das, D. Feakins, W. A. Last, S. N. Nabi, S. K. Ray, R. A. Shaw, and B. C. Smith, *J. Chem. Soc., A* p. 616 (1970).
129. H. R. Allcock and W. M. Thomas, U.S. Pat. 3,329,663 (1967).
130. H. R. Allcock, P. S. Forgione, and K. J. Valan, *J. Org. Chem.* **30**, 947 (1965).
131. E. T. McBee, P. Johncock, S. French, and H. P. Braendlin, *U.S. Govt. Res. Rep.* AD 254,985 (1961).
132. S. K. Das, R. Keat, R. A. Shaw, and B. C. Smith, *J. Chem. Soc., London* p. 5032 (1965).
133. R. Keat and R. A. Shaw, *Angew. Chem., Int. Ed. Engl.* **7**, 212 (1968).
134. R. Rätz, E. Kober, C. Grundmann, and G. Ottmann, *Inorg. Chem.* **3**, 757 (1964).
135. W. Bull, Ph.D. Dissertation, University of Illinois (1957).
136. A. W. Hofmann, *Chem. Ber.* **17**, 1905 (1884).
137. R. Schenck, *Chem. Ber.* **60A**, 160 (1927).
138. R. Schenck and G. Römer, *Chem. Ber.* **57B**, 1343 (1924).
139. H. Bode and H. Clausen, *Z. Anorg. Allg. Chem.* **258**, 99 (1949).
140. H. Lederle, G. Ottmann, and E. Kober, *Inorg. Chem.* **5**, 1818 (1966).
141. W. Couldridge, *J. Chem. Soc., London* **53**, 398 (1888).
142. W. Couldridge, *Bull. Soc. Chim. Fr.* **50**, 534 (1888).
143. Y. Kobayashi, L. A. Chasin, and L. B. Clapp, *Inorg. Chem.* **2**, 212 (1963).
144. H. R. Allcock and R. L. Kugel, *Inorg. Chem.* **5**, 1016 (1966).
145. A. A. Kropacheva, N. M. Kashnikova, and V. A. Parshina, *Zh. Obshch. Khim.* **34**, 530 (1964); *J. Gen. Chem. USSR* **34**, 532 (1964).
146. A. A. Kropacheva, L. E. Mukhina, N. M. Kashikova, and V. A. Parshina, *Zh. Obshch. Khim.* **31**, 1036 (1961); *J. Gen. Chem. USSR* **31**, 957 (1961).
147. A. A. Kropacheva and N. M. Kashnikova, *Zh. Obshch. Khim.* **35**, 2229 (1965).
148. R. Keat, S. K. Ray, and R. A. Shaw, *J. Chem. Soc., London* p. 7193 (1965).
149. H. McD. McGeachin and F. R. Tromans, *Chem. Ind. (London)* p. 1131 (1960).
150. V. B. Desai, R. A. Shaw, B. C. Smith, and D. Taylor, *Chem. Ind. (London)* p. 1177 (1969)
151. G. Allen, D. J. Oldfield, N. L. Paddock, F. Rallo, J. Serregi, and S. M. Todd, *Chem. Ind. (London)* p. 1032 (1965).
152. B. Green and D. B. Sowerby, *J. Chem. Soc., A* p. 987 (1970).
153. B. Green and D. B. Sowerby, *Inorg. Nucl. Chem. Lett.* **5**, 989 (1969).
154. R. Keat, R. A. Shaw, and C. Stratton, *J. Chem. Soc., London* p. 2223 (1965).
155. R. Stahlberg and E. Steger, *Spectrochim. Acta, Part A* **23**, 2005 (1967).
156. R. Keat and R. A. Shaw, *Chem. Ind. (London)* p. 1232 (1964).
157. R. Keat and R. A. Shaw, *J. Chem. Soc., London* p. 4067 (1965).
158. H. Koopman, F. J. Spruit, F. van Deursen, and J. Bakker, *Rec. Trav. Chim. Pays-Bas* **84**, 341 (1965).
159. A. Schmidpeter, C. Weingand, and E. Hafner-Roll, *Z. Naturforsch. B* **24**, 799 (1969).
160. C. T. Ford, F. E. Dickson, and I. I. Bezman, *Inorg. Chem.* **6**, 1594 (1967).
161. W. Lehr, *Naturwissenschaften* **56**, 214 (1969).
162. R. Stahlberg and E. Steger, *J. Inorg. Nucl. Chem.* **29**, 961 (1967).
163. S. N. Nabi, R. A. Shaw, and C. Stratton, *Chem. Ind. (London)* p, 166 (1969).
164. R. Stahlberg and E. Steger, *J. Inorg. Nucl. Chem.* **28**, 684 (1966).
165. N. I. Shvetsov, K. A. Nuridzhanyan, A. Ya. Yakubovich, and F. F. Sukhov, *Zh. Obshch. Khim.* **33**, 3936 (1963); *J. Gen. Chem. USSR* **33**, 3874 (1963).
166. I. I. Bezman and C. T. Ford, *Chem. Ind. (London)* p. 163 (1963).
167. B. Grushkin, J. L. McClanahan, and R. G. Rice, *J. Amer. Chem. Soc.* **86**, 4204 (1964).
168. J. H. Smalley, F. E. Dickson, and I. I. Bezman, *Inorg. Chem.* **3**, 1780 (1964).
169. R. Keat and R. A. Shaw, *J. Chem. Soc., A* p. 908 (1966).
170. V. B. Desai, R. A. Shaw, and B. C. Smith, *Angew. Chem., Int. Ed. Engl.* **7**, 887 (1968).

171. M. Becke-Goehring and K. John, *Z. Anorg. Allg. Chem.* **304**, 126 (1960).
172. A. A. Kropacheva and L. E. Mukhina, *Zh. Obshch. Khim.* **31**, 2437 (1961).
173. E. T. McBee, L. Brinkmann, and H. P. Braendlin, *U.S. Govt. Res. Rep.* AD 254,983 (1961).
174. G. F. Ottmann, R. P. Alexander, and E. Kober, U.S. Pat. 3,157,638 (1964).
175. G. Ottmann, H. Agahigian, H. Hooks, G. D. Vickers, E. Kober, and R. Rätz, *Inorg. Chem.* **3**, 753 (1964).
176. A. A. Kropacheva and N. M. Kashnikova, *Zh. Obshch. Khim.* **35**, 1988 (1965).
177. A. A. Kropacheva and N. M. Kashnikova, *Zh. Obshch. Khim.* **32**, 652 (1962).
178. A. A. Kropacheva and N. M. Kashnikova, *Zh. Obshch. Khim.* **33**, 1046 (1963).
179. A. A. Kropacheva and N. M. Kashnikova, *Zh. Obshch. Khim.* **38**, 136 (1968).
180. B. Capon, K. Hills, and R. A. Shaw, *J. Chem. Soc., London* p. 4059 (1965).
181. R. A. Shaw and F. B. G. Wells, private communication (1971).
182. A. J. Berlin, B. Grushkin, and L. R. Moffett, *Inorg. Chem.* **7**, 589 (1968).
183. A. A. Kropacheva and L. E. Mukhina, *Zh. Obshch. Khim.* **32**, 521 (1962).
184. A. A. Kropacheva and L. E. Mukhina, *Zh. Obshch. Khim.* **33**, 706 (1963).
185. L. E. Mukhina and A. A. Kropacheva, *Zh. Obshch. Khim.* **38**, 313 (1966).
186. R. J. A. Otto and L. F. Audrieth, *J. Amer. Chem. Soc.* **80**, 3575 (1958).
187. B. Dishon, *J. Amer. Chem. Soc.* **71**, 2251 (1949).
188. B. W. Fitzsimmons and R. A. Shaw, *Chem. Ind.* (*London*) p. 109 (1961).
189. B. W. Fitzsimmons and R. A. Shaw, *J. Chem. Soc., London* p. 1735 (1964).
190. M. Yokoyama, *Nippon Kagaku Zasshi* **80**, 1192 (1959); *Chem. Abstr.* **55**, 4341 (1961).
191. B. W. Fitzsimmons, C. Hewlett, K. Hills, and R. A. Shaw, *J. Chem. Soc., A* p. 679 (1967)
192. M. R. Pitina, T. M. Ivanova, and N. I. Shvetsov-Shilov, *Zh. Obshch. Khim.* **37**, 2076 (1967); *J. Gen. Chem. USSR* **37**, 1968 (1967).
193. B. W. Fitzsimmons and R. A. Shaw, *Proc. Chem. Soc., London* p. 258 (1961).
194. B. W. Fitzsimmons and R. A. Shaw, *Inorg. Syn.* **8**, 77 (1966).
195. R. Rätz and M. Hess, *Chem. Ber.* **84**, 889 (1951).
196. H. R. Allcock and R. L. Kugel, *J. Amer. Chem. Soc.* **87**, 4216 (1965).
197. V. I. Spitsyn, N. A. Afans'eva, A. K. Pikaev, I. D. Kolli, and P. Ya. Glazunov, *Dokl. Akad. Nauk SSSR* **131**, 1106 (1960); *Chem. Abstr.* **54**, 22128 (1960).
198. M. F. Sorokin and V. K. Latov, *Zh. Obshch. Khim.* **35**, 1471 (1965).
199. A. J. Matuszko and M. S. Chang, *J. Org. Chem.* **31**, 2004 (1966).
200. R. Pornin, *Bull. Soc. Chim. Fr.* [9] p. 2861 (1966).
201. M. S. Chang and A. J. Matuszko, *Chem. Ind.* (*London*) p. 410 (1962).
202. R. Rätz and C. Grundmann, U.S. Pat. 2,876,247 (1959).
203. R. Rätz, H. Schroeder, H. Ulrich, E. Kober, and C. Grundmann, *J. Amer. Chem. Soc.* **84**, 551 (1962).
204. E. T. McBee and H. R. Allcock, unpublished work (1956).
205. E. T. McBee, L. Brinkmann, and H. P. Braendlin, *U.S. Govt. Res. Rep.* AD 254,982 (1961).
206. M. V. Lenton and B. Lewis, *J. Chem. Soc., A* p. 665 (1966).
207. G. Ottmann, H. Lederle, and E. Kober, *Ind. Eng. Chem., Prod. Res. Develop.* **5**, 202 (1966).
208. E. Kober, H. Lederle, and G. Ottmann, *Amer. Soc. Lub. Eng. Trans.* **7**, 389 (1964).
209. S. H. Rose, *J. Polym. Sci., Part B* **6**, 837 (1968).
210. T. J. Mao, R. D. Dresdner, and J. A. Young, *J. Inorg, Nucl. Chem.* **24**, 53 (1962).
211. R. Rätz and C. Grundmann, U.S. Pat. 2,876,248 (1959).
212. H. R. Allcock, *J. Amer. Chem. Soc.* **86**, 2591 (1964).
213. E. T. McBee, K. Okuhara, and C. J. Morton, *Inorg. Chem.* **5**, 450 (1966).

214. E. T. McBee, P. Johncock, and H. P. Braendlin, *U.S. Govt. Res. Rep.* AD 254,984 (1961).
215. G. M. Nichols, *ASD Tech. Rep.* **61–2**, Contract No. AF 33(616)7158 (1961).
216. D. Dell, B. W. Fitzsimmons, and R. A. Shaw, *J. Chem. Soc., London* p. 4070 (1965).
217. D. Dell, B. W. Fitzsimmons, R. Keat, and R. A. Shaw, *J. Chem. Soc., A* p. 1680 (1966).
218. V. B. Tolstoguzov, V. V. Pisarenko, and V. V. Kireev, *Russ. J. Inorg. Chem.* **10**, 382 (1965).
219. C. T. Ford, J. M. Barr, F. E. Dickson, and I. I. Bezman, *Inorg. Chem.* **5**, 351 (1966).
220. C. T. Ford, F. E. Dickson, and I. I. Bezman, *Inorg. Chem.* **4**, 419 (1965).
221. J. Parrod and R. Pornin, *C. R. Acad. Sci.* **7**, 3022 (1964).
221a. H. R. Allcock and E. J. Walsh, *Inorg. Chem.* **10**, 1643 (1971).
222. G. M. Nichols, *ASD Tech. Rept.* **61–2**, Part II, Contract No. AF 33(616)7158 (1962).
223. E. Kober, H. Lederle, and G. Ottmann, *Inorg. Chem.* **5**, 2239 (1966).
224. E. Kober, H. Lederle, and G. Ottmann, Tech. Rep. Contract NObs-90092 (1964).
224a. H. R. Allcock and E. J. Walsh, unpublished work (1969).
225. H. Lederle, E. Kober, and G. Ottmann, *J. Chem. Eng. Data* **11**, 221 (1966).
226. G. M. Nichols, Ger. Pat. **1**,182,660 (1964).
227. G. M. Nichols, U.S. Pat. 3,316,330 (1967).
228. E. Kober, H. Lederle, and G. Ottmann, *U.S. Govt. Res. Rep.* AD 430,861 (1964).
229. A. Schmidpeter and C. Weingand, *Z. Naturforsch. B* **24**, 177 (1969).
230. A. P. Carroll and R. A. Shaw, *Inorg. Syn.* **8**, 84 (1966).
231. A. P. Carroll and R. A. Shaw, *J. Chem. Soc., A* p. 914 (1966).
232. A. P. Carroll and R. A. Shaw, *Chem. Ind.* (*London*) p. 1908 (1962).
233. B. Yanik, V. Zheshutko, and T. Pel'char, *J. Gen. Chem. USSR* **36**, 1451 (1966).
234. M. Feldt and T. Moeller, *J. Inorg. Nucl. Chem.* **30**, 2351 (1968).
235. M. Biddlestone and R. A. Shaw, *Chem. Commun.* p. 320 (1969).
236. S. M. Zhivukhin and V. V. Kireev, *Zh. Obshch. Khim.* **34**, 3126 (1964).
237. J. N. Rapko and G. R. Feistel, *Chem. Commun.* p. 474 (1968).
238. M. F. Lappert and G. Srivastava, *J. Chem. Soc., A* p. 210 (1966).
239. G. E. Coxon and D. B. Sowerby, *J. Chem. Soc., A* p. 3012 (1969).
240. J. Dyson and N. L. Paddock, *Chem. Commun.* p. 191 (1966).
241. K. L. Paciorek and R. H. Kratzer, *Inorg. Chem.* **5**, 538 (1966).
242. K. Denny and S. Lanoux, *J. Inorg. Nucl. Chem.* **31**, 1531 (1969).
243. R. Schmutzler, *Chem. Commun.* p. 19 (1965).
244. G. C. Demitras, R. A. Kent, and A. G. MacDiarmid, *Chem. Ind.* (*London*) p. 1712 (1964).
245. A. J. Downs, *Chem. Commun.* p. 628 (1967).
246. M. P. Yagupsky, *Inorg. Chem.* **6**, 1770 (1967).
247. M. Becke-Goehring and H. Weber, *Z. Anorg. Allg. Chem.* **365**, 185 (1969).
248. M. Becke-Goehring and H. J. Wald, *Z. Anorg. Allg. Chem.* **371**, 88 (1969).
249. M. Becke-Goehring, L. Leichner, and B. Scharf, *Z. Anorg. Allg. Chem.* **343**, 154 (1966).
250. E. Fluck, *Z. Anorg. Allg. Chem.* **320**, 64 (1963).
251. M. Green, R. N. Haszeldine, and G. S. A. Hopkins, *J. Chem. Soc., A* p. 1766 (1966).
252. H.-G. Horn and M. Becke-Goehring, *Z. Anorg. Allg. Chem.* **367**, 165 (1969).
253. T. Moeller and A. H. Westlake, *J. Inorg. Nucl. Chem.* **29**, 957 (1967).
254. M. G. Barlow, M. Green, R. N. Haszeldine, and H. G. Higson, *J. Chem. Soc., C* p. 1592 (1966).
255. H. Bock and W. Wiegräbe, *Chem. Ber.* **99**, 377 (1966).
256. E. W. Abel and G. Willey, *Proc. Chem. Soc., London* p. 308 (1962).

257. B. W. Fitzsimmons, C. Hewlett, and R. A. Shaw, *J. Chem. Soc., London* p. 7432 (1965).
258. H. R. Allcock and E. J. Walsh, unpublished work (1967).
259. M. Lustig, *Inorg. Chem.* **8**, 443 (1969).
260. E. S. Kozlov, A. A. Kisilenko, A. I. Sedlov, and A. V. Kirsanov, *Zh. Obshch. Khim.* **37**, 1611 (1967).
261. E. Fluck and F. L. Goldmann, *Z. Anorg. Allg. Chem.* **356**, 307 (1968).
262. I. N. Zhmurova and A. V. Kirsanov, *Zh. Obshch. Khim.* **32**, 2576 (1962); *J. Gen. Chem. USSR* **32**, 2540 (1962).
263. V. A. Shokol, A. A. Kisilenko, and G. I. Derkach, *Zh. Obshch. Khim.* **39**, 874 (1969); *J. Gen. Chem. USSR* **39**, 839 (1969).
264. W. Wiegräbe and H. Bock, *Chem. Ber.* **101**, 1414 (1968).
265. H. W. Roesky, *Z. Anorg. Allg. Chem.* **367**, 151 (1969).
266. I. N. Zhmurova and A. V. Kirsanov, *Zh. Obshch. Khim.* **31**, 3685 (1961); *J. Gen. Chem. USSR* **31**, 3440 (1961).
267. H. Bock and H. Dieck, *Z. Naturforsch. B* **21**, 739 (1966).
268. C. Walker and H. Shechter, *J. Amer. Chem. Soc.* **90**, 5626 (1968).
269. A. F. Vandi and T. Moeller, *Chem. Ind. (London)* p. 221 (1962).
270. B. I. Stepanov, T. G. Edelman, and I. M. Kolbasova, *Zh. Obshch. Khim.* **37**, 1883 (1967); *J. Gen. Chem. USSR* **37**, 1793 (1967).
271. I. N. Zhmurova, I. Yu. Voitsekhovskaya, and A. V. Kirsanov, *Zh. Obshch. Khim.* **31**, 3741 (1961); *J. Gen. Chem. USSR* **31**, 3495 (1961).
272. G. I. Derkach, E. S. Gubnitskaya, and A. V. Kirsanov, *Zh. Obshch. Khim.* **31**, 3746 (1961); *J. Gen. Chem. USSR* **31**, 3500 (1961).
273. H. Schmidbaur and W. Wolfsberger, *Chem. Ber.* **100**, 1000 (1967).
274. M. Becke-Goehring and E. Fluck, *Inorg. Syn.* **8**, 92 (1966).
275. E. Fluck, *Chem. Ber.* **94**, 1388 (1961).
276. J. Emsley and P. B. Udy, *J. Chem. Soc., A* p. 3025 (1970).
277. M. Becke-Goehring and E. Fluck, *Angew. Chem., Int. Ed. Engl.* **1**, 281 (1962).
278. A. Schmidpeter and J. Ebeling, *Chem. Ber.* **101**, 815 (1968).
279. H. G. Mardersteig, L. Meinel, and H. Nöth, *Z. Anorg. Allg. Chem.* **368**, 254 (1969).
280. H. W. Roesky and L. F. Grimm, *Chem. Ber.* **102**, 2319 (1969).
281. H. W. Roesky and L. F. Grimm, *Chem. Ber.* **103**, 3114 (1970).
282. M. Becke-Goehring and B. Scharf, *Z. Anorg. Allg. Chem.* **353**, 320 (1967).
283. I. M. Filatova, E. L. Zaitseva, A. P. Simonov, and A. Ya. Yakubovich, *Zh. Obshch. Khim.* **38**, 1304 (1968); *J. Gen. Chem. USSR* **38**, 1256 (1968).
284. A. Schmidpeter, H. Brecht, and J. Ebeling, *Chem. Ber.* **101**, 3902 (1968).
285. V. V. Sheluchenko, I. M. Filatova, E. L. Zaitseva, and A. Ya. Yakubovich, *Zh. Obshch. Khim.* **39**, 194 (1969); *J. Gen. Chem. USSR* **39**, 178 (1969).
286. V. A. Gilyarov, R. V. Kudryavtsev, and M. I. Kabachnik, *Zh. Obshch. Khim.* **38**, 352 (1968); *J. Gen. Chem. USSR* **38**, 351 (1968)
287. E. Fluck, *Z. Anorg. Allg. Chem.* **315**, 181 (1962).
288. E. F. Moran, *J. Inorg. Nucl. Chem.* **30**, 1405 (1968).
289. K. L. Paciorek, *Inorg. Chem.* **3**, 96 (1964).
290. R. H. Kratzer and K. L. Paciorek, *J. Org. Chem.* **32**, 853 (1967).

AUTHOR INDEX

Numbers in parentheses are reference numbers and indicate that an author's work is referred to, although his name is not cited in the text. Numbers in italics show the page on which the complete reference is listed.

A

B

E

F

H

I

J

K

L

M

T

Y

Z

SUBJECT INDEX

Cyclo- or polyphosphazenes which contain both organic and halogeno substituents can be located most readily in the index under the organic component.

D

E

F

T

U